# Lecture Notes in Artificial Intelligence 7647

Subseries of Lecture Notes in Computer Science

LNAI Series Editors

Randy Goebel
  *University of Alberta, Edmonton, Canada*
Yuzuru Tanaka
  *Hokkaido University, Sapporo, Japan*
Wolfgang Wahlster
  *DFKI and Saarland University, Saarbrücken, Germany*

LNAI Founding Series Editor

Joerg Siekmann
  *DFKI and Saarland University, Saarbrücken, Germany*

Vicenç Torra   Yasuo Narukawa
Beatriz López   Mateu Villaret (Eds.)

# Modeling Decisions
# for Artificial Intelligence

9th International Conference, MDAI 2012
Girona, Catalonia, Spain, November 21-23, 2012
Proceedings

 Springer

Series Editors

Randy Goebel, University of Alberta, Edmonton, Canada
Jörg Siekmann, University of Saarland, Saarbrücken, Germany
Wolfgang Wahlster, DFKI and University of Saarland, Saarbrücken, Germany

Volume Editors

Vicenç Torra
IIIA-CSIC, Bellaterra, Catalonia, Spain
E-mail: vtorra@iiia.csic.es

Yasuo Narukawa
Toho Gakuen, Kunitachi, Tokyo, Japan
E-mail: nrkwy@ybb.ne.jp

Beatriz López
Universitat de Girona, Catalonia, Spain
E-mail: beatriz.lopez@udg.edu

Mateu Villaret
Universitat de Girona, Catalonia, Spain
E-mail: villaret@ima.udg.edu

ISSN 0302-9743                                e-ISSN 1611-3349
ISBN 978-3-642-34619-4                         e-ISBN 978-3-642-34620-0
DOI 10.1007/978-3-642-34620-0
Springer Heidelberg Dordrecht London New York

Library of Congress Control Number: Applied for

CR Subject Classification (1998): I.2.3, I.2.6, I.5.2-3, K.6.5, H.3.4-5, G.1.6, G.1.9,
F.2.1, H.2.8, K.4.4

LNCS Sublibrary: SL 7 – Artificial Intelligence

*Typesetting:* Camera-ready by author, data conversion by Scientific Publishing Services, Chennai, India

Printed on acid-free paper

Springer is part of Springer Science+Business Media (www.springer.com)

# Preface

This volume contains papers presented at the 9th International Conference on Modeling Decisions for Artificial Intelligence (MDAI 2012), held in Girona, Catalonia, Spain, November 21–23, 2012. This conference followed MDAI 2004 (Barcelona, Catalonia), MDAI 2005 (Tsukuba, Japan), MDAI 2006 (Tarragona, Catalonia), MDAI 2007 (Kitakyushu, Japan), MDAI 2008 (Sabadell, Catalonia), MDAI 2009 (Awaji Island, Japan), MDAI 2011 (Perpinyà, Catalonia, France), and MDAI 2012 (Changsha, China) with proceedings also published in the LNAI series (Vols. 3131, 3558, 3885, 4617, 5285, 5861, 6408, and 6820).

The aim of this conference was to provide a forum for researchers to discuss theory and tools for modeling decisions, as well as applications that encompass decision-making processes and information-fusion techniques.

The organizers received 49 papers from 17 different countries, from Europe, Asia, and Australia, 32 of which are published in this volume. Each submission received at least two reviews from the Program Committee and a few external reviewers. We would like to express our gratitude to them for their work. The plenary talks presented at the conference are also included in this volume.

The conference was supported by the Institut d'Informàtica i Aplicacions de la Universitat de Girona, the Fundació Privada: Girona, Universitat i Futur, the Universitat de Girona, the Catalan Association for Artificial Intelligence (ACIA), the European Society for Fuzzy Logic and Technology (EUSFLAT), the Japan Society for Fuzzy Theory and Intelligent Informatics (SOFT), the UNESCO Chair in Data Privacy, the China Computer Federation, the Spanish MINECO (TIN2011-15580-E), and the Spanish MEC (ARES - CONSOLIDER INGENIO 2010 CSD2007-00004).

September 2012

Vicenç Torra
Yasuo Narukawa
Beatriz López
Mateu Villaret

# Organization

## General Chairs

Beatriz López — Universitat de Girona, Catalonia, Spain
Mateu Villaret — Universitat de Girona, Catalonia, Spain

## Program Chairs

Vicenç Torra — IIIA-CSIC, Bellaterra, Catalonia, Spain
Yasuo Narukawa — Toho Gakuen, Tokyo, Japan

## Advisory Board

Bernadette Bouchon-Meunier — Computer Science Laboratory of the University Paris 6 (LiP6), CNRS, France
Didier Dubois — Institut de Recherche en Informatique de Toulouse (IRIT), CNRS, France
Lluis Godo — IIIA-CSIC, Catalonia, Spain
Kaoru Hirota — Tokyo Institute of Technology, Japan
Janusz Kacprzyk — Systems Research Institute, Polish Academy of Sciences, Poland
Sadaaki Miyamoto — University of Tsukuba, Japan
Michio Sugeno — European Centre for Soft Computing, Spain
Ronald R. Yager — Machine Intelligence Institute, Iona Collegue, NY, USA

## Program Committee

Gleb Beliakov — Deakin University Australia
Gloria Bordogna — Consiglio Nazionale delle Ricerche, Italia
Tomasa Calvo — Universidad Alcala de Henares, Spain
Marc Daumas — Université de Perpignan, France
Susana Díaz — Universidad de Oviedo, Spain
Josep Domingo-Ferrer — Universitat Rovira i Virgili, Catalonia
Jozo Dujmovic — San Francisco State University, California
Michel Grabisch — Université Paris I Panthéon-Sorbonne, France
Enrique Herrera-Viedma — Universidad de Granada, Spain
Masahiro Inuiguchi — Osaka University, Japan

| | |
|---|---|
| Ivan Kojadinovic | Université de Pau et des Pays de l'Adour, France |
| Xinwang Liu | Southeast University, China |
| Jun Long | National University of Defense Technology, China |
| Jean-Luc Marichal | University of Luxembourg, Luxembourg |
| Rosa Meo | Università di Torino, Italia |
| Radko Mesiar | Slovak University of Technology, Slovakia |
| Tetsuya Murai | Hokkaido University, Japan |
| Toshiaki Murofushi | Tokyo Institute of Technology, Japan |
| Guillermo Navarro-Arribas | Universitat Autònoma de Barcelona, Catalonia, Spain |
| Michael Ng | Hong Kong Baptist University |
| Gabriella Pasi | Università di Milano Bicocca, Italia |
| Susanne Saminger-Platz | Jihannes Kepler University, Austria |
| Roman Słowiński | Poznan University of Technology, Poland |
| László Szilágyi | Sapientia-Hungarian Science University of Transylvania, Hungary |
| Aida Valls | Universitat Rovira i Virgili, Catalonia, Spain |
| Zeshui Xu | Southeast University, China |
| Yuji Yoshida | University of Kitakyushu, Japan |
| Gexiang Zhang | Southwest Jiaotong University, China |

## Local Organizing Committee Chairs

Beatriz López
Mateu Villaret

## Local Organizing Committee

Miquel Bofill
Sonia Buxeda
Pablo Gay
Xavier Manyer
Miquel Palahí

## Additional Referees

José Luis Vivas
Montserrat Batet
Damià Castellà
David Sànchez
Constantinos Patsakis

# Supporting Institutions

Institut d'Informàtica i Aplicacions de la Universitat de Girona
Fundació Privada Girona, Universitat i Futur
Universitat de Girona
The Catalan Association for Artificial Intelligence (ACIA)
The European Society for Fuzzy Logic and Technology (EUSFLAT)
The Japan Society for Fuzzy Theory and Intelligent Informatics (SOFT)
The UNESCO Chair in Data Privacy
The China Computer Federation
The Spanish MEC (ARES - CONSOLIDER INGENIO 2010 CSD2007-00004)
The Spanish MINECO (TIN2011-15580-E)

# Table of Contents

## Invited Papers

An Overview of Hierarchical and Non-hierarchical Algorithms of
Clustering for Semi-supervised Classification ........................ 1
  Sadaaki Miyamoto

Fuzzy Programming Approaches to Robust Optimization ............. 11
  Masahiro Inuiguchi

Do We Know How to Integrate? ................................. 13
  Radko Mesiar

Combination and Soft-Normalization of Belief Functions
on MV-Algebras ............................................... 23
  Tommaso Flaminio, Lluís Godo, and Tomáš Kroupa

## Regular Papers

## Aggregation Operators

Using Linear Programming for Weights Identification of Generalized
Bonferroni Means in R ......................................... 35
  Gleb Beliakov and Simon James

An Ordered Weighted Average with a Truncation Weight
on Intervals .................................................. 45
  Yuji Yoshida

Choquet Integral on the Real Line as a Generalization of the OWA
Operator...................................................... 56
  Yasuo Narukawa

On WOWA Rank Reversal ...................................... 66
  Wlodzimierz Ogryczak, Patrice Perny, and Paul Weng

Extending Concordance and Discordance Relations to Hierarchical Sets
of Criteria in ELECTRE-III Method ............................. 78
  Luis Del Vasto-Terrientes, Aida Valls, Roman Slowinski, and
  Piotr Zielniewicz

An Extended LibQUAL+ Model Based on Fuzzy Linguistic
Information ................................................... 90
  Francisco J. Cabrerizo, Ignacio J. Pérez, Javier López-Gijón, and
  Enrique Herrera-Viedma

Measure of Inconsistency for the Potential Method ................... 102
  *Lavoslav Čaklović*

# Integrals

Hierarchical Bipolar Sugeno Integral Can Be Represented as
Hierarchical Bipolar Choquet Integral ............................. 115
  *Katsushige Fujimoto and Michio Sugeno*

Qualitative Integrals and Desintegrals - Towards a Logical View ....... 127
  *Didier Dubois, Henri Prade, and Agnés Rico*

A Quantile Approach to Integration with Respect to Non-additive
Measures ......................................................... 139
  *Marta Cardin*

# Data Privacy and Security

Analysis of On-Line Social Networks Represented as Graphs –
Extraction of an Approximation of Community Structure Using
Sampling .......................................................... 149
  *Néstor Martínez Arqué and David F. Nettleton*

Using Profiling Techniques to Protect the User's Privacy in Twitter .... 161
  *Alexandre Viejo, David Sánchez, and Jordi Castellà-Roca*

Detecting Sensitive Information from Textual Documents: An
Information-Theoretic Approach .................................... 173
  *David Sánchez, Montserrat Batet, and Alexandre Viejo*

A Study of Anomaly Detection in Data from Urban Sensor Networks ... 185
  *Christoffer Brax and Anders Dahlbom*

Comparing Random-Based and $k$-Anonymity-Based Algorithms for
Graph Anonymization .............................................. 197
  *Jordi Casas-Roma, Jordi Herrera-Joancomartí, and Vicenç Torra*

Heuristic Supervised Approach for Record Linkage ................... 210
  *Javier Murillo, Daniel Abril, and Vicenç Torra*

Sampling Attack against Active Learning in Adversarial
Environment ...................................................... 222
  *Wentao Zhao, Jun Long, Jianping Yin, Zhiping Cai, and Geming Xia*

Dynamic Credit-Card Fraud Profiling............................... 234
  *Marc Damez, Marie-Jeanne Lesot, and Adrien Revault d'Allonnes*

# Reasoning

Representing Fuzzy Logic Programs by Graded Attribute
Implications . . . . . . . . . . . . . . . . . . . . . . . . . . . . . . . . . . . . . . . . . . . . . . . . . . . . . .   246
    Tomas Kuhr and Vilem Vychodil

Refining Discretizations of Continuous-Valued Attributes . . . . . . . . . . . . .   258
    Eva Armengol and Àngel García-Cerdaña

Linear Programming with Graded Ill-Known Sets . . . . . . . . . . . . . . . . . . . .   270
    Shizuya Kawamura and Masahiro Inuiguchi

Sharing Online Cultural Experiences: An Argument-Based Approach . . .   282
    Leila Amgoud, Roberto Confalonieri, Dave de Jonge,
    Mark d'Inverno, Katina Hazelden, Nardine Osman,
    Henri Prade, Carles Sierra, and Matthew Yee-King

Simple Proof of Basic Theorem for General Concept Lattices by
Cartesian Representation . . . . . . . . . . . . . . . . . . . . . . . . . . . . . . . . . . . . . . . . .   294
    Radim Belohlavek, Jan Konecny, and Petr Osicka

On Some Properties of the Negative Transitivity Obtained from
Transitivity . . . . . . . . . . . . . . . . . . . . . . . . . . . . . . . . . . . . . . . . . . . . . . . . . . . . . . .   306
    Susana Díaz, Susana Montes, and Bernard De Baets

# Applications

Multi Criteria Operators for Multi-attribute Auctions . . . . . . . . . . . . . . . .   318
    Albert Pla, Beatriz Lopez, and Javier Murillo

Finding Patterns in Large Star Schemas at the Right Aggregation
Level . . . . . . . . . . . . . . . . . . . . . . . . . . . . . . . . . . . . . . . . . . . . . . . . . . . . . . . . . . . . .   329
    Andreia Silva and Cláudia Antunes

Application of Quantitative MCDA Methods for Parameter Setting
Support of an Image Processing System . . . . . . . . . . . . . . . . . . . . . . . . . . . . .   341
    Lionel Valet and Vincent Clivillé

# Clustering and Similarity

Inductive Clustering and Twofold Approximations in Nearest Neighbor
Clustering . . . . . . . . . . . . . . . . . . . . . . . . . . . . . . . . . . . . . . . . . . . . . . . . . . . . . . . . .   355
    Sadaaki Miyamoto and Satoshi Takumi

Marginality: A Numerical Mapping for Enhanced Exploitation of
Taxonomic Attributes . . . . . . . . . . . . . . . . . . . . . . . . . . . . . . . . . . . . . . . . . . . . . .   367
    Josep Domingo-Ferrer

Introducing Incomparability in Modeling Qualitative Belief
Functions ............................................................ 382
   *Amel Ennaceur, Zied Elouedi, and Eric Lefevre*

On Rough Set Based Non Metric Model .......................... 394
   *Yasunori Endo, Ayako Heki, and Yukihiro Hamasuna*

An Efficient Reasoning Method for Dependencies over Similarity and
Ordinal Data ....................................................... 408
   *Radim Belohlavek, Pablo Cordero, Manuel Enciso, Angel Mora, and
Vilem Vychodil*

**Author Index** .................................................... 421

# An Overview of Hierarchical
# and Non-hierarchical Algorithms
# of Clustering for Semi-supervised Classification

Sadaaki Miyamoto

Department of Risk Engineering,
Faculty of Systems and Information Engineering
University of Tsukuba, 1-1-1 Tennodai, Tsukuba, Ibaraki 305-8573, Japan
miyamoto@risk.tsukuba.ac.jp

**Abstract.** An overview of a variety of methods of agglomerative hierarchical clustering as well as non-hierarchical clustering for semi-supervised classification is given. Two different formulations for semi-supervised classification are introduced: one is with pairwise constraints, while the other does not use constraints. Two methods of the mixture of densities and fuzzy c-means are contrasted and their theoretical properties are discussed. A number of agglomerative hierarchical algorithms are then discussed. It will be shown that the single linkage has different characteristics when compared with the complete linkage and average linkage. Moreover the centroid method and the Ward method are discussed. It will also be shown that the must-link constraints and the cannot-link constraints are handled in different ways in these methods.

**Keywords:** agglomerative hierarchical clustering, semi-supervised classification, pairwise constraints, $K$-means.

## 1 Introduction

Recently, semi-supervised learning [5, 24] has extensively been studied and relation to cluster analysis has also been considered. Special attention has been paid to constrained clustering and various methods have been developed [2–4, 6, 7, 10, 11, 17, 18, 21–23].

An extensive survey is not aimed at in this paper, but it focuses on two relatively unknown topics. First, the way how the method of fuzzy $c$-means [1, 16] is related to the mixture of distributions [14, 24] in semi-supervised classification is considered. Second, a class of agglomerative hierarchical algorithms with constraints [17, 18] are discussed. Throughout the paper, non-probabilistic models are mostly discussed, while the standard model of the mixture of densities are briefly referred to. Moreover, the focus is more on agglomerative hierarchical clustering than a family of non-hierarchical techniques such as the $K$-means.

V. Torra et al. (Eds.): MDAI 2012, LNAI 7647, pp. 1–10, 2012.
© Springer-Verlag Berlin Heidelberg 2012

## 2  Non-hierarchical Clustering and Semi-supervised Classification

Let $X = \{x_1, \ldots, x_N\}$ be the set of objects for clustering, and an object $x \in X$ be a point of $p$-dimensional Euclidean space ($X \subset \mathbf{R}^p$). The squared Euclidean distance is denoted by

$$D(x, y) = \|x - y\|^2 = \sum_{j=1}^{p} (x^j - y^j)^2 \tag{1}$$

where $x = (x^1, \ldots, x^p), y = (y^1, \ldots, y^p) \in \mathbf{R}^p$. Moreover the squared Mahalanobis distance is denoted by

$$D(x, y; S) = (x - y)^\top S^{-1}(x - y) \tag{2}$$

where $S$ is a positive-definite matrix.

Zhu and Goldberg [24] distinguish the concepts of semi-supervised classification and constrained clustering, contrary to our usage that semi-supervised classification includes constrained clustering. Let us distinguish these two concepts for convenience hereafter.

Semi-supervised classification by their definition uses a set of labeled samples $\{(x_k, y_k)\}$ $k = 1, \ldots, N$, where $y_k$ is the class label of $x_k$ and another set of *unlabeled* samples $\{x'_\ell\}$, $\ell = N + 1 \ldots, N + L$. The purpose is to have the labels on $x'_\ell$ or that for all $\mathbf{R}^p$.

In contrast, constrained clustering uses two sets of constraints: $ML = \{(x_k, x_l)\}$ and $CL = \{(x_i, x_j)\}$. $ML$ is called must-link and $CL$ is called cannot-link. A pair in $ML$ such as $(x_k, x_l)$ should be in the same cluster, while a pair $(x_i, x_j)$ in $CL$ should be in different clusters.

Let us first consider semi-supervised classification.

### 2.1  EM Solution for Semi-supervised Classification

Zhu and Goldberg [24] show an iterative solution by the EM algorithm [14] for the mixture of Gaussian distributions, abbreviated GMM, in the case of semi-supervised classification. The formulation is omitted here, but the iterative solution is as follows. The number of clusters is assumed to be $c$.

**EM Algorithm for GMM for Semi-supervised Classification ([24], p.27)**

1. Set initial values $\theta^{(t)} = (\pi_j^{(t)}, \mu_j^{(t)}, \Sigma_j^{(t)})$ for $t = 0$, where $\pi_j$ is the prior probability, $\mu_j$ is the mean value, and $\Sigma_j$ is the covariance matrix for $j$-th Gaussian distribution.

2. Calculate the probability of allocating $x_k$ to cluster $j$:

$$\gamma_{kj} = P(y_j | x_k, \theta^{(t)}) = \frac{\pi_j^{(t)} \mathcal{N}(x_k; \mu_j^{(t)}, \Sigma_j^{(t)})}{\sum_{l=1}^{c} \pi_l^{(t)} \mathcal{N}(x_k; \mu_l^{(t)}, \Sigma_l^{(t)})}, \quad k = N + 1 \ldots, N + L,$$

where $\mathcal{N}(x_k; \mu_l^{(t)}, \Sigma_l^{(t)})$ is the $l$-th Gaussian density function. Define $\gamma_{kj}$ for labeled samples:

$$\gamma_{kj} = \begin{cases} 1 & (y_k = j), \\ 0 & (otherwise), \end{cases} \quad k = 1, \ldots, N.$$

3. Calculate parameters $\theta^{(t+1)}$:

$$L_j = \sum_{k=1}^{N+L} \gamma_{kj}, \tag{3}$$

$$\mu_j^{(t+1)} = \frac{1}{L_j} \sum_{k=1}^{N+L} \gamma_{kj} x_k, \tag{4}$$

$$\Sigma_j^{(t+1)} = \frac{1}{L_j} \sum_{k=1}^{N+L} \gamma_{kj}(x_k - \mu_j^{(t+1)})(x_k - \mu_j^{(t+1)})^\top, \tag{5}$$

$$\pi_j^{(t+1)} = \frac{L_j}{N + L}. \tag{6}$$

## 2.2 Fuzzy $c$-Means Clustering for Semi-supervised Classification

We proceed to consider fuzzy $c$-means clustering that is closely related to the above solution by the EM algorithm. For this purpose it is adequate to use the KL-information based objective function [12, 13, 16]:

$$J(U, V, A, S) = \sum_{i=1}^{c} \sum_{k=1}^{N} [\tilde{u}_{ki} D(x_k, v_i; S_i) - \lambda^{-1} \tilde{u}_{ki} \log \alpha_i + \tilde{u}_{ki} \log |S_i|$$
$$+ W(u_{ki} - \tilde{u}_{ki})^2]$$
$$+ \sum_{i=1}^{c} \sum_{k=N+1}^{N+L} [u_{ki} D(x_k, v_i; S_i) + \lambda^{-1} u_{ki} \log \frac{u_{ki}}{\alpha_i} + u_{ki} \log |S_i|], \tag{7}$$

where $\{\tilde{u}_{ki}\}$ $(k = 1, \ldots, N)$ are membership values given beforehand, while $U = (u_{ki})$ $(k = 1, \ldots, N + L)$ are variables with the constraint $\sum_i u_{ki} = 1$ and $u_{ki} \geq 0$; $v_i$ is a cluster center and $V = (v_1, \ldots, v_c)$. Other two variables are $A = (\alpha_1, \ldots, \alpha_c)$ and $S = (S_1, \ldots, S_c)$. Variable $\alpha_i$ controls cluster sizes and satisfies $\sum_i \alpha_i = 1$, $\alpha_i \geq 0$. $S_i$ is a positive definite matrix used in

$$D(x, v_i; S_i) = (x - v_i)^\top S_i^{-1}(x - v_i).$$

Other constants of $\lambda$ and $W$ are positive.

An alternate minimization of $J(U, V, A, S)$ is used: after initial values $(U, V, A, S)$ are set, the following steps are repeated until convergence.

Step 1: *Minimize* $J(U, V, A, S)$ *w.r.t.* $U$ while $(V, A, S)$ is fixed.
Step 2: *Minimize* $J(U, V, A, S)$ *w.r.t.* $V$ while $(U, A, S)$ is fixed.
Step 3: *Minimize* $J(U, V, A, S)$ *w.r.t.* $A$ while $(U, V, S)$ is fixed.
Step 4: *Minimize* $J(U, V, A, S)$ *w.r.t.* $S$ while $(U, V, A)$ is fixed.

We first have

$$u_{ki} = \tilde{u}_{ki} \quad for \ k = 1, \ldots, N, \ i = 1, \ldots c. \tag{8}$$

Substituting this into the above function, we have the following solutions.

$$u_{ki} = \frac{\frac{\alpha_i}{|S_i|^{\frac{1}{2}}} \exp(-\lambda D(x_k, v_i; S_i))}{\sum_{j=1}^{c} \frac{\alpha_j}{|S_j|^{\frac{1}{2}}} \exp(-\lambda D(x_k, v_j; S_j))}, \quad k = N+1, \ldots, N+L, \ i = 1, \ldots, c, \tag{9}$$

$$v_i = \frac{\sum_{k=1}^{N+L} u_{ki} x_k}{\sum_{k=1}^{N+L} u_{ki}}, \quad i = 1, \ldots, c, \tag{10}$$

$$\alpha_i = \frac{\sum_{k=1}^{N+L} u_{ki}}{N+L}, \quad i = 1, \ldots, c, \tag{11}$$

$$S_i = \frac{1}{\sum_{k=1}^{N+L} u_{ki}} \sum_{k=1}^{N+L} u_{ki}(x_k - v_i)(x_k - v_i)^\top, \quad i = 1, \ldots, c. \tag{12}$$

It is easy to see that the solutions from the objective function $J(U, V, A, S)$ are the same as those by the EM algorithm by taking $\lambda = \frac{1}{2}$. We thus have a result of the equivalence of the both method, and also the fuzzy $c$-means with the KL information method is more general by changing $\lambda$ in the semi-supervised classification.

## 3    Constrained Clustering and Agglomerative Hierarchical Algorithms

The agglomerative hierarchical clustering starts from each object as an initial cluster, a pair of clusters is then merged at a time, and finally it ends with the one cluster of the whole set. It uses an inter-cluster dissimilarity denoted by $D(G_i, G_j)$.

Many non-hierarchical algorithms for constrained clustering have been proposed [2, 3, 11, 21–23]. but fewer studies are on agglomerative hierarchical clustering: although theoretical considerations on some methods have been done [6, 7], there is no comprehensive discussion that covers all well-known linkage methods.

We note that handling $ML$ and $CL$ should be different, since $ML$ defines connected components of a graph which should be initial clusters in an agglomerative hierarchical algorithm, while $CL$ cannot be handled by the initial setting but should be considered while clusters are generated.

We hence describe the general procedure first and then discuss how the constraints are included. There are two options for handling $CL$:

**Option 1:** Modify distance $D(x, y)$ for $(x, y) \in CL$.
**Option 2:** Put penalty on $CL$ when clusters are merged.

The precise definition is given after the next algorithm called **AHCC**.

## AHCC: Agglomerative Hierarchical Clustering with Constraints

1) Let each initial cluster $G_i = \hat{G}_i$ be a connected component of the graph generated from $ML$, $i = 1, \ldots, K_0$, and the number of clusters $K = K_0$. If option 1 is used, then

$$D(G_i, G_j) = D(\hat{G}_i, \hat{G}_j) + \sum_{x \in \hat{G}_i, y \in \hat{G}_j, (x,y) \in CL} D(x_i, x_j)$$

else $D(G_i, G_j) = D(\hat{G}_i, \hat{G}_j)$. Note that $D(\hat{G}_i, \hat{G}_j)$ uses the definition of dissimilarity in each linkage method.

2) Find the pair of clusters with minimum distance:

$$(G_p, G_q) = \arg\min_{i,j}\{D(G_i, G_j) + \mathcal{P}(G_i, G_j)\} \tag{13}$$

where $\mathcal{P}(G_i, G_j)$ is the penalty term when option 2 is used; if option 1 is used, $\mathcal{P}(G_i, G_j) = 0$.
Merge $G_r = G_p \cup G_q$ at the level of dissimilarity

$$m_K = D(G_p, G_q) = \min_{i,j} D(G_i, G_j). \tag{14}$$

Reduce the number of clusters: $K = K - 1$.

3) If $K = 1$, stop, else update distance $D(G_r, G_j)$, for all other clusters $G_j$. Go to step 2).

**End AHCC.**

**Option 1:** Let *Const.* be a constant large enough to prevent linkage. Put

$$\mathcal{D}(x_i, x_j) = Const., \quad \text{for } (x_i, x_j) \in CL.$$

**Option 2:** Let *Const.* be a constant large enough to prevent linkage. Put

$$\mathcal{P}(G_i, G_j) = \sum_{x \in \hat{G}_i, y \in \hat{G}_j, (x,y) \in CL} Const.$$

The two options are mutually exclusive: if option 1 is used, then we do not use option 2, and vice versa. Note also the difference between the two options. The modification of dissimilarity reflects the level $m_K$, while the penalty not.

Five linkage methods for the definition of the distance between clusters are well-known: they are the single linkage, complete linkage, average linkage, centroid method, and the Ward method [9]. They have different definitions for dissimilarity between clusters, and hence the formulas for updating $D(G_r, G_j)$ are different.

**The Single Linkage:** The single linkage alias the nearest neighbor method has the definition of dissimilarity between clusters given by

$$D(G_i, G_j) = \min_{x \in G_i, y \in G_j} D(x, y). \tag{15}$$

We can use an updating formula in step **3)**:

$$D(G_r, G_j) = \min\{D(G_p, G_j), D(G_q, G_j)\}.$$

**The Complete Linkage:** The complete linkage alias the furthest neighbor method has the definition of dissimilarity between clusters given by

$$D(G_i, G_j) = \max_{x \in G_i, y \in G_j} D(x, y). \tag{16}$$

We can use an updating formula

$$D(G_r, G_j) = \max\{D(G_p, G_j), D(G_q, G_j)\}.$$

**The Average Linkage:** The average linkage alias the group average method has the definition of dissimilarity between clusters given by

$$D(G_i, G_j) = \frac{1}{|G_i||G_j|} \sum_{x \in G_i, y \in G_j} D(x, y), \tag{17}$$

where $|G_i|$ is the number of elements in $G_i$. We can use an updating formula

$$D(G_r, G_j) = \frac{|G_p|}{|G_r|} D(G_p, G_j) + \frac{|G_q|}{|G_r|} D(G_q, G_j),$$

Note that $|G_r| = |G_p| + |G_q|$.

**Centroid Method:** Let the centroid of a cluster $G$ be

$$M(G) = \frac{1}{|G|} \sum_{x_k \in G} x_k.$$

The distance between two clusters is then defined by the squared distance between the two centroids:

$$D(G_i, G_j) = \|M(G_i) - M(G_j)\|^2. \tag{18}$$

The formula for updating in step **3)** of the above algorithm is omitted here.

**Ward Method:** The Ward method is also based on the Euclidean distance. Let us define

$$E(G) = \sum_{x \in G} \|x - M(G)\|^2,$$

which is the sum of distances between each $x \in G$ and the centroid for $G$. Using $E(G)$, we put

$$D(G_i, G_j) = E(G_i \cup G_j) - E(G_i) - E(G_j) \tag{19}$$

which is always positive. The formula for updating in step **3)** of the above algorithm is omitted here. Note also that the initial distance is given by the following:

$$D(x_i, x_j) = \frac{1}{2}\|x_i - x_j\|^2.$$

**Table 1.** Linkage methods and the two options: 'Yes' means constraints $CL$ work without a problem, while 'No' means constraints $CL$ do not work generally. 'Distance' means that constraints $CL$ work, but a problem of carefully handling distance is to be considered. 'Reversal' means that $CL$ works, but reversals in dendrograms will occur.

|  | single linkage | complete linkage | average linkage | centroid method | Ward method |
|---|---|---|---|---|---|
| Option 1 | No | Yes | Yes | Distance | Distance |
| Option 2 | Reversal | Reversal | Reversal | Reversal | Yes |

## 3.1  Problems in Linkage Methods with Pairwise Constraints

When these methods are used with pairwise constraints, we have problems to be considered. Table 1 summarizes problems in the options and the linkage methods. We note the single linkage with the constraints has problems; the centroid method and the Ward method have another problem.

**Difficulties in the Single Linkage Method:** The single linkage method with option 1 (modification of dissimilarity) does not work well, since even if a cannot-link $(x_h, x_l)$ exists between $G$ and $G'$, the cannot-link does not prevent a link between $G$ and $G'$. If $D(x_h, x_l) > \min_{x \in G, y \in G'} D(x, y)$, then the dissimilarity $D(G, G')$ remain the same; if $D(x_h, x_l) = \min_{x \in G, y \in G'} D(x, y)$, then $G$ and $G'$ will be merged at the level of second smallest dissimilarity of $\{D(x, y) : x \in G, y \in G'\}$. Thus the cannot-link is not effective unless all pairs between $G$ and $G'$ have cannot-link, which is unrealistic.

On the other hand, the single linkage method with penalty (option 2) seems to work. However, there is another problem of 'reversal in dendrogram' [15]. If the level $m_K$ is monotonic:

$$m_{K_0} \le m_{K_0-1} \le \cdots \le m_2 \tag{20}$$

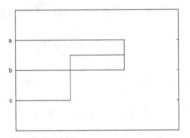

**Fig. 1.** A simple example of reversal, where $m_3 = D(a, b) > m_2 = D(\{a, b\}, c)$

then the output of dendrogram does not have a reversal. Figure 1 shows a simple example of a reversal. The above linkage methods are known to be without a reversal, except the centroid method. The single linkage with the constraints handled by the penalty has reversals. To construct a simple example of a reversal with $CL$ is easy, but omitted here.

*Note 1.* For the complete linkage and the average linkage, the cannot-link works well and reversals do not occur when option 1 is used, while reversals may occur when option 2 is used. We omit the detail.

**Problems in the Centroid Method and the Ward Method:** The centroid method and the Ward method are based on the squared Euclidean distance. Although dissimilarity can be rather freely modified in other three methods, the modification of dissimilarity using the large constant *Const.* is unacceptable in a Euclidean space. Let us mitigate the situation by introducing a kernel space [20] by using

$$D_K(x, y) = \|\Phi(x) - \Phi(y)\|_H^2$$

with a high-dimensional mapping $\Phi \colon R^p \to H$ instead of the original data space. It is also known that the centroid method and the Ward method can be used in a kernel space [8] with appropriate changes of the initial values. Using the Gaussian kernel, $D_K(x, y)$ with $\|x - y\| \to \infty$ leads to $D_K(x, y) \to 2$. It thus seems acceptable to use $D_K(x, y) = 2$ instead of $D(x, y) = Const.$ However, another problem occurs: when such a modification is introduced, the kernel is no longer positive-definite in general. The authors tried option 1 by this method [17], and the results seem acceptable in typical numerical examples. However, the penalty (option 2) for the centroid and Ward methods seems to work still better and it is without such a theoretical problem.

## 4   Conclusions

Semi-supervised classification by a variation of fuzzy $c$-means and constrained clustering using different linkage methods of agglomerative hierarchical clustering have been overviewed. The discussion in this paper does not seem to be found elsewhere. Numerical examples were omitted to save space, but the author with his colleagues has been investigating, in particular, constrained clustering using the centroid method and the Ward method [17–19]. The results suggest that these methods are useful in typical examples and real data sets.

We thus showed new possibilities of fuzzy c-means and agglomerative hierarchical clustering; these methods will be useful in various real applications.

**Acknowledgment.** This work has partly been supported by the Grant-in-Aid for Scientific Research, Japan Society for the Promotion of Science, No. 23500269.

# References

1. Bezdek, J.C.: Pattern Recognition with Fuzzy Objective Function Algorithms. Plenum Press, New York (1981)
2. Basu, S., Bilenko, M., Mooney, R.J.: A Probabilistic Framework for Semi-Supervised Clustering. In: Proc. of the Tenth ACM SIGKDD (KDD 2004), pp. 59–68 (2004)
3. Basu, S., Banerjee, A., Mooney, R.J.: Active Semi-Supervision for Pairwise Constrained Clustering. In: Proc. of the SIAM International Conference on Data Mining (SDM 2004), pp. 333–344 (2004)
4. Basu, S., Davidson, I., Wagstaff, K.L.: Constrained Clustering. CRC Press, Boca Raton (2009)
5. Chapelle, O., Schölkopf, B., Zien, A. (eds.): Semi-Supervised Learning. MIT Press, Cambridge (2006)
6. Davidson, I., Ravi, S.S.: Agglomerative Hierarchical Clustering with Constraints: Theoretical and Empirical Results. In: Jorge, A.M., Torgo, L., Brazdil, P.B., Camacho, R., Gama, J. (eds.) PKDD 2005. LNCS (LNAI), vol. 3721, pp. 59–70. Springer, Heidelberg (2005)
7. Davidson, I., Ravi, S.S.: Using instance-level constraints in agglomerative hierarchical clustering: theoretical and empirical results. Data Min., Knowl., Disc. 18, 257–282 (2009)
8. Endo, Y., Haruyama, H., Okubo, T.: On some hierarchical clustering algorithms using kernel functions. In: Proc. of FUZZ-IEEE 2004, CD-ROM Proc., Budapest, Hungary, July 25-29, pp. 1–6 (2004)
9. Everitt, B.S.: Cluster Analysis, 3rd edn. Arnold, London (1993)
10. Klein, D., Kamvar, S.D., Manning, C.: From Instance-level Constraints to Space-level Constraints: Making the Most of Prior Knowledge in Data Clustering. In: Proc. of the Intern. Conf. on Machine Learning, Sydney, Australia, pp. 307–314 (2002)
11. Kulis, B., Basu, S., Dhillon, I., Mooney, R.: Semi-supervised graph clustering: a kernel approach. Mach. Learn. 74, 1–22 (2009)
12. Ichihashi, H., Honda, K., Tani, N.: Gaussian mixture PDF approximation and fuzzy c-means clustering with entropy regularization. In: Proc. of Fourth Asian Fuzzy Systems Symposium, vol. 1, pp. 217–221 (2000)
13. Ichihashi, H., Miyagishi, K., Honda, K.: Fuzzy c-means clustering with regularization by K-L information. In: Proc. of 10th IEEE International Conference on Fuzzy Systems, vol. 2, pp. 924–927 (2001)

14. McLachlan, G.J., Krishnan, T.: The EM algorithms and Extensions. Wiley, New York (1997)
15. Miyamoto, S.: Fuzzy Sets in Information Retrieval and Cluster Analysis. Kluwer, Dordrecht (1990)
16. Miyamoto, S., Ichihashi, H., Honda, K.: Algorithms for Fuzzy Clustering. Springer (2008)
17. Miyamoto, S., Terami, A.: Semi-Supervised Agglomerative Hierarchical Clustering Algorithms with Pairwise Constraints. In: Proc. of WCCI 2010 IEEE World Congress on Computational Intelligence, CCIB, Barcelona, Spain, July, 18-23, pp. 2796–2801 (2010)
18. Miyamoto, S., Terami, A.: Constrained Agglomerative Hierarchical Clustering Algorithms with Penalties. In: Proc. of 2011 IEEE International Conference on Fuzzy Systems, Taipei, Taiwan, June 27-30, pp. 422–427 (2011)
19. Miyamoto, S., Terami, A.: Inductive vs. Transductive Clustering Using Kernel Functions and Pairwise Constraints. In: Proc. of 11th Intern. Conf. on Intelligent Systems Design and Applications (ISDA 2011), Cordoba, Spain, November 22-24, pp. 1258–1264 (2011)
20. Schölkopf, B., Smola, A.J.: Learning with Kernels. MIT Press, Cambridge (2002)
21. Shental, N., Bar-Hillel, A., Hertz, T., Weinshall, D.: Computing Gaussian Mixture Models with EM using Equivalence Constraints. In: Thrun, S., Saul, L.K., Schölkopf, B. (eds.) Advances In Neural Information Processing Systems 16, pp. 465–472 (2004)
22. Wagstaff, K., Cardie, C., Rogers, S., Schroedl, S.: Constrained K-means Clustering with Background Knowledge. In: Proc. of the 9th ICML, pp. 577–584 (2001)
23. Wang, N., Li, X., Luo, X.: Semi-supervised Kernel-based Fuzzy c-Means with Pairwise Constraints. In: Proc. of WCCI 2008, pp.1099-1103 (2008)
24. Zhu, X., Goldberg, A.B.: Introduction to Semi-Supervised Learning. Morgan and Claypool (2009)

# Fuzzy Programming Approaches to Robust Optimization

Masahiro Inuiguchi

Graduate School of Engineering Science, Osaka University, Toyonaka, Japan

**Abstract.** In the real world problems, we may face the cases when parameters of linear programming problems are not known exactly. In such cases, parameters can be treated as random variables or possibilistic variables. The probability distribution which random variables obey are not always easily obtained because they are assumed to be obtained by strict measurement owing to the cardinality of the probability. On the other hand, the possibility distribution restricting possibilistic variables can be obtained rather easily because they are assumed to be obtained from experts' perception owing to the ordinality of possibility. Then possibilistic programming approach would be convenient as an optimization technique under uncertainty.

In this talk, we review possibilistic linear programming approaches to robust optimization. Possibilistic linear programming approaches can be classified into three cases: optimizing approach, satisficing approach and two-stage approach [1]. Because the third approach has not yet been very developed, we focus on the other two approaches. First we review the optimization approach. We describe a necessarily optimal solution [2] as a robust optimal solution in the optimization approach. Because a necessarily optimal solution do not always exist, necessarily soft optimal solutions [3] have been proposed. In the necessarily soft optimal solutions, the optimality conditions is relaxed to an approximate optimality conditions. The relation to minimax regret solution [4,5] is shown and a solution procedure for obtaining a best necessarily soft optimal solution is briefly described.

Next we talk about the modality constrained programming approach [6]. A robust treatment of constraints are introduced. Then the necessity measure optimization model and necessity fractile optimization model are described as treatments of an objective function. They are models from the viewpoint of robust optimization. The simple models can preserve the linearity of the original problems. We describe how much we can generalize the simple models without great loss of linearity. A modality goal programming approach [7] is briefly introduced. By this approach, we can control the distribution of objective function values by a given goal.

Finally, we conclude this talk by giving future topics in possibilistic linear programming [8,9,10].

**Keywords:** robust optimization, fuzzy programming, necessity measure, optimization approach, satisficing approach.

V. Torra et al. (Eds.): MDAI 2012, LNAI 7647, pp. 11–12, 2012.

# References

[1] Inuiguchi, M., Ramik, J.: Possibilistic Linear Programming: A Brief Review of Fuzzy Mathematical Programming and a Comparison with Stochastic Programming in Portfolio Selection Problem. Fuzzy Sets and Systems 111, 29–45 (2000)

[2] Inuiguchi, M., Sakawa, M.: Possible and Necessary Optimality Tests in Possibilistic Linear Programming Problems. Fuzzy Sets and Systems 67, 29–46 (1994)

[3] Inuiguchi, M., Sakawa, M.: Robust Optimization under Softness in a Fuzzy Linear Programming Problem. International Journal of Approximate Reasoning 18, 21–34 (1998)

[4] Inuiguchi, M., Sakawa, M.: Minimax Regret Solutions to Linear Programming Problems with an Interval Objective Function 86, 526–536 (1995)

[5] Inuiguchi, M., Tanino, T.: On Computation Methods for a Minimax Regret Solution Based on Outer Approximation and Cutting Hyperplanes. International Journal of Fuzzy Systems 3, 548–557 (2001)

[6] Inuiguchi, M., Ichihashi, H., Kume, Y.: Modality Constrained Programming Problems: A Unified Approach to Fuzzy Mathematical Programming Problems in the Setting of Possibility Theory. Information Sciences 67, 93–126 (1993)

[7] Inuiguchi, M., Tanino, T.: Modality Goal Programming Models with Interactive Fuzzy Numbers. In: Proceedings of the Eighth International Fuzzy Systems Association World Congress, vol. 1, pp. 519–523 (1999)

[8] Inuiguchi, M.: Approaches to Linear Programming Problems with Interactive Fuzzy Numbers. In: Lodwick, W.A., Kacprzyk, J. (eds.) Fuzzy Optimization. STUDFUZZ, vol. 254, pp. 145–161. Springer, Heidelberg (2010)

[9] Inuiguchi, M.: Possibilistic Linear Programming Using General Necessity Measures Preserves the Linearity. In: Torra, V., Narakawa, Y., Yin, J., Long, J. (eds.) MDAI 2011. LNCS (LNAI), vol. 6820, pp. 186–197. Springer, Heidelberg (2011)

[10] Inuiguchi, M., Higuchi, T., Tsurumi, M.: Linear Necessity Measures and Their Applications to Possibilistic Linear Programming. In: Yao, J., Ramanna, S., Wang, G., Suraj, Z. (eds.) RSKT 2011. LNCS (LNAI), vol. 6954, pp. 280–289. Springer, Heidelberg (2011)

# Do We Know How to Integrate?

Radko Mesiar

Slovak University of Technology, Bratislava, Slovakia,
and Institute of Theory of Information and Automation
Czech Academy of Sciences, Prague, Czech Republic
radko.mesiar@stuba.sk
http://www.math.sk/mesiar

**Abstract.** After a short history of integration on real line, some examples of optimization tasks are given to illustrate the philosophy behind some types of integrals with respect to monotone measures and related to the standard arithmetics on real line. Basic integrals are then described both in discrete case and general case. A general approach to integration known as universal integrals is recalled, and introduced types of integrals as universal integrals are discussed. A special stress is given to copula-based universal integrals. Several types of integrals based on arithmetics different from the standard one are given, too. Finally, some concluding remarks are added.

**Keywords:** Choquet integral, monotone measure, Sugeno integral, universal integral.

## 1 Introduction and Historical Remarks

When asking a randomly chosen person whether he/she knowns something about integrals, almost all positively reacting persons have in mind the Riemann integral. This integral is a background of the classical natural sciences, and it acts on (possibly n–dimensional) real line equipped with the standard Lebesgue measure. Obviously, the history of integration, at least till 1925, is related to the Riemann integral and its genuine generalizations. As a first trace of constructive approaches to integration can be considered a formula for the volume of a frustum of a square pyramid proposed in ancient Egypt around 1850 BC (the Moscow Mathematical Papyrus, Problem 14). The first documented systematic technique allowing to determine integrals is the exhaustion method of the ancient Greek astronomer Eudoxus (around 370 BC). This method was further developed by several Greek mathematicians, including Archimedes. Similar methods were independently developed in China (Liu Hui around the third century, father Zu Chogzhi and son Zu Geng in the fifth century describing the volume of a sphere) and in India (Aryabhata in the fifth century). Only more than 1000 years later, several European scientists have done next important steps in the integration area. We recall J. Kepler (his approach to computation of the volume of barrels in now known as Simpson rule), Cavalieri (with his method of indivisibles he was able to integrate polynomials till order 9), J. Wallis (algebraic law

V. Torra et al. (Eds.): MDAI 2012, LNAI 7647, pp. 13–22, 2012.

for integration), P. de Fermat (his was the first to use infinite series in his integration method). Modern notation for (indefinite) integral was introduced by G. Leibniz in 1675. He adapted the integral symbol $\int$ from the letter known as long $\int$, standing for "summa". The modern notation for the definite integral was first used by J. Fourier around 1820. In this period, A. Cauchy developed a method for integration of continuous functions. All the roots and backgrounds for the "integral", including the fundamental work of I. Newton and G. Leibniz, were known in the middle of the 19th century. It was B. Riemann in his Habilitation Thesis at University of Göttingen [15] in 1854 who gave the first indubitable access to integration. This integral, now called the Riemann integral, is the best known integral, taught in each Calculus course. Its limitations (real line, standard Lebesgue measure) were challenging several scholars to generalize it. We recall here H. Lebesgue [7] who in 1904 introduced a rather general integral, acting on an arbitrary measurable space $(X, \mathcal{A})$, and defined for an $\sigma$–additive measure $m : \mathcal{A} \to [0, \infty]$. Observe that this integral is a background of the probability theory, among others. The next words bring a quotation from H. Lebesgue lecture held in May 8, 1926, in Copenhagen and entitled "The development of the notion of the integral": "... a generalization mode not for the vain pleasure of generalizing, but rather for the solution of problems previously posed, is always a fruitful generalization. The diverse applications which have already taken the concepts which we have just examined prove this superabundantly" (for the full text see [18]). Note that there is no concept of improper Lebesgue integral as it is the case of Riemann integral. Therefore there is no guarantee that a Riemann integrable function is also Lebesgue integrable. As a typical example consider the function $f : R \to R$ given by

$$f(x) = \begin{cases} 1 & \text{if } x = 0 \\ \frac{\sin x}{x} & \text{else.} \end{cases}$$

Then the Lebesgue integral $\int_R f(x) \, d\mu$ with $\mu$ being the standard Lebesgue measure on Borel subsets of $R$ does not exist (indeed, $\int_R |f(x)| \, d\mu = +\infty$), but the (improper) Riemann integral can be computed to be finite.

All till now mentioned integrals were additive (as functionals) and defined with respect to an $(\sigma-)$ additive measure. The first known approach to integration based on a monotone but not necessarily additive measure is due to Vitali [21] from 1925. Approach of G. Vitali (dealing with inner and outer measures) is a predecessor of the fundamental work of G. Choquet [4] yielding the Choquet integral. Another fundamental integral defined for monotone measures is due to M. Sugeno [20] in 1974.

All mentioned integrals consider the (non-negative) real values of both functions and measures. There are numerous kinds of integrals defined on more general structures. In this contribution, we consider only the framework of already mentioned integrals, i.e., we will deal with measurable spaces $(X, \mathcal{A})$ from the class $\mathcal{S}$ of all measurable spaces ($X$ is a non–empty set, universe, and $\mathcal{A} \subseteq 2^X$ is a $\sigma$–algebra of subsets of $X$), with $\mathcal{A}$–measurable functions

$f : X \rightarrow [0, \infty]$ from the class $\mathcal{F}_{(X,\mathcal{A})}$ of all such functions, and with monotone $m : \mathcal{A} \rightarrow [0, \infty]$, $m(\emptyset) = 0, m(X) > 0$, from the class $\mathcal{M}_{(X,\mathcal{A})}$ of all such measures. Integral is then a mapping

$$I : \bigcup_{(X,\mathcal{A}) \in \mathcal{S}} (\mathcal{M}_{(X,\mathcal{A})} \times \mathcal{F}_{(X,\mathcal{A})}) \rightarrow [0, \infty]$$

with some special properties we will discuss in the next sections.

The aim of this paper is to discuss some approaches to integration with respect to monotone measures. In the next section, we bring an optimisation problem under different constraints, illustrating the philosophy of several such integrals linked to the standard summation and multiplication of reals. These integrals are then properly defined and further discussed, including the study of their relationship. Section 3 is devoted to the introduction of a framework of universal integrals recently proposed in [5]. We introduce here some universal integrals, including copula based integrals (here we restrict our considerations to the unit interval $[0, 1]$). In Section 4, pseudo-arithmetical operations based integrals are discussed. Finally, several concluding remarks are added.

## 2    Optimisation of a Global Performance and Integrals

Consider a group $X = \{a, b, c\}$ of three workers with working capacity $f : X \rightarrow [0, \infty]$ given in hours by $f(a) = 5$, $f(b) = 4$, $f(c) = 3$. A performance per hour of a group of our workers is given by a set function $m : 2^X \rightarrow [0, \infty]$,

$m(\emptyset) = 0$, $m(\{a\}) = 2$, $m(\{b\}) = 3$, $m(\{c\}) = 4$,
$m(\{a, b\}) = 7$, $m(\{b, c\}) = 5$, $m(\{a, c\}) = 4$, $m(\{a, b, c\}) = 8$.

Our aim is to find a strategy to reach the optimal total performance of our workers under given work constraints:

**(1)** only one group can work for a fixed time period;
**(2)** several disjoint groups can work (fixed working time in each group may differ);
**(3)** one group starts to work, once a worker stops to work, he cannot start to work again;
**(4)** several disjoint groups can start to work, in each group after some working time we can split a working group into smaller groups, and a worker after stopping to work cannot start again;
**(5)** there are no constraints.

We formalize the optimal total performances under these five constraints settings and give the solution for our example. Hence the optimal total performance $T_i$ under constraints (i) is:

$$T_1 = \max \{k \cdot m(A) \mid k \cdot 1_A \leq f\} = \min \{f(a), f(b)\} \cdot m(\{a, b\}) = 4 \cdot 7 = 28;$$

$$T_2 = \max \left\{ \sum k_i \cdot m(A_i) \mid \sum k_i \cdot 1_{A_i} \leq f, (A_i)_i \text{ is disjoint system} \right\} =$$
$$= \min \{f(a), f(b)\} \cdot m(\{a, b\}) + f(c) \cdot m(\{c\}) = 4 \cdot 7 + 3 \cdot 4 = 40;$$

$$T_3 = \max \left\{ \sum k_i \cdot m(A_i) \mid \sum k_i \cdot 1_{A_i} \leq f, (A_i)_i \text{ is a chain} \right\} =$$

$$= \min \{f(a), f(b), f(c)\} \cdot m(\{a, b, c\}) + \min \{f(a) - f(c), f(b) - f(c)\} \cdot m(\{a, b\}) +$$

$$+ (f(a) - f(b)) \cdot m(\{a\}) = 3 \cdot 8 + 1 \cdot 7 + 1 \cdot 2 = 33;$$

$$T_4 = \max \left\{ \sum k_i \cdot m(A_i) \mid \sum k_i \cdot 1_{A_i} \leq f, A_i \cap A_j \in \{\emptyset, A_i, A_j\} \text{ for each i,j} \right\} =$$

$$= \min \{f(a), f(b)\} \cdot m(\{a, b\}) + (f(a) - f(b)) \cdot m(\{a\}) + f(c) \cdot m(\{c\}) =$$

$$= 4 \cdot 7 + 1 \cdot 2 + 3 \cdot 4 = 42;$$

$$T_5 = \max \left\{ \sum k_i \cdot m(A_i) \mid \sum k_i \cdot 1_{A_i} \leq f \right\} = T_4 = 42.$$

From the constraints settings it is obvious that the following inequalities always hold, independently of $f$ and $m$:

$$T_1 \leq T_i, \quad i \in \{1, 2, 3, 4, 5\};$$

$$T_5 \geq T_i, \quad i \in \{1, 2, 3, 4, 5\};$$

$$T_4 \geq T_i, \quad i \in \{1, 2, 3, 4\},$$

i.e., we have the following Hasse diagram (see Figure 1).

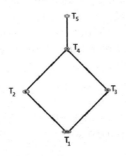

**Fig. 1.** Hasse diagram for relationships between functionals $T_1 - T_5$

As another example, consider $f : X \to [0, \infty]$ given by $f(a) = 8$, $f(b) = 3$, $f(c) = 6$, and $m : 2^X \to [0, \infty]$ given by,
$m(\emptyset) = 0$, $m(\{a\}) = 2$, $m(\{b\}) = 3$, $m(\{c\}) = 4$ and $m(A) = 10$ in all other cases. Then:

$$T_1 = 60, \quad T_2 = 69, \quad T_3 = 64, \quad T_4 = 73, \quad T_5 = 84,$$

(for more details see [19]).

All introduced functionals can be seen as special instances of decomposition integral proposed recently by Event and Lehrer [2], and some of them are, in fact, famous integrals introduced in past decades. We recall them now in a general setting, considering an arbitrary measurable space $(X, \mathcal{A}) \in \mathcal{S}$, as a mappings

$$I_i : \bigcup_{(X,\mathcal{A}) \in \mathcal{S}} (\mathcal{M}_{(X,\mathcal{A})} \times \mathcal{F}_{(X,\mathcal{A})}) \to [0, \infty], \ i \in \{1, 2, 3, 4, 5\}.$$

The first optimal performance $T_1$ is linked to the Shilkret integral [17],

$$I_1(m, f) = \sup \{k \cdot m(A) \mid k \cdot 1_A \leq f\}.$$

Note that all sets considered in this paper are supposed to be measurable, $A \in \mathcal{A}$. Evidently, our $T_1 = I_1(m, f)$ for $m, f$ given on $X = \{a, b, c\}$ and $\mathcal{A} = 2^X$.

Concerning the second optimal performance, $T_2$ is linked to the PAN–integral introduced by Yang in [23],

$$I_2(m, f) =$$

$$= \sup \left\{ \sum_{i=1}^{n} k_i \cdot m(A_i) \mid n \in N, \sum_{i=1}^{n} k_i \cdot 1_{A_i} \leq f, (A_i)_{i=1}^{n} \text{ is a disjoint system} \right\}.$$

The third optimization task describes the philosophy of the Choquet integral [4],

$$I_3(m, f) = \sup \left\{ \sum_{i=1}^{n} k_i \cdot m(A_i) \mid n \in N, \sum_{i=1}^{n} k_i \cdot 1_{A_i} \leq f, (A_i)_{i=1}^{n} \text{ is a chain} \right\}.$$

Note that due to the definition of the classical Riemann integral it holds

$$I_3(m, f) = \int_0^\infty m(\{f \geq t\}) \, dt.$$

The fourth approach to optimization constraints brings a new integral $I_4$ proposed recently by Stupňanová [19],

$$I_4(m, f) = \sup \left\{ \sum_{i=1}^{n} k_i \cdot m(A_i) \mid n \in N, \sum_{i=1}^{n} k_i \cdot 1_{A_i} \leq f, A_i \cap A_j \in \{\emptyset, A_i, A_j\} \right.$$

$$\text{for any } i, j \in \{1, \cdots, n\} \}.$$

Finally, another recent integral is linked to $T_5$, namely the concave integral introduced by Lehrer [8],

$$I_5(m, f) = \sup \left\{ \sum_{i=1}^{n} k_i \cdot m(A_i) \mid n \in N, \sum_{i=1}^{n} k_i \cdot 1_{A_i} \leq f \right\}.$$

The Hasse diagram in Figure 2 depicts the relationships between these integrals. Each of these integrals is linked to the standard arithmetical operations on real line, and for each of them it holds

$$I_i(k \cdot 1_{\{x\}}) = k \cdot m(\{x\}), k \in [0, \infty[,$$

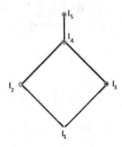

**Fig. 2.** Hasse diagram for relationships between integrals $I_1 - I_5$

if the singleton $\{x\} \in \mathcal{A}$. However, considering a general $A \in \mathcal{A}$, $I_i(k \cdot 1_A) = k \cdot m(A)$ holds only for $i \in \{1,3\}$, in general. Consequently, integrals $I_2, I_4$ and $I_5$ have a failure admitting the existence of $m_1 \neq m_2$ such that $I(m_1, f) = I(m_2, f)$ for each $f \in \mathcal{F}_{(X,\mathcal{A})}$. Note also that each introduced integral is homogeneous, i.e., $I_i(k \cdot f) = k \cdot I_i(f)$ for each $k \in [0, \infty[$ and $i \in \{1,2,3,4,5\}$.

Considering special types of monotone measures, we have the next equalities valid for any measurable function $f \in \mathcal{F}_{(X,\mathcal{A})}$:

- if $m \in \mathcal{M}_{(X,\mathcal{A})}$ is supermodular then

$$I_5(m, \cdot) = I_3(m, \cdot),$$

i.e., then the concave integral coincide with the Choquet integral, see [8];
- if $m$ is subadditive then

$$I_5(m, \cdot) = I_2(m, \cdot),$$

i.e., then the concave integral coincide with the PAN–integral, see [19];
- if $m$ is an unaminity measure, i.e., there is $A \in \mathcal{A}, A \neq \emptyset$, so that $m(B) = \begin{cases} 1 \text{ if } A \subseteq B, \\ 0 \text{ else} \end{cases}$, then all introduced integrals coincide, and then

$$I_i(m, f) = \inf \{f(x) | x \in A\}, \ i = 1, \cdots, 5,$$

see [19].

Obviously, if $X$ is finite and $m$ is additive then

$$I_i(m, f) = \sum_{x \in X} f(x) \cdot m(\{x\}), \ i \in \{2,3,4,5\}.$$

Moreover, if $m$ is a $\sigma$–additive measure, then integrals $I_i(m, \cdot), i \in \{2,3,4,5\}$ coincide with the standard Lebesgue integral,

$$I_i(m, f) = \int_X f \, dm.$$

# 3   Universal Integrals

To capture the idea of the majority of integrals proposed as functionals on abstract measurable spaces, Klement et al. [5] have recently proposed the concept of universal integrals.

**Definition 1.** *A mapping* $I : \bigcup_{(X,\mathcal{A}) \in \mathcal{S}} \left( \mathcal{M}_{(X,\mathcal{A})} \times \mathcal{F}_{(X,\mathcal{A})} \right) \to [0, \infty]$ *is called a universal integral whenever it satisfies the next properties:*

**(UI1)** *$I$ is nondecreasing in both components, i.e., $I(m_1, f_1) \leq I(m_2, f_2)$ whenever there is $(X, \mathcal{A}) \in \mathcal{S}$ such that $m_1, m_2 \in \mathcal{M}_{(X,\mathcal{A})}$ and $m_1 \leq m_2$, $f_1, f_2 \in \mathcal{F}_{(X,\mathcal{A})}$ and $f_1 \leq f_2$;*

**(UI2)** *there is an operation $\otimes : [0, \infty]^2 \to [0, \infty]$ (called a pseudo–multiplication) with annihilator $0$ (i.e., $a \otimes 0 = 0 \otimes a = 0$ for each $a \in [0, \infty]$) and a neutral element $e \in ]0, \infty]$ (i.e., $a \otimes e = e \otimes a = a$ for each $a \in [0, \infty]$) so that*

$$I(m, k \cdot 1_A) = k \otimes m(A)$$

*for any $(X, \mathcal{A}) \in \mathcal{S}, m \in \mathcal{M}_{(X,\mathcal{A})}, A \in \mathcal{A}$ and $k \in [0, \infty]$;*

**(UI3)** *for any two pairs $(m_1, f_1) \in (X_1, \mathcal{A}_1), (m_2, f_2) \in (X_2, \mathcal{A}_2)$ such that $m_1 (\{f_1 \geq t\}) = m_2 (\{f_2 \geq t\})$ for each $t \in ]0, \infty]$ (such pairs are called integral equivalent) it holds*

$$I(m_1, f_1) = I(m_2, f_2).$$

Similarly we can introduce the concept of universal integrals on the unit interval $[0, 1]$ (compare the concepts of measure theory and probability theory). In such a case, we deal with normed monotone measures, $m(X) = 1$ (these measures are also called fuzzy measures or capacities), measurable functions $f : X \to [0, 1]$, and the considered pseudo-multiplication $\otimes$ is defined on $[0, 1]^2$, $\otimes : [0, 1]^2 \to [0, 1]$, with neutral element $e = 1$ (then $\otimes$ is called a semicopula, or conjuctor, or weak $t$–norm, depending on the literature). For more details we recommend [5]. Here we recall only two distinguished classes of universal integrals.

**Proposition 1.** *Let $\otimes : [0, 1]^2 \to [0, 1]$ be a fixed pseudo-multiplication. Then the mapping $I_{\otimes} : \bigcup_{(X,\mathcal{A}) \in \mathcal{S}} \left( \mathcal{M}_{(X,\mathcal{A})} \times \mathcal{F}_{(X,\mathcal{A})} \right) \to [0, \infty]$ given by*

$$I_{\otimes}(m, f) = \sup \{ t \otimes m (\{f \geq t\}) \, | t \in [0, \infty] \}$$

*is a universal integral which is the smallest one linked to $\otimes$ through the axiom (UI2).*

Note that the Shilkret integral $I_1$ is related to the standard product $\cdot$, $I_1 = I_{\cdot}$, while $\otimes = \wedge$ (min) yields the famous Sugeno integral [20],

$$I_1(m, f) = Su(m, f) = \sup \{ t \wedge m (\{f \geq t\}) \, | t \in [0, \infty] \}.$$

The second type of universal integrals we recall is defined on $[0, 1]$ and it is linked to copulas.

Observe that a copula $C : [0, 1]^2 \to [0, 1]$ is a pseudo-multiplication on $[0, 1]$ which is supermodular, i.e., for all $\mathbf{x}, \mathbf{y} \in [0, 1]^2$ it holds $C(\mathbf{x} \wedge \mathbf{y}) + C(\mathbf{x} \vee \mathbf{y}) \geq C(\mathbf{x}) + C(\mathbf{y})$. Copulas are in a one-to-one correspondence with probability measures on the Borel subsets of $[0, 1]^2$ with uniformly distributed margins. This link is fully characterized by the equality

$$P_C\left([0, u] \times [0, v]\right) = C(u, v)$$

valid for all $u, v \in [0, 1]$. For more details we recommend [14].

**Proposition 2.** *Let* $C : [0, 1]^2 \to [0, 1]$ *be a fixed copula. Then the mapping* $I_{(C)} : \bigcup_{(X, \mathcal{A}) \in \mathcal{S}} \left(\mathcal{M}^1_{(X, \mathcal{A})} \otimes \mathcal{F}^1_{(X, \mathcal{A})}\right) \to [0, 1]$, *where* $\mathcal{M}^1_{(X, \mathcal{A})} = \{m \in \mathcal{M}_{(X, \mathcal{A})} | m(X) = 1\}$, $\mathcal{F}^1_{(X, \mathcal{A})} = \{f \in \mathcal{F}_{(X, \mathcal{A})} | \operatorname{Ran} f \subseteq [0, 1]\}$, *given by*

$$I_{(C)}(m, f) = P_C\left(\{(u, v) \in [0, 1]^2 | v \leq m(\{f \geq u\})\}\right)$$

*is a universal integral on* $[0, 1]$.

Observe that for the product copula $\Pi$ one have $I_\Pi = T_3$ (restricted to $[0, 1]$), i.e., the Choquet integral is obtained. Similarly, for the greatest copula Min (i.e., for $\wedge$), the Sugeno integral on $[0, 1]$ is obtained, $I_{(Min)} = Su$.

Finally note that integrals $I_2$ (PAN-integral), $I_4$ (Stupňanová integral) and $I_5$ (concave integral of Lehrer) introduced in the previous section are not universal integrals.

## 4    Integrals and Pseudo-Arithmetical Operations

In Section 2, we have tried to answer the question how to integrate under different constraint settings, utilizing as a basic tool for our processing the standard arithmetical operations on the real line. There are possible several modifications of these operations, yielding new types of integrals. First of all, we can rescale our original scale $[0, \infty]$ by means of some automorphism $\varphi : [0, \infty] \to [0, \infty]$ (i.e., $\varphi$ is an increasing bijection). Then the standard addition $+$ becomes a pseudo-addition $\oplus : [0, \infty]^2 \to [0, \infty]$ given by

$$u \oplus v = \varphi^{-1}\left(\varphi(u) + \varphi(v)\right).$$

Similarly, pseudo-multiplication $\otimes : [0, \infty]^2 \to [0, \infty]$ is given by

$$u \otimes v = \varphi^{-1}\left(\varphi(u) \cdot \varphi(v)\right).$$

Modifying $I_1$ (Shilkret integral) into

$$I_{1, \varphi}(m, f) = \sup\left\{k \otimes m(A) \mid k \cdot 1_A \leq f\right\}$$

one gets the universal integral $I_\otimes$. However, $I_\otimes(m, f) = \varphi^{-1}\left(I_1(\varphi \cdot m, \varphi \cdot f)\right)$, i.e., we have a $\varphi$-transform of $I_1$ only. Similarly, the remaining integrals $I_i, i =$

2, 3, 4, 5, can be transformed. Note that the transformed Choquet integral $I_{3,\varphi}$ is a special instance of Choquet-like integrals introduced by Mesiar [9].

Pseudo-addition $\oplus$ and pseudo-multiplication $\otimes$ can be introduced axiomatically, see e.g. [1]. Not going more deeply into details, recall only that for $\oplus = \vee$ (max, supremum), and any pseudo-multiplication $\otimes$ as given in Definition 1, when replacing $+$ by $\vee$ and $\cdot$ by $\otimes$, in the definition of integrals $I_i, i = 1, \cdots, 5$, all of them collapse into the universal integral $I_\otimes$ characterized in Proposition 1. For a deeper overview of integrals based on pseudo-arithmetical operations (on finite spaces) we recommend [11].

## 5 Concluding Remarks

We have discussed the integrals, first from historical point of view, and then as optimization procedures when considering different constraints settings. The concept of universal integrals on $[0, \infty]$ and on $[0, 1]$ was also given, and several positive and negative examples were added.

Note that the axiomatic approach to several of introduced integrals was introduced several years after their constructive introduction. This is not the case of Riemann integral only, but for example the Choquet integral was axiomatized by Schmeidler in 1986 [16]. For an overview of axiomatic approaches to integrals we recommend [6].

Adding some constraints on monotone measures, one can get some distinguished aggregation functions. So, for example, when considering universal integrals on $[0, 1]$ and symmetric monotone measure on finite space $X$ (i.e., $m(A)$ depends on the cardinality of $A$ only), then the Choquet integral becomes OWA operator [22], [3], and copula-based integral $I_{(C)}(m, \cdot)$ becomes OMA operator [10] (i.e., ordered modular average).

Integrals can be also combined. So, for example, any convex combination $I = \lambda I_{(1)} + (1 - \lambda) I_{(2)}$, of two universal integrals related to pseudo-multiplications $\otimes_1$ and $\otimes_2$ with the same neutral element $e$ is a universal integral related to the pseudo-multiplication $\otimes = \lambda \otimes_1 + (1 - \lambda) \otimes_2$, independently of $\lambda \in [0, 1]$. For two copulas $C_1, C_2$, also $C = \lambda C_1 + (1 - \lambda) C_2$ is a copula, and then $I_{(C)} = \lambda I_{C_1} + (1 - \lambda) I_{C_2}$. Another approach to combine integrals was proposed by Narukawa and Torra [12], and multidimensional integrals were introduced and discussed by the same authors in [13].

As we see, though we have touched the problem how to integrate, this area is an expanding field attracting an intensive research and we believe to see not only many new theoretical results soon, but first of all numerous applications in several engineering and human reasoning connected branches.

**Acknowledgement.** The work on this contribution was supposed by grants VEGA 1/0171/12 and GAČR P–402–11–0378.

# References

1. Benvenuti, P., Mesiar, R., Vivona, D.: Monotone set functions-based integrals. In: Pap, E. (ed.) Handbook of Measure Theory, vol. II, ch. 33, pp. 1329–1379. Elsevier Science, Amsterdam (2002)
2. Event, Y., Lehrer, E.: Decomposition-Integral: Unifying Choquet and the Concave Integrals (preprint)
3. Grabisch, M.: Fuzzy integral in multicriteria decision making. Fuzzy Sets and Systems 69, 279–298 (1995)
4. Choquet, G.: Theory of capacities. Ann. Inst. Fourier 5, 131–295 (1953 - 1954)
5. Klement, E.P., Mesiar, R., Pap, E.: A Universal Integral as Common Frame for Choquet and Sugeno Integral. IEEE Transactions on Fuzzy Systems 18(1), 178–187 (2010)
6. Klement, E.P., Mesiar, R.: Discrete integrals and axiomatically defined functionals. Axioms 1 (2012), doi:10.3390/axioms 1010009, 9–20
7. Lebesgue, H.: Leçons sur l'intégration et la recherche des fonctions primitives. Gauthier–Villars, Paris (1904)
8. Lehrer, E.: A new integral for capacities. Econom. Theory 39, 157–176 (2009)
9. Mesiar, R.: Choquet-like integrals. J. Math. Anal. Appl. 194, 477–488 (1995)
10. Mesiar, R., Mesiarová-Zemánková, A.: The ordered modular averages. IEEE Trans. Fuzzy Systems 19, 42–50 (2011)
11. Mesiar, R., Li, J., Pap, E.: Discrete pseudo-integrals. Int. J. Approximate Reasoning (in press)
12. Narukawa, Y., Torra, V.: Generalized transformed t–conorm and multifold integral. Fuzzy Sets and Systems 157, 1354–1392 (2006)
13. Narukawa, Y., Torra, V.: Multidimensional generalized fuzzy integral. Fuzzy Sets and Systems 160, 802–815 (2009)
14. Nelsen, R.B.: An introduction to copulas, 2nd edn. Springer Series in Statistics. Springer, New York (2006)
15. Riemann, B.: Über die Darstellbarkeit einer Function durch eine trigonometrische Reihe. Habilitation Thesis, University of Göttingen (1854)
16. Schmeidler, D.: Integral representation without additivity. Proc. Amer. Math. Soc. 97, 255–261 (1986)
17. Shilkret, N.: Maxitive measure and integration. Indag. Math. 33, 109–116 (1971)
18. Chae, S.B.: Lebesgue integration. Marcel Dekker, Inc., New York (1980)
19. Stupňanová, A.: A Note on Decomposition Integrals. In: Greco, S., Bouchon-Meunier, B., Coletti, G., Fedrizzi, M., Matarazzo, B., Yager, R.R. (eds.) IPMU 2012, Part IV. CCIS, vol. 300, pp. 542–548. Springer, Heidelberg (2012)
20. Sugeno, M.: Theory of fuzzy integrals and its applications. PhD thesis, Tokyo Institute of Technology (1974)
21. Vitali, G.: Sulla definizione di integrale delle funzioni di una variabile. Ann. Mat. Pura ed Appl. IV 2, 111–121 (1925)
22. Yager, R.R.: On ordered weighted averaging aggregation operators in multicriteria decisionmaking. IEEE Trans. Systems Man Cybernet. 18, 183–190 (1988)
23. Yang, Q.: The pan-integral on fuzzy measure space. Fuzzy Mathematics 3, 107–114 (1985) (in Chinese)

# Combination and Soft-Normalization of Belief Functions on MV-Algebras

Tommaso Flaminio[1], Lluís Godo[1], and Tomáš Kroupa[2]

[1] IIIA Artificial Intelligence Research Institute (CSIC)
Campus UAB s/n, Bellaterra 08193, Spain
{tommaso,godo}@iiia.csic.es
[2] Institute of Information Theory and Automation of the ASCR
Pod Vodárenskou věží 4, 182 08 Prague, Czech Republic
kroupa@utia.cas.cz

**Abstract.** Extending the notion of belief functions to fuzzy sets leads to the generalization of several key concepts of the classical Dempster-Shafer theory. In this paper we concentrate on characterizing normalized belief functions and their fusion by means of a generalized Dempster rule of combination. Further, we introduce soft-normalization that arises by either rising up the usual level of contradiction above 0, or by decreasing the classical level of normalization below 1.

## 1 Introduction

The *Dempster-Shafer theory of evidence* [4,13] is a generalization of Bayesian probability theory that allows to combine all the available informations about a given event $E$ into a unique one. The theory shows how all the available evidences can be used to evaluate the degree of belief of $E$ via a *belief function bel*. In fact, in the classical setting, pieces of evidence are encoded by means of subsets of a fixed domain $X$ called the *frame of discernment*. To each piece of evidence (i.e. to each subset of $X$) is attached a weight (called *mass* in Dempster-Shafer theory) that is given by a probability distribution $m$ defined over the powerset $2^X$. If a subset is assigned a strictly positive mass, it is called a *focal element*.

Specifically, our belief is encoded by a mass assignment $m : 2^X \to [0,1]$, that is, $\sum_{B \in 2^X} m(B) = 1$ and $m(\emptyset) = 0$. Its associated *belief function bel* $: 2^X \to [0,1]$ attaches to each $A \in 2^X$ the sum of the masses of those pieces of evidence supporting $A$, i.e.

$$bel(A) = \sum_{B \subseteq A} m(B). \tag{1}$$

It is worth noticing that, since every mass $m$ is a probability distribution over $2^X$, the belief of $A$ can be equivalently defined as

$$bel(A) = P_m(\beta_A) \tag{2}$$

V. Torra et al. (Eds.): MDAI 2012, LNAI 7647, pp. 23–34, 2012.

where $P_m$ is the probability measure defined over $2^{(2^X)}$ and $\beta_A$ is the character-istic function[1] of the inclusion set $\{B \in 2^X : B \subseteq A\}$.

Recently, several generalizations of belief function theory to the algebraic setting of MV-algebras of continuous fuzzy sets have been proposed [9,5]. The soft-computational setting of fuzzy sets and the related algebraic framework open the door to the generalization of the key concepts that form the basis of classical Dempster-Shafer theory. In this paper, after some needed preliminaries on MV-algebras of fuzzy sets and finitely additive measures on them, called *states*, we first recall those generalized notions of belief functions. For the particular class of belief functions whose focal elements are crisp, we also study their Möbious transform. Then, always in the generalized setting of MV-algebras of continuous fuzzy sets, we discuss the notion of normalized belief function and characterize it in terms of the support of the state underlying it. Finally, after recaling some generalized forms of the Dempster rule of combination (not only conjunctive), we consider a notion of *soft-normalization* that arises by either rising up above 0 the usual levels of contradiction, or by decreasing the classical level of normalization below 1.

## 2   MV-Algebras of Fuzzy Sets and States

MV-algebras were introduced by Chang [1] as the equivalent algebraic seman-tics for the infinite-valued Łukasiewicz calculus. They are algebraic structures $M = (M, \oplus, \neg, 0)$ of type $(2, 1, 0)$ satisfying the following requirements: the reduct $(M, \oplus, 0)$ is a commutative monoid, and for every $a, b \in M$, the following equations hold: $\neg\neg a = a$, $a \oplus \neg 0 = \neg 0$ and $\neg(\neg a \oplus b) \oplus b = \neg(\neg b \oplus a) \oplus a$.

It is well known [2] that the class of MV-algebras forms an algebraic variety. Moreover, in every MV-algebra the following operations are definable: $a \odot b$ is $\neg(\neg a \oplus \neg b)$; $a \Rightarrow b$ is $\neg a \oplus b$; $a \vee b$ is $(a \Rightarrow b) \Rightarrow b$, $a \wedge b$ is $\neg(\neg a \wedge \neg b)$, and the constant 1 stands for $\neg 0$. In every MV-algebra $M$, a partial order relation is defined as follows: for every $a, b \in M$, $a \leq b$ iff $a \Rightarrow b = 1$. An MV-algebra is said to be linearly ordered (or an MV-chain), if the order $\leq$ is linear.

*Example 1.* (1) Every Boolean algebra is an MV-algebra. Moreover, for every MV-algebra $M$, the set of its idempotent elements $B(M) = \{a \in M : a \oplus a = a\}$ is the domain of the largest Boolean subalgebra of $M$, the so called *Boolean skeleton* of $M$.

(2) Consider the real unit interval $[0,1]$ equipped with Łukasiewicz operations: for every $a, b \in [0,1]$,

$$a \oplus b = \min\{1, a + b\}, \qquad \neg a = 1 - a.$$

Then the structure $[0,1]_{MV} = ([0,1], \oplus, \neg, 0)$ is an MV-chain. Chang theorem [1,2] says that an equation holds in $[0,1]_{MV}$ iff it holds in every MV-algebra.

---

[1] Throughout the paper, we make no formal distinction between a set and its charac-teristic function.

It is worth noticing that in $[0,1]_{MV}$ the above introduced operations have the following form:

$$a \odot b = \max\{0, a + b - 1\}, \, a \Rightarrow b = \min\{1, 1 - a + b\},$$
$$a \vee b = \max\{a, b\}, \qquad a \wedge b = \min\{a, b\}.$$

(3) For every $n \in \mathbb{N}$, consider the class $F_n$ of $n$-place McNaughton functions, i.e. functions from $[0,1]^n$ into $[0,1]$ which are continuous, piecewise linear, each piece having integer coefficient. The algebra $(F_n, \oplus, \neg, \overline{0})$ with operations $\oplus$ and $\neg$ defined pointwise, and where $\overline{0}$ here denotes the zero-constant function, is an MV-algebra that coincides with the free MV-algebra over $n$ generators. We will henceforth denote this algebra by Free$(n)$.

An *MV-clan over a set* $X$ is a collection of functions from $X$ into $[0,1]$ (i.e. a set of fuzzy subsets of $X$) that contains the zero-constant function and that is closed under the finitary pointwise application of $\oplus$ and $\neg$ as defined in $[0,1]_{MV}$. We will denote by $[0,1]^X$ the clan of all functions from $X$ into $[0,1]$. A clan $M \subseteq [0,1]^X$ is said to be *separating* if for every $x_1, x_2 \in X$ with $x_1 \neq x_2$, there exists a function $f \in M$ such that $f(x_1) \neq f(x_2)$. Clearly, $[0,1]^X$ is separating, and it is well known that for every $n \in \mathbb{N}$, Free$(n)$ is a separating MV-clan as well (cf. [2, §3.6]).

Whenever $X$ is finite, we will call $[0,1]^X$ a *finite domain* MV-clan. Finite domain MV-clans will play a central role in this paper. The following notion of *state* is the MV-counterpart of the notion of a finitely-additive probability measure on a Boolean algebra.

**Definition 1 ([11]).** *Let $M$ be an MV-algebra. A* state *on $M$ is a map* $\mathbf{s} : M \to [0,1]$ *satisfying* $\mathbf{s}(1) = 1$, *and* $\mathbf{s}(a \oplus b) = \mathbf{s}(a) + \mathbf{s}(b)$ *whenever* $a \odot b = 0$. *A state* $\mathbf{s}$ *is said to be* faithful *if* $\mathbf{s}(x) = 0$ *implies* $x = 0$.

The following theorem, independently proved in [8] and [12], shows an integral representation of states by Borel probability measures defined on the $\sigma$-algebra $\mathfrak{B}(X)$ of Borel subsets of $X$, where $X$ is any compact Hausdorff topological space.

**Theorem 1.** *Let $M \subseteq [0,1]^X$ be a separating clan of continuous functions over a compact Hausdorff space $X$. Then there is a one-to-one correspondence between the states on $M$ and the regular Borel probability measures on $\mathfrak{B}(X)$. In particular, for every state $\mathbf{s}$ on $M$, there exists a unique regular Borel probability measure $\mu$ on $\mathfrak{B}(X)$ such that for every $f \in M$,*

$$\mathbf{s}(f) = \int_X f \, d\mu. \tag{3}$$

## 3   Belief Functions on MV-Algebras of Fuzzy Sets

In what follows we will assume $X$ to be a finite set.

## 3.1   Crisp-Focal Belief Functions

In [9], the author proposes the following generalization of belief functions. Let $M = [0,1]^X$ be a finite domain MV-clan and consider, for every $f : X \to [0,1]$, the map $\hat{\rho}_f : 2^X \to [0,1]$ defined as follows: for every $B \subseteq X$,

$$\hat{\rho}_f(B) = \min\{f(x) : x \in B\}. \tag{4}$$

*Remark 1.* Notice that $\hat{\rho}_f$ generalizes $\beta_A$ in the following sense: whenever $A \in B(M) = 2^X$, then $\hat{\rho}_A = \beta_A$. Namely, for every $A \in B(M)$, $\hat{\rho}_A(B) = 1$ if $B \subseteq A$, and $\hat{\rho}_A(B) = 0$, otherwise.

**Definition 2.** *A map* $\hat{\mathbf{b}} : M \to [0,1]$ *is called a* crisp-focal belief function *whenever there is a state* $\hat{\mathbf{s}} : [0,1]^{(2^X)} \to [0,1]$ *such that* $\hat{\mathbf{s}}(\{\emptyset\}) = 0$ *and, for every* $f \in M$,

$$\hat{\mathbf{b}}(f) = \hat{\mathbf{s}}(\hat{\rho}_f). \tag{5}$$

With $X$ being finite, Theorem 1 yields a unique probability measure $\mu : 2^{(2^X)} \to [0,1]$ such that $\hat{\mathbf{s}}(\hat{\rho}_f) = \sum_{C \in 2^X} \hat{\rho}_f(C) \cdot \mu(\{C\})$. Moreover, it is easy to see that, for every $C \subseteq 2^X$, $\mu(\{C\}) = \hat{\mathbf{s}}(\{C\})$. Since $\mu(\{\emptyset\}) = 0$, probability measure $\mu$ induces a mass assignment $m$ such that $m(C) = \mu(\{C\})$.

In Dempster-Shafer theory, given a belief function $bel : 2^X \to [0,1]$, the mass $m$ that defines $bel$ can be recovered from $bel$ by Möbius transform:

$$m(A) = \sum_{B \subseteq A}(-1)^{|A \setminus B|} bel(B).$$

In case of crisp-focal belief functions, the situation is analogous.

**Proposition 1.** *Let* $\hat{\mathbf{b}} : [0,1]^X \to [0,1]$ *be a crisp-focal belief function, defined as* $\hat{\mathbf{b}}(f) = \hat{\mathbf{s}}(\rho_f)$ *for some state* $\hat{\mathbf{s}}$ *on* $[0,1]^{2^X}$ *such that* $\hat{\mathbf{s}}(\{\emptyset\}) = 0$ *and* $\hat{\mathbf{s}}(\{C\}) > 0$ *iff* $C(x) \in \{0,1\}$, *where* $C \neq \emptyset$. *Then*

$$\hat{\mathbf{s}}(\{A\}) = m(A) = \sum_{B \subseteq A}(-1)^{|A \setminus B|}\hat{\mathbf{b}}(B)$$

*for each* $A \subseteq X$.

*Proof.* Definition (5) directly gives that $\hat{\rho}_A(C), \in \{0,1\}$ for each pair of crisp sets $A, C \subseteq X$ and thus

$$\hat{\mathbf{b}}(A) = \sum_{C \in 2^X} \hat{\rho}_A(C) \cdot \hat{\mathbf{s}}(\{C\}) = \sum_{B \subseteq A} \hat{\mathbf{s}}(\{B\}) = \sum_{B \subseteq A} m(B).$$

This implies that the restriction of $\hat{\mathbf{b}}$ to $2^X$ is a classical belief function. See [10] for further details. $\qquad \square$

As a corollary, observe that, in the hypothesis of the above proposition, the values $\hat{\mathbf{b}}(f)$ for non-crisp $f \in [0,1]^X$ are necessarily determined by the values of $\hat{\mathbf{b}}$ over the crisp sets of $2^X$. Indeed, in [9] it shown that, for any $f \in [0,1]^X$, $\hat{\mathbf{b}}(f)$ is in fact the Choquet integral of $f$ with respect to the restriction of $\hat{\mathbf{b}}$ over $2^X$.

Moreover, this shows another characterization of crisp-focal belief functions. Indeed, a function $bel : [0,1]^X \to [0,1]$ is a crisp-focal belief function iff its restriction on crisp sets $2^X$ is a total monotone function, i.e., for every natural $n$ and every $A_1, \ldots, A_n \in 2^X$, the following inequality holds:

$$bel\left(\bigvee_{i=1}^{n} A_i\right) \geq \sum_{\emptyset \neq I \subseteq \{1,\ldots,n\}} (-1)^{|I|+1} \cdot bel\left(\bigwedge_{k \in I} A_k\right).$$

## 3.2   General Belief Functions

The definition introduced in [5] generalizes crisp-focal belief function by introducing, for every $f \in M$, a map $\rho_f$ associating with each *fuzzy set* $g \in M$ the degree of inclusion of $g$ into $f$. Specifically, let $M = [0,1]^X$ be a finite domain MV-clan, and consider, for every $f \in M$, the map $\rho_f : M \to [0,1]$ defined as follows: for every $g \in M$,

$$\rho_f(g) = \min\{g(x) \Rightarrow f(x) : x \in X\}. \tag{6}$$

The choice of $\Rightarrow$ in the above definition is due to the MV-algebraic setting, but other choices could be made in other fuzzy logics.

Those mappings $\rho_f$ can be regarded as *generalized inclusion operators* between fuzzy sets (cf. [5] for further details). For every $f \in \{0,1\}^X$ (i.e. whenever $f$ is identified with a vector in $[0,1]^X$ with *integer* components), the map $\rho_f : [0,1]^X \to [0,1]$ is a pointwise minimum of finitely many linear functions with integer coefficients, and hence $\rho_f$ is a non-increasing McNaughton function [2].

**Lemma 1.** *The MV-algebra $\mathcal{R}_2$ generated by the set $\varrho_2 = \{\rho_a : a \in \{0,1\}^X\}$ coincides with* Free$(n)$, *where $n$ is the cardinality of $X$.*

*Proof.* By [3, Theorem 3.13], if a variety $\mathbb{V}$ of algebras is generated by an algebra $A$, then the free algebra over a cardinal $n > 0$ is, up to isomorphisms, the subalgebra of $A^{A^X}$ generated by the projection functions $\theta_i : A^X \to A$. Therefore, in order to prove our claim it suffices to show that the projection functions $\theta_1, \ldots, \theta_n$ belong to $\varrho_2$.

Consider, for every $i = 1, \ldots, n$ the point $\bar{i} \in \{0,1\}^X$ such that

$$\bar{i}(j) = \begin{cases} 0, & \text{if } j = i \\ 1, & \text{otherwise.} \end{cases}$$

Then $\rho_{\bar{i}} = 1 - \theta_i$. In fact, for every $b \in [0,1]^X$, and for every $i, j \in X$ such that $j \neq i$, we have $b(j) \to \bar{i}(j) = 1$, and $b(i) \to \bar{i}(i) = 1 - b(i)$, so that $1 - \rho_{\bar{i}(b)} = \theta_i(b) = b(i)$. This actually shows that the MV-algebra $\mathcal{R}_2^-$ generated by the set $\neg \varrho_2 = \{1 - \rho_a : a \in \{0,1\}^X\}$ is isomorphic to Free$(n)$. Clearly $\mathcal{R}_2$ and $\mathcal{R}_2^-$ are isomorphic through the map $g : a \in \mathcal{R}_2 \mapsto 1 - a \in \mathcal{R}_2^-$.   □

Therefore, if we consider the MV-algebra $\mathcal{R}$ generated by $\varrho = \{\rho_f : f \in [0,1]^X\}$ we obtain a semisimple MV-algebra that properly extends Free$(n)$, and whose elements are continuous functions from $[0,1]^X$ into $[0,1]$. This implies, in particular, that $\mathcal{R}$ is separating.

**Definition 3.** *Let $X$ be a finite set and let $M = [0,1]^X$. A map $\mathbf{b} : M \to [0,1]$ will be called a* belief function *on the finite domain MV-clan $M$ provided there exists a state $\mathbf{s} : \mathcal{R} \to [0,1]$ such that for every $a \in M$,*

$$\mathbf{b}(a) = \mathbf{s}(\rho_a). \tag{7}$$

*We will denote by $Bel(M)$ the class of all the belief functions over a finite domain MV-clan $M$.*

Note that if $\mathbf{s}$ is such that the set $\{f \in M \mid \mathbf{s}(\{f\}) > 0\}$ is countable, then the above expression yields

$$\mathbf{b}(a) = \sum_{f \in M} \rho_a(f) \cdot \mathbf{s}(\{f\}).$$

As in the previous section, we will identify the mass of a belief function $\mathbf{b}$ with the unique Borel regular probability measure $\mu$ over $\mathfrak{B}([0,1]^X)$ that represents the state $\mathbf{s}$ via Theorem 1.

Since belief functions on $[0,1]^X$ are defined as states on $\mathcal{R}$ and different states $\mathbf{s}_1$ and $\mathbf{s}_2$ determine different belief functions $\mathbf{b}_1$ and $\mathbf{b}_2$, the set $Bel([0,1]^X)$ of belief functions on $[0,1]^X$ is in 1-1 correspondence with the set $\mathcal{S}(\mathcal{R})$ of all states on $\mathcal{R}$. Hence $Bel([0,1]^X)$ is a compact convex subset of $[0,1]^{[0,1]^X}$. Therefore Krein-Mil'man theorem shows that $Bel([0,1]^X)$ is in the closed convex hull of its extremal points. The following result characterizes the extremal points of $Bel([0,1]^X)$.

**Proposition 2.** *For every $x \in [0,1]^X$, the belief function $\mathbf{b}_x$ defined by*

$$\mathbf{b}_x(f) = \mathbf{s}_x(\rho_f) = \rho_f(\{x\}), \quad f \in [0,1]^X, \tag{8}$$

*is an extremal point of $Bel([0,1]^X)$.*

*Proof.* A belief function $\mathbf{b} \in Bel([0,1]^X)$ is extremal iff its state assignment is extremal in $\mathcal{S}(\mathcal{R})$. In fact $\mathbf{s}$ is not extremal iff there exist $\mathbf{s}_1, \mathbf{s}_2 \in \mathcal{S}(\mathcal{R})$ and a real number $\lambda \in (0,1)$ such that $\mathbf{s} = \lambda \mathbf{s}_1 + (1-\lambda)\mathbf{s}_2$. In particular, for every $a \in [0,1]^X$,

$$\mathbf{b}(a) = \mathbf{s}(\rho_a) = \lambda \mathbf{s}_1(\rho_a) + (1-\lambda)\mathbf{s}_2(\rho_a) = \lambda \mathbf{b}_1(a) + (1-\lambda)\mathbf{b}_2(a),$$

whence $\mathbf{b}$ would not be extremal as well.

□

As we recalled above, $\mathcal{R}$ is separating. Therefore from Proposition 2 the extreme points of its state space are MV-homomorphisms $\mathbf{s}_x$, for each $x \in [0,1]^X$. Hence the following holds due to (8).

**Theorem 2.** *Every belief function is a pointwise limit of a convex combination of some elements $\rho_.(x^1), \ldots, \rho_.(x^k)$, where $x^1, \ldots, x^k \in [0,1]^X$.*

### 3.3    On Normalized Belief Functions

In classical Dempster-Shafer theory, the notion of focal element is crucial for classifying belief functions. Whenever $X = \{1, \ldots, n\}$ is a finite set, the Boolean algebra $2^X$ is finite, and hence the mass assignment $m : 2^X \to [0,1]$ defines obviously only finitely many focal elements. On the other hand, the MV-algebra $[0,1]^X$ has uncountably many elements, and hence we cannot find in general a mass assignment $\mu$ defined over $\mathfrak{B}([0,1]^X)$ that defines a belief function $\mathbf{b}$ through (10) which is supported by a finite set. This observation leads to the following definition.

**Definition 4.** *Let $\mathcal{K}$ be the set of all compact subsets of a finite domain MV-clan $[0,1]^X$. For every regular Borel probability measure $\mu$ defined on $\mathfrak{B}([0,1]^X)$, we call the set*

$$\mathrm{spt}\mu = \bigcap \{K | K \in \mathcal{K}, \mu(K) = 1\}$$

*the support of $\mu$.*

By Theorem 1 we can regard $\mathrm{spt}\mu$ as the support of the state $\mathbf{s}$ defined from $\mu$ via (3). In particular, the following holds:

$$\mathbf{b}(a) = \int_{[0,1]^X} \rho_a \, \mathrm{d}\mu = \int_{\mathrm{spt}\mu} \rho_a \, \mathrm{d}\mu. \tag{9}$$

Therefore, for a belief function $\mathbf{b}$ on $[0,1]^X$ whose state assignment $\mathbf{s}$ is characterized through (3) by a regular Borel probability measure $\mu$, we will henceforth refer to $\mathrm{spt}\mu$ as the set of focal elements of $\mathbf{b}$. We restrict our attention to those belief functions on $[0,1]^X$ such that their state assignment $\mathbf{s}$ on Free$(n)$ satisfies the condition

$$\mathbf{b}(\overline{0}) = \mathbf{s}(\rho_{\overline{0}}) = 0. \tag{10}$$

**Proposition 3.** *The set $\mathcal{S}_0$ of all states on $\mathcal{R}$ satisfying (10) is a nonempty compact convex subset of $[0,1]^{\mathcal{R}}$ considered with its product topology.*

*Proof.* $\mathcal{S}_0$ is nonempty: let $\mathbf{s}_1$ be defined by

$$\mathbf{s}_1(f) = f(1, \ldots, 1),$$

for every $f \in \mathcal{R}$. This gives $\mathbf{s}_1(\rho_{\overline{0}}) = \rho_{\overline{0}}(1, \ldots, 1) = 0$ and thus $\mathbf{s}_1 \in \mathcal{S}_0$. Let $\mathbf{s}, \mathbf{s}' \in \mathcal{S}_0$ and $\alpha \in (0,1)$. Then the function $\mathcal{R} \to [0,1]$ given by

$$\alpha\mathbf{s} + (1 - \alpha)\mathbf{s}'$$

is a state on $\mathcal{R}$ which clearly satisfies (10). Hence $\mathcal{S}_0$ is a convex subset of the product space $[0,1]^{\mathcal{R}}$. Since the space $[0,1]^{\mathcal{R}}$ is compact, we only need to show that $\mathcal{S}_0$ is closed (in its subspace product topology). To this end, consider a convergent sequence $(\mathbf{s}_m)_{m \in \mathbb{N}}$ in $\mathcal{S}_0$ whose limit is $\mathbf{s}$. As the set of all states on $\mathcal{R}$ is closed, $\mathbf{s}$ is a state. That $\mathbf{s}$ satisfies (10) follows from $\mathbf{s}(\rho_{\overline{0}}) = \lim_{m \to \infty} \mathbf{s}_m(\rho_{\overline{0}}) = 0$. $\qquad\square$

The family of states $\mathcal{S}_0$ can be characterized by employing integral representation of states. Namely, we will show that a state assignment $\mathbf{s} \in \mathcal{S}_0$ iff $\mathbf{s}$ is "supported" by normal fuzzy sets in $[0,1]^X$, i.e. fuzzy sets $f \in [0,1]^X$ such that $f(x) = 1$ for some $x \in X$. We will denote by $\mathcal{NF}(X)$ the set of normalized fuzzy sets from $[0,1]^X$, i.e.

$$\mathcal{NF}(X) = \{f \in [0,1]^X \mid f(x) = 1 \text{ for some } x \in X\}.$$

**Proposition 4.** *Let $\mathbf{s}$ be a state assignment on $\mathcal{R}$ and $\mu$ be the regular Borel probability measure associated with $\mathbf{s}$. Then $\mathrm{spt}\mu \subseteq \mathcal{NF}(X)$ if and only if $\mathbf{s} \in \mathcal{S}_0$.*

*Proof.* Let $\mu$ be a probability measure on Borel subsets of $[0,1]^X$ such that $\mathrm{spt}\mu \subseteq \mathcal{NF}(X)$. Put

$$\mathbf{s}(f) = \int_{[0,1]^X} f \, d\mu, \quad f \in \mathcal{R}. \tag{11}$$

Since $\rho_{\overline{0}}(x) = 0$ for each $x \in \mathrm{spt}\mu$, it follows that

$$\mathbf{s}(\rho_{\overline{0}}) = \int_{\mathrm{spt}\mu} \rho_{\overline{0}} \, d\mu = 0,$$

hence $\mathbf{s} \in \mathcal{S}_0$. Conversely, assume that

$$\mathbf{s}(\rho_{\overline{0}}) = \int_{[0,1]^X} \rho_{\overline{0}} \, d\mu = 0,$$

which implies $\rho_{\overline{0}} = 0$ $\mu$-almost everywhere over $[0,1]^X$. Since $\rho_{\overline{0}}(x) = 0$ iff $x \in [0,1]^X$ is such that $x_i = 1$, for some $i \in X$, we obtain $\mu(\mathcal{NF}(X)) = 1$. $\qquad\square$

In particular, every state assignment of a generalized belief function in the sense of [9] belongs to $\mathcal{S}_0$.

## 4   Generalized Dempster Rule of Combination

In [5] the authors present a way to generalize the well-known Dempster rule to combine the information carried by two belief functions $\mathbf{b}_1, \mathbf{b}_2 \in Bel(M)$, into a belief function $\mathbf{b}_{1,2} \in Bel(M)$. In this section we will recall the basic steps of that construction, and we also add some remarks. We start with an easy result about the definition of states in a product space.

**Proposition 5.** *For every MV-algebra $N$, and for every pair of states $\mathbf{s}_1, \mathbf{s}_2 : N \to [0,1]$, there exists a state $\mathbf{s}_{1,2}$ defined on the direct product $N \times N$ such that for every $(b,c) \in N \times N$, $\mathbf{s}_{1,2}(b,c) = \mathbf{s}_1(b) \cdot \mathbf{s}_2(c)$.*

Let now $M = [0,1]^X$, and let $\mathcal{R}$ be as defined in Section 3. Also let $\mathbf{s}_1, \mathbf{s}_2$ be two states on $\mathcal{R}$ such that $\mathbf{b}_1(f) = \mathbf{s}_1(\rho_f)$ and $\mathbf{b}_2(f) = \mathbf{s}_2(\rho_f)$ for all $f \in M$. Furthermore, let $\mu_1, \mu_2 : \mathfrak{B}(M) \to [0,1]$ be the two regular probability measures of support $\mathrm{spt}\mu_i$ (for $i = 1,2$), such that for $i = 1,2$,

$$\mathbf{s}_i(f) = \int_{\mathrm{spt}\mu_i} f \, d\mu_i.$$

Take the mapping $\mu_{1,2} : \mathfrak{B}(M \times M) \to [0,1]$ to be the product measure on Borel subsets generated by $M \times M$. Let $\mathbf{s}_{1,2}$ be a state on $[0,1]^{M \times M}$ defined by integrating measurable functions $M \times M \to [0,1]$ with respect to $\mu$. If there exist $g, h : M \to [0,1]$ and $f$ such that $f(x,y) = g(x) \cdot h(y)$, then Proposition 5 yields $\mathbf{s}_{1,2}(f) = \mathbf{s}_1(g) \cdot \mathbf{s}_2(h)$.

Finally, for every $f \in M$, consider the map $\rho_f^{\wedge} : M \times M \to [0,1]$ defined by $\rho_a^{\wedge}(b,c) = \rho_a(b \wedge c)$. Then we are ready to define the following combination of belief functions.

**Definition 5 (Generalized Dempster rule).** *Given $\mathbf{b}_1, \mathbf{b}_2 \in Bel(M)$ as above, define its* min*-conjunctive combination $\mathbf{b}_{1,2} : M \to [0,1]$ as follows: for all $a \in M$,*

$$\mathbf{b}_{1,2}(a) = \mathbf{s}_{1,2}(\rho_a^{\wedge}). \tag{12}$$

Regarding the support of the combined measure, it is worth noticing that by [6, Theorem 417C (v)], $\mathrm{spt}\mu_{1,2} = \mathrm{spt}\mu_1 \times \mathrm{spt}\mu_2$, and hence, whenever $\mu_1$ and $\mu_2$ are normalized in the sense that their support is included into $\mathcal{NF}(X)$, $\mathrm{spt}\mu_{1,2} \subseteq \mathcal{NF}(X)$ as well. Therefore, by Proposition 4 one might deduce that, if $\mathbf{b}_1$ and $\mathbf{b}_2$ are normalized belief functions, then $\mathbf{b}_{1,2}$ is normalized as well. The following example shows that it is not the case, since in the definition of $\mathbf{b}_{1,2}$, together with the product measure $\mu_{1,2}$ we also use the map $\rho^{\wedge}$ which, in fact, is not a genuine fuzzy-inclusion operator.

*Example 2.* Consider two belief functions $\mathbf{b}_1$ and $\mathbf{b}_2$ on $[0,1]^2$ with masses concentrated as follows:

$$\mu_1(1,0) = 1/4;\ \mu_1(1,1) = 3/4;\ \mu_2(0,1) = 1/2;\ \mu_2(1,1) = 1/2.$$

Then, the product measure $\mu_{1,2}$ has support in the cartesian product of the supports of the two masses, $\{((1,0),(0,1)), ((1,0),(1,1)), ((1,1),(0,1)), ((1,1), (1,1))\}$, and it takes values

$$\mu_{1,2}((1,0),(0,1)) = 1/8,\ \mu_{1,2}((1,0),(1,1)) = 1/8,\ \mu_{1,2}((1,1),(0,1)) = 3/8,$$
$$\mu_{1,2}((1,1),(1,1)) = 3/8.$$

So, $\mu_{1,2}$ is normalized in the sense that each of its focal elements can be regarded as a normalized vector of $[0,1]^4$. On the other hand, $\mathbf{b}_{1,2}$ is not normalized because $(0,0) = (1,0) \wedge (0,1)$, $\rho_{(0,0)}(0,0) = 1$, and hence

$$\mathbf{b}(0,0) = \sum_{b \wedge c = (0,0)} \rho_{(0,0)}((0,0)) \cdot \mu_1(b) \cdot \mu_2(c) = \rho_{(0,0)}(0,0) \cdot \mu_1(1,0) \cdot \mu_2(0,1) = 1/8 > 0.$$

The above min-conjunctive combination can easily be extended to well-known MV-operations on fuzzy sets, such as max-disjunction $\vee$, strong conjunction $\odot$ and strong disjunction $\oplus$, by defining $(b_1 \circledast b_2)(a) = s_{1,2}(\rho_a^{\circledast})$, for $\circledast$ being one of these operations, and defining $\rho_a^{\circledast}(b,c) = \rho_a(b \circledast c)$. In this generalized case, the map $\mathbf{b}_{1,2}^{\circledast}$ resulting from the respective combination rule will be called the $\circledast$-*combination of* $\mathbf{b}_1$ *and* $\mathbf{b}_2$.

Whenever the supports of $\mu_1$ and $\mu_2$ are countable, it is easy to prove that $\mathbf{b}_{1,2}^{\circledast}$ is a belief function in the sense of Definition 3. In fact, in this case Definition 5 yields $\mathbf{b}_{1,2}^{\circledast}(a) = \sum_{b,c \in M} \rho_a(b \circledast c) \cdot \mu_1(\{b\}) \cdot \mu_2(\{c\})$. Notice that the above expression reduces to

$$\mathbf{b}_{1,2}(a) = \sum_{d \in M} \sum_{b,c \in M, b \circledast c = d} \rho_a(d) \cdot (\mu_1(\{b\}) \cdot \mu_2(\{c\})) = \sum_{d \in M} \rho_a(d) \cdot \mu^*(\{d\}),$$

where

$$\mu^*(\{d\}) = \sum_{b,c \in M, b \circledast c = d} \mu_1(\{b\}) \cdot \mu_2(\{c\})$$

is indeed a mass assignment and hence $\mathbf{b}_{1,2}^{\circledast} \in Bel(M)$.

Therefore, turning back to the above Example 2 and Proposition 4, there exists a mass $\mu \neq \mu_{1,2}$ for $\mathbf{b}_{1,2}^{\circledast}$ such that $\mathrm{spt}\mu \not\subseteq \mathcal{NF}(X)$.

## 5    Soft Normalization for Mass Assignments

The *height* of a fuzzy set $f \in [0,1]^X$ is defined in the literature as

$$h(f) = \max\{f(x) : x \in X\}. \tag{13}$$

The value $h(f)$ can be interpreted as the degree of normalization of $f$. As a matter of fact, a fuzzy set $f$ is called normalized whenever $h(f) = 1$, otherwise it is called non-normalized. A non-normalized fuzzy set represents a partially inconsistent information.

Consider now a belief function $\mathbf{b}$ defined by a state with support $\mathrm{spt}\mu$. Assume there exists a focal element $f \in \mathrm{spt}\mu$ with $\mu(\{f\}) > 0$ that is a non-normalized fuzzy set.[2] This means that $\mathbf{b}$ is associating a positive degree of evidence to a (partially) inconsistent information, which is reflected on the value that $\mathbf{b}$ assigns to the $\overline{0}$. Indeed, in such a case we have $\rho_{\overline{0}}(f) > 0$, and hence $\mathbf{b}(\overline{0}) \geq \rho_{\overline{0}}(f) \cdot \mu(\{f\}) > 0$. And in fact it is easy to see that the more inconsistent are the focal elements of $\mathbf{b}$, the greater is the value $\mathbf{b}(\overline{0})$. When events and focal elements are crisp sets (and hence the unique possible not-normalized focal element is $\overline{0}$), normalization consists in redistributing the mass that $\mu$ assigns to $\overline{0}$ to the other focal elements of $\mu$ (if any).

Dealing with fuzzy focal elements, allows us to introduce a notion of *soft normalization* for belief functions. In particular, it allows a softer redistribution of the masses, depending on two thresholds.

**Definition 6.** *A mass assignment $\mu : [0,1]^X \to [0,1]$ is said to be $\alpha$-normalized provided that $\inf\{h(f) : f \in \mathrm{spt}\mu\} = \alpha$.*

In other words, a mass is $\alpha$-normalized provided that each focal element of $\mu$ has at least height $\alpha$. In particular, for a belief function $\mathbf{b}$ we define the *degree of normalization* of $\mathbf{b}$ as the value

---

[2] Notice that if $\mathrm{spt}\mu$ is not countable, the condition $f \in \mathrm{spt}\mu$ does not guarantee $\mu(\{f\}) > 0$.

$$\inf\{h(f) : f \in \mathrm{spt}\mu\},$$

where $\mu$ is the mass associated to $\mathbf{b}$.

In what follows we assume masses $\mu$ such that their supports $\mathrm{spt}\mu$ are countable. Let now $\mu : [0,1]^X \to [0,1]$ be an $\alpha$-normalized mass assignment, and assume that there exists a focal element $g$ for $m$ such that $h(g) = \beta > \alpha$.

The mass $\mu$ can be renormalized to the higher degree $\beta$ by defining a new mass $\mu^\beta$ as follows: for every $f \in \mathcal{F}(X)$,

$$\mu^\beta(\{f\}) = \begin{cases} 0, & \text{if } h(f) < \beta \\ \frac{\mu(\{g\})}{1-K}, & \text{otherwise.} \end{cases} \tag{14}$$

where $K = \sum_{h(l)<\beta} \mu(\{l\})$.

The idea of this $\beta$-normalization, similarly to the classical normalization, consists in fixing the value $\beta$ as a new level of consistency for the mass we are considering. Since $\alpha < \beta \le 1$, the class of focal elements of height lower then $\beta$ is not empty. Then the process of $\beta$-normalization consists in redistributing all the mass $K = \sum_{h(l)<\beta} \mu(\{l\})$, which $\mu$ assigns to the fuzzy sets of height lower than $\beta$, to those focal elements of height greater of (or equal to) $\beta$.

Clearly a mass $\mu$ can be renormalized, up to the maximum value

$$\beta_{max} = \sup\{h(f) : f \in \mathrm{spt}(\mu)\}.$$

Consider two belief functions $\mathbf{b}_1$ and $\mathbf{b}_2$ with associated masses $\mu_1$ and $\mu_2$ respectively, also we assume for simplicity $\mathrm{spt}\mu_1$ and $\mathrm{spt}\mu_2$ to be countable. Let $\mathbf{b}_{1,2}^\circledast$ the belief function defined by the $\circledast$-combination of $\mathbf{b}_1$ and $\mathbf{b}_2$ as we introduced in Section 4. Then the focal elements of $\mathbf{b}_{1,2}^\circledast$ forms the following set:

$$\{f \circledast g : f \in \mathrm{spt}\mu_1 \text{ and } g \in \mathrm{spt}\mu_2\}.$$

Therefore for each focal element $f \circledast g$ of $\mathbf{b}_{1,2}^\circledast$, its height is easily calculated as $h(f \circledast g) = \max\{f(x) \circledast g(x) : x \in X\}$. Therefore the level to which a $\circledast$ combined belief function $\mathbf{b}_{1,2}^\circledast$ allows to be normalized can be similarly calculated by the height of the focal elements of the combining functions $\mathbf{b}_1$ and $\mathbf{b}_2$. It is worth to point out that, whenever $\circledast$ is a conjunctive operation (like a t-norm for instance), $h(f \circledast g) \le \min\{h(f), h(g)\}$.

**Acknowledgments.** Flaminio and Godo acknowledge partial support from the Spanish projects TASSAT (TIN2010-20967-C04-01) and ARINF (TIN2009-14704-C03-03). The work of Tomáš Kroupa was supported by the grant P402/12/1309 of Czech Science Foundation.

# References

1. Chang, C.C.: Algebraic Analysis of Many-valued Logics. Trans. Am. Math. Soc. 88, 467–490 (1958)
2. Cignoli, R., D'Ottaviano, I.M.L., Mundici, D.: Algebraic Foundations of Many-valued Reasoning. Kluwer, Dordrecht (2000)

3. Cohn, P.M.: Universal Algebra. Revisited Edition. D. Reidel Pub. Co., Dordrecht (1981)
4. Dempster, A.P.: Upper and lower probabilities induced by a multivalued mapping. The Annals of Mathematical Statistics 38(2), 325–339 (1967)
5. Flaminio, T., Godo, L., Marchioni, E.: Belief Functions on MV-Algebras of Fuzzy Events Based on Fuzzy Evidence. In: Liu, W. (ed.) ECSQARU 2011. LNCS (LNAI), vol. 6717, pp. 628–639. Springer, Heidelberg (2011)
6. Fremlin, D.H.: Measure theory, vol. 4. Torres Fremlin, Colchester (2006), Topological measure spaces. Part I, II, Corrected second printing of the 2003 original
7. Goodearl, K.R.: Partially Ordered Abelian Group with Interpolation. AMS Math. Survey and Monographs 20 (1986)
8. Kroupa, T.: Every state on semisimple MV-algebra is integral. Fuzzy Sets and Systems 157, 2771–2782 (2006)
9. Kroupa, T.: From Probabilities to Belief Functions on MV-Algebras. In: Borgelt, C., González-Rodríguez, G., Trutschnig, W., Lubiano, M.A., Gil, M.Á., Grzegorzewski, P., Hryniewicz, O. (eds.) Combining Soft Computing and Statistical Methods in Data Analysis. AISC, vol. 77, pp. 387–394. Springer, Heidelberg (2010)
10. Kroupa, T.: Extension of Belief Functions to Infinite-valued Events. Soft Computing (to appear), doi:10.1007/s00500-012-0836-2
11. Mundici, D.: Averaging the truth-value in Łukasiewicz logic. Studia Logica 55(1), 113–127 (1995)
12. Panti, G.: Invariant measures in free MV-algebras. Communications in Algebra 36(8), 2849–2861 (2008)
13. Shafer, G.: A Mathematical Theory of Evidence. Princeton University Press, Princeton (1976)

# Using Linear Programming
# for Weights Identification
# of Generalized Bonferroni Means in R

Gleb Beliakov and Simon James

School of Information Technology, Deakin University
221 Burwood Hwy, Burwood 3125, Australia
{gleb,sjames}@deakin.edu.au

**Abstract.** The generalized Bonferroni mean is able to capture some interaction effects between variables and model mandatory requirements.
We present a number of weights identification algorithms we have developed in the R programming language in order to model data using
the generalized Bonferroni mean subject to various preferences. We then
compare its accuracy when fitting to the journal ranks dataset.

**Keywords:** Aggregation functions, means, generalized Bonferroni mean,
weights identification, least absolute deviation (LAD) fitting.

## 1 Introduction

In decision-making and information processing contexts, the need often arises to
merge multiple inputs into a single representative output. For more sophisticated
aggregation functions to find use in everyday applications, ways of interpreting
their behavior and implementation tools need to be developed to make them
accessible to practitioners. In recent years, such developments include the *Kappalab* R package by Grabisch et al. [13], and *AOTool* and *fmtools* by Beliakov [1].
These tools allow the parameters and weights of aggregation functions to be automatically learned from data and used to predict unknown values or analyze
the datasets.

The Bonferroni mean [11] is an aggregation function with the ability to model
mandatory requirements, i.e. we can ensure that some criteria are at least partially satisfied for a high overall score. Since it was generalized by Yager in [21] a
number of publications have followed, with generalizations refined in [8, 15, 22],
extensions to higher level fuzzy sets in [7, 18–20] and lattices [6]. As well as
modeling mandatory requirements, the Bonferroni mean could also be useful as
a non-linear function which is able to capture interaction effects. Indeed, in its
original form the terms of the function are similar to those in statistics used to
model interaction between pairs of variables in regression models.

There are a number of ways to construct aggregation functions for applications. Sometimes the parameters can be specified by experts while in other cases
we may have an existing dataset and we want a model that reflects the relationship between the inputs and outputs. In the latter case, we can use optimization

V. Torra et al. (Eds.): MDAI 2012, LNAI 7647, pp. 35–44, 2012.

techniques and perform fitting in order to learn the parameters or weights of the function we wish to use in the model. Due to the composition of the generalized Bonferroni mean, however, a number of issues arise in attempting to learn its weights from data. In general, the problem is not one that can be framed as a linear or quadratic program. In this paper we give an overview of some approaches we have taken and implemented in the R programming language [16]. The fitting algorithms we have developed, as well as our datasets and preprocessing techniques are available as R source files at our website[1].

We investigate three simplifications that allow the problem to be formulated as a linear programming problem and compare the accuracy of the resulting Bonferroni means to other functions with a real data set.

The paper will be structured as follows. In Section 2, we give an overview of aggregation functions and how they can be fit to data. It is here that we also provide the definition of the generalized Bonferroni mean. In Section 3 we show how the weights identification problem for the Bonferroni mean can be formulated linearly, using the same techniques as are employed in fmtool. We then show how the Bonferroni mean compares to other functions when fit to some journal rankings datasets in Section 4. As well as fitting to each full journal set, we also use 10-fold cross-validation tests to show the robustness of the fitting process. We give a brief summary in Section 5.

## 2     Preliminaries

We will give an overview here of the definitions and methods that will be used throughout the rest of the paper. In particular, we are concerned with Aggregation functions and techniques for learning their parameters from data.

### 2.1     Aggregation Functions

The study of aggregation functions for decision making and information processing applications has become increasingly widespread. A number of recent monographs give an overview of their use and properties, [9,14,17]. We will consider aggregation functions defined over the unit interval.

**Definition 1.** *An aggregation function* $f : [0,1]^n \to [0,1]$ *is a function non-decreasing in each argument and satisfying* $f(0,\ldots,0) = 0$ *and* $f(1,\ldots,1) = 1$.

**Definition 2.** *An aggregation function is considered to be:* averaging *where* $\min(\mathbf{x}) \leq f(\mathbf{x}) \leq \max(\mathbf{x})$, conjunctive *where* $f(\mathbf{x}) \leq \min(\mathbf{x})$, disjunctive *where* $f(\mathbf{x}) \geq \max(\mathbf{x})$, *and* mixed *otherwise.*

Due to the monotonicity of aggregation functions, averaging behavior is equivalent to idempotency, i.e. $f(t,t,...,t) = t$.

---

[1] http://aggregationfunctions.wordpress.com

In this paper, we are primarily concerned with averaging aggregation functions, although the Bonferroni mean uses the product $f(x,y) = xy$ in its composition which is one of the archetypical conjunctive functions.

Well known means include the arithmetic mean (also commonly referred to as the average) and the median. The arithmetic mean, as well as geometric means and power means can be expressed as special cases of the weighted quasi-arithmetic mean. We provide its definition here.

**Definition 3.** *For a strictly monotone continuous generating function $\phi : [0, 1] \rightarrow [-\infty, \infty]$ and weighting vector* **w**, *the weighted quasi-arithmetic mean is given by,*

$$QAM_\mathbf{w}(\mathbf{x}) = \phi^{-1}\left(\sum_{i=1}^{n} w_i \phi(x_i)\right). \tag{1}$$

Special cases include weighted arithmetic means, where $\phi(t) = t$, weighted power means where $\phi(t) = t^p$ and weighted geometric means (i.e. $G(\mathbf{x}) = \prod_{i=1}^{n} x_i^{w_i}$) if $\phi(t) = -\ln t$. The weights $w_i$ are usually non-negative and sum to one.

On the other hand, the median, maximum and minimum operators can be expressed as special cases of the ordered weighted averaging (OWA) operator. Rather than weight arguments according to their position or source, the OWA allocates a weight depending on the relative size of the input. It was formally defined by Yager in 1988 [23].

**Definition 4.** *For a weighting vector* **w**, *the ordered weighted averaging (OWA) operator is given by,*

$$OWA_\mathbf{w}(\mathbf{x}) = \sum_{i=1}^{n} w_i x_{(i)}, \tag{2}$$

*where the parentheses (.) indicate a reordering of the inputs such that $x_{(1)} \geq x_{(2)} \geq \ldots \geq x_{(n)}$.*

Special cases include the maximum when $\mathbf{w} = (1, 0, \ldots, 0)$, the minimum when $\mathbf{w} = (0, \ldots, 0, 1)$ and the median if $w_i = 1$ for $i = \frac{n+1}{2}$ and 0 otherwise where $n$ is odd, and $w_i = 0.5$ for $i = \frac{n}{2}, \frac{n}{2} + 1$ and 0 otherwise where $n$ is even.

The Bonferroni mean was defined in 1950 [11] and later generalized by Yager and others in the computational intelligence and decision making field. In its original form, it is defined as follows.

**Definition 5.** *Let $p, q \geq 0$ and $x_i \geq 0, i = 1, \ldots, n$. The Bonferroni mean is the function*

$$B^{p,q}(\mathbf{x}) = \left(\frac{1}{n(n-1)} \sum_{i,j=1, i \neq j}^{n} x_i^p x_j^q\right)^{\frac{1}{p+q}}. \tag{3}$$

In the case of $p = q$ for $n = 2$ the Bonferroni mean is equivalent to the geometric mean. For $q = 0$ (or $p = 0$), it will reduce to a power mean and can therefore express functions such as the arithmetic mean ($p = 1$), quadratic mean ($p = 2$)

and the limiting case of the geometric mean $p = 0$. As the ratio $\frac{p}{q}$ approaches infinity (or 0), the mean approaches the maximum operator. When $n > 2$, there must exist at least one pair $(i, j)$ such that $x_i, x_j > 0$, for the Bonferroni mean to return a non-zero output $B^{p,q}(\mathbf{x}) > 0$. It is this property that makes it possible for the generalizations of the Bonferroni mean to express mandatory requirements.

In [8], the Bonferroni mean was expressed as a composed aggregation function, generalizing it in terms of two means and a conjunctive function. With this construction, the function is able to model partial conjunction [12] with respect to any number of arguments, i.e. we can specify mandatory requirements that must at least partially be fulfilled for the function to have a non-zero output.

The notation $\mathbf{x}_{j \neq i}$ is used to denote the vector in $[0, 1]^{n-1}$ that includes the arguments from $\mathbf{x} \in [0, 1]^n$ in each dimension except the $i$-th, $\mathbf{x}_{j \neq i} = (x_1, \ldots, x_{i-1}, x_{i+1}, \ldots, x_n)$.

**Definition 6.** *[8]. Let $\mathbb{M}$ denote a 3-tuple of aggregation functions $< M_1, M_2, C >$, with $M_1 : [0, 1]^n \to [0, 1]$, $M_2 : [0, 1]^{n-1} \to [0, 1]$ and $C : [0, 1]^2 \to [0, 1]$, with the diagonal of $C$ denoted by $C_*(t) = C(t, t)$ and inverse diagonal $C_*^{-1}$. The generalized Bonferroni mean is given by,*

$$B_{\mathbb{M}}(\mathbf{x}) = C_*^{-1}\Big(M_1\big(C(x_1, M_2(\mathbf{x}_{j \neq 1})), \ldots, C(x_n, M_2(\mathbf{x}_{j \neq n}))\big)\Big). \tag{4}$$

The original Bonferroni mean is returned where $M_1 = WAM(\mathbf{x})$, $M_2 = PM_q(\mathbf{x})$ and $C = x^p y^q$ (with all weights equal).

Since $M_1$ is an averaging function of $n$ arguments while $M_2$ is a function of $n - 1$ arguments, they will have weighting vectors of different dimension. In order to choose the weights appropriately, so that they are consistent with the application and inputs, the following convention is used for the weighting vector of $M_2$ [8].

Given $\mathbf{u} \in [0, 1]^n$, the vectors $\mathbf{u}^i \in [0, 1]^{n-1}$, $i = 1, \ldots, n$ are defined by

$$u_j^i = \frac{u_j}{\sum_{k \neq i} u_k} = \frac{u_j}{1 - u_i}, \quad u_i \neq 1. \tag{5}$$

Note that for every $i$, $\mathbf{u}^i$ sum to one.

This allows one to either use the same weighting vector or differing vectors if each stage of aggregation requires it.

## 2.2   Fitting Aggregation Functions to Data

The usual framework for fitting a function $f$ to data involves an objective equation that minimizes the difference between the observed values $y_k$ and predicted values $f(\mathbf{x}_k)$ in some norm. In particular, we have the $L_2$ norm or least squares approach,

$$\sum_{k=1}^{K} \big(f(\mathbf{x}_k) - y_k\big)^2, \tag{6}$$

and $L_1$ or least absolute deviation (LAD) approach,

$$\sum_{k=1}^{K} \left| f(\mathbf{x}_k) - y_k \right|. \tag{7}$$

For our algorithms, we are interested in the latter approach, which can be converted into a linear program [2, 10].

Given a dataset with $K$ rows $(x_{k1}, x_{k2}, ..., x_{kn}, y_k)$ where each $k$ represents an observed value, we firstly represent each residual in terms of its positive and negative components (one of which will be zero), i.e. $r_k = |f(\mathbf{x})_k - y_k| = r_k^+ + r_k^-$.

We then minimize the sum of these residuals subject to equality constraints ensuring the $y_k$ are equal to the predicted function value and the residual.

$$\text{Minimize} \quad \sum_{k=1}^{K} r_k^+ + r_k^-,$$
$$\text{s.t.} \quad f(\mathbf{x}_k) + r_k^+ - r_k^- = y_k, k = 1 \ldots K \tag{8}$$
$$r_k^+, r_k^- \geq 0.$$

With suitable transformations or rearrangements of the data, many interesting aggregation functions can be represented in this way, for example, to fit a weighted quasi-arithmetic mean with generating function $g$, we can use the constraints:

$$\left( \sum_{i=1}^{n} w_i g(x_{ki}) \right) + r_k^+ - r_k^- = g(y_k), k = 1 \ldots K$$
$$w_i \geq 0, \forall \, i,$$
$$\sum_{i=1}^{n} w_i = 1.$$

Note that the residuals in this case are the differences between the transformed data - not the actual data itself.

For ordered functions such as the OWA, the data can be transformed so that the weights are learned from the reordered data. In both cases, although the functions themselves are not linear, the weights are only fit to linear data.

In our AggWaFit.R source file, the commands ordfit.GenOWA and ordfit.QAM can be used to find the weighting vector $\mathbf{w}$ from a given data set where the generator and its inverse are specified. These commands are designed for fitting to data where the outputs are ordinal and will return a stats file with root mean squared error (RMSE), average $L_1$ loss, prediction accuracy (for predicting the ordinal classes), a confusion matrix and the resulting $w$. A file with the predicted values from the function and corresponding classes is also returned with the original data.

# 3   Formulating Weights Identification Problems

The generalized Bonferroni mean is defined with respect to the two weighting vectors, **w** and **u**. Due to its composition, however, we cannot transform the data and fit the weights as we do for the quasi-arithmetic mean and OWA. We look at three simplifications that will enable us to fit generalized Bonferroni means to data.

## 3.1   Fitting $v_{ij}$ Weights to Product Pairs

We can firstly consider the case of $M_1, M_2$ weighted arithmetic means and $C$ the product operation with powers $p, q$. This leads to the following expression and simplification.

$$\left( \sum_{i=1}^{n} w_i x_i^p \left( \sum_{j \neq i} \frac{u_j}{1 - u_i} x_j^q \right) \right)^{\frac{1}{p+q}} = \left( \sum_{i=1, j=1, i \neq j}^{n} \frac{w_i u_j}{1 - u_i} x_i^p x_j^q \right)^{\frac{1}{p+q}}$$

Although we still cannot separate the weights linearly, we can consider each $x_i x_j$ term and consider coefficients $v_{ij}$. We hence transform the instances of the dataset $(x_{k1}, x_{k2}, \ldots, x_{kn}, y_k)$ such that we fit to $(x_{k1}^p x_{k2}^q, x_{k1}^p x_{k3}^q, \ldots, x_{k,(n-1)}^p x_{kn}^q, y_k^{p+q})$ and introduce the following linear constraints.

$$\left( \sum_{i=1, j=1, i \neq j}^{n} v_{ij} x_{ki}^p x_{kj}^q \right) + r_k^+ - r_k^- = y_k^{p+q}, k = 1 \ldots K$$

$$v_{ij} \geq 0, \forall \; ij,$$

$$\sum_{i=1}^{n} v_{ij} = 1.$$

The resulting $v_{ij}$ will not be separable into the $w_i, u_i, u_j$ etc, however we can gain an idea of the rough contribution of $w_i$ which is associated with the $p$ index and $u_i$ associated with the $q$ index by summing the rows and columns of the $v_{ij}$ matrix respectively.

In our BonFit.R source file, this fitting is done to ordinal data using the ordfit.bonf.vij command. Different $p, q$ can be specified and further restrictions placed on the $v_{ij}$ if desired.

## 3.2   Fitting $w_i$ Weights with Fixed u

An alternative to fitting to the pairs $x_i^p x_j^q$ is to fix the weighting vector **u**. This way, we can use alternative means for $M_1, M_2$ (in particular, any QAM) whereas before we were limited to weighted arithmetic means. We hence perform fitting by transforming each of the input terms $x_{ki}$ by combining with the mean of the $\mathbf{x}_{k,j\neq i}$ and using the generator functions. Denoting the generator of $M_1$ by $m_1$, each of the terms will be given by

$$m_1\left(x_{ki}^p(M_2(\mathbf{x}_{k,j\neq i}))^q\right),$$

where the weighting vector for $M_2$ is defined separately for each $i$ using each of the $\mathbf{u}^i$ determined from the supplied vector $\mathbf{u}$. In this case, we introduce the following constraints.

$$\left(\sum_{i=1,j=1,i\neq j}^n w_i m_1(x_{ki}^p(M_2(\mathbf{x}_{k,j\neq i}))^q)\right) + r_k^+ - r_k^- = (m_1(y_k))^{p+q},$$

$$k = 1\ldots K,$$

$$w_i \geq 0, \qquad \forall\, i,$$

$$\sum_{i=1}^n w_i = 1.$$

This fitting is done with BonFit.R using `ordfit.bonf.quasi` where the generators of both $M_1, M_2$ can be specified, as well as the weighting vector $\mathbf{u}$ associated with $M_2$ and the indices $p, q$.

## 3.3  Enforcing Mandatory Requirements

In some applications, it may be desirable to define a model with one or two mandatory requirements, but which still fits the data as well as it can. In this case, we can use projections on $\mathbf{w}$. Denoting the generator of $M_2$ (a weighted quasi-arithmetic mean) by $m_2$, this will result in the following expression,

$$\left(x_i^p m_2^{-1}\left(\sum_{j\neq i}\frac{u_j}{1-u_i}m_2(x_j)\right)^q\right)^{\frac{1}{p+q}}.$$

As we can see, the $u_i, u_j$ do not occur as linear cofactors, however since the $i$-th variable will always be mandatory, we can transform the dataset such that we only fit the weighting vector $\mathbf{u}^i$. We hence will not obtain a weight for the relative importance of $i$ with respect to the other variables, however this would usually be acceptable as it is not needed in the model to calculate new values. By rearranging the function, we then introduce the following linear constraints into the fitting problem.

$$\left(\sum_{j=1,j\neq i}^n u_j m_2(x_{kj})\right) + r_k^+ - r_k^- = m_2\left(\frac{y_k^{(p+q)/q}}{x_{ki}^{p/q}}\right), \quad k = 1\ldots K,$$

$$u_j \geq 0, \forall\, j,$$

$$\sum_{j=1,j\neq i}^n u_j = 1.$$

To fit a function this way in BonFit.R, we can use the command `ordfit.bonf.proj`, where the variable that is to be mandatory is specified, as well as the desired generator for $M_2$.

## 4   Modeling the Journal Rankings Dataset

In our previous work [3–5], we have used a dataset synthesized from the Australian journal rankings, which pairs the indices provided by Thomson and Reuters' ISI Web of Knowledge database with the quality ranks allocated by the Australian Research Council (ARC). The motivation for using such a dataset is that the relationship between the indices and the quality rank should be roughly monotone, while there are also likely to exist correlations between the inputs.

Before the rankings were disbanded, we collected the 2011 data for journals in all disciplines with both ARC and ISI data. This gave us a list of over 5000 journals spread across different fields of research (FoR) categories. For comparing the accuracy of the Bonferroni mean, we used 17 FoR categories, each with more than 80 journals and one (1103 Clinical Sciences) with 706 journals. The data first had to be transformed so that each of the indices ranged between $[0, 1]$ and so that the distribution of the scores was such that idempotency could be obtained. The algorithm from this is also available at the previously mentioned website. We used the algorithms in BonFit.R for the Bonferroni means ($m_2 = t^3$ for `ordfit.bonf.quasi`, and $m_2 = t$, $i = 2$, for `ordfit.bonf.proj` i.e. the Impact Factor is made mandatory), AggWaFit.R for the WAM, OWA, power means ($p = 2, 3$) and geometric mean, and $fmtools$ for the Choquet integrals (2-additive and general). Table 1 shows the overall standard and 10-fold cross validation accuracy when fitting to the journals data, averaged across the 17 FoR codes. As we can see, the Bonferroni mean performs reasonably well in fitting to each of the datasets. Weighting each pair using $v_{ij}$ could be interpreted similarly to modeling interaction between pairs as is done with the 2-additive Choquet integral, and it is worth noting that their performance is similar for the 10-fold tests. Enforcing the impact factor as a mandatory requirement in this case did not lead to good accuracy, however this may be necessary in some cases for reflecting the decision maker's preferences.

**Table 1.** Overall classification and L1-accuracy for various aggregation functions

|  | $B.v_{ij}$ | $B.qam$ | $B.proj$ | $WAM$ | $OWA$ | $PM_2$ | $PM_3$ | $GM$ | $Ch_{2-add}$ | $Ch_{gen}$ |
|---|---|---|---|---|---|---|---|---|---|---|
| *All* | | | | | | | | | | |
| L1 | 0.124 | 0.123 | 0.150 | 0.123 | 0.125 | 0.117 | 0.117 | 0.149 | 0.113 | **0.106** |
| Acc. | 0.676 | 0.662 | 0.576 | 0.661 | 0.642 | 0.672 | 0.673 | 0.621 | 0.691 | **0.715** |
| *10fold* | | | | | | | | | | |
| L1 | 0.132 | 0.126 | 0.170 | 0.129 | 0.133 | **0.123** | 0.124 | 0.216 | 0.126 | 0.126 |
| Acc. | 0.652 | 0.654 | 0.440 | 0.645 | 0.616 | **0.655** | 0.646 | 0.485 | 0.649 | 0.654 |

# 5 Conclusion

We have introduced some methods for fitting the generalized Bonferroni mean to data. To date, such methods have not been investigated for the Bonferroni mean. As well as making these available, we also draw attention to the datasets and R-code at our website, which can be used to further the study of aggregation functions and their use in decision making. We found that the generalized Bonferroni mean offers comparable performance to other means when modeling data. Although it is not possible to develop linear or quadratic programs in general, it is possible to write efficient algorithms for various special cases.

# References

1. Beliakov, G.: fmtools package, version 1.0 (2007), http://www.deakin.edu.au/~gleb/aotool.html
2. Beliakov, G.: Construction of aggregation functions from data using linear programming. Fuzzy Sets and Systems 160, 65–75 (2009)
3. Beliakov, G., James, S.: Citation based journal rankings: an application of fuzzy measures. In: Proc. of the 5th Intl Summer School of Aggregation Operators AGOP 2009, Palma de Mallorca, Spain, July 6-11 (2009)
4. Beliakov, G., James, S.: Using Choquet integrals for evaluating citation indices in journal ranking. In: Proc. Eurofuse Workshop 2009, Pamplona, Spain, September 16-18 (2009)
5. Beliakov, G., James, S.: Citation-based journal ranks: The use of fuzzy measures. Fuzzy Sets and Systems 167, 101–119 (2011)
6. Beliakov, G., James, S.: Defining Bonferroni means over lattices. In: Proc. of FUZZIEEE, Brisbane, Austarlia (2012)
7. Beliakov, G., James, S.: On extending generalized Bonferroni means to Atanassov orthopairs in decision making contexts. Fuzzy Sets and Systems (in press, 2012), doi: 10.1016/j.fss.2012.03.018
8. Beliakov, G., James, S., Mordelová, J., Rückschlossová, T., Yager, R.R.: Generalized Bonferroni mean operators in multi-criteria aggregation. Fuzzy Sets and Systems 161, 2227–2242 (2010)
9. Beliakov, G., Pradera, A., Calvo, T.: Aggregation Functions: A Guide for Practitioners. Springer, Heidelberg (2007)
10. Bloomfield, P., Steiger, W.L.: Least Absolute Deviations. Theory, Applications and Algorithms. Birkhauser, Boston (1983)
11. Bonferroni, C.: Sulle medie multiple di potenze. Bollettino Matematica Italiana 5, 267–270 (1950)
12. Dujmovic, J.J.: Continuous preference logic for system evaluation. IEEE Trans. on Fuzzy Systems 15, 1082–1099 (2007)
13. Grabisch, M., Kojadinovic, I., Meyer, P.: A review of methods for capacity identification in Choquet integral based multi-attribute utility theory: Applications of the Kappalab R package. European Journal of Operational Research 186, 766–785 (2008)
14. Grabisch, M., Marichal, J.-L., Mesiar, R., Pap, E.: Aggregation Functions. Cambridge University press, Cambridge (2009)
15. Mordelová, J., Rückschlossová, T.: ABC-aggregation functions. In: Proc. of the 5th Intl. Summer School on Aggregation Operators, Palma de Mallorca, Spain (2009)

16. R Development Core Team. R: A Language and Environment for Statistical Computing. R Foundation for Statistical Computing, Vienna, Austria (2011), http://www.R-project.org/
17. Torra, Y., Narukawa, V.: Modeling Decisions. Information Fusion and Aggregation Operators. Springer (2007)
18. Xia, M., Xu, Z., Zhu, B.: Generalized intuitionistic fuzzy Bonferroni means. International Journal of Intelligent Systems 27, 23–47 (2011)
19. Xu, Z.: Uncertain Bonferroni mean operators. International Journal of Computational Intelligence Systems 3(6), 761–769 (2010)
20. Xu, Z., Yager, R.R.: Intuitionistic fuzzy Bonferroni means. IEEE Transactions on Systems, Man, and Cybernetics, Part B: Cybernetics 41, 568–578 (2011)
21. Yager, R.: On generalized Bonferroni mean operators for multi-criteria aggregation. International Journal of Approximate Reasoning 50, 1279–1286 (2009)
22. Yager, R., Beliakov, G., James, S.: On generalized Bonferroni means. In: Proc. Eurofuse 2009 Conference, Pamplona, Spain (2009)
23. Yager, R.R.: On ordered weighted averaging aggregation operators in multicriteria decision making. IEEE Trans. on Systems, Man and Cybernetics 18, 183–190 (1988)

# An Ordered Weighted Average
# with a Truncation Weight on Intervals

Yuji Yoshida

Faculty of Economics and Business Administration, University of Kitakyushu
4-2-1 Kitagata, Kokuraminami, Kitakyushu 802-8577, Japan
yoshida@kitakyu-u.ac.jp

**Abstract.** This paper deals with ordered weighted averages on a closed interval, and their fundamental properties are investigated. In this paper we focus on ordered weighted average with a truncation weight, and the sub-additivity of a top-concentrated average is derived. Several examples are given to understand the idea. Further we deal with ordered weighted average from the bottom, and their relations are investigated. Finally, ordered weighted averages based on a probability are discussed and value-at-risks are explained as their example.

## 1 Introduction

Weighted averages are fundamental tools in decision making and they are represented by aggregated operators ([4,3]), and ordered weighted averages have been studied by Yager [13,14,16]. Ordered weighted averages have a lot of applications and Beliakov et al. [3] has discussed how to choose the weights in practice. Ordered weighted averages are described as follows: Let $n$ be a positive integer, and let $\{w_1, w_2, \cdots, w_n\}$ be a weighting sequence such that $w_i \in [0,1](i = 1, 2, \cdots, n)$ and $\sum_{i=1}^{n} w_i = 1$. An *ordered weighted average* of a finite sequence $\{x_1, x_2, \cdots, x_n\}$ is defined by

$$\xi(x_1, x_2, \cdots, x_n) := \sum_{i=1}^{n} w_i x_{(i)}, \tag{1}$$

where $x_{(i)}$ is the $i$-th largest element in $\{x_1, x_2, \cdots, x_n\}$, i.e. $\{(1), (2), \cdots, (n)\}$ is a permutation of the index set $\{1, 2, \cdots, n\}$ and $x_{(i-1)} \geq x_{(i)}$ for $i = 2, 3, \cdots, n$. Extending the ordered weighted averages, Yager [15] and Yager and Xu [17] have studied continuous weighted ordered weighted averages with weights defined by a fuzzy quantifier, and Torra [10], Torra and Narukawa [12], Torra and Godo [11], Narukawa and Torra [7] and Narukawa et al. [8] have demonstrated the relation between weighted ordered weighted averages and Choquet integral. This paper constructs ordered weighted averages on a closed interval and investigates their various properties. We demonstrate a top-concentrated average as one of ordered weighted averages with a truncation weight, and we discuss its sub-additivity.

In Section 2, ordered weighted averages on a closed interval are introduced extending the ordered weighted averages (1), and we investigate their fundamental properties. Next, in Section 3, we discuss ordered weighted averages with

V. Torra et al. (Eds.): MDAI 2012, LNAI 7647, pp. 45–55, 2012.
© Springer-Verlag Berlin Heidelberg 2012

a truncation weight, and we focus on a top-concentrated average. We derive the sub-additivity of the top-concentrated average, representing the average by the supremum of weighted averages. Several examples are given to understand the idea. Further, in Section 4, we deal with ordered weighted averages from the bottom, and we investigate their relations by a dual representation. Finally, in Section 5, we discuss ordered weighted averages based on a probability and we explain value-at-risks as its examples in finance.

## 2   Ordered Weighted Averages on an Interval

In this section we construct ordered weighted averages on a closed interval from the concept of (1). Let $\mathbb{R}$ be the set of all real numbers, and let $m$ be the Lebesgue measure on $\mathbb{R}$. For real numbers $a, b \in \mathbb{R}$ satisfying $a < b$, $\mathcal{C}([a, b])$ denotes the class of all real-valued continuous functions on $[a, b]$. Let $w : [a, b] \mapsto [0, \infty)$ be a nonnegative measurable function such that $0 < \int_a^b w(x)\,dx < \infty$. Then $w$ is called a *weighting function*. Let $f \in \mathcal{C}([a, b])$ be a function to be estimated. The range of $f$ is given by a closed interval $R := \{f(x) \mid x \in [a, b]\} = [\underline{f}, \overline{f}]$ with the lower bound $\underline{f} := \inf_{x \in [a,b]} f(x)$ and the upper bound $\overline{f} := \sup_{x \in [a,b]} f(x)$. Now we introduce a map $m^f : [\underline{f}, \overline{f}] \mapsto [a, b]$ as follows:

$$m^f(\alpha) := a + m(\{x \in [a, b] \mid f(x) \geq \alpha\}) \tag{2}$$

for $\alpha \in [\underline{f}, \overline{f}]$. Then a map $\alpha \mapsto m^f(\alpha)$ is left-continuous and non-increasing on $(\underline{f}, \overline{f}]$ since $\bigcap_{\beta \in (\underline{f}, \overline{f}]: \beta < \alpha}\{x \in [a, b] \mid f(x) \geq \beta\} = \{x \in [a, b] \mid f(x) \geq \alpha\}$ for $\alpha \in (\underline{f}, \overline{f}]$. Hence we define

$$m^f(\alpha_+) := \begin{cases} \lim_{\beta \downarrow \alpha} m^f(\beta) & \text{if } \alpha \in [\underline{f}, \overline{f}) \\ a & \text{if } \alpha = \overline{f}, \end{cases} \tag{3}$$

$$m^f(\alpha_-) := \begin{cases} \lim_{\beta \uparrow \alpha} m^f(\beta) & \text{if } \alpha \in (\underline{f}, \overline{f}] \\ b & \text{if } \alpha = \underline{f}. \end{cases} \tag{4}$$

Then we have $m^f(\alpha_-) = m^f(\alpha)$ for $\alpha \in [\underline{f}, \overline{f}]$. Since $\bigcup_{\alpha \in [\underline{f}, \overline{f}]}[m^f(\alpha_+), m^f(\alpha)] = [a, b]$, we can define a quasi-inverse of $m^f(\alpha)$ by

$$F(x) := \alpha \quad \text{if } x \in [m^f(\alpha_+), m^f(\alpha)] \tag{5}$$

for $x \in [a, b]$. $F$ is non-increasing and it is called an *ordered function* of $f$. Now we define an *ordered weighted average* (from the top) $\text{OWA}_w^{[a,b]}(f)$ of the function $f$ with the weighting function $w$ as follow:

$$\text{OWA}_w^{[a,b]}(f) := \int_a^b F(x)\,w(x)\,dx \Big/ \int_a^b w(x)\,dx. \tag{6}$$

Here we note that $F$ is defined by (5), which is based on the set $\{x \in [a, b] \mid f(x) \geq \alpha\}$. Then it is trivial that $\text{OWA}_w^{[a,b]}(f) \in [\underline{f}, \overline{f}]$. As a weighting function

$w$, we use an element of $\mathcal{W}_0$, where $\mathcal{W}_0$ is the class of nonnegative measurable functions on $[a, b]$ satisfying $0 < \int_a^b w(x)\, dx < \infty$. Instead of it, taking into account of $w(x)/\int_a^b w(x)\, dx$, we may choose an element of $\mathcal{W}$, where $\mathcal{W}$ is the class of nonnegative measurable functions on $[a, b]$ satisfying $\int_a^b w(x)\, dx = 1$. Next we need the following concepts to describe a comonotonic property of ordered weighted averages.

**Definition 1.** (Dellacherie [5], Renneberg [9]).

(i) Two functions $f, g(\in \mathcal{C}([a, b]))$ are called *comonotonic* if $(f(x) - f(y))(g(x) - g(y)) \geq 0$ for all $x, y \in [a, b]$.
(ii) A map $I : \mathcal{C}([a, b]) \mapsto \mathbb{R}$ is called *comonotonically additive* if $I(f + g) = I(f) + I(g)$ holds for all comonotonic $f, g \in \mathcal{C}([a, b])$.

Then, the following results are known.

**Lemma 1.** ([5,9]). *For functions $f, g \in \mathcal{C}([a, b])$, the following three conditions (a) - (c) are equivalent.*

(a) *There exist increasing functions $\phi, \psi : \mathbb{R} \mapsto \mathbb{R}$ and a function $h : [a, b] \mapsto \mathbb{R}$ such that $f = \phi(h)$ and $g = \psi(h)$.*
(b) *For all $x, y \in [a, b]$, it holds that $f(x) < f(y)$ then $g(x) \leq g(y)$.*
(c) *For all $x, y \in [a, b]$, it holds that $(f(x) - f(y))(g(x) - g(y)) \geq 0$.*

For ordered weighted averages $\mathrm{OWA}_w^{[a,b]}(f)$, we have the following fundamental results.

**Theorem 1.** *Let $w(\in \mathcal{W}_0)$ be a weighting function on $[a, b]$. Then the ordered weighted average $\mathrm{OWA}_w^{[a,b]}$ has the following properties:*

(i) $\mathrm{OWA}_w^{[a,b]}(f) \leq \mathrm{OWA}_w^{[a,b]}(g)$ *for $f, g \in \mathcal{C}([a, b])$ satisfying $f \leq g$. (monotonicity)*
(ii) $\mathrm{OWA}_w^{[a,b]}(f + \theta) = \mathrm{OWA}_w^{[a,b]}(f) + \theta$ *for $f \in \mathcal{C}([a, b])$ and real numbers $\theta$. (translation invariance)*
(iii) $\mathrm{OWA}_w^{[a,b]}(\lambda f) = \lambda\, \mathrm{OWA}_w^{[a,b]}(f)$ *for $f \in \mathcal{C}([a, b])$ and $\lambda \geq 0$. (positive homogeneity)*
(iv) $\mathrm{OWA}_w^{[a,b]}$ *is comonotonically additive. (comonotonical additivity)*
(v) *Let $\varphi : \mathbb{R} \mapsto \mathbb{R}$ be a strictly increasing continuous convex (concave) function. Then $\mathrm{OWA}_w^{[a,b]}(\varphi(f)) \geq \varphi(\mathrm{OWA}_w^{[a,b]}(f))$ $(\mathrm{OWA}_w^{[a,b]}(\varphi(f)) \leq \varphi(\mathrm{OWA}_w^{[a,b]}(f))$ resp.).*

In Theorem 1(iv), ordered weighted averages $\mathrm{OWA}_w^{[a,b]}$ have comonotonical additivity, however it is not additive in general (Example 1(i)). In next section, we discuss the sub-additive property regarding a special ordered weighted average with a truncation weight. In the rest of this section, we give several examples for ordered weighted averages on an interval to understand the idea in this section.

## Example 1

(i) Let a domain $[0,1]$ and let $f(x) = x(1-x)$, $g(x) = x^2$ and a weighting function $w(x) = 1-x$. Then from (2) we have $m^f(\alpha) = \sqrt{1-4\alpha}$, $m^g(\alpha) = 1-\sqrt{\alpha}$, and their inverse functions are $F(x) = \frac{1}{4}(1-x^2)$ and $G(x) = (1-x)^2$ respectively. Thus by (6) we get $\text{OWA}_w^{[0,1]}(f) = \frac{5}{24}$ and $\text{OWA}_w^{[0,1]}(g) = \frac{1}{2}$. On the other hand, we also have $f(x) + g(x) = x$. Then $m^{f+g}(\alpha) = 1-\alpha$ and we get $\text{OWA}_w^{[0,1]}(f+g) = \frac{2}{3}$. We note that $\text{OWA}_w^{[0,1]}(f) + \text{OWA}_w^{[0,1]}(g) = \frac{5}{24} + \frac{1}{2} \neq \frac{2}{3} = \text{OWA}_w^{[0,1]}(f+g)$.

(ii) Let a domain $[0,\pi]$ and let $f(x) = \sin x$. Then from (2) we have $m^f(\alpha) = \pi - \arcsin\alpha$ and its inverse function is $F(x) = \sin(\frac{\pi-x}{2})$. For weighting functions $w(x) = \pi - x$ and $w(x) = \sqrt{\pi - x}$, by (6) we obtain $\text{OWA}_w^{[0,\pi]}(f) = \frac{8}{\pi^2} = 0.810569\cdots$ and $\text{OWA}_w^{[0,\pi]}(f) = 0.744743\cdots$ respectively.

(iii) Let a domain $[0,\pi]$ and let $f(x) = 1+\sin 2x$. Then in the same way we have its ordered function $F(x) = 1 + \sin(x + \frac{\pi}{2})$ (Fig.1). For weighting functions $w(x) = \pi - x$ and $w(x) = \sqrt{\pi - x}$, we obtain $\text{OWA}_w^{[0,\pi]}(f) = 1 + \frac{4}{\pi^2} = 1.40528\cdots$ and $\text{OWA}_w^{[0,\pi]}(f) = 1.24105\cdots$ respectively.

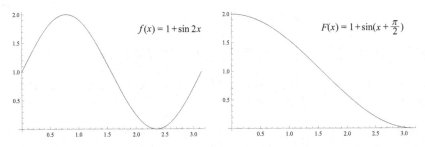

**Fig. 1.** A function $f(x) = 1 + \sin 2x$ on $[0,\pi]$ and its ordered function $F(x)$

(iv) Let a domain $[0,2]$ and let $f(x) = e^{-(x-1)^2}$. Then in the same way we have its ordered function $F(x) = e^{-x^2/4}$. For weighting functions $w(x) = 2 - x$ and $w(x) = e^{-x}$, we obtain $\text{OWA}_w^{[0,2]}(f) = 0.861528\cdots$ and $\text{OWA}_w^{[0,2]}(f) = 0.850428\cdots$ respectively.

(v) Let a domain $[0,2]$ and let $f(x) = \frac{1}{1+(x-1)^2}$. Then in the same way we have its ordered function $F(x) = \frac{4}{4+x^2}$. For weighting functions $w(x) = 2 - x$ and $w(x) = e^{-x}$, we obtain $\text{OWA}_w^{[0,2]}(f) = \frac{\pi - \log 4}{2} = 0.877649\cdots$ and $\text{OWA}_w^{[0,2]}(f) = 0.869833\cdots$ respectively.

(vi) Let a domain $[0,2]$ and let

$$f(x) = \begin{cases} 2 - 4x & \text{if } 0 \leq x < 1/4 \\ 1 & \text{if } 1/4 \leq x < 3/4 \\ -2 + 4x & \text{if } 3/4 \leq x < 5/4 \\ 3 & \text{if } 5/4 \leq x < 7/4 \\ 10 - 4x & \text{if } 7/4 \leq x \leq 2. \end{cases} \tag{7}$$

From (2) we have

$$m^f(\alpha) = \begin{cases} 2 & \text{if } 0 \le \alpha \le 1 \\ 3 - \alpha & \text{if } 1 < \alpha \le 3 \\ 0 & \text{if } 3 < \alpha. \end{cases} \qquad (8)$$

We note that (8) is a left-continuous and mon-increasing function. By (5) we have its ordered function

$$F(x) = \begin{cases} 3 & \text{if } 0 \le x < 1/2 \\ 4 - 2x & \text{if } 1/2 \le x < 3/2 \\ 1 & \text{if } 3/2 \le x \le 2. \end{cases} \qquad (9)$$

For weighting functions $w(x) = 2 - x$ and $w(x) = \sqrt{2 - x}$, we obtain $\text{OWA}_w^{[0,2]}(f) = 2.45833 \cdots$ and $\text{OWA}_w^{[0,2]}(f) = 2.27058 \cdots$ respectively.

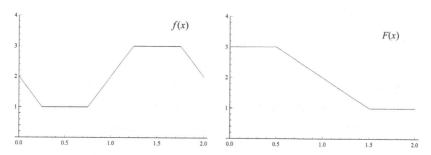

**Fig. 2.** A function $f(x)$ in (7) and its ordered function $F(x)$ in (9)

## 3  Ordered Weighted Average with a Truncation Weight

Let $n$ be a positive integer. A weighting sequence $\{w_1, w_2, \cdots, w_n\}$ is called a *truncation weight* if it is given by

$$w_i := \begin{cases} \frac{1}{l} & \text{if } i = 1, 2, \cdots, l \\ 0 & \text{otherwise} \end{cases} \qquad (10)$$

for a positive integer $l$ satisfying $l < n$. Extending this weight to one on an interval, we can concentrate the weight on the top values and truncate the lower values. In this section, we discuss the sub-additivity of ordered weighted averages on an interval using this kind of weights. We put $a = 0$ and $b = 1$ for simplicity. Regarding weighting functions, we have $\mathcal{W} \subset \mathcal{W}_0$, which are complete with respect to $L^1$-norm. Now we give a lemma to establish sub-additivity for an ordered weighted average with a truncation weight.

**Lemma 2.** *Let $c$ be a constant satisfying $0 < c < 1$ and let $f \in \mathcal{C}([0,1])$. Define a weighting function $w^* \in \mathcal{W}_0$ by*

$$w^*(x) := \begin{cases} \frac{1}{c} & \text{if } 0 \le x \le c \\ 0 & \text{otherwise.} \end{cases} \qquad (11)$$

*Then it holds that*

$$\text{OWA}_{w^*}^{[0,1]}(f) = \sup_{w:0 \le w \le \frac{1}{c}, \int_0^1 w(x)\,dx=1} \int_0^1 f(x)w(x)\,dx. \tag{12}$$

*Hence, the supremum is attained by the weight $w^*$ in (11).*

The weight $w^*$ defined by (11) is a *truncation weight*, and we have $w^* \in \mathcal{W}$ since $\int_0^1 w^*(x)\,dx = 1$. For this weight $w^*$, we call $\text{OWA}_{w^*}^{[0,1]}(f)$ a *top-concentrated average*. Lemma 2 shows that a top-concentrated average is represented by the supremum of weighted averages. We define a class of weighting functions $\mathcal{W}(c) := \{w \in \mathcal{W} \mid 0 \le w \le \frac{1}{c}\}$ for $c \in (0,1)$, and then we get the following result since $w^*$ does not depend on $f$.

**Theorem 2.** *Let $f, g \in \mathcal{C}([0,1])$. The top-concentrated average $\text{OWA}_{w^*}^{[0,1]}$ has the following sub-additivity:*

$$\text{OWA}_{w^*}^{[0,1]}(f+g) \le \text{OWA}_{w^*}^{[0,1]}(f) + \text{OWA}_{w^*}^{[0,1]}(g). \tag{13}$$

**Corollary 1.** *Let $a, b \in \mathbb{R}$ satisfy $a < b$ and let $f, g \in \mathcal{C}([a,b])$. Then the top-concentrated average $\text{OWA}_{w^*}^{[a,b]}$ has the sub-additivity:*

$$\text{OWA}_{w^*}^{[a,b]}(f+g) \le \text{OWA}_{w^*}^{[a,b]}(f) + \text{OWA}_{w^*}^{[a,b]}(g) \tag{14}$$

*and*

$$\text{OWA}_{w^*}^{[a,b]}(f) = \sup_{w \in \mathcal{W}(c-a)} \int_a^b f(x)w(x)\,dx, \tag{15}$$

*where the supremum is attained by*

$$w^*(x) = \begin{cases} \frac{1}{c-a} & \text{if } a \le x \le c \\ 0 & \text{otherwise} \end{cases} \tag{16}$$

*for $c \in (a,b)$.*

Finally we give a example for ordered weighted averages with a truncation weight.

**Example 2.** Let a domain $[0, \pi]$ and let $f(x) = 1 + \sin 2x$ in Example 1(iii). Then its ordered function is $F(x) = 1 + \sin(x + \frac{\pi}{2})$. Let $c = 1$ and let $\Gamma := \{x \in [0,\pi] \mid f(x) \ge 1 + \sin(1 + \frac{\pi}{2})\}$ (Fig.3). For a truncation weight

$$w^*(x) = \begin{cases} 1 \text{ if } 0 \le x \le 1 \\ 0 \text{ otherwise,} \end{cases}$$

we obtain $\text{OWA}_{w^*}^{[0,\pi]}(f) = \int_0^\pi F(x)w^*(x)\,dx = \int_0^1 F(x) \cdot 1\,dx = 1 + \sin(1) = 1.84147\cdots$. This value is larger than the result $\text{OWA}_w^{[0,\pi]}(f) = 1.40528\cdots$ in Example 1(iii).

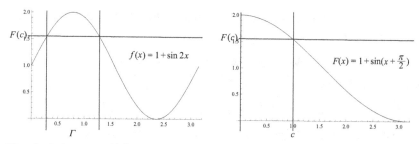

**Fig. 3.** A function $f(x) = 1 + \sin 2x$, its ordered function $F(x)$ and $\Gamma$ at $c = 1$

## 4    Ordered Weighted Averages from the Bottom

Let $a, b \in \mathbb{R}$ satisfy $a < b$, and let $f \in \mathcal{C}([a, b])$. In this section, we investigate ordered weighted averages based on the set $\{x \in [a, b] \mid f(x) \leq \alpha\}$. The range of $f$ is given by $R := \{f(x) \mid x \in [a, b]\} = [\underline{f}, \overline{f}]$ with the lower bound $\underline{f} := \inf_{x \in [a, b]} f(x)$ and the upper bound $\overline{f} := \sup_{x \in [a, b]} f(x)$. We introduce a map $\underline{m}^f : [\underline{f}, \overline{f}] \mapsto [a, b]$ by

$$\underline{m}^f(\beta) := a + m(\{x \in [a, b] \mid f(x) \leq \beta\}) \tag{17}$$

for $\beta \in [\underline{f}, \overline{f}]$. Then we have $\underline{m}^f(\beta) = a + m(\{x \in [a, b] \mid f(x) \leq \beta\}) = a + m(\{x \in [a, b] \mid \underline{f} + \overline{f} - f(x) \geq \underline{f} + \overline{f} - \beta\}) = m^{\underline{f} + \overline{f} - f}(\underline{f} + \overline{f} - \beta)$. In a similar way to (3) and (4), we have $\underline{m}^f(\beta_+) = \underline{m}^f(\beta)$ and the corresponding ordered function is

$$\underline{F}(x) := \beta \quad \text{if } x \in [\underline{m}^f(\beta_-), \underline{m}^f(\beta)] \tag{18}$$

for $x \in [a, b]$. Thus the ordered weighted averages based on the set $\{x \in [a, b] \mid f(x) \leq \beta\}$, which we call an *ordered weighted average from the bottom* is

$$\underline{\mathrm{OWA}}_w^{[a,b]}(f) := \int_a^b \underline{F}(x)\, w(x)\, dx \Big/ \int_a^b w(x)\, dx. \tag{19}$$

Putting $\alpha = \underline{f} + \overline{f} - \beta$, we get $\underline{m}^f(\beta) = m^{\underline{f} + \overline{f} - f}(\underline{f} + \overline{f} - \beta) = m^{\underline{f} + \overline{f} - f}(\alpha)$ and $\underline{m}^f(\beta_-) = m^{\underline{f} + \overline{f} - f}(\alpha_+)$. Eq. (18) is reduced to

$$\underline{F}(x) := \underline{f} + \overline{f} - \alpha \quad \text{if } x \in [m^{\underline{f} + \overline{f} - f}(\alpha_+), m^{\underline{f} + \overline{f} - f}(\alpha)] \tag{20}$$

for $x \in [a, b]$. Therefore from (5),(6),(18),(19) and Theorem 1(ii) we obtain the following lemma.

**Lemma 3.** *Let $a, b \in \mathbb{R}$ satisfy $a < b$ and let $f \in \mathcal{C}([a, b])$.*

$$\underline{\mathrm{OWA}}_w^{[a,b]}(f) = \underline{f} + \overline{f} - \mathrm{OWA}_w^{[a,b]}(\underline{f} + \overline{f} - f) = -\overline{\mathrm{OWA}}_w^{[a,b]}(-f). \tag{21}$$

For ordered weighted averages from the bottom $\underline{\mathrm{OWA}}_w^{[a,b]}(f)$, from (21) and the results in Section 3 we have the following theorem.

**Theorem 3.** *Let $w(\in \mathcal{W}_0)$ be a weighting function on $[a,b]$. Then the ordered weighted average from the bottom $\underline{\mathrm{OWA}}_w^{[a,b]}$ has the following properties:*

(i) $\underline{\mathrm{OWA}}_w^{[a,b]}(f) \le \underline{\mathrm{OWA}}_w^{[a,b]}(g)$ *for $f,g \in \mathcal{C}([a,b])$ satisfying $f \le g$. (monotonicity)*

(ii) $\underline{\mathrm{OWA}}_w^{[a,b]}(f + \theta) = \underline{\mathrm{OWA}}_w^{[a,b]}(f) + \theta$ *for $f \in \mathcal{C}([a,b])$ and real numbers $\theta$. (translation invariance)*

(iii) $\underline{\mathrm{OWA}}_w^{[a,b]}(\lambda f) = \lambda \underline{\mathrm{OWA}}_w^{[a,b]}(f)$ *for $f \in \mathcal{C}([a,b])$ and $\lambda \ge 0$. (positive homogeneity)*

(iv) $\underline{\mathrm{OWA}}_w^{[a,b]}$ *is comonotonically additive. (comonotonical additivity)*

(v) *Let $f,g \in \mathcal{C}([a,b])$. A bottom-concentrated average $\underline{\mathrm{OWA}}_{w^*}^{[a,b]}$ has the super-additivity:*

$$\underline{\mathrm{OWA}}_{w^*}^{[a,b]}(f + g) \ge \underline{\mathrm{OWA}}_{w^*}^{[a,b]}(f) + \underline{\mathrm{OWA}}_{w^*}^{[a,b]}(g) \qquad (22)$$

*and*

$$\underline{\mathrm{OWA}}_{w^*}^{[a,b]}(f) = \inf_{w \in \mathcal{W}(c-a)} \int_a^b f(x)w(x)\,dx, \qquad (23)$$

*where the infimum is attained a truncation weight $w^* \in \mathcal{W}$ by*

$$w^*(x) = \begin{cases} \frac{1}{c-a} & \text{if } a \le x \le c \\ 0 & \text{otherwise} \end{cases} \qquad (24)$$

*for $c \in (a,b)$.*

We give an example for a bottom-concentrated average, using Example 1(iii).

**Example 3.** Let a domain $[0,\pi]$ and let $f(x) = 1 + \sin 2x$ in Examples 1(iii) and 3. Then its ordered function is $\underline{F}(x) = 1 - \sin(x + \frac{\pi}{2})$. Let $c = 1$ and let $\Gamma := \{x \in [0,\pi] \mid f(x) \le 1 - \sin(1 + \frac{\pi}{2})\}$ (Fig.4). For a truncation weight

$$w^*(x) = \begin{cases} 1 \text{ if } 0 \le x \le 1 \\ 0 \text{ otherwise,} \end{cases}$$

we obtain $\underline{\mathrm{OWA}}_{w^*}^{[0,\pi]}(f) = \int_0^\pi \underline{F}(x)w^*(x)\,dx = \int_0^1 \underline{F}(x) \cdot 1\,dx = 1 - \sin(1) = 0.158529\cdots$.

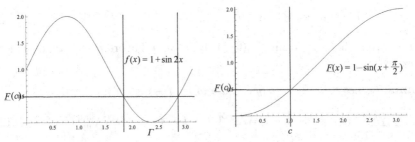

**Fig. 4.** A function $f(x) = 1 + \sin 2x$, its ordered function $\underline{F}(x)$ and $\Gamma$ at $c = 1$

## 5    Ordered Weighted Averages Based on a Probability

In this section we discuss ordered weighted averages based on a probability. Let $(\Omega, P)$ be a probability space, where $P$ is non-atomic. Let $\mathcal{X}$ be the set of all integrable real random variables $X$ on $\Omega$ with a continuous distribution function $x \mapsto F_X(x) := P(X < x)$ for which there exists a non-empty open interval $I$ such that $F_X(\cdot) : I \mapsto (0,1)$ is strictly increasing and onto. Then there exists a strictly increasing and continuous inverse function $F_X^{-1} : (0,1) \mapsto I$. We note that $F_X(\cdot) : I \mapsto (0,1)$ and $F_X^{-1} : (0,1) \mapsto I$ are one-to-one and onto, and we put $\lim_{x \downarrow \inf I} F_X(x) = 0$ and $\lim_{x \uparrow \sup I} F_X(x) = 1$. Let the closure of $I$ by $R$, which is the range of $X$ and $R := [\underline{X}, \overline{X}]$ with the lower bound $\underline{X} = \operatorname{ess\,inf}_{\omega \in \Omega} X(\omega)$ and the upper bound $\overline{X} = \operatorname{ess\,sup}_{\omega \in \Omega} X(\omega)$. Let $w(\in \mathcal{W}_0) : [0.1] \mapsto [0, \infty)$ be a *weighting function*. In this section, it is not needed generally that the domain of $X$ coincides with the domain of $w$, i.e. we may take them as $\Omega \neq [0,1]$. We introduce a threshold probability $\underline{m}^X : [\underline{X}, \overline{X}] \mapsto [0,1]$ by

$$\underline{m}^X(\alpha) := P(\{\omega \in \Omega \mid X(\omega) \leq \alpha\}) = F_X(\alpha) \tag{25}$$

for $\alpha \in [\underline{X}, \overline{X}]$. In this case, since the map $\alpha \mapsto \underline{m}^X(\alpha)$ is continuous and increasing on $I = (\underline{X}, \overline{X})$, its inverse function is $F_X^{-1}(x)$ for $x \in [0,1]$. Now we define an *ordered weighted average from the bottom* $\underline{\mathrm{OWA}}_w^{[0,1]}(X)$ of the random variable $X(\in \mathcal{X})$ with the weighting function $w$ as follows.

$$\underline{\mathrm{OWA}}_w^{[0,1]}(X) := \int_0^1 F_X^{-1}(x)\, w(x)\, dx \Big/ \int_0^1 w(x)\, dx. \tag{26}$$

Then it holds that $\underline{\mathrm{OWA}}_w^{[0,1]}(X) \in [\underline{X}, \overline{X}]$. Here we give an example of criteria which is used generally in financial engineering.

**Example 4.** We introduce an average value-at-risk for real random variables ([1,2]). Let $p$ be a positive probability. The *value-at-risk (VaR)* at a risk probability $p$ is given by the percentiles of the distribution function $F_X$.

$$\mathrm{VaR}_p(X) := \begin{cases} \underline{X} & \text{if } p = 0 \\ \sup\{x \in R \mid F_X(x) \leq p\} & \text{if } 0 < p < 1 \\ \overline{X} & \text{if } p = 1. \end{cases} \tag{27}$$

Then we have $F_X(\mathrm{VaR}_p(X)) = p$ and $\mathrm{VaR}_p(X) = F_X^{-1}(p)$ for $0 < p < 1$ (Fig.5). VaR is a risk-sensitive criterion based on percentiles, and it is one of the standard criteria in asset management ([6,18,19]). VaR is a kind of risk values of the asset prices at a specified risk-level probability and it is useful for selecting portfolios to get rid of bad scenarios in investment. Further, the *average value-at-risk (AVaR)* at a probability level $p$ is also used for a risk criterion in asset management, and it is given as follows.

$$\mathrm{AVaR}_p(X) := \begin{cases} \underline{X} & \text{if } p = 0 \\ \dfrac{1}{p} \int_0^p \mathrm{VaR}_q(X)\, dq = \dfrac{1}{p} \int_0^p F_X^{-1}(q)\, dq & \text{if } 0 < p \leq 1. \end{cases} \tag{28}$$

From (26) and (28), we obtain

$$\text{AVaR}_p(X) = \underline{\text{OWA}}_{w^*}^{[0,1]}(X) \tag{29}$$

with a truncation weight

$$w^*(x) = \begin{cases} \frac{1}{p} & \text{if } 0 \le x \le p \\ 0 & \text{otherwise.} \end{cases} \tag{30}$$

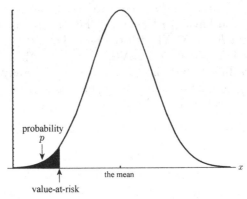

**Fig. 5.** Value-at-risk $\text{VaR}_p(X)$ at a probability $p$

The following results hold for the average value-at-risks from the results in Theorem 3 and Example 4.

**Corollary 2.** *Let $p$ be a positive probability. The average value-at-risk $\text{AVaR}_p$ has the following properties:*

(i) $\text{AVaR}_p(X) \le \text{AVaR}_p(Y)$ *for $X, Y \in \mathcal{X}$ satisfying $X \le Y$. (monotonicity)*
(ii) $\text{AVaR}_p(X+\theta) = \text{AVaR}_p(X)+\theta$ *for $X \in \mathcal{X}$ and real numbers $\theta$. (translation invariance)*
(iii) $\text{AVaR}_p(\lambda X) = \lambda\,\text{AVaR}_p(X)$ *for $X \in \mathcal{X}$ and $\lambda \ge 0$. (positive homogeneity)*
(iv) $\text{AVaR}_p$ *is comonotonically additive. (comonotonical additivity)*
(v) $\text{AVaR}_p(X+Y) \ge \text{AVaR}_p(X)+\text{AVaR}_p(Y)$ *for $X, Y \in \mathcal{X}$. (super-additivity)*

## Concluding Remarks

(i) Narukawa and Torra [7] and Narukawa et al. [8] have introduced continuous ordered weighted averages in a general form by Choquet integral, and they have studied its mathematical analysis. The ordered weighted averages (6) are the same one as those however in this paper we took a constructive approach to discuss the sub-additivity, which will be applicable to decision making (Section 5).
(ii) One of further researches is to find more general conditions for weighting functions to preserve the sub-additivity for the ordered weighted averages for applications in decision making.

# References

1. Artzner, P., Delbaen, F., Eber, J.-M., Heath, D.: Thinking coherently. Risk 10, 68–71 (1997)
2. Artzner, P., Delbaen, F., Eber, J.-M., Heath, D.: Coherent measures of risk. Mathematical Finance 9, 203–228 (1999)
3. Beliakov, G., Pradera, A., Calvo, T.: Aggregation Functions: A Guide for Practioners. Springer (2007)
4. Calvo, T., Kolesárová, A., Komorníková, M., Mesiar, R.: Aggregation operators: Basic concepts, issues and properties. In: Calvo, T., Gmayor, Mesiar, R. (eds.) Aggregation Operators: New Trends and Applications, pp. 3–104. Phisica-Verlag, Springer (2002)
5. Dellacherie, C.: Quelques commentarires sur les prolongements de capacités, Séminare de Probabilites 1969/1970, Strasbourg. LNAI, vol. 191, pp. 77–81. Springer(1971)
6. Meucci, A.: Risk and Asset Allocation. Springer, Heidelberg (2005)
7. Narukawa, Y., Torra, V.: Continuous OWA Operator and its Calculation. In: (CD-ROM Proceedings, IFSA-EUSFLAT 2009), pp. 1132–1134 (2009)
8. Narukawa, Y., Torra, V., Sugeno, M.: Choquet integral of a function on the real line. In: Torra, V., Narukawa, Y., Daumas, M. (eds.) Modeling Decisions for Artificial Intelligence 2010, CD-ROM Proceedings, MDAI 2010, pp. 24–33 (2010)
9. Renneberg, D.: Non Additive Measure and Integral. Kluwer Academic Publ., Dordrecht (1994)
10. Torra, V.: The weighted OWA operator. Int. J. of Intel. Syst. 12, 153–166 (1997)
11. Torra, V., Godo, L.: Continuous WOWA operators with application to defuzzification. In: Aggregation Operators: New Trends and Applications, pp. 159–176. Physica-Verlag, Springer (2002)
12. Torra, V., Narukawa, Y.: Modeling Decisions - Information Fusion and Aggregation Operators. Springer (2002)
13. Yager, R.R.: Ordered weighted averaging aggregation operators in multi-criteria decision making. IEEE Trans. on Systems, Man and Cybernetics. Int. J. of Intel. Syst. 18, 183–190 (1988)
14. Yager, R.R.: Families of OWA operators. Fuzzy Sets and Systems 59, 125–148 (1993)
15. Yager, R.R.: OWA aggregation over a continuous interval argument with application to decision making. IEEE Trans. on Systems, Man, and Cybern. - Part B: Cybernetics 34, 1952–1963 (2004)
16. Yager, R.R., Filev, D.P.: Parameterized and-like and or-like OWA opera-tors. Int. J. of General Systems 22, 297–316 (1994)
17. Yager, R.R., Xu, Z.: The continuous order weighted geometric operator and its application to decision making. Fuzzy Sets and Systems 157, 1393–1402 (2006)
18. Yoshida, Y.: A Perception-Based Portfolio Under Uncertainty: Minimization of Average Rates of Falling. In: Torra, V., Narukawa, Y., Inuiguchi, M. (eds.) MDAI 2009. LNCS (LNAI), vol. 5861, pp. 149–160. Springer, Heidelberg (2009)
19. Yoshida, Y.: An average value-at-risk portfolio model under uncertainty: A perception-based approach by fuzzy random variables. Journal of Advanced Computational Intelligence and Intelligent Informatics 15, 56–62 (2011)

# Choquet Integral on the Real Line
# as a Generalization of the OWA Operator*

Yasuo Narukawa

Toho Gakuen,
3-1-10, Naka, Kunitachi, Tokyo, 186-0004, Japan
Department of Computational Intelligence and Systems Science
Tokyo Institute of Technology
4259 Nagatuta, Midori-ku, Yokohama 226-8502, Japan
narukawa@d4.dion.ne.jp

**Abstract.** The Choquet integral is one of the operators that can be used
for aggregation and synthesis of information. It integrates a function with
respect to a fuzzy measure. In this paper we study the Choquet integral
with respect to a symmetric fuzzy measure, which is a generalization of
the OWA operator. We present some results about the approximation of
Choquet integral for the calculation. We also present the inequalities for
Choquet integral with respect to a symmetric fuzzy measure.

**Keywords:** Fuzzy measure, Non additive measure, Choquet integral,
Aggregation operator, OWA operator.

## 1 Introduction

Choquet integral [3] is one of the approaches used in aggregation operators [7, 8]
to combine information from several sources. Formally speaking, the integral
integrates a function with respect to a fuzzy measure, where a fuzzy measure is
a monotone set function. The Choquet integral with respect to a fuzzy measure
can be applied to the decision modeling with uncertainty and risk [2, 23].

Among the aggregation operators, the Choquet integral is well known as a
generalization of other operators as the weighted mean, the Ordered Weighted
Averaging (OWA) operator [24–26] as well as the arithmetic mean.

Due to these properties, the Choquet integral is a flexible operator that can
be used in different applications, and this has caused the interest of several
researchers for its properties.

There are a lot of papers studying the theory and applications of Choquet
integral. Most of them assume the discrete set as universal set [6, 5, 12]. Other
papers are on the abstract space [22, 13]. There are very few papers for the
Choquet integral of a function on real line [15–17].

* Partial support by the Spanish MEC (projects ARES – CONSOLIDER INGENIO
2010 CSD2007-00004 –, eAEGIS – TSI2007-65406-C03-02 –, co-Privacy TIN2011-
27076-C03-03) is acknowledged.

V. Torra et al. (Eds.): MDAI 2012, LNAI 7647, pp. 56–65, 2012.

This paper is one of the first attempts for the calculation of Choquet integral with respect to a fuzzy measure of a function on real line.

This paper is devoted to the study of the Choquet integral of a function on the real line. Especially we study the Choquet integral with respect to a symmetric fuzzy measure, which is the geralization of OWA operator. We introduce some approximation of such integrals for particular types of fuzzy measures. Then, we study the inequalities of the Choquet integral with respect to a special fuzzy measure, including symmetric fuzzy measure.

The structure of the paper is organized as follows. In Section 2 some preliminaries needed in the rest of the paper are given. In Section 3, we present the results of the Choquet integral with respect to a symmetric fuzzy measure. In Section 4, we present the inequalities of the Choquet integral. The paper finishes with a conclusion.

## 2    Preliminaries

In this section, we define fuzzy measures, the Choquet integral and the OWA operator, and show their basic properties.

Let $X$ be a unit interval or a subset of natural numbers and $\mathcal{B}$ be a class of its Borel sets, that is, the smallest $\sigma-$algebra which includes the class of all closed sets. We say that $(X, \mathcal{B})$ is a measurable space.

**Definition 1.** [21] Let $(X, \mathcal{B})$ be a measurable space. A fuzzy measure (or a non-additive measure) $\mu$ is a real valued set function, $\mu : \mathcal{B} \longrightarrow [0, 1]$ with the following properties;

1. $\mu(\emptyset) = 0$
2. $\mu(A) \leq \mu(B)$ whenever $A \subset B$, $A, B \in \mathcal{B}$.

We say that the triplet $(X, \mathcal{B}, \mu)$ is a fuzzy measure space if $\mu$ is a fuzzy measure.

A fuzzy measure is said to be continuous if $A_n \uparrow A$ implies $\mu(A_n) \uparrow \mu(A)$ and $A_n \downarrow A$ implies $\mu(A_n) \downarrow \mu(A)$.

We assume that $\mu$ is continuous if $X$ is a unit interval.

**Definition 2.** Let $(X, \mathcal{B}, \mu)$ be a fuzzy measure space.

1. $\mu$ is said to be submodular, if

$$\mu(A) + \mu(B) \geq \mu(A \cup B) + \mu(A \cap B).$$

2. $\mu$ is said to be supermodular if

$$\mu(A) + \mu(B) \leq \mu(A \cup B) + \mu(A \cap B).$$

**Definition 3.** Let $(X, \mathcal{B})$ be a measurable space. A function $f : X \to R$ is said to be measurable if $\{x | f(x) \geq \alpha\} \in \mathcal{B}$ for all $\alpha \in R$.

$\mathcal{F}(X)$ denotes the class of non-negative measurable functions, that is,

$$\mathcal{F}(X) = \{f | f : X \to R^+, f : \text{measurable}\}$$

**Definition 4.** [3, 18] Let $(X, \mathcal{B}, \mu)$ be a fuzzy measure space. The Choquet integral of $f \in \mathcal{F}(X)$ with respect to $\mu$ is defined by

$$(C) \int f d\mu = \int_0^\infty \mu_f(r) dr,$$

where $\mu_f(r) = \mu(\{x | f(x) \geq r\})$.

Let $A \subset X$. The Choquet integral restricted on $A$ is defined by

$$(C) \int_A f d\mu := (C) \int f \cdot 1_A d\mu.$$

Let $\mu$ be a fuzzy measure on $(X, \mathcal{B})$. $\mathcal{F}_\mu(X)$ denotes the class of non-negative measurable functions with Choquet integrable, that is,

$$\mathcal{F}_\mu(X) = \{f | f \in \mathcal{F}_\mu(X), (C) \int f d\mu < \infty\}.$$

The next proposition is obvious from the definition of the Choquet integral.

**Proposition 1.** *Let $\mu$ and $\nu$ be a fuzzy measure on $(X, \mathcal{B})$ and $a, b$ be a real number. We have*

$$(C) \int_A f d(a\mu + b\nu) = a(C) \int_A f d\mu + b(C) \int_A f d\nu$$

*for $f \in \mathcal{F}_\mu(X) \cap \mathcal{F}_\nu(X)$.*

In relation to Choquet integral with respect to a submodular (a super-modular) non-additive measure, we have the next famous theorem.

**Theorem 1.** [3, 4, 19] Let $\mu$ be a non-additive measure in $(X, \mathcal{B})$ and $f, g \in \mathcal{M}^+$.

1. If $\mu$ is submodular, then

$$(C) \int (f + g) d\mu \leq (C) \int f d\mu + (C) \int g d\mu.$$

2. If $\mu$ is supermodular, then

$$(C) \int (f + g) d\mu \geq (C) \int f d\mu + (C) \int g d\mu.$$

Next we will introduce the general definition of aggregation operator (aggregation function [8]).

**Definition 5.** *Let $\mathcal{I}$ be a non empty real interval. An aggregation operator $Ag$ is a function $Ag : \mathcal{I}^{(N)} \to \mathcal{I}$ with the following properties;*

1. *(Monotonicity)*
   *If $a_i \leq b_i$ for all $i = 1, \ldots, N$, $\mathbf{a} = (a_1, \ldots, a_N), \mathbf{b} = (b_1, \ldots, b_N)$ $\mathbf{a}, \mathbf{b} \in D$, then $Ag(\mathbf{a}) \leq Ag(\mathbf{b})$.*
2. *(boundary conditions)*
   $\inf_{\mathbf{x} \in \mathcal{I}^{(N)}} Ag(\mathbf{x}) = \inf \mathcal{I}$ *and* $\sup_{\mathbf{x} \in \mathcal{I}^{(N)}} Ag(\mathbf{x}) = \sup \mathcal{I}$.

We say that an aggregation operator $Ag$ satisfies an idempotency (or unanimity) if $Ag(a, \ldots, a) = a$ if $a \in \mathcal{I}$.

In the following we assume that $\mathcal{I}$ be a unit interval $[0, 1]$ and $Ag$ is idempotent.

Yager introduced the Ordered Weighted Averaging operator in [24], which is one of the most famous aggregation operator with idempotency.

A weighting vector $\mathbf{w}$ with weights $(w_1, \ldots, w_N)$ is a vector $\mathbf{w} \in R^N$ satisfying $\sum_{i=1}^{N} w_i = 1$ and $w_i \geq 0$ for all $i = 1, 2, \ldots, N$.

**Definition 6.** *[24] Given a weighting vector $\mathbf{w}$ with weights $(w_1, \ldots, w_N)$, the Ordered Weighted Averaging operator is defined as follows:*

$$OWA_{\mathbf{w}}(\mathbf{a}) = \sum_{i=1}^{N} w_i a_{\sigma(i)}$$

*where $\sigma$ defines a permutation of $\{1, \ldots, N\}$ such that $a_{\sigma(i)} \geq a_{\sigma(i+1)}, \mathbf{a} = (a_1, \ldots, a_N)$.*

**Definition 7.** *Let $X := \{1, \ldots, N\}$. A fuzzy measure $\mu$ on $\mathcal{B}$ is said to be symmetric [9] if $\mu(A) = \mu(B)$ for $|A| = |B|$, $A, B \in \mathcal{B}$.*

Symmetric fuzzy measures on $(X, \mathcal{B})$ can be represented in terms of $N$ weights $w_i$ for $i = 1, \ldots, N$ so that $\mu(A) = \sum_{i=1}^{|A|} w_i$. Using a symmetric fuzzy measure, we can represent any $OWA$ operator as a Choquet integral.

**Proposition 2.** *Let $X := \{1, 2, \ldots, N\}$; then, for every $OWA_{\mathbf{w}}$, there exists a symmetric fuzzy measure satisfying $\mu(\{1\}) := w_1$ and $\mu(\{1, \ldots, i\}) := w_1 + \cdots + w_i$ for $i = 1, 2, \ldots, N$, such that*

$$OWA_{\mathbf{w}}(\mathbf{a}) = (C) \int \mathbf{a} d\mu$$

*for $\mathbf{a} \in R_+^N$.*

## 3 Choquet Integral with Respect to a Symmetric Fuzzy Measure

In the following we consider the Choquet integral of a monotone increasing function on the real line. We assume that $X$ is a unit interval, that is, $X = [0, 1]$.

Let $\lambda$ be a Lebesgue measure on $C$, that is, a measure generated by $\lambda([a,b]) = b - a$ for $[a,b] \subset X$. Let $\mathcal{F}_c(X)$ be a class of continuous functions on $X$.

We will define a continuous version of symmetric fuzzy measure.

**Definition 8.** *Let $\mu$ be a fuzzy measure on $(X, \mathcal{B})$ and $\mu(X) = 1$. $\mu$ is said to be symmetric, if $\lambda(A) = \lambda(B)$ implies $\mu(A) = \mu(B)$.*

**Remark.** Let $X := \{1, \ldots, N\}$ and $m(A) := |A|$ for $A \subset X$. Then $m$ is an additive measure on $2^X$. If $\mu$ is symmetric, we have $m(A) = m(B)$ implies $\mu(A) = \mu(B)$. The symmetry in Definition 8 is essentially same as one in Definition 7.

Let $\mu$ be a symmetric fuzzy measure on $([0,1], \mathcal{B})$. Define a function $\varphi : [0,1] \to [0,1]$ by $\varphi(x) := \mu([0,x])$. Suppose that $x < y$. Since $[0,x] \subset [0,y]$, we have $\varphi(x) \leq \varphi(y)$.

Let $\lambda(A) := x$ for arbitrary $A \in \mathcal{B}$. Then we have $\lambda(A) = \lambda([0,x])$.

Since $\mu$ is symmetric, we have

$$\mu(A) = \mu([0,x]) = \varphi(x) = \varphi(\lambda(A))$$

for arbitrary $A \in \mathcal{B}$. Therefore we have the next proposition.

**Proposition 3.** *Let $\mu$ be a symmetric fuzzy measure on $(X, \mathcal{B})$. There exists a monotone increasing function $\varphi : [0,1] \to [0,1]$ such that $\mu = \varphi \circ \lambda$.*

Let $x_n \to x$ for $x_n \in [0,1], n = 1, 2, \ldots$. Define $\{a_n\}$ and $\{b_n\}$ by

$$a_n = \inf_{k \geq n} x_k, \quad b_n = \sup_{k \geq n} x_k.$$

Since $a_n \uparrow x$, we have $\cup_n [0, a_n] = [0, x]$. It follows from the continuity of $\mu$ that

$$\lim_{n \to \infty} \mu([0, a_n]) = \mu([0, x]),$$

that is,

$$\lim_{n \to \infty} \varphi(a_n) = \varphi(x).$$

In the same way, we have

$$\lim_{n \to \infty} \varphi(b_n) = \varphi(x).$$

Since $\varphi$ is monotone and $a_n \leq x_n \leq b_n$, we have

$$\lim_{n \to \infty} \varphi(x_n) = \varphi(x).$$

Therefore we have the next proposition.

**Proposition 4.** *Let $\mu$ be a symmetric fuzzy measure on $(X, \mathcal{B})$. $\varphi$ in Proposition 3 is continuous.*

We say that a function $\varphi$ in Proposition 1 is a weight function for a symmetric fuzzy measure $\mu$. Since $\varphi$ is continuous, it follows from Weierstrass approximation theorem that $\varphi$ can be approximated by the polynomial, that is , for any $\epsilon > 0$, there exists real numbers $a_1, \cdots, a_N$ such that $|\varphi(x) - \sum_{k=1}^{N} a_k x^k| < \epsilon$ for $x \in X$. Therefore every symmetric fuzzy measure $\mu$ can be approximated by a fuzzy measure $\sum_{k=1}^{N} a_k \lambda^k$.

**Proposition 5.** *Let $\mu$ be a symmetric fuzzy measure on $(X, \mathcal{B})$. For any $\epsilon > 0$, there exist real numbers $a_1, \ldots, a_N$ such that*

$$|(C) \int f d\mu - \sum_{k=1}^{N} a_k (C) \int f d\lambda^k| < \epsilon.$$

*for $f \in \mathcal{F}_\mu(X)$.*

*Example 1.* Let $\mu := \lambda^{1/2}$, that is, $\varphi(x) = x^{1/2}$. We have a sequence $\varphi_n$ such that $\varphi_n \to \varphi$ uniformly. In fact, $\varphi_1(x) = x$,
$\varphi_2(x) = f(\frac{1}{2})_2 C_1 x(1 - x) + f(1)_2 C_2 x^2 = \sqrt{2} x - (\sqrt{2} - 1) x^2$.
. . .

Therefore we have

$$(C) \int f d\lambda^{1/2} \approx \sqrt{2}(C) \int f d\lambda - (\sqrt{2} - 1)(C) \int f d\lambda^2.$$

Moreover suppose that $\varphi$ in Proposition 1 be analytic. Then we can express $w$ by

$$\varphi(x) := \sum_{k=1}^{\infty} a_k x^k.$$

Then we have

$$(C) \int_{[0,x]} f d\mu = \sum_{k=1}^{\infty} a_k (C) \int f d\lambda^k$$

for $x \in [0, 1]$.

*Example 2.* Let $\mu(A) := \log_2(\lambda(A) + 1)$, that is, $\varphi(x) = \log_2(x + 1)$. Since we have

$$\varphi(x) = \frac{1}{\log 2} \sum_{k=1}^{\infty} \frac{(-1)^{k-1}}{k} x^k.$$

Therefore

$$(C) \int f d\log_2(\lambda + 1) = \frac{1}{\log 2} \sum_{k=1}^{\infty} \frac{(-1)^{k-1}}{k} \int f d\lambda^k.$$

# 4   Inequalities

In this section, we will present some basic inequalities for the Choquet integral with respect to a fuzzy measure generated by a convex function or a concave function.

**Definition 9.** *Let $\varphi$ be a real valued function on closed interval $[c,d]$. $\varphi$ is said to be convex if*

$$\varphi(\lambda x + (1-\lambda)y) \leq \lambda\varphi(x) + (1-\lambda)\varphi(y)$$

*for $x, y \in [c,d]$, $0 < \lambda < 1$.*
   *$\varphi$ is said to be concave if*

$$\varphi(\lambda x + (1-\lambda)y) \geq \lambda\varphi(x) + (1-\lambda)\varphi(y)$$

*for $x, y \in [c,d]$, $0 < \lambda < 1$.*

We have the next Jensen's inequality from the definition [4].

**Proposition 6.** *Let $\mu$ be a fuzzy measure on $(X, \mathcal{B})$ with $\mu(X) = 1$.*

1. *If $\varphi$ is convex, then*

$$(C)\int \varphi(f)d\mu \geq \varphi((C)\int f d\mu).$$

2. *If $\varphi$ is concave, then*

$$(C)\int \varphi(f)d\mu \leq \varphi((C)\int f d\mu).$$

Suppose that $\varphi$ is convex or concave and monotone. Applying the theorem above to classical Lebesgue integral, we have

$$\int_0^1 \varphi(\mu(\{x|f(x) \geq a\}))da \geq \varphi(\int_0^1 (\mu(\{x|f(x) \geq a\}))da).$$

Therefore we have the next inequalities.

**Proposition 7.** *Let $\mu$ be a fuzzy measure on $(X, \mathcal{B})$.*

1. *If $\varphi : [0,1] \to [0,1]$ is a non-decreasing convex function on closed interval with $\varphi(0) = 0$ and $\varphi(1) = 1$.*

$$(C)\int f d(\varphi \circ \mu) \geq \varphi((C)\int f d\mu).$$

2. *If $\varphi$ is a non-decreasing concave function on closed interval with $\varphi(0) = 0$ and $\varphi(1) = 1$,*

$$(C)\int f d(\varphi \circ \mu) \leq \varphi((C)\int f d\mu).$$

Using subadditivity of Choquet integral with respect to a submodular fuzzy measure, we have the next proposition [14, 11].

**Proposition 8.** Let $\mu$ be a submodular fuzzy measure on $(X, \mathcal{B})$ and $p \geq 1$, $q \geq 1$, $1/p + 1/q = 1$.

1.

$$(C) \int fg d\mu \leq ((C) \int f^p d\mu)^{1/p} ((C) \int g^q d\mu)^{1/q}$$

2.

$$(C) \int (f + g)^p d\mu \leq ((C) \int f^p d\mu)^{1/p} + ((C) \int g^p d\mu)^{1/p}$$

Note that if $\varphi : [0, 1] \to [0, 1]$ be concave and continuous, and $\mu := \varphi \circ \lambda$. Then $\mu$ is submodular [19].

In the following we suppose that $\varphi : [0, 1] \to [0, 1]$ be concave and continuous, and $\mu := \varphi \circ \lambda$.

**Definition 10.** *We say that a continuous function $\varphi : [0, 1] \to [0, 1]$ is semi convex if there exists $C > 0$ such that for all $x, y \in [0, 1]$ and $0 \leq a \leq 1$*

$$\varphi(ax + (1 - a)y) \leq C\{(a\varphi(x)) + (1 - a\varphi(y))\}.$$

*We say that a continuous function $\varphi : [0, 1] \to [0, 1]$ is strongly semi convex if there exists $C > 0$ such that for all $x_i \in [0, 1]$, $0 \leq a_i \leq 1$ and $\sum_i a_i = 1$,*

$$\varphi(\sum_i a_i x_i) \leq C \sum_i a_i \varphi(x_i).$$

Suppose that $\varphi$ is continuous and concave with $\varphi(0) = 0$ and $\varphi(1) = 1$. Then for all $x \in [0, 1]$ we have $x \leq \varphi(x)$. Therefore we have the next proposition.

**Proposition 9.** *Suppose that $\varphi$ is continuous and concave with $\varphi(0) = 0$ and $\varphi(1) = 1$. $\varphi$ is strongly semi convex if there exists $C > 1$ such that for all $x$, $\varphi(x) \leq Cx$.*

*Example 3.* Let $\varphi(x) = x(2 - x)$. $\varphi$ is concave with $\varphi(0) = 0$ and $\varphi(1) = 1$. Then we have $x \leq \varphi(x) \leq 2x$.

Next we will define a maximal function with respect to a fuzzy measure.

**Definition 11.** *Let $\mu$ be a fuzzy measure on $(X, B)$ and $f \in \mathcal{F}_\mu(X)$.*
*A maximal function $M_\mu f$ with respect to $\mu$ of $f$ is defined by*

$$M_\mu f(x) := \sup_r \frac{1}{\mu([x - r, x + r])} (C) \int_{[x-r, x+r]} f d\mu.$$

If $\mu$ is a classical measure, $M_\mu f$ is Hardy Littlewood maximal function.
If $\mu$ is symmetric with some special conditions, we have the next theorem that is similar to classical one.

**Theorem 2.** *Let $\varphi$ be continuous and concave with $\varphi(0) = 0$ and $\varphi(1) = 1$ and $\varphi$ be strongly semi convex.*
*Let $\mu = \varphi \circ \lambda$ and $f \in \mathcal{F}_\mu(X)$.*
*There exists a constant $C$ such that for all $\alpha > 0$*

$$\mu(\{x \mid M_\mu f(x) > \alpha\}) \leq \frac{C}{\alpha}(C) \int f d\mu.$$

## 5    Conclusion

In this paper we have studied some properties of the Choquet integral. We have given some new expressions to compute the Choquet integral with respect to a function on the real line, We have given some inequalities.

As future work we will consider the extension of the space to multi dimensional Euclidean space, and also the application of these results.

## References

1. Calvo, T., Mayor, G., Mesiar, R. (eds.): Aggregation Operators. Physica-Verlag (2002)
2. Chateauneuf, A.: Modeling attitudes towards uncertainty and risk through the use of Choquet integral. Annals of Operations Research 52, 1–20 (1994)
3. Choquet, G.: Theory of capacities. Ann. Inst. Fourier, Grenoble. 5, 131–295 (1955)
4. Denneberg, D.: Non additive measure and integral. Kluwer Academic Publishers, Dordorecht (1994)
5. Faigle, U., Grabisch, M.: A discrete Choquet integral for ordered systems. Original Research Article Fuzzy Sets and Systems 168(1), 3–17 (2011)
6. Gilboa, I., Schmeidler, D.: Additive representations of non-additive measures and the Choquet integral. Annals of Operations Research 52(1), 43–65 (1994)
7. Grabisch, M., Murofushi, T., Sugeno, M. (eds.): Fuzzy Measures and Integrals: Theory and Applications. Physica-Verlag (2000)
8. Grabisch, M., Marichal, J.-L., Mesiar, R., Pap: Aggregation Functions. In: Encyclopedia of Mathematics and its Applications, vol. 127. Cambridge University Press (2009)
9. Miranda, P., Grabisch, M.: $p$-symmetric fuzzy measures. In: Proc. of the IPMU 2002 Conference, Annecy, France, pp. 545–552 (2002)
10. Grabisch, M., Labreuche, C.: A decade of application of the Choquet and Sugeno integrals in multi-criteria decision aid. Annals of Operations Research 175(1), 247–286 (2010)
11. Mesiar, R., Li, J., Pap, E.: The Choquet integral as Lebesgue integral and related inequalities. Kybernetika 46(6), 1098–1107 (2010)
12. Mesiar, R., Mesiarova-Zemankova, A., Ahmad, K.: Discrete Choquet integral and some of its symmetric extensions. Fuzzy Sets and Systems 184(1), 148–155 (2011)
13. Narukawa, Y., Murofushi, T., Sugeno, M.: Regular fuzzy measure and representation of comonotonically additive functional. Fuzzy Sets and Systems 112(2), 177–186 (2000)
14. Narukawa, Y.: Distances defined by Choquet integral. In: IEEE International Conference on Fuzzy Systems, London, England CD-ROM, July 24-26 (2007)

15. Narukawa, Y., Torra, V.: Continuous OWA operator and its calculation. In: Proc. IFSA-EUSFLAT, Lisbon, Portugal, July 20-24, pp. 1132–1135 (2009) ISBN:978-989-95079-6-8
16. Narukawa, Y., Torra, V.: Aggregation operators on the real line. In: Proc. 3rd International Workshop on Soft Computing Applications (SOFA 2009), Szeged, Hungary and Arad, Romania, pp. 185–188 (August 2009) ISBN: 978-1-4244-5056-5
17. Narukawa, Y., Torra, V.: Choquet Integral on Locally Compact Space: A Survey. In: Huynh, V.-N., Nakamori, Y., Lawry, J., Inuiguchi, M. (eds.) Integrated Uncertainty Management and Applications. AISC, vol. 68, pp. 71–81. Springer, Heidelberg (2010)
18. Murofushi, T., Sugeno, M.: An interpretation of fuzzy measures and the Choquet integral as an integral with respect to a fuzzy measure. Fuzzy Sets and Systems 29, 201–227 (1989)
19. Pap, E.: Null-Additive set functions. Kluwer Academic Publishers, Dordorecht (1995)
20. Ralescu, A.L., Ralescu, D.A.: Extensions of fuzzy aggregation. Fuzzy Sets and Systems 86, 321–330 (1997)
21. Sugeno, M.: Theory of fuzzy integrals and its applications, Doctoral Thesis, Tokyo Institute of Technology (1974)
22. Sugeno, M., Narukawa, Y., Murofushi, T.: Choquet integral and fuzzy measures on locally compact space. Fuzzy Sets and Systems 99(2), 205–211 (1998)
23. Torra, V., Narukawa, Y.: Modeling decisions: information fusion and aggregation operators. Springer (2007)
24. Yager, R.R.: On ordered weighted averaging aggregation operators in multi-criteria decision making. IEEE Trans. on Systems, Man and Cybernetics 18, 183–190 (1988)
25. Yager, R.R., Filev, D.P.: Parameterized and-like and or-like OWA operators. Int. J. of General Systems 22, 297–316 (1994)
26. Yager, R.R.: Families of OWA operators. Fuzzy Sets and Systems 59, 125–148 (1993)

# On WOWA Rank Reversal

Wlodzimierz Ogryczak[1], Patrice Perny[2], and Paul Weng[2]

[1] ICCE, Warsaw University of Technology, Warsaw, Poland
wogrycza@elka.pw.edu.pl
[2] LIP6 - UPMC, Paris, France
{patrice.perny,paul.weng}@lip6.fr

**Abstract.** The problem of aggregating multiple criteria to form an overall measure is of considerable importance in many disciplines. The ordered weighted averaging (OWA) aggregation, introduced by Yager, uses
weights assigned to the ordered values rather than to the specific criteria. This allows one to model various aggregated preferences, preserving
simultaneously the impartiality (neutrality) with respect to the individual criteria. However, importance weighted averaging is a central task in
multicriteria decision problems of many kinds. It can be achieved with
the Weighted OWA (WOWA) aggregation, introduced by Torra, covering both the weighted means and the OWA averages as special cases. In
this paper we analyze the monotonicity properties of the WOWA aggregation with respect to changes of importance weights. In particular, we
demonstrate that a rank reversal phenomenon may occur in the sense
that increasing the importance weight for a given criterion may enforce
the opposite WOWA ranking than that imposed by the criterion values.

**Keywords:** OWA, WOWA, Multicriteria Optimization, Rank Reversal.

## 1 Introduction

Consider a decision problem defined by $m$ criteria. That means the decisions are
characterized by $m$-dimensional outcome vectors $\eta = (\eta_1, \eta_2, \dots, \eta_m)$. In order
to make the multicriteria model operational for the decision support process,
one needs to assume some aggregation function $a : R^m \to R$. The aggregated
value can then be optimized (maximized or minimized).

The most commonly used aggregation is based on the weighted mean where
positive importance weights $p_i$ $(i = 1, \dots, m)$ are allocated to several criteria

$$A_{\mathbf{p}}(\eta) = \sum_{i=1}^{m} p_i \eta_i \qquad (1)$$

The weights are typically normalized to the total 1 ($\sum_{i=1}^{m} p_i = 1$). However, the
weighted mean while being able to define the importance of criteria is not able to
model the decision maker's preferences regarding the distribution of outcomes.
The latter is crucial when aggregating (normalized) uniform achievement criteria
like those used in the fuzzy optimization methodologies [18] as well as in the goal

V. Torra et al. (Eds.): MDAI 2012, LNAI 7647, pp. 66–77, 2012.

programming and the reference point approaches to multiple criteria decision support [7]. In stochastic problems uniform objectives may represent various possible values of the same (uncertain) outcome under several scenarios [6].

The preference weights can be effectively introduced with the so-called Ordered Weighted Averaging (OWA) aggregation function developed by Yager [16]. In the OWA aggregation the weights are assigned to the ordered values (i.e. to the smallest value, the second smallest and so on) rather than to the specific criteria. Since its introduction, the OWA aggregation has been successfully applied to many fields of decision making [18,19].

The OWA operator is able to model various aggregation functions from the maximum through the arithmetic mean to the minimum. Thus, it enables modeling of various preferences from the optimistic to the pessimistic one. On the other hand, the OWA is not able to allocate any importance weights to specific criteria. Actually, the weighted mean (1) cannot be expressed in terms of the OWA aggregations.

Importance weighted averaging is a central task in multicriteria decision problems of many kinds, such as selection, classification, object recognition, and information retrieval. Therefore, several attempts have been made to incorporate importance weighting into the OWA operator [17,2]. Finally, Torra [13] has introduced the Weighted OWA (WOWA) aggregation defined by two weighting vectors: the preferential weights $\mathbf{w}$ and the importance weights $\mathbf{p}$. It covers both the weighted means (defined with $\mathbf{p}$) and the OWA averages (defined with $\mathbf{w}$) as special cases. Actually, the WOWA average is reduced to the weighted mean in the case of equal preference weights and it becomes the standard OWA average in the case of equal importance weights. Since its introduction, the WOWA operator has been successfully applied to many fields of decision making [15,9,10,7,8] including metadata aggregation problems [1,5].

While considering the importance weighting of the criteria one may expect some monotonicity properties of the aggregation with respect to the (relative) increase of a given importance weight. The basic stability requirements with respect to a given importance weight can be formalized as two properties: rank stability and asymptotic monotonicity. We say that an aggregation satisfies the rank stability property if whenever the aggregation ranks two vectors consistently with the inequality on a given criterion it preserves this ranking for any positive increase of importance weight for the given criterion. If despite that for some importance weights the aggregation ranks two vectors consistently with the relation on a given criterion, a positive increase of importance weight for the given criterion may result in an opposite inequality, we say that the rank reversal phenomenon occurs. We say that an aggregation satisfies the asymptotic monotonicity property if for any importance weights independently from the relation between the aggregation values for two vectors, a sufficiently large increase of the importance weight for a given criterion enforces the aggregation ranking consistently with the inequality on the given criterion values. Both stability properties are satisfied by the weighted mean (1). We analyze how the WOWA aggregation models the importance weighting stability properties. Unfortunately, we are able to show a possible rank reversal phenomenon which may be considered a serious flaw of the WOWA importance weighting scheme. However, the WOWA aggregation fulfills the asymptotic monotonicity property.

The paper is organized as follows. In the next section we formally introduce the WOWA operator and recall some alternative computational formula based on the Lorenz curves. In Section 3 we show some examples of the rank reversal phenomenon for the WOWA aggregation. Next, in Section 4 we prove the asymptotic monotonicity showing required levels for sufficiently large increase of the importance weight for various special cases of the WOWA aggregation.

## 2   WOWA Aggregation

Let $\mathbf{w} = (w_1, \ldots, w_m)$ be a weighting vector of dimension $m$ such that $w_i \geq 0$ for $i = 1, \ldots, m$ and $\sum_{i=1}^{m} w_i = 1$. The corresponding OWA aggregation of outcomes $\eta = (\eta_1, \ldots, \eta_m)$ can be mathematically formalized as follows [16]. Let $\langle \eta \rangle = (\eta_{\langle 1 \rangle}, \eta_{\langle 2 \rangle}, \ldots, \eta_{\langle m \rangle})$ denote the vector obtained from $\eta$ by rearranging its components in the non-increasing order. That means $\eta_{\langle 1 \rangle} \geq \eta_{\langle 2 \rangle} \geq \cdots \geq \eta_{\langle m \rangle}$ and there exists a permutation $\tau$ of set $I = \{1, \ldots, m\}$ such that $\eta_{\langle i \rangle} = \eta_{\tau(i)}$ for $i = 1, \ldots, m$. Further, we apply the weighted sum aggregation to ordered outcome vectors $\langle \eta \rangle$, i.e. the OWA aggregation has the following form:

$$OWA_{\mathbf{w}}(\eta) = \sum_{i=1}^{m} w_i \eta_{\langle i \rangle} \tag{2}$$

Due to the strict monotonicity of the OWA aggregation with positive weighting vectors [4], the OWA optimization generates a Pareto optimal solution.

The OWA aggregation (2) allows to model various aggregation functions from the maximum ($w_1 = 1$, $w_i = 0$ for $i = 2, \ldots, m$) through the arithmetic mean ($w_i = 1/m$ for $i = 1, \ldots, m$) to the minimum ($w_m = 1$, $w_i = 0$ for $i = 1, \ldots, m - 1$). However, the weighted mean (1) cannot be expressed as an OWA aggregation. Actually, the OWA aggregations are symmetric (impartial, neutral) with respect to the individual criteria and it does not allow to represent any importance weights allocated to specific criteria.

Importance weighted averaging is a central task in multicriteria decision problems of many kinds and the ordered averaging model enables one to introduce importance weights to affect criteria importance by rescaling accordingly its measure within the distribution of achievements as defined in the so-called Weighted OWA (WOWA) aggregation [13]. Let $\mathbf{w} = (w_1, \ldots, w_m)$ be OWA weights and let $\mathbf{p} = (p_1, \ldots, p_m)$ be an additional importance weighting vector such that $p_i \geq 0$ for $i = 1, \ldots, n$ and $\sum_{i=1}^{m} p_i = 1$. The corresponding Weighted OWA aggregation of achievements $\eta = (\eta_1, \ldots, \eta_m)$ is defined as follows [13]:

$$WOWA_{\mathbf{w}, \mathbf{p}}(\eta) = \sum_{i=1}^{m} v_i(\mathbf{p}, \eta) \eta_{\langle i \rangle} \tag{3}$$

where weights $v_i$ are defined as

$$v_i(\mathbf{p}, \eta) = \varphi(\sum_{k \leq i} p_{\tau(k)}) - \varphi(\sum_{k < i} p_{\tau(k)}) \tag{4}$$

with $\varphi$ a monotone increasing function that interpolates points $(\frac{i}{m}, \sum_{k \leq i} w_k)$ together with point $(0.0)$ and $\tau$ representing the ordering permutation for $\eta$ (i.e., $\eta_{\tau(i)} = \eta_{\langle i \rangle}$). Moreover, function $\varphi$ is required to be a straight line when the point can be interpolated in this way, thus allowing the WOWA to cover the standard weighted mean with weights $p_i$ as a special case of equal OWA weights ($w_i = 1/m$ for $i = 1, \ldots, m$). Indeed, the WOWA defined by (3)–(4) as OWA aggregation with modified preferential weights may be rewritten as the weighted mean with modified weights:

$$WOWA_{\mathbf{w}, \mathbf{p}}(\eta) = \sum_{i=1}^{m} \pi_i(\mathbf{p}, \eta)\eta_i \tag{5}$$

where the weights $\pi_i$ are defined as

$$\pi_i(\mathbf{p}, \eta) = \varphi(p_i + \sum_{k < \tau(i)} p_{\tau(k)}) - \varphi(\sum_{k < \tau(i)} p_{\tau(k)}). \tag{6}$$

Actually, the WOWA aggregation is a special case of the rank dependent utility [12] with a piecewise linear probability weighting function $\varphi$ defined by the importance weights.

The WOWA may be expressed with a more direct formula where preferential (OWA) weights $w_i$ are applied to the averages of the corresponding portions of ordered outcomes (quantile intervals) according to the distribution defined by importance weights $p_i$ [9]. Note that one may alternatively compute the WOWA values by using rational importance weights to replicate the corresponding achievements and then calculate the OWA aggregations. This approach can be generalized to any real importance weights and the WOWA aggregation can be equivalently defined as follows [9]:

$$WOWA_{\mathbf{w}, \mathbf{p}}(\eta) = \sum_{i=1}^{m} w_i m \int_{\frac{i-1}{m}}^{\frac{i}{m}} \overline{F}_\eta^{(-1)}(\xi) \, d\xi \tag{7}$$

where $\overline{F}_\eta^{(-1)}$ is the stepwise function $\overline{F}_\eta^{(-1)}(\xi) = \eta_{\langle k \rangle}$ for $\sum_{j < k} p_{\tau(j)} < \xi \leq \sum_{j \leq k} p_{\tau(j)}$, for $k = 1, \ldots, m$ with $\tau$ representing the ordering permutation for $\eta$ (i.e., $\eta_{\tau(k)} = \eta_{\langle k \rangle}$). It can also be mathematically formalized as the quantile function defined as the left-continuous inverse of the decumulative distribution function, i.e., $\overline{F}_\eta^{(-1)}(\xi) = \sup \{z : \overline{F}_\eta(z) \geq \xi\}$ for $0 < \xi \leq 1$ with $\overline{F}_\eta(z) = \sum_{i=1}^{n} p_i \zeta_i(z)$ where $\zeta_i(z) = 1$ if $\eta_i \geq z$ and 0 otherwise.

Formula (7), defining the WOWA value by applying preferential weights $w_i$ to importance weighted averages within quantile intervals, may be reformulated with the tail averages (Lorenz components):

$$WOWA_{\mathbf{w}, \mathbf{p}}(\eta) = \sum_{k=1}^{m} \overline{w}_k m L(\eta, \mathbf{p}, \frac{k}{m}) \quad \text{where} \quad L(\eta, \mathbf{p}, \xi) = \int_0^\xi \overline{F}_\eta^{(-1)}(\zeta) d\zeta \tag{8}$$

and differential weights

$$\overline{w}_k = w_k - w_{k+1} \text{ for } k = 1, \ldots, m-1 \quad \text{and} \quad \overline{w}_m = w_m \qquad (9)$$

Note that the differential weights $\overline{w}_i$ are positive in the case of positive and strictly decreasing preferential (OWA) weights $w_1 > w_2 > \ldots > w_m > 0$. Graphs of functions $L(\eta, \mathbf{p}, \xi)$ (with respect to $\xi$) take the form of concave piecewise linear curves, the so-called (upper) absolute Lorenz curves. Moreover, values of function $L(\eta, \mathbf{p}, \xi)$ for any $0 \leq \xi \leq 1$ can be given by linear programming (LP) optimization which enables the WOWA minimization to be implemented with a LP model [9], in the case of positive and decreasing preferential (OWA) weights $w_1 \geq w_2 \geq \ldots \geq w_m > 0$.

Applying the WOWA aggregation to a multiple criteria optimization problem we get the WOWA optimization model. For any positive weights $\mathbf{w}$ and $\mathbf{p}$, the WOWA aggregation is strictly monotonic [7]. Therefore, the WOWA optimal solutions are then Pareto-optimal.

## 3   Rank Reversal

When considering the importance weighting of the criteria one may expect some monotonicity properties of the aggregation with respect to changes of the importance weights. Note that for any vector of importance weights $\mathbf{p}$ any positive increase of a given importance weight must be accompanied by decrease of some other weights. We will focus on weights changes represented by a positive increase of a given importance weight $p_{i_o}$ with proportional decrease of other weights, i.e., we will consider a parameterized importance weight modification

$$\mathbf{p}(\varepsilon) = \frac{1}{1+\varepsilon}(\mathbf{p} + \varepsilon \mathbf{e}_{i_o}) \quad \text{with} \quad \varepsilon > 0 \qquad (10)$$

where $\mathbf{e}_i$ denotes the $i$th unit vector. The basic stability requirements with respect to a given importance weight can be formalized as two properties: rank stability and asymptotic monotonicity.

*Rank stability and rank reversal.* Let $\eta'$ and $\eta''$ be vectors such that $\eta'_{i_o} < \eta''_{i_o}$ for a criterion $i_o \in I$. We say that an aggregation satisfies the rank stability property if whenever for any importance weights $\mathbf{p}$, the aggregation of $\eta'$ is less or equal to that for $\eta''$, then this inequality remains valid for any positive increase of importance weight $p_{i_o}$ with proportional decrease of other weights. If despite that for some importance weights $\mathbf{p}$ the aggregation of $\eta'$ is less than that for $\eta''$, a positive increase of importance weight $p_{i_o}$ with proportional decrease of other weights may result in opposite inequality, we say that the rank reversal phenomenon occurs.

*Asymptotic monotonicity.* Let $\eta'$ and $\eta''$ be vectors such that $\eta'_{i_o} < \eta''_{i_o}$ for a criterion $i_o \in I$. We say that an aggregation satisfies the asymptotic monotonicity property if for any importance weights $\mathbf{p}$ independently from the relation between the aggregation values of $\eta'$ and $\eta''$, a sufficiently large increase of importance weight $p_{i_o}$ with proportional decrease of other weights enforces the aggregation ranking consistently with inequality $\eta'_{i_o} < \eta''_{i_o}$.

One may notice that both the stability properties are satisfied by the weighted mean (1). Indeed, for any vectors $\eta'$, $\eta''$ and importance weights $\mathbf{p}$, while increasing importance weight $p_{i_o}$ with proportional decrease of other weights, following (10) one gets

$$
\begin{aligned}
A_{\mathbf{p}(\varepsilon)}(\eta') - A_{\mathbf{p}(\varepsilon)}(\eta'') &= \sum_{i=1}^{m} p_i(\varepsilon)(\eta_i' - \eta_i'') \\
&= \frac{1}{1+\varepsilon} \sum_{i=1}^{m} p_i(\eta_i' - \eta_i'') + \frac{\varepsilon}{1+\varepsilon}(\eta_{i_o}' - \eta_{i_o}'') \qquad (11) \\
&= \frac{1}{1+\varepsilon}(A_{\mathbf{p}}(\eta') - A_{\mathbf{p}}(\eta'')) + \frac{\varepsilon}{1+\varepsilon}(\eta_{i_o}' - \eta_{i_o}'')
\end{aligned}
$$

This leads to the following statements.

**Proposition 1.** *Let $\eta'$ and $\eta''$ be outcome vectors such that $\eta_{i_o}' < \eta_{i_o}''$ for a criterion $i_o \in I$. If $A_{\mathbf{p}}(\eta') \le A_{\mathbf{p}}(\eta'')$ for some importance weights $\mathbf{p}$, then any positive increase of importance weight $p_{i_o}$ with proportional decrease of other weights, following (10), results in strict inequality on averages $A_{\mathbf{p}(\varepsilon)}(\eta') < A_{\mathbf{p}(\varepsilon)}(\eta'')$.*

**Proposition 2.** *Let $\eta'$ and $\eta''$ be outcome vectors such that $\eta_{i_o}' < \eta_{i_o}''$ for a criterion $i_o \in I$. For any vector of importance weights $\mathbf{p}$, a sufficiently large increase of importance weight $p_{i_o}$ with proportional decrease of other weights, following (10) with*

$$
\varepsilon > \frac{\max\{A_{\mathbf{p}}(\eta') - A_{\mathbf{p}}(\eta''), 0\}}{\eta_{i_o}'' - \eta_{i_o}'}
$$

*results in strict inequality on averages $A_{\mathbf{p}(\varepsilon)}(\eta') < A_{\mathbf{p}(\varepsilon)}(\eta'')$.*

Unfortunately, the WOWA aggregation does not guarantee the rank stability. We will show that the rank reversal phenomenon may occur for the WOWA aggregation even in a simple case of ordered vectors. Consider two vectors $\eta' = (1000, 102, 10)$ and $\eta'' = (1000, 100, 12)$. While introducing preferential weights $\mathbf{w} = (0.8, 0.1, 0.1)$ and assuming an equal importance of all the criteria, i.e. $\mathbf{p} = (1/3, 1/3, 1/3)$, one gets:

$$
\begin{aligned}
WOWA_{\mathbf{w},\mathbf{p}}(\eta') &= OWA_{\mathbf{w}}(\eta') = 0.8 \cdot 1000 + 0.1 \cdot 102 + 0.1 \cdot 10 = 811.2 \\
WOWA_{\mathbf{w},\mathbf{p}}(\eta'') &= OWA_{\mathbf{w}}(\eta'') = 0.8 \cdot 1000 + 0.1 \cdot 100 + 0.1 \cdot 12 = 811.2
\end{aligned}
$$

Thus with equally important criteria $WOWA_{\mathbf{w},\mathbf{p}}(\eta') = WOWA_{\mathbf{w},\mathbf{p}}(\eta'')$ and according to the ordered aggregation both the vectors are equally good.

Suppose one wish to consider criterion $\eta_3$ as much more important, say 4 times more important than those related to the first or second criterion. For this purpose, importance weights $\bar{\mathbf{p}} = (1/6, 1/6, 2/3)$ are introduced. Note that $\bar{\mathbf{p}}$ may be understood as a result of increasing $p_3$ by 1 and renormalizing all weights, i.e., $\bar{\mathbf{p}} = \frac{1}{1+\varepsilon}(\mathbf{p} + \varepsilon \mathbf{e}_3)$ with $\varepsilon = 1$. Since $\eta_3' < \eta_3''$, one may expect $WOWA_{\mathbf{w},\bar{\mathbf{p}}}(\eta') < WOWA_{\mathbf{w},\bar{\mathbf{p}}}(\eta'')$. However this is not the case, as we show now.

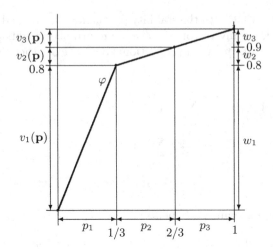

**Fig. 1.** Definition of function $\varphi$ for $\mathbf{w} = (0.8, 0.1, 0.1)$ and WOWA weights $v_i$ for equally important attributes $\mathbf{p} = (1/3, 1/3, 1/3)$ and vectors $\eta$ with ordered coefficients $\eta_1 \geq \eta_2 \geq \eta_3$

To take into account the importance weights in the WOWA aggregation (3) we introduce the piecewise linear function $\varphi$ (Fig. 1):

$$\varphi(\xi) = \begin{cases} 2.4\xi & \text{for } 0 \leq \xi \leq 1/3 \\ 0.8 + 0.3(\xi - 1/3) & \text{for } 1/3 < \xi \leq 2/3 \\ 0.9 + 0.3(\xi - 2/3) & \text{for } 2/3 < \xi \leq 1.0 \end{cases} \qquad (12)$$

Actually, since vectors $\eta'$ and $\eta''$ are both already ordered, the ordered weights $v_i$ are identical for both of them $v_i(\mathbf{p}, \eta') = v_i(\mathbf{p}, \eta'') = v_i(\mathbf{p})$. In the case of equal importance weights $\mathbf{p} = (1/3, 1/3, 1/3)$, obviously, $v_i(\mathbf{p}) = w_i$ (as presented in Fig. 1). Calculating weights $v_i$ according to formula (4) with function $\varphi$ given by (12), as illustrated in Fig. 2, one gets $v_1(\bar{\mathbf{p}}) = \varphi(1/6) = 0.4$, $v_2(\bar{\mathbf{p}}) = \varphi(1/3) - \varphi(1/6) = 0.4$ and $v_3(\bar{\mathbf{p}}) = 1 - \varphi(1/3) = 0.2$. Hence,

$$WOWA_{\mathbf{w}, \bar{\mathbf{p}}}(\eta') = 0.4 \cdot 1000 + 0.4 \cdot 102 + 0.2 \cdot 10 = 442.8$$
$$WOWA_{\mathbf{w}, \bar{\mathbf{p}}}(\eta'') = 0.4 \cdot 1000 + 0.4 \cdot 100 + 0.2 \cdot 12 = 442.4$$

Thus, despite $\eta_3''$ is 20% larger than $\eta_3'$ while $\eta_2''$ is only 2% smaller than $\eta_2'$, an increase of the importance weight for the third criterion results in a lower WOWA evaluation of $\eta''$ in comparison to $\eta'$.

Compare with the same weights vector $\eta' = (1000, 102, 10)$ with $\eta''' = (1000, 100, 13)$. Assuming an equal importance of all criteria, i.e. $\mathbf{p} = (1/3, 1/3, 1/3)$, one gets:

$$WOWA_{\mathbf{w}, \mathbf{p}}(\eta') = OWA_{\mathbf{w}}(\eta') = 0.8 \cdot 1000 + 0.1 \cdot 102 + 0.1 \cdot 10 = 811.2$$
$$WOWA_{\mathbf{w}, \mathbf{p}}(\eta''') = OWA_{\mathbf{w}}(\eta''') = 0.8 \cdot 1000 + 0.1 \cdot 100 + 0.1 \cdot 13 = 811.3$$

Thus with equally important criteria $WOWA_{\mathbf{w}, \mathbf{p}}(\eta')$ is a little bit smaller than $WOWA_{\mathbf{w}, \mathbf{p}}(\eta''')$ similar to inequality on the third criterion $\eta_3' < \eta_3'''$. Consider

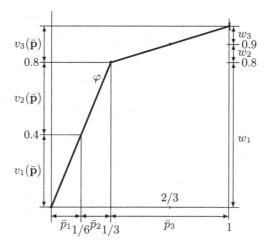

**Fig. 2.** Definition of WOWA weights $v_i$ with $\mathbf{w} = (0.8, 0.1, 0.1)$ and $\bar{\mathbf{p}} = (1/6, 1/6, 2/3)$ for vectors $\eta$ with ordered coefficients $\eta_1 \geq \eta_2 \geq \eta_3$

now criterion $\eta_3$ as 4 times more important than those related to the first or second criterion, i.e., importance weights $\bar{\mathbf{p}} = (1/6, 1/6, 2/3)$. Since, the vectors are already ordered, the corresponding ordered weights calculation remains the same as for the earlier comparison of vectors $\eta'$ and $\eta''$ (see Fig. 2). Hence,

$$WOWA_{\mathbf{w}, \bar{\mathbf{p}}}(\eta''') = 0.4 \cdot 1000 + 0.4 \cdot 100 + 0.2 \cdot 13 = 442.6 < 442.8 = WOWA_{\mathbf{w}, \bar{\mathbf{p}}}(\eta')$$

Thus, despite that for equal importance weights the WOWA aggregation ranks vectors $\eta'$ and $\eta'''$ consistently with the inequality on the third criterion, increasing the importance weight for this criterion results in a rank reversal. Note that this phenomenon occurs despite $\eta_3'''$ is 30% larger than $\eta_3'$ while $\eta_2'''$ is only 2% smaller than $\eta_2'$.

Since our examples are built on ordered vectors, the WOWA rank reversal phenomenon can easily be explained with an analysis of the graph of function $\varphi$. Note that in the case of equal importance weights (Fig. 1), weight $v_1$ is defined by an interval on a high slope segment of $\varphi$ whereas both $v_2$ and $v_3$ are defined on a lower slope segment. While increasing the importance weight for $\eta_3$ one gets increased $v_3$ due to expanded interval. Intervals defining $v_1$ and $v_2$ are appropriately decreased. However, while $v_1$ is indeed decreased, $v_2$ is actually increased since a smaller interval is applied to a higher slope, as the expansion of $\bar{p}_3$ pushes $\bar{p}_2$ on the high slope segment of function $\varphi$.

## 4    Asymptotic Monotonicity

In the previous section, we have given a counterexample illustrating that rank stability may not hold for the WOWA aggregation. Now, we show that nonetheless it satisfies asymptotic monotonicity and we give the required levels of importance weight change to guarantee it.

The WOWA aggregation is continuous with respect to importance weights. Therefore, it obviously fulfills the property of asymptotic monotonicity. Note that for any outcome vectors $\eta'$ and $\eta''$ such that $\eta'_{i_o} < \eta''_{i_o}$ for a criterion $i_o \in I$, one gets $WOWA_{\mathbf{w},\bar{\mathbf{p}}}(\eta') < WOWA_{\mathbf{w},\bar{\mathbf{p}}}(\eta'')$ with $\bar{\mathbf{p}} = \mathbf{e}_{i_o} = \lim_{\varepsilon \to \infty} \mathbf{p}(\varepsilon)$, following (10). Indeed, the following statement can be directly proven.

**Proposition 3.** *Let $\eta'$ and $\eta''$ be outcome vectors such that $\eta'_{i_o} < \eta''_{i_o}$ for a criterion $i_o \in I$. For any positive preferential weights $w_i \geq 0$ and any vector of importance weights $\mathbf{p}$, a sufficiently large increase of importance weight $p_{i_o}$ with proportional decrease of other weights, following (10) with $\varepsilon > \Delta$*

$$\Delta = \max \left\{ \frac{\max\limits_{i \neq i_o}(\eta'_i - \eta''_i)\max\limits_{k \in I} w_k}{(\eta''_{i_o} - \eta'_{i_o})\min\limits_{k \in I} w_k}, 0 \right\} + \frac{2\max\limits_{i \in I}|\eta''_i|(\max\limits_{k \in I} w_k - \min\limits_{k \in I} w_k)}{(\eta''_{i_o} - \eta'_{i_o})\min\limits_{k \in I} w_k}$$

*results in strict inequality $WOWA_{\mathbf{w},\mathbf{p}(\varepsilon)}(\eta') < WOWA_{\mathbf{w},\mathbf{p}(\varepsilon)}(\eta'')$.*

*Proof.* Note that following (5), one gets

$$WOWA_{\mathbf{w},\mathbf{p}(\varepsilon)}(\eta') - WOWA_{\mathbf{w},\mathbf{p}(\varepsilon)}(\eta'') = \sum_{i=1}^{m}\pi_i(\mathbf{p}(\varepsilon),\eta')\eta'_i - \sum_{i=1}^{m}\pi_i(\mathbf{p}(\varepsilon),\eta'')\eta''_i$$

$$= \sum_{i=1}^{m}\pi_i(\mathbf{p}(\varepsilon),\eta')(\eta'_i - \eta''_i) + \sum_{i=1}^{m}(\pi_i(\mathbf{p}(\varepsilon),\eta') - \pi_i(\mathbf{p}(\varepsilon),\eta''))\eta''_i$$

and from (6), one gets for any outcome vector $\eta$

$$mp_i(\varepsilon)\min_{k \in I} w_k \leq \pi_i(\mathbf{p}(\varepsilon),\eta) \leq mp_i(\varepsilon)\max_{k \in I} w_k,$$

$$1 - m(1 - p_i(\varepsilon))\max_{k \in I} w_k \leq \pi_i(\mathbf{p}(\varepsilon),\eta) \leq 1 - m(1 - p_i(\varepsilon))\min_{k \in I} w_k.$$

Hence, $|\pi_i(\mathbf{p}(\varepsilon),\eta') - \pi_i(\mathbf{p}(\varepsilon),\eta'')| \leq m(\max\limits_{k \in I} w_k - \min\limits_{k \in I} w_k)\min\{p_i(\varepsilon), 1 - p_i(\varepsilon)\}$, $\pi_{i_o}(\mathbf{p}(\varepsilon),\eta') \geq mp_{i_o}(\varepsilon)\min\limits_{k \in I} w_k$, $\pi_i(\mathbf{p}(\varepsilon),\eta') \leq mp_i(\varepsilon)\max\limits_{k \in I} w_k$ for $i \neq i_o$ and

$$WOWA_{\mathbf{w},\mathbf{p}(\varepsilon)}(\eta') - WOWA_{\mathbf{w},\mathbf{p}(\varepsilon)}(\eta'')$$
$$\leq m[\min_{k \in I} w_k p_{i_o}(\varepsilon)(\eta'_{i_o} - \eta''_{i_o}) + \max_{k \in I} w_k \sum_{i \neq i_o} p_i(\varepsilon)\max\{\max_{i \neq i_o}\eta'_i - \eta''_i, 0\}]$$
$$+ m(\max_{k \in I} w_k - \min_{k \in I} w_k)[(1 - p_{i_o}(\varepsilon))|\eta''_{i_o}| + \sum_{i \neq i_o} p_i(\varepsilon)\max_{i \neq i_o}|\eta''_i|]$$
$$\leq m\min_{k \in I} w_k[p_{i_o}(\varepsilon) - (1 - p_{i_o}(\varepsilon))\Delta](\eta'_{i_o} - \eta''_{i_o}).$$

Thus, for large enough $\varepsilon > \Delta$ one gets $p_{i_o}(\varepsilon) = (p_{i_o} + \varepsilon)/(1 + \varepsilon) > \Delta/(1 + \Delta)$ and thereby $WOWA_{\mathbf{w},\mathbf{p}(\varepsilon)}(\eta') < WOWA_{\mathbf{w},\mathbf{p}(\varepsilon)}(\eta'')$.

Proposition 3 states that when having a WOWA optimal solution with a non satisfactory achievement for criterion $i_o$, one may increase the importance of this criterion, e.g., by setting new importance weights $p(\varepsilon)_{i_o} = (p_{i_o} + \varepsilon)/(1 + \varepsilon)$ and

$p(\varepsilon)_i = p_i/(1 + \varepsilon)$ for all $i \neq i_o$. For a sufficiently large increment $\varepsilon$, following Proposition 3 it will exclude solutions with worse resulting values for criterion $i_o$. However, the required amount of the weight increase for a general case, following $\Delta$ in Proposition 3 is impracticably large. It can be reduced for special types of the WOWA operators like for the case of monotonic preferential weights which is well suited for decisions under risk [10] or fair optimization [11].

Note that, following (8), we have

$$WOWA_{\mathbf{w},\mathbf{p}}(\eta') - WOWA_{\mathbf{w},\mathbf{p}}(\eta'') = \sum_{i=1}^{m} \overline{w}_k m[L(\eta', \mathbf{p}, \frac{k}{m}) - L(\eta'', \mathbf{p}, \frac{k}{m})]$$

where $\overline{w}_k$ are positive differential OWA weights defined as (9) and

$$L(\eta', \mathbf{p}, \xi) - L(\eta'', \mathbf{p}, \xi) = \max_{\mathbf{u} \in U(\mathbf{p},\xi)} \sum_{i=1}^{m} \eta'_i u_i - \max_{\mathbf{u} \in U(\mathbf{p},\xi)} \sum_{i=1}^{m} \eta''_i u_i$$

with $U(\mathbf{p}, \xi) = \{\mathbf{u} = (u_1, \ldots, u_m) : \sum_{i=1}^{m} u_i = \xi, \quad 0 \le u_i \le p_i \quad i \in I\}$. Hence,

$$L(\eta', \mathbf{p}, \xi) - L(\eta'', \mathbf{p}, \xi) \le \sum_{i=1}^{m} \eta'_i \bar{u}_i(\xi) - \sum_{i=1}^{m} \eta''_i \bar{u}_i(\xi) = \sum_{i=1}^{m} (\eta'_i - \eta''_i)\bar{u}_i(\xi) \quad (13)$$

where $\bar{u}(\xi)$ is an optimal solution to the problem $\max_{\mathbf{u} \in U(\mathbf{p},\xi)} \sum_{i=1}^{m} \eta'_i u_i$.

**Proposition 4.** *Let $\eta'$ and $\eta''$ be outcome vectors such that $\eta'_{i_o} < \eta''_{i_o}$ for a criterion $i_o \in I$ and $\eta'_{i_o} \ge \eta'_i$ for all $i \in I$. For any positive and decreasing preferential weights $w_1 \ge w_2 \ge \ldots \ge w_m > 0$ and any vector of importance weights $\mathbf{p}$, a sufficiently large increase of importance weight $p_{i_o}$ with proportional decrease of other weights, following (10) with $\varepsilon > \Delta$*

$$\Delta = \max\left\{\max_{i \neq i_o} \frac{\eta'_i - \eta''_i}{\eta''_{i_o} - \eta'_{i_o}}, 0\right\}$$

*results in strict inequality $WOWA_{\mathbf{w},\mathbf{p}(\varepsilon)}(\eta') < WOWA_{\mathbf{w},\mathbf{p}(\varepsilon)}(\eta'')$.*

*Proof.* Applying inequality (13) to importance weights $\mathbf{p}(\varepsilon)$ one gets

$$L(\eta', \mathbf{p}(\varepsilon), \xi) - L(\eta'', \mathbf{p}(\varepsilon), \xi) \le [\bar{u}_{i_o}(\xi) - \Delta \sum_{i \neq i_o} \bar{u}_i(\xi)](\eta'_{i_o} - \eta''_{i_o})$$

where, due to $\eta'_{i_o} \ge \eta'_i$ for all $i$, $\bar{u}_{i_o}(\xi) = \min\{\xi, p_{i_o}(\varepsilon)\}$ and $\bar{u}_i(\xi) \le \min\{\xi - \bar{u}_{i_o}(\xi), p_i(\varepsilon)\}$ for all $i \neq i_o$. Hence,

$$L(\eta', \mathbf{p}(\varepsilon), \xi) - L(\eta'', \mathbf{p}(\varepsilon), \xi) \le \begin{cases} \xi(\eta'_{i_o} - \eta''_{i_o}) & \xi \le p_{i_o}(\varepsilon) \\ (p_{i_o}(\varepsilon) - \Delta \sum_{i \neq i_o} p_i(\varepsilon))(\eta'_{i_o} - \eta''_{i_o}) & \xi > p_{i_o}(\varepsilon) \end{cases}$$

Therefore, for a large enough $\varepsilon > \Delta$ one gets $p_{i_o}(\varepsilon) > \Delta/(1 + \Delta)$ and $p_{i_o}(\varepsilon) - \Delta(1 - p_{i_o}(\varepsilon)) > 0$. Thus $L(\eta', \mathbf{p}(\varepsilon), \xi) < L(\eta'', \mathbf{p}(\varepsilon), \xi)$ for any $0 < \xi \leq 1$ and, due to nonnegative differential weights $\bar{w}_k$, inequality $WOWA_{\mathbf{w}, \mathbf{p}(\varepsilon)}(\eta') < WOWA_{\mathbf{w}, \mathbf{p}(\varepsilon)}(\eta'')$ is valid.

**Proposition 5.** *Let $\eta'$ and $\eta''$ be outcome vectors such that $\eta'_{i_o} < \eta''_{i_o}$ for a criterion $i_o \in I$. For any positive and decreasing preferential weights $w_1 \geq w_2 \geq \ldots \geq w_m > 0$ and any vector of importance weights $\mathbf{p}$, a sufficiently large increase of importance weight $p_{i_o}$ with proportional decrease of other weights, following (10) with $\varepsilon > \Delta$*

$$\Delta = \max\left\{ \max_{i \neq i_o} \frac{(\eta'_i - \eta''_i)w_1}{(\eta''_{i_o} - \eta'_{i_o})w_m}, 0 \right\}$$

*results in strict inequality $WOWA_{\mathbf{w}, \mathbf{p}(\varepsilon)}(\eta') < WOWA_{\mathbf{w}, \mathbf{p}(\varepsilon)}(\eta'')$.*

*Proof.* Let $\delta = \max\{\max_{i \neq i_o}(\eta'_i - \eta''_i), 0\}/(\eta''_{i_o} - \eta'_{i_o})$. Applying inequality (13) to importance weights $\mathbf{p}(\varepsilon)$ one gets

$$WOWA_{\mathbf{w}, \mathbf{p}(\varepsilon)}(\eta') - WOWA_{\mathbf{w}, \mathbf{p}(\varepsilon)}(\eta'')$$
$$\leq \sum_{k=1}^{m} \bar{w}_k m [\bar{u}_{i_o}(\frac{k}{m}) - \delta \sum_{i \neq i_o} \bar{u}_i(\frac{k}{m})](\eta'_{i_o} - \eta''_{i_o})$$
$$\leq m[w_m p_{i_o}(\varepsilon) - \delta w_1 \sum_{i \neq i_o} p_i(\varepsilon)](\eta'_{i_o} - \eta''_{i_o})$$

since $\bar{u}_{i_o}(\frac{k}{m}) \geq 0$ for all $k$, $\bar{u}_{i_o}(\frac{m}{m}) = p_{i_o}(\varepsilon)$, and $\bar{u}_i(\frac{k}{m}) \leq p_i(\varepsilon)$ for all $i$. Thus, for a large enough $\varepsilon > \Delta$ one gets $p_{i_o}(\varepsilon) > \Delta/(1 + \Delta) = \delta w_1/(\delta w_1 + w_m)$ and thereby $WOWA_{\mathbf{w}, \mathbf{p}(\varepsilon)}(\eta') < WOWA_{\mathbf{w}, \mathbf{p}(\varepsilon)}(\eta'')$.

## 5   Concluding Remarks

In this paper, we have investigated the monotonicity of WOWA with respect to weight perturbations in favor of a single criterion. Contrary to intuition, there exist configurations where such an improvement in favor of a criterion $i_o$ impact negatively the performance of the optimal solution on that criterion. This may reduce the controllability of WOWA when used as a scalarizing function in interactive exploration of feasible solutions. Hopefully, we also have established positive results showing that some controllability can be ensured for sufficiently large weight improvements. Our results show that the WOWA importance weighting mechanism alone is insufficient for effective multiple criteria preference modeling. For this purpose the WOWA aggregation should be supported by additional control parameters like aspiration levels in the reference point methods [7]. We think similar studies are worth investigating for a more general class of aggregation operators, such as Choquet integrals.

**Acknowledgements.** The research by W. Ogryczak was partially supported by the Polish National Budget Funds 2010–2013 for science under the grant N N514 044438.

The research by P. Perny and P. Weng was supported by the project ANR-09-BLAN-0361 GUaranteed Efficiency for PAReto optimal solutions Determination (GUEPARD).

# References

1. Damiani, E., De Capitani di Vimercati, S., Samarati, P., Viviani, M.: A WOWA-based aggregation technique on trust values connected to metadata. Electronic Notes in Theoretical Computer Science 157(3), 131–142 (2006)
2. Larsen, H.L.: Importance weighted OWA aggregation of multicriteria queries. In: North American Fuzzy Information Processing Society (NAFIPS), pp. 740–744 (1999)
3. Liu, X.: Some properties of the weighted OWA operator. IEEE Transactions on Systems, Man, and Cybernetics, Part B: Cybernetics 368(1), 118–127 (2006)
4. Llamazares, B.: Simple and absolute special majorities generated by OWA operators. European Journal of Operational Research 158(3), 707–720 (2004)
5. Nettleton, D.F., Muñiz, J.: Processing and representation of meta-data for sleep apnea diagnosis with an artificial intelligence approach. Int. J. Medical Informatics 63(1-2), 77–89 (2001)
6. Ogryczak, W.: Multiple criteria optimization and decisions under risk. Control & Cybernetics 31(4), 975–1003 (2002)
7. Ogryczak, W.: Ordered weighted enhancement of preference modeling in the reference point method for multicriteria optimization. Soft Comp. 14(5), 435–450 (2010)
8. Ogryczak, W., Perny, P., Weng, P.: On Minimizing Ordered Weighted Regrets in Multiobjective Markov Decision Processes. In: Brafman, R. (ed.) ADT 2011. LNCS (LNAI), vol. 6992, pp. 190–204. Springer, Heidelberg (2011)
9. Ogryczak, W., Śliwiński, T.: On Optimization of the Importance Weighted OWA Aggregation of Multiple Criteria. In: Gervasi, O., Gavrilova, M.L. (eds.) ICCSA 2007, Part I. LNCS, vol. 4705, pp. 804–817. Springer, Heidelberg (2007)
10. Ogryczak, W., Śliwiński, T.: On efficient WOWA optimization for decision support under risk. Int. Journal of Approximate Reasoning 50(6), 915–928 (2009)
11. Ogryczak, W., Wierzbicki, A., Milewski, A.: A multi-criteria approach to fair and efficient bandwidth allocation. Omega 36(3), 451–463 (2008)
12. Quiggin, J.: Generalized Expected Utility Theory. The Rank-Dependent Model. Kluwer Academic, Dordrecht (1993)
13. Torra, V.: The weighted OWA operator. Int. J. Intell. Syst. 12(2), 153–166 (1997)
14. Torra, V., Narukawa, Y.: Modeling Decisions Information Fusion and Aggregation Operators. Springer, Berlin (2007)
15. Valls, A., Torra, V.: Using classification as an aggregation tool for MCDM. Fuzzy Sets Systems 115(1), 159–168 (2000)
16. Yager, R.R.: On ordered weighted averaging aggregation operators in multicriteria decision making. IEEE Trans. on Syst., Man and Cyber. 18(1), 183–190 (1988)
17. Yager, R.R.: Including Importances in OWA Aggegations Using Fuzzy Systems Modeling. IEEE Transactions on Fuzzy Systems 6(2), 286–294 (1998)
18. Yager, R.R., Filev, D.P.: Essentials of Fuzzy Modeling and Control. Wiley (1994)
19. Yager, R.R., Kacprzyk, J., Beliakov, G. (eds.): Recent Developments in the Ordered Weighted Averaging Operators: Theory and Practice. STUDFUZZ, vol. 265. Springer, Heidelberg (2011)

# Extending Concordance and Discordance Relations to Hierarchical Sets of Criteria in ELECTRE-III Method

Luis Del Vasto-Terrientes[1], Aida Valls[1], Roman Slowinski[2], and Piotr Zielniewicz[2]

[1] Universitat Rovira i Virgili. Departament d'Enginyeria Informàtica i Matemàtiques,
Av Països Catalans, 26. 43007 Tarragona, Catalonia, Spain
{luismiguel.delvasto,aida.valls}@urv.cat
[2] Institute of Computing Science, Poznan University of Technology,
Piotrowo 2, 60-965 Poznan, Poland
{roman.slowinski,piotr.zielniewicz}@cs.put.poznan.pl

**Abstract.** In many real-world multiple criteria decision problems, the family of criteria has a hierarchical structure presented in the form of a tree. The leaves of the tree correspond to elementary criteria on which a finite set of alternatives is directly evaluated. Evaluations of alternatives on elementary criteria are aggregated to form a sub-criterion at an upper level of the tree. Then, evaluations of alternatives on sub-criteria having the same predecessor in the hierarchy tree are aggregated again in a sub-criterion of a higher level, and so on, until the aggregation at the general goal criterion, which is the root of the tree, where the alternatives are finally ranked from the best to the worst. At each node of the tree, above the leaves, we are considering the aggregation of multiple criteria evaluations using the ELECTRE-III method, based on building and exploiting outranking relations for each pair of alternatives. As the result of ELECTRE-III is a partial preorder of alternatives, the sub-criteria are ordering the alternatives just partially. Therefore, in this paper we propose a new way of calculating concordance and discordance indices that take part in the definition of an outranking relation aggregating the (partially ordered) evaluations on sub-criteria. A robustness analysis has been performed to analyze the behavior of the proposed method in different settings.

**Keywords:** Hierarchical criteria, Multiple Criteria Decision Making, Partial Preorder, Outranking method.

## 1   Introduction

Ranking a finite set of alternatives evaluated on multiple conflicting criteria is not a trivial task. Many different methods have been proposed to support decision makers (DM) in this task. They are based on different models for representing the preferences of the DM in order to rank the alternatives taking into account a value system of the DM. The decision aiding process is based on three elements: (1) a finite set of alternatives $A$ (also called actions if they can be put in operation simultaneously), (2) a finite set of criteria $G=\{g_1, g_2,..,g_n\}$, where each criterion $g_j$ provides a performance score

V. Torra et al. (Eds.): MDAI 2012, LNAI 7647, pp. 78–89, 2012.
© Springer-Verlag Berlin Heidelberg 2012

for each alternative in $A$, and (3) a preference system that for each possible pair of alternatives assigns one of the following types of relations: indifference, preference or incomparability.

The preferences over the alternatives depend on how well they perform according to a number of criteria. Initially the alternatives are evaluated with respect to each criterion separately. To aggregate the scores of the alternatives on individual criteria into an overall preference order of the alternatives, many methods have been proposed, each with its own informational requirements and mathematical properties [1].

We will focus on outranking-based methods. These methods have developed rapidly during the last few decades because of their adaptability to the poor structure of many real decision problems. An interesting characteristic of this approach is the treatment of heterogeneous criteria scales (numerical, qualitative and ordinal), and a non-compensatory character of the aggregation. This heterogeneity of scales is usually an inconvenience for many decision support systems, which often require a common scale of measurement for all the criteria.

A well-known family of methods based on outranking is the ELECTRE family [2],[3]. Since the first proposal of the basic ELECTRE-I method, several parent methods have been developed, e.g., ELECTRE-Is for the selection of best alternatives, ELECTRE-II and ELECTRE-III/IV for constructing a ranking, and ELECTRE-TRI for sorting problems. ELECTRE methods have been widely considered as an effective and efficient decision aiding tool with successful applications in different domains [4],[5],[6].

In some real-world decision problems, criteria are naturally defined in a hierarchical structure, distinguishing different levels of generality that model the implicit taxonomical relations between the criteria [7],[8],[9][10]. Such a hierarchical structure has the form of a tree, where the root is a general goal criterion, the nodes of the tree descending from the goal are sub-criteria, and nodes descending from these sub-criteria are the sub-criteria of the lower level, and so on, until the leaves that correspond to elementary criteria on which the alternatives are directly evaluated. The current versions of ELECTRE methods, apart from a recent proposal presented in [11], require, however, that all criteria are defined on a common level and do not accept a hierarchy of criteria. Thus, to apply ELECTRE methods in case of hierarchically structured criteria, one needs to generalize its crucial concepts, being concordance and discordance indices, to take into account aggregation of sub-criteria that rank the alternatives in a partial way. This is precisely the goal of our work. Let us add that the generalization of ELECTRE to a hierarchical set of criteria presented in [11] follows a different principle based on robust ordinal regression from examples.

In this paper we focus on decision problems with the goal of finding a partial pre-order of the alternatives, as in [6],[12]. Therefore, the ELECTRE-III method for ranking will be taken as basis. To extend ELECTRE-III to a hierarchy of criteria, we propose to follow the hierarchical organization of the criteria to aggregate the information at each node of the tree, according to the corresponding sub-criteria. First, the evaluations of alternatives on elementary criteria are aggregated to form a sub-criterion at the lowest level of the hierarchy tree. Then, evaluations of alternatives on sub-criteria having the same predecessor in the hierarchy tree are aggregated again to form a sub-criterion of an upper level, and so on, until the aggregation at the general goal criterion, where the alternatives are finally ranked from the best to the worst.

At the nodes of the tree above the leaves we are considering the classic ELECTRE-III method. As the result of ELECTRE-III is a partial preorder, the sub-criteria are ordering the alternatives just partially. This is why we propose a new way of calculating concordance and discordance indices that take part in the definition of an outranking relation aggregating the (partially ordered) evaluations of alternatives on sub-criteria.

The paper is organized as follows. First, in section 2, some basics of ELECTRE method are reminded, with a focus on the construction of outranking relations involving concordance and discordance indices. Section 3 presents the extension of the concordance and discordance indices to the case of a hierarchical structure of the set of criteria. The results shown in section 4 illustrate the behavior of the new indices. Finally, some conclusions and directions of a future work are given.

## 2     Some Basics on ELECTRE-III Method

The ELECTRE-III method has two steps: first, the construction of an outranking relation over all the possible pairs of alternatives, and second, the exploitation of this outranking relation to solve the ranking decision problem [3].

### 2.1     Concordance and Discordance Indices

Given two alternatives $a, b \in A$, alternative $a$ **outranks** alternative $b$ if $a$ outperforms $b$ on enough criteria of sufficient importance, and $a$ is not outperformed by $b$ by having a significantly inferior performance on any single criterion. Each alternative in the set of alternatives is compared to all the other members of the set in a pairwise manner to determine their credibility of outranking. So, the outranking relation $aSb$ is built on the basis of two tests: the *concordance test*, sometimes referred to as "the respect of the majority", involving the calculation of a *concordance index* $c(a,b)$ measuring the strength of the coalition of criteria that support the hypothesis "$a$ is at least as good as $b$"; and the *discordance test*, sometimes referred to as "the respect of minorities", involving the calculation of *discordance indices* $d_j(a,b)$ measuring the strength of evidence provided by the $j$-th criterion against this hypothesis.

Each alternative $a \in A$ is evaluated on set $G$ of $n$ criteria $g_j$, $j=1,...,n$, where each criterion $g_j$ is assigned by a DM a weight $w_j$, and three thresholds: indifference threshold $q_j$, preference threshold $p_j$ and veto threshold $v_j$, which, in general, can depend on the evaluations $g_j(a)$. For consistency: $0 \leq q_j \leq p_j \leq v_j$. We assume, moreover, without loss of generality that all the criteria are of the gain type, i.e. the greater the value, the better. The weights represent the relative importance of each criterion, which can be interpreted as a voting power of each criterion. The weights of criteria do not mean substitution rates as in the case of compensatory aggregation operators.

Given an ordered pair of alternatives $(a,b) \in A \times A$, the outranking relation $aSb$ means: "$a$ is at least as good as $b$". Calculation of the credibility of outranking $\rho(aSb)$ involves a partial concordance index $c_j(a,b)$, and a partial discordance index $d_j(a,b)$ for each criterion $g_j$. The overall concordance index is computed for each ordered pair $(a,b)$ of alternatives as follows:

$$c(a,b) = \frac{1}{w}\sum_{j=1}^{n} w_j c_j(a,b) \qquad (1)$$

where $W = \sum_{j=1}^{n} w_j$, and the partial concordance index $c_j(a,b)$ is defined as:

$$c_j(a,b) = \begin{cases} 1 & \text{if } g_j(a) \geq g_j(b) - q_j \\ 0 & \text{if } g_j(a) \leq g_j(b) - p_j \\ \dfrac{g_j(a) - g_j(b) + p_j}{p_j - q_j} & \text{otherwise.} \end{cases} \qquad (2)$$

The computation of the discordance index takes into account the criteria that disagree with the assertion $aSb$. In this case, each criterion is assigned a veto threshold $v_j$. The veto is the maximum difference allowed between the values of a pair of alternatives when $g_j(a) < g_j(b)$. The partial discordance index is defined as follows:

$$d_j(a,b) = \begin{cases} 1 & \text{if } g_j(a) - g_j(b) \leq -v_j \\ 0 & \text{if } g_j(a) - g_j(b) \geq -p_j \\ \dfrac{g_j(a) - g_j(b) + p_j}{p_j - v_j} & \text{otherwise.} \end{cases} \qquad (3)$$

The overall concordance and partial discordance indices are then combined to obtain a valued outranking relation with credibility $\rho(aSb) \in [0, 1]$ defined by:

$$\rho(aSb) = \begin{cases} c(a,b) & \text{if } d_j(a,b) \leq c(a,b), \ \forall j \\ c(a,b) \prod_{j \in J(a,b)} \dfrac{1 - d_j(a,b)}{1 - c(a,b)}, & \text{otherwise,} \end{cases} \qquad (4)$$

where $J(a,b)$ is the set of criteria for which $d_j(a,b) > c(a,b)$, and the credibility of outranking is equal to the overall concordance index when there is no discordant criterion.

## 2.2    ELECTRE-III Ranking Procedure

The exploitation procedure proposed for ELECTRE-III is an iterative distillation algorithm that selects at each step a subset of alternatives, taking into account the previously established outranking relations. It starts with a value $\lambda = \max_{a,b \in A} \rho(a,b)$ to compute such a binary relation in $A$ that it is true for a credibility of outranking greater than $\lambda$, and false for a credibility of outranking smaller or equal to $\lambda$. This yields a crisp outranking relation for which the qualification $Q(a)$ of each alternative is computed (i.e. the number of alternatives which are outranked by $a$ minus the number of alternatives which outrank $a$). This leads to the generation of a set of alternatives with the greatest qualification called the *first distillate* and denoted by $D_1$.

If $D_1$ contains only one alternative, the procedure is repeated in $A \backslash D_1$. Otherwise, the same procedure but with a smaller $\lambda$ is applied inside $D_1$ to obtain the *second distillate* $D_2$; if $D_2$ result in a singleton, the procedure is repeated again in $D_1 \backslash D_2$ (except if the latter is empty); otherwise, it is applied inside $D_2$ repeatedly until $D_1$ is completely used up, before starting with $A \backslash D_1$. Notice that it may happen that two or more alternatives belong to one distillate because they have the same qualification and neither of them can be ranked better or worse than others. In this case, the alternatives are said to be *indifferent* and are assigned to the same ranking position. This procedure, which yields a first complete preorder $O_\downarrow$ is called the descending distillation chain.

A second complete preorder, $O_\uparrow$, is obtained by an ascending distillation chain, in which the alternatives having the smallest qualification are first retained. A final partial pre-order $O$ is then built as the intersection of the two complete pre-orders, $O_\downarrow$ and $O_\uparrow$. The resulting ranking is a partial preorder, i.e. for any two alternatives from set $A$, one may be preferred over the other, or they may be indifferent, or they may be incomparable. The incomparability of two alternatives occurs when one of these alternatives, say $a$, is ranked higher than the other alternative, say $b$, in $O_\downarrow$ (or $O_\uparrow$), and $b$ is ranked higher than $a$ in $O_\uparrow$ (or $O_\downarrow$).

The result of this exploitation stage is partial preorder $O$ that establishes a preference structure on the set of alternatives $A$. For each possible pair of alternatives, it assigns one of the following four binary relations $\{P, P^-, I, R\}$, having the following meaning:

- $a\,P\,b$: $a$ is preferred to $b$
- $a\,P^-\,b$: $b$ is preferred to $a$
- $a\,I\,b$: $a$ is indifferent to $b$
- $a\,R\,b$: $a$ is incomparable to $b$

# 3    Concordance and Discordance Indices for a Hierarchical Structure of Criteria

We are considering a hierarchical structure of criteria. An example of this structure is shown in Figure 1. The main goal criterion of the DM is placed in the *root* of the hierarchy tree. The leaves of the hierarchy tree, at the lowest level, are called *elementary criteria* and correspond to the most specific criteria. The tree nodes between the root criterion and the elementary criteria are called *intermediate criteria* or *sub-criteria* defined on the basis of other intermediate or elementary criteria directly descending from them. Different levels of intermediate criteria can be considered. Each criterion from a subset of criteria having a common predecessor in the tree has a weight to indicate its relative importance with respect to other criteria in the subset.

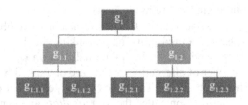

**Fig. 1.** A hierarchy tree of criteria

We assume that the original ELECTRE-III method is applied to aggregate evaluations of alternatives on the elementary criteria. For example, taking into account the hierarchy tree from Fig. 1, the original ELECTRE-III it is applied for subsets of elementary criteria separately: $\{g_{1.1.1}, g_{1.1.2}\}$ and $\{g_{1.2.1}, g_{1.2.2}, g_{1.2.3}\}$. The results of this application are partial preorders $O_{1.1}$ and $O_{1.2}$, which correspond to evaluations of the considered alternatives on the parent criteria $g_{1.1}$ and $g_{1.2}$, respectively. Then, in order to get the ranking of alternatives with respect to the goal criterion $g_1$, one has to aggregate their evaluations on criteria $g_{1.1}$ and $g_{1.2}$, i.e. to aggregate the partial preorders $O_{1.1}$ and $O_{1.2}$. The original ELECTRE-III can only deal with evaluations of alternatives on criteria, that are complete preorders, and thus it has to be adapted to aggregation of partial preorders.

In the following sub-sections, we propose a new method for calculation of partial concordance and discordance indices for a partial preorder induced by a $j$-th sub-criterion belonging to a subset of sub-criteria with a common predecessor. This partial preorder is composed of the four binary relations $P$, $P^-$, $I$, $R$. Thus, considering any ordered pair of alternatives $(a,b) \in A \times A$ in this preorder, we will have to answer the question how each of these binary relations translates to values of $c_j(a,b)$ and $d_j(a,b)$.

**Preference and Indifference Relations, P and I.** The first situation we consider is when $a$ is strictly preferred or indifferent to $b$ in a partial preorder $O_j$. Remember that the concordance index measures the support to the outranking relation defined as "$a$ is at least as good as $b$" ($a\ S\ b$). Since $S = I \lor P$, both preference $aPb$ and indifference $aIb$ relations in $O_j$, give evidence that $O_j$ clearly supports this claim. Therefore, the value of partial concordance index is set to 1:

$$c_j(a\ P\ b) = 1 \tag{5}$$

$$c_j(a\ I\ b) = 1. \tag{6}$$

According to the previous rationale, when $aPb$ and $aIb$ in $O_j$, we set the partial discordance index to 0:

$$d_j(a\ P\ b) = 0 \tag{7}$$

$$d_j(a\ I\ b) = 0 \tag{8}$$

**Inverse Preference Relation, P–.** When $b$ is preferred over $a$ in a partial preorder $O_j$, then this fact contradicts the outranking relation $aSb$. In other words, $aP^-b$ is incompatible with $aSb$, and thus the partial concordance index is set to 0:

$$c_j(a\ P^-b) = 0. \tag{9}$$

However, the statement of discordance is not so direct. Having $aP^-b$ in $O_j$, the strength of the opposition of $O_j$ against $aSb$ depends on the number of alternatives "separating" $b$ from $a$ in $O_j$. According to this rationale, the partial discordance index $d_j(a,b)$ will be equal to 0 when the difference between the number of alternatives preferred to $a$ and to $b$ is smaller or equal than a veto threshold $v$ (Eq. 10), and $d_j(a,b)$ will be equal to 1 when this difference is greater than the veto $v$ (Eq. 11):

$$if\ \Gamma(a) - \Gamma(b) \le v,\ then\ d_j(a\ P^-b) = 0, \tag{10}$$

$$if\ \Gamma(a) - \Gamma(b) > v,\ then\ d_j(a\ P^-\ b) = 1, \tag{11}$$

where $\Gamma(a)$ is the number of alternatives preferred to $a$, and $\Gamma(b)$ is the number of alternatives preferred to $b$ in $O_j$. Let us consider an example of $O_j$ shown in Figure 2. Based on this partial preorder, one can state the relations for all possible pairs of alternatives, as shown in Table 1.

Considering $n$ alternatives in the partial preorder, and taking into account that each alternative is indifferent with itself, the veto threshold must be in the range $0 < v < n - 1$.

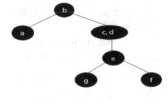

**Table 1.** Matrix of preference relations in preorder $O_1$

|   | a | b | c | d | e | f | g | Γ |
|---|---|---|---|---|---|---|---|---|
| a | I | P | R | R | R | R | R | 1 |
| b | P | I | P | P | P | P | P | 0 |
| c | R | P | I | I | P | P | P | 1 |
| d | R | P | I | I | P | P | P | 1 |
| e | R | P | P | P | I | P | P | 3 |
| f | R | P | P | P | P | I | R | 4 |
| g | R | P | P | P | P | R | I | 4 |

**Fig. 2.** Partial preorder $O_1$

**Incomparability Relation, R.** When in partial preorder $O_j$, alternative $a$ is incomparable with alternative $b$, then it is impossible to state whether $aPb$ or $aIb$ or $aP^-b$, thus, the partial preorder gives no clear support to the outranking hypothesis. In this case, we will take into account additional information about the alternatives $a$ and $b$, given by the function $\Gamma(\cdot)$, which indicates how many other alternatives are preferred to $a$ or $b$. If the difference between the values of $\Gamma(a)$ and $\Gamma(b)$ was negative or close to 0, then this should enforce the conviction that $aRb$ could rather turn to $aPb$ or $aIb$, than to $aP^-b$. Otherwise, if the difference between the values of $\Gamma(a)$ and $\Gamma(b)$ was positive, then this should enforce the conviction that $aRb$ could rather turn to $aP^-b$, than to $aPb$ or $aIb$. According to this rationale, and relating the difference $\Gamma(a) - \Gamma(b)$ to the veto threshold $v$, two situations can be distinguished:

$$if \ \Gamma(a) - \Gamma(b) \le v, \ then \ \begin{cases} c_j(a \ R \ b) = k^c + \delta^c(a,b) \\ d_j(a \ R \ b) = 0 \end{cases} \tag{12}$$

$$if \ \Gamma(a) - \Gamma(b) > v, \ then \ \begin{cases} d_j(a \ R \ b) = k^d + \delta^d(a,b) \\ c_j(a \ R \ b) = 0 \end{cases} \tag{13}$$

The first situation (Eq. 12) corresponds to the belief that $aRb$ could turn to $aPb$ or $aIb$, because the difference between the number of alternatives preferred to $a$ and the number of alternatives preferred to $b$ is not as big as the veto threshold $v$. In this situation, the partial concordance index is set to a constant value $k^c \in [0,1]$ plus a correcting factor $\delta^c(a,b)$, while keeping the partial discordance equal to 0. In the example from Figure 2, assuming $v>1$, this situation happens to alternative $a$ compared to alternatives $c, d, e, f, g$ (see the values of $\Gamma(\cdot)$ in Table 1).

The second situation (Eq. 13) corresponds to the belief that $aRb$ could turn to $aP^-b$, because the difference between the number of alternatives preferred to $a$ and the number of alternatives preferred to $b$ is bigger than the veto threshold $v$. In this situation, the partial discordance index is set to a constant value $k^d \in [0,1]$ plus a correcting factor $\delta^d(a,b)$, while keeping the partial concordance index equal to 0. In the example from Figure 2, assuming $v>1$, this situation happens to alternatives $e, f, g$ compared to alternative $a$.

Let us pass to explanation of $k^c, k^d, \delta^c$ and $\delta^d$. Considering that $aRb$ could turn with an equal probability to $aPb, aIb, aP^-b$, and that $S=I\vee P$, we conclude that only two of the three possible relations confirm $S$. Thus, there is 2/3 chance that $aRb$ would confirm $aSb$. Therefore, we propose to assign $k^c$ the value 2/3 (Eq. 14). Analogously, we propose to assign $k^d$ the value 1/3 (Eq. 15).

$$c_j(a\ R\ b) = 2/3 + \delta^c(a, b), \tag{14}$$

$$d_j(a\ R\ b) = 1/3 + \delta^d(a, b). \tag{15}$$

The rationale for adding correcting factors $\delta^c$ and $\delta^d$ to the above formulas is founded on a premise that the partial concordance and discordance indices in these two situations should depend on the magnitude of the difference $\Gamma(a) - \Gamma(b)$.

In the example shown in Figure 2, alternative $a$ is incomparable with alternatives $c$, $d$, $e$, $f$ and $g$, but the number of alternatives preferred to $a$ is much less than the number of alternatives preferred to $f$ and $g$ (1 vs. 4) while the number of alternatives preferred to $a$ is the same as the number of alternatives preferred to $c$ and $d$ (1 vs. 1). This fact should reinforce $c_j(aRf)$ and $c_j(aRg)$, compared to $c_j(aRc)$ and $c_j(aRd)$.

For this reason, the correcting factors $\delta^c$ and $\delta^d$ are introduced to slightly increase or decrease the base partial concordance $k^c$ or discordance $k^d$ of $aRb$, depending on the value of the difference $\Gamma(a) - \Gamma(b)$.

**$\delta^c$ for the Partial Concordance Index.** We establish the following logical condition for concordance $c_j(aRb)$: "If alternative $a$ is incomparable to alternatives $b$ and $d$ in a partial preorder $O_j$, and $\Gamma(a) - \Gamma(b) < \Gamma(a) - \Gamma(d) \leq v$, then $c_j(aRb)$ should be greater than $c_j(aRd)$". According to this condition, for each pair $(a,b) \in A \times A$, such that $aRb$ and $\Gamma(a) - \Gamma(b) \leq v$ in $O_j$, we propose:

$$\delta^c(a, b) = \frac{\Gamma(b) - \Gamma(a)}{n - 2} \times \alpha, \tag{16}$$

where $n$ is the number of alternatives in set $A$, and 2 is subtracted from $n$ in the denominator because there are 2 alternatives ($a$ and $b$) that are indifferent with $a$ and $b$, respectively.

The value $\alpha$ has been introduced to control the maximum degree of change permitted to the original partial concordance index for incomparability. We set $\alpha$ to 0.25, so that $-0.25 \leq \delta^c(a, b) \leq 0.25$ and $0.42 \leq c_j(aRb) \leq 0.92$.

**$\delta^d$ for the Partial Discordance Index.** We establish the following logical condition for $d_j(aRb)$: "If alternative $a$ is incomparable to alternatives $b$ and $d$ in a partial preorder $O_j$, and $\Gamma(a) - \Gamma(b) > \Gamma(a) - \Gamma(d) > v$, then $d_j(aRb)$ should be greater than $d_j(aRd)$". According to this condition, for each pair $(a,b) \in A \times A$, such that $aRb$ and $\Gamma(a) - \Gamma(b) > v$ in $O_j$, we propose:

$$\delta^d(a, b) = \frac{\Gamma(a) - \Gamma(b) - v}{n - 2} \times \alpha \tag{17}$$

Again, the value $\alpha$ is controlling the maximum degree of change permitted to the original partial discordance index for incomparability. Setting $\alpha$ to 0.25, we get $-0.25 \leq \delta^d(a, b) \leq 0.25$ and $0.08 \leq d_j(aRb) \leq 0.58$.

# 4    Analysis of Partial Concordance and Discordance Indices

In this section, we investigate how the partial concordance and discordance indices depend on the values of the parameters $\alpha$ and $v$. Two kinds of analysis have been performed: (1) fixing the parameter $\alpha$ to 0.25 and observing the changes due to different veto thresholds $v$; (2) fixing the veto threshold $v$ to 2 and observing the changes due to $\alpha$.

To facilitate the comparison between different configurations of $\alpha$ and $v$, we have considered a set of consecutive cutting levels for the values of the partial concordance and discordance indices. Values equal or above the threshold are changed to 1 and those below to 0. In this way, we get 0-1 matrices showing which pairs of alternatives are in partial concordance or discordance with $aSb$. A graphical representation of the number of pairs getting 1 will permit to compare different configurations. This kind of analysis permits to draw some preliminary conclusions on behavior of each of the measures proposed. A similar study of the integral behavior of the generalized ELECTRE-III method is left for a future work.

The tests have been performed for three different case studies, including a different number of alternatives with a different organization of the partial preorder (different relations of incomparability, preference and indifference).

The first case study concerns set $A = \{ a, b, c, d, e, f, g \}$ of alternatives, structured in a partial preorder $O_2$ with two branches, as shown in Figure 3. Suppose that this partial preorder has been generated with ELECTRE-III for a subset of elementary criteria. The corresponding preference relations are given in Table 2.

**Fig. 3.** Partial preorder $O_2$ for case 1

**Table 2.** Matrix of preference relations in preorder $O_2$

|   | a | b | c | d | e | f | g | Γ |
|---|---|---|---|---|---|---|---|---|
| a | I | R | P⁻ | R | R | P | R | 1 |
| b | R | I | P⁻ | P | P⁻ | P | P | 2 |
| c | P | P | I | P | P | P | P | 0 |
| d | R | P⁻ | P⁻ | I | P⁻ | P | P | 3 |
| e | R | P | P⁻ | P | I | P | P | 1 |
| f | P⁻ | P⁻ | P⁻ | P⁻ | P⁻ | I | P⁻ | 6 |
| g | R | P⁻ | P⁻ | P⁻ | P⁻ | P | I | 4 |

## 4.1    Veto Threshold Test

In this test we study the behavior of partial concordance and discordance indices when the veto threshold $v$ changes, while the value of $\alpha = 0.25$.

Figure 4 and Figure 5 present the number of pairs with a partial concordance (resp. discordance) above a cutting level. Four veto thresholds are considered, taking into consideration different cutting levels.

These results indicate that the concordance index increases when veto is increasing. This is due to the fact that a higher veto threshold involves a higher number of incomparability relations interpreted as $P$ or $I$, increasing the support to concordance, instead of discordance. For the same reason, the discordance index decreases with the increase of the veto.

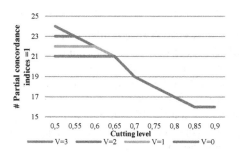

**Fig. 4.** $v$ test for partial concordance in $O_2$

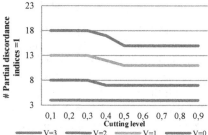

**Fig. 5.** $v$ test for partial discordance in $O_2$

Comparing both figures, we observe that the effect on the number of pairs of alternatives with discordance is higher than on the number of pairs with concordance when the veto is increasing. Veto is not only affecting the incomparability relations $R$, but also the situations of $P^-$. In case of incomparability (Eq. 12), differences in the partial concordance values are observed for different veto thresholds until the cutting level equal to 0.65. Above this value, the behavior is independent of the veto because we use (Eq. 13). This is not the case of discordance, where there is a fixed difference due to $P^-$ (Eq. 10-11).

### 4.2    Test for the Parameter α

In this test, we compare the values of partial concordance and discordance indices for three different values of α. The veto threshold is fixed to $v = 0$, because for higher values of veto the number of pairs of alternatives that enter the situation of discordance is low in this case study. From the results displayed in Figure 6 and Figure 7, we observe that the number of pairs reaching value 1 is more diversified when alpha is high (α = 0.25), having a smooth decrease in function of the cutting level. For α = 0, as the partial concordance for incomparability is always set to $k^c = 0.67$, the partial concordance index has only three possible values (0, 0.67 and 1); hence, the decrease of this function is much more steep. This tendency is also observed in the partial discordance index.

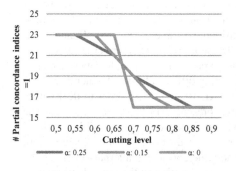

**Fig. 6.** α test for partial concordance in $O_2$

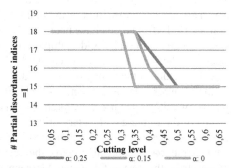

**Fig. 7.** α test for partial discordance in $O_2$

### 4.3    Comparative Analysis for 3 Case Studies

To complete this analysis, two other case studies have been considered. The partial preorders ($O_3$ and $O_4$) with different depth and structure, as shown in Figures 8 and 9. To check if the behavior of the indices is stable in the different case studies, a correlation of the results has been measured. We have taken the number of pairs of alternatives with the partial concordance index above a cutting level (changing from 0.5 to 0.9) for $\alpha = 0.25$ and $v = 2$. The Pearson correlation on the values obtained for the three case studies has been calculated. The same analysis has been done for the partial discordance. The correlation values obtained are shown in Tables 3 and 4.

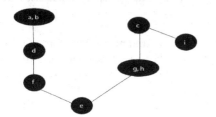

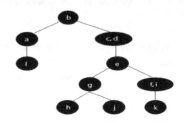

**Fig. 8.** $O_3$ for case 2          **Fig. 9.** $O_4$ for case 3

**Table 3.** Partial concordances correlations          **Table 4.** Partial discordances correlations

| Corr. | $O_1$ | $O_2$ | $O_3$ | Corr. | $O_1$ | $O_2$ | $O_3$ |
|-------|-------|-------|-------|-------|-------|-------|-------|
| $O_1$ | 1 | 0.99 | 0.98 | $O_1$ | 1 | 0.99 | 1 |
| $O_2$ | 0.99 | 1 | 0.99 | $O_2$ | 0.99 | 1 | 0.99 |
| $O_3$ | 0.98 | 0.99 | 1 | $O_3$ | 1 | 0.99 | 1 |

We can observe very high correlation factors for the three case studies. This is an indication of the high robustness of the measures proposed, regarding the structure and number of alternatives in the partial preorder.

## 5    Conclusions and Future Work

The well-known ELECTRE family of MCDA methods based on outranking consists of two main steps: construction of the outranking relation and the exploitation of this relation. ELECTRE methods have been widely considered as an effective and efficient decision aiding tool in real-world multiple criteria decision problems where criteria are considered at the same level. Many real-world decision problems involve, however, a hierarchical structure of the family of criteria, distinguishing different levels of generality.

In this work, we have generalized the ELECTRE-III method in order to deal with a family of criteria having a hierarchical structure presented in the form of a tree. The leaves of the tree are elementary criteria considered as pseudo-criteria, and higher level criteria are sub-criteria aggregating some lower level (sub- or elementary) criteria, until the general goal which is the root of the tree, where the considered alternatives should be ranked from the best to the worst. To manage the propagation of the outranking relation from the leaves to the root, we adapted the partial concordance and discordance indices to consideration of partial preorders induced by sub-criteria on the considered set of alternatives. They are used in concordance and non-discordance tests when constructing outranking relations at different levels of the tree. To check the robustness of the proposed definitions with respect to different structures of the partial preorders and some adopted parameters, we performed a simulation study which gave a positive response. This result motivates us to continue the study of the behavior of the generalized ELECTRE-III method in the context of real-world applications, where the goal is not obtaining an overall measure of the performance but to discover the preference relations between the objects.

**Acknowledgments.** The first author is supported by a FI predoctoral grant from Generalitat de Catalunya. This work is supported by the Spanish Ministry of Science and Innovation (DAMASK project, TIN2009-11005) and the Spanish Government (PlanE, Spanish Economy and Employment Stimulation Plan). The last two authors wish to acknowledge financial support from the Polish National Science Centre, grant no. N N519 441939.

# References

1. Torra, V., Nakurawa, Y.: Modeling decisions – information fusion and aggregation operators. Springer, Berlin (2007)
2. Figueira, J., Greco, S., Ehrgott, M. (eds.): Multicriteria decision analysis: State of the Art Surveys. Springer, Berlin (2005)
3. Figueira, J., Greco, S., Roy, B., Slowinski, R.: ELECTRE Methods: Main Features and Recent Developments. In: Zopounidis, C., Pardalos, P.M. (eds.) Handbook of Multicriteria Analysis, ch.3, pp. 51–89. Springer, Berlin (2010)
4. Arondel, C., Girando, P.: Sorting Cropping Systems on the Basis of Impact of Groundwater Quality. European J. Operational Research 127, 467–482 (2000)
5. Colson, G.: The OR's prize winner and the software ARGOS: How a multijudge and a multicriteria ranking GDSS helps a jury to attribute a scientific award. Computers & Operations Research 27, 741–755 (2000)
6. Damaskos, X., Kalkafakou, G.: Application of ELECTRE III and DEA Methods in the BRP of a Bank Branch Network. Yugoslav Journal of Operation Research 15(2), 259–276 (2005)
7. Wang, G., Huang, S.H., Dismukes, J.P.: Production-driven supply chain selection using integrated multi-criteria decision-making methodology. International Journal of Production Economics 91, 1–15 (2004)
8. Ahsan, M.K., Bartlema, J.: Monitoring healthcare performance by analytic hierarchy process: a developing-country perspective. Int. Transactions in Operational Research 11, 465–478 (2004)
9. Valls, A., Schuhmacher, M., Pijuan, J., Passuello, A., Nadal, M., Sierra, J.: Preference assessment for the management of sewage sludge application on agricultural soils. Int. Journal of Multicriteria Decision Making 1(1), 4–24 (2010)
10. Aydin, S., Kahraman, C., Kaya, I.: A new fuzzy multicriteria decision making approach: an application for European Quality Award assessment. Knowledge-based Systems 32, 37–46 (2012)
11. Corrente, S., Greco, S., Slowinski, R.: Multiple Criteria Hierarchy Process with ELECTRE and PROMETHEE. Submitted to OMEGA (2011)
12. Papadopoulos, A., Karagiannidis, A.: Application of the multi-criteria analysis method ELECTRE III for the optimisation of decentralised energy systems. OMEGA: The Int. Journal of Management Science 36, 766–776 (2008)

# An Extended LibQUAL+ Model
# Based on Fuzzy Linguistic Information

Francisco J. Cabrerizo[1], Ignacio J. Pérez[2],
Javier López-Gijón[3], and Enrique Herrera-Viedma[2]

[1] Dept. of Software Engineering and Computer Systems,
Distance Learning University of Spain (UNED), 28040 Madrid, Spain
`cabrerizo@issi.uned.es`
[2] Dept. of Computer Science and Artificial Intelligence,
University of Granada, 18071 Granada, Spain
`{ijperez,viedma}@decsai.ugr.es`
[3] Dept. of Library Sciences,
University of Granada, 18071 Granada, Spain
`jgijon@ugr.es`

**Abstract.** LibQUAL+ model is a web-based survey to measure the library service quality according to the users' perceptions. Although it is the most popular method, it presents two major drawbacks: (i) it is devised on cardinal scale to measure the library service quality (from 1 to 9), but, due to the subjectivity, impression and vagueness of the human beings when attempting to qualify phenomena related to human perception, it seems natural that they use words in natural language (linguistic terms) instead of numerical values to provide their preferences, and (ii) it considers that all the users' opinions on the library services are equally important, however, some users should be more influential than others in some questions as they do not play equal roles in measuring library service quality. The aim of this paper is to present an extended LibQUAL+ model based on fuzzy linguistic information overcoming the above drawbacks.

**Keywords:** Academic library, quality evaluation, LibQUAL+, fuzzy linguistic modeling.

## 1 Introduction

The evaluation of academic libraries is a topic and an activity of importance in all countries with established library services. Academic libraries play a notable role in the educational progress and their evaluation is essential to improve their services as an important part of a learning environment [1]. Since the mission of an academic library is to resolve user's expectations, one of the appropriate evaluation methods is the based on the feedback or comments point of the users. Academic libraries are service institutions, and better service will be provided if the nature and needs of users are known. According to user comments, observed weaknesses and strengths can be understood and, in order to eliminate

V. Torra et al. (Eds.): MDAI 2012, LNAI 7647, pp. 90–101, 2012.

defects and develop strengths, proposals can be provided to this end. Furthermore, focusing on users in the academic libraries and the efforts to resolve their expectations, it makes the academic libraries more dynamic [2].

One of the most important methods for evaluating the library service quality using user's perceptions is the survey method, because detailed information is provided about user comments, it makes clear the concept of service, it shows the problems, and it offers possible solutions [2]. Among the survey methods, LibQUAL+ model [3] is the most popular and the best-known one. It was developed in the US in order to collect data on the quality of library services. The aim of its designers was to develop a tool that would help libraries better understand their user's perceptions of service quality and to use this information in planning their operations. The survey data allow identification of areas in which service levels should be improved, and they have also been used to identify best practices and reallocate resources accordingly [5]. Since 2000, more than 1.100 libraries have participated in LibQUAL+, including college and university libraries, community college libraries, health sciences libraries, academic law libraries, and public libraries [2,4,5,6,7,8].

However, the LibQUAL+ model has some drawbacks that should be addressed. On the one hand, users answer each question giving a score from one to nine on a 9-points Likert scale [9]. Most of the criticism about scale based on measurement is that scores do not necessarily represent user's preference. This is because respondents have to internally convert preference to scores and the conversion may introduce misrepresentation of the preference being captured [10]. In view of the fact that user service evaluation depends largely on what users perceived, linguistics judgement is a good option in avoiding such inconvenience. The use of words in natural language rather than numerical values is, in general, a less specific, more flexible, direct, realistic, and adequate form to express judgments. Thus, linguistic terms as for example, "satisfied", "fair", "dissatisfied", are regarded as the natural representation of the preference or judgment. These characteristics indicate the applicability of fuzzy set theory [11] in capturing the user's preference, which aids in measuring the ambiguity of concepts that are associated with human being's subjective judgment. Since the evaluation is resulted from the different evaluator's view of linguistic variables, its evaluation must therefore be conducted in an uncertain, fuzzy environment. On the other hand, LibQUAL+ model considers that all the users have the same importance in evaluating each question on the library service levels, but in the quality evaluation of library services, the information handled is not equally important, i.e., the framework is heterogeneous. For example, when a group of users expresses its opinions on the community space for group learning, its assessments must not be considered with equal relevance, given that, there will be users, such as students, with more knowledge on the community space for group learning than others, such as professors, and therefore, all the opinions shall not be equally reliable; although a final and global assessment must be made using the initial and individual assessments.

The aim of this paper is to present an extended LibQUAL+ model based on fuzzy linguistic information, which overcomes the above drawbacks of the LibQUAL+ model, to evaluate the quality of academic libraries according to user's satisfaction. To do so, our proposed model uses the ordinal fuzzy linguistic modeling [12] to represent the user's perceptions and takes into account that the users' opinions on the library service levels are not equally important. To do so, tools of computing with words based on the linguistic aggregation operators LOWA [12] and LWA [13] to compute the quality assessments are used.

The rest of this paper is set out as follows. In Section 2, the tools used for developing our model are presented. Section 3 describes the extended LibQUAL+ model based on fuzzy linguistic information to evaluate the quality of academic libraries. Finally, some conclusions are drawn in Section. 4.

## 2    Preliminaries

In this section, the LibQUAL+ model is described and the fuzzy linguistic approach for computing with words, which is used to design our fuzzy linguistic extended LibQUAL+ model, is presented.

### 2.1    The LibQUAL+ Model

In 1999, a major project to develop a standardized measure of library service quality was undertaken by the Association of Research Libraries (ARL) in collaboration with Texas A&M University. The result of this project was LibQUAL+ [3], which is an extension of the SERVQUAL (for SERVice QUALity) tool [14]. SERVQUAL has been carefully tested and widely accepted after a dozen years of application in the private sector and elsewhere. Grounded in the gap theory of service quality, the singular percept of SERVQUAL is that "only customers judge quality; all other judgments are essentially irrelevant" [15]. According to the gap model, service quality is the gap between customer's expectations and perceptions. When experiences exceed expectations, the quality of the service is high, and vice versa [5]. Service quality is conceptualized as a gap between customers' minimum/desired expectations of service quality and their perceptions of the service quality actually received. A positive gap indicates that the service performance has exceeded customers' expectations, whereas a negative gap indicates that the service performance has fallen short of the expected service. Gap models are intuitively appealing to many research consumers [16] since its interpretation is straightforward. For instance, if the perceived rating on a item is below the minimum, it clearly indicates that the subject the item evaluates needs improvement. On the other hand, if the perceived rating on an item is very above the desired level of service, it may suggest that the item is not a concern to consumers.

Following that idea, LibQUAL+ is a survey administered by the ARL to measure library user's perception of library service quality and to help libraries identify service areas needing improvement [3]. To do so, the LibQUAL+ survey is composed of 22 core questions that measure perceptions concerning three dimensions of library service quality [4]:

- *Affect of service.* This dimension assesses empathy, responsiveness, assurance, and reliability of library employees. It includes the following nine questions:
  - $q_1$: Employees who instill confidence in users.
  - $q_2$: Giving users individual attention.
  - $q_3$: Employees who are consistently courteous.
  - $q_4$: Readiness to respond to users' questions.
  - $q_5$: Employees who have the knowledge to answer user questions.
  - $q_6$: Employees who deal with users in a caring fashion.
  - $q_7$: Employees who understand the needs of their users.
  - $q_8$: Willingness to help users.
  - $q_9$: Dependability in handling users' service problems.
- *Library as place.* This dimension measures the usefulness of space, the symbolic value of the library, and the library as a refuge for word of study. It includes the following five questions:
  - $q_{10}$: Library space that inspires study and learning.
  - $q_{11}$: Quiet space for individual activities.
  - $q_{12}$: A comfortable and inviting location.
  - $q_{13}$: A getaway for study, learning, or research.
  - $q_{14}$: Community space for group learning and group study.
- *Information control.* This dimension measures how users want to interact with the modern library and include scope, timeliness and convenience, ease of navigation, modern equipment, and self-reliance. It includes the following eight questions:
  - $q_{15}$: Making electronic resources accessible from my home or office.
  - $q_{16}$: A library Web site enabling me to locate information on my own.
  - $q_{17}$: Printed library materials I need for my work.
  - $q_{18}$: The electronic information resources I need.
  - $q_{19}$: Modern equipment that lets me easily access needed information.
  - $q_{20}$: Easy-to-use access tools that allow me to find things on my own.
  - $q_{21}$: Making information easily accessible for independent use.
  - $q_{22}$: Print and/or electronic journal collections I require for my work.

For each question, respondents are asked to indicate their minimum acceptable service level, their desired service level, and the perception of the actual service provided by the library by giving a score from one to nine. The minimum service level and the desired service level reflect the importance of that service to the user: a low level means that it is not considered very important, and vice versa – when the minimum or desired service level receive high scores, the issue is important. An adequacy gap (the perceived quality in relation to the accepted minimum level) and a superiority gap (the perceived quality in relation to the desired service) are determined based on the answers [5].

## 2.2    A Fuzzy Linguistic Approach for Computing with Words

Many problems present fuzzy and vague qualitative aspects (decision making, risk assessment, information retrieval, etc.). In such problems, the information cannot be assessed precisely in a quantitative form, but it may be done in a qualitative one, and thus, the use of a linguistic approach is necessary. For example, when attempting to qualify phenomena related to human perception, we are often led to use words in natural language instead of numerical values. The fuzzy linguistic approach is an approximate technique appropriate to deal with fuzzy and vague qualitative aspects of problems. It models linguistic information by means of linguistic terms supported by linguistic variables [17,18,19], whose values are not numbers but words or sentences in a natural or artificial language.

The ordinal fuzzy linguistic approach [12,13] is a very useful kind of fuzzy linguistic approach used for modeling the computing with words process as well as linguistic aspects of problems. It facilitates the fuzzy linguistic modeling very much because it simplifies the definition of the semantic and syntactic rules. It is defined by considering a finite and totally ordered label set $S = \{s_i\}$, $i \in \{0, \ldots, \mathcal{T}\}$, in the usual sense, i.e., $s_i \geq s_j$ if $i \geq j$, and with odd cardinality. Typical values of cardinality used in the linguistic models are odd values, such as 7 or 9, with an upper limit of granularity of 11 or no more than 13, where the mid term represents an assessment of "approximately 0.5", and the rest of the terms being placed symmetrically around it. The semantics of the linguistic term set is established from the ordered structure of the label set by considering that each linguistic term for the pair $(s_i, s_{\mathcal{T}-i})$ is equally informative. For example, we can use the following set of nine labels to provide the user evaluations: $\{N = None, EL = Extremely\ Low, VL = Very\ Low, L = Low, M = Medium, H = High, VH = Very\ High, EH = Extremely\ High, T = Total\}$.

An advantage of the ordinal fuzzy linguistic approach is the simplicity and quickness of its computational model. It is based on the symbolic computation [12,13] and acts by direct computation on labels by taking into account the order of such linguistic assessments in the ordered structure of linguistic terms. This symbolic tool seems natural when using the fuzzy linguistic approach, because the linguistic assessments are simply approximations which are given and handled when it is impossible or unnecessary to obtain more accurate values. Usually, the ordinal fuzzy linguistic model for computing with words is defined by establishing (i) a negation operator, $Neg(s_i) = s_j \mid j = \mathcal{T} - i$, (ii) comparison operators based on the ordered structure of linguistic terms: Maximization operator: $MAX(s_i, s_j) = s_i$ if $s_i \geq s_j$; and Minimization operator: $MIN(s_i, s_j) = s_i$ if $s_i \leq s_j$, and (iii) adequate aggregation operators. In the following, we present two aggregation operators based on symbolic computation to complete the ordinal fuzzy linguistic computational model.

**The LOWA Operator.** An important aggregation operator of ordinal linguistic values based on symbolic computation is the LOWA operator [12]. The *Linguistic Ordered Weighted Averaging* (LOWA) is an operator used to aggregate non-weighted ordinal linguistic information, i.e., linguistic information values with equal importance [12].

**Definition 1.** *Let $A = \{a_1, \ldots, a_m\}$ be a set of labels to be aggregated, then the LOWA operator, $\phi$, is defined as:*

$$\phi(a_1, \ldots, a_m) = W \cdot B^T = \mathcal{C}^m\{w_k, \ b_k, \ k = 1, \ldots, m\}$$
$$= w_1 \odot b_1 \oplus (1 - w_1) \odot \mathcal{C}^{m-1}\{\beta_h, \ b_h, \ h = 2, \ldots, m\} \,, \tag{1}$$

*where $W = [w_1, \ldots, w_m]$ is a weighting vector, such that, $w_i \in [0, 1]$ and $\Sigma_i w_i = 1$. $\beta_h = w_h / \Sigma_2^m w_k$, $h = 2, \ldots, m$, and $B = \{b_1, \ldots, b_m\}$ is a vector associated to $A$, such that, $B = \sigma(A) = \{a_{\sigma(1)}, \ldots, a_{\sigma(m)}\}$, where, $a_{\sigma(j)} \leq a_{\sigma(i)} \ \forall \ i \leq j$, with $\sigma$ being a permutation over the set of labels $A$. $\mathcal{C}^m$ is the convex combination operator of $m$ labels and if $m = 2$, then it is defined as:*

$$\mathcal{C}^2\{w_i, \ b_i, \ i = 1, 2\} = w_1 \odot s_j \oplus (1 - w_1) \odot s_i = s_k \,, \tag{2}$$

*such that, $k = min\{\mathcal{T}, \ i + round(w_1 \cdot (j - i))\}$, $s_j, s_i \in S$, $(j \geq i)$, where "round" is the usual round operation, and $b_1 = s_j$, $b_2 = s_i$. If $w_j = 1$ and $w_i = 0$, with $i \neq j$, $\forall i$, then the convex combination is defined as: $\mathcal{C}^m\{w_i, \ b_i, \ i = 1, \ldots, m\} = b_j$.*

The LOWA operator is an "or-and" operator [12] and its behavior can be controlled by means of $W$. In order to classify OWA operators with regards to their localization between "or" and "and", Yager [20] introduced a measure of *orness*, associated with any vector $W$ : $orness(W) = \frac{1}{m-1}\sum_{i=1}^{m}(m - i)w_i$. This measure characterizes the degree to which the aggregation is like an "or" (MAX) operation. Note that an OWA operator with $orness(W) \geq 0.5$ will be an *orlike*, and with $orness(W) < 0.5$ will be an *andlike* operator.

An important question of the LOWA operator is the determination of the weighting vector $W$. In [20], it was defined an expression to obtain $W$ that allows to represent the concept of fuzzy majority [21] by means of a fuzzy linguistic nondecreasing quantifier $Q$ [22]:

$$w_i = Q(i/n) - Q((i - 1)/n), \ i = 1, \ldots, n \,. \tag{3}$$

When a fuzzy linguistic quantifier $Q$ is used to compute the weights of LOWA operator $\phi$, it is symbolized by $\phi_Q$.

**The LWA Operator.** Another important aggregation operator of ordinal linguistic values is the *Linguistic Weighted Averaging* (LWA) operator [13]. It is based on the LOWA operator and is defined to aggregate weighted ordinal fuzzy linguistic information, i.e., linguistic information values with not equal importance.

**Definition 2.** *The aggregation of a set of weighted linguistic opinions, $\{(c_1, a_1), \ldots, (c_m, a_m,)\}$, $c_i, a_i \in S$, according to the LWA operator, $\Phi$, is defined as:*

$$\Phi[(c_1, a_1), \ldots, (c_m, a_m)] = \phi(h(c_1, a_1), \ldots, h(c_m, a_m)) \,, \tag{4}$$

*where $a_i$ represents the weighted opinion, $c_i$ the importance degree of $a_i$, and $h$ is the transformation function defined depending on the weighting vector $W$ used for the LOWA operator $\phi$, such that, $h = MIN(c_i, a_i)$ if $orness(W) \geq 0.5$, and $h = MAX(Neg(c_i), a_i)$ if $orness(W) < 0.5$.*

We should point out that the LOWA and LWA operators are the basis of the fuzzy linguistic extended LibQUAL+ model. We have chosen these operators due to the following reasons:

- Both operators are complementary (the LWA operator is defined from the LOWA operator) and this simplifies the design of the evaluation model.
- Both operators act by symbolic computation and, therefore, linguistic approximation processes are unnecessary and this simplifies the processes of computing with words.
- The concept of fuzzy majority represented by linguistic quantifiers acts in their processes of computation and, in such a way, the assessments on academic libraries are obtained according to the majority of evaluations provided by the users.

# 3 An Extended LibQUAL+ Model Based on Fuzzy Linguistic Information

In this section, we present the extended LibQUAL+ model based on fuzzy linguistic information to evaluate the quality of academic libraries according to user's satisfaction. It is developed with the aim of solving the drawbacks of the LibQUAL+ model shown in Section 1. To do so, we define a quality evaluation model of academic libraries which presents two elements: (i) an evaluation scheme that contains the twenty-two questions relating to three dimensions of library service quality, and (ii) a computation method to generate quality assessments of academic libraries and to obtain their weaknesses and strengths.

## 3.1 Evaluation Scheme

The evaluation scheme is based on a set of twenty-two questions relating to three dimensions of library service quality: (i) affect of service, which relates to user interactions with, and the general helpfulness and competency of academic library staff, (ii) library as a place, which deals with the physical environment of the academic library as a place for individual study, group work, and inspiration, and (iii) information control, which relates to whether users are able to find the required information in the academic library in the format of their choosing. It presents the following characteristics:

- *It is user driven.* The evaluation scheme necessarily requires the inclusion of questions about library service quality easily understandable to any user rather than questions that can be measures objectively independently of users. As the basis of our model is the LibQUAL+ model, we use the same twenty-two questions relating to the three dimensions of library service quality (see Section 2.1), which are easily understandable to any user. This number of questions is not excessive in order to help users in understanding it and avoiding confusion. It is due to long and complex evaluation schemes cause user idleness and limit their own application possibilities.

- *It is weighted.* The users of the academic library do not play equal roles in measuring library service quality: i.e., some users should be more influential than others in some questions as it is not always valid that all group of users have equal importance with respect to the decision being made. This is because the degree of relevancy, knowledge, and experience may not be equal among them. For example, the student's opinion on the community space for group learning and group study should be more important that the professor's opinion. Therefore, there must be an allowance for such differences in weight or importance as the framework is heterogeneous.

## 3.2   Computation Method

We have designed a computation method to generate quality assessment in academic libraries that has two main characteristics:

- *It is a user-centered computation method.* The quality assessment in academic libraries is obtained from individual linguistic judgments provided by their users rather than from assessments obtained objectively by means of the direct observation of the academic libraries characteristics.
- *It is a majority guided computation method.* The quality assessments are values representative of the majority of individual judgments provided by the users of the academic library. The aggregation to compute the quality assessments is developed by means of the LOWA and LWA operators.

First, a quality evaluation questionnaire based on the LibQUAL+ model is defined, which consists of twenty-two questions relating to three dimensions of library service quality described in Section 2.1. Users are asked for impressions about the twenty-two questions according to minimum service level they are willing to accept, desired service level they would like to receive, and perceived performance level, that is, actual level of service they perceive to have been rendered. The minimum service level, the desired service level, and the perceived performance level behind each question are rated on a linguistic term set $S$. For instance, the linguistic term set presented in Section 2 can be used. We use fuzzy linguistic variables to represent user's opinions by mean of linguistic labels because they are more easily understood by the users than numerical ones. In addition, we assume that each user does not have the same importance in the evaluation scheme. It is assigned a relative linguistic importance degree, $I(e_l, q_i) \in S$, for each user, $e_l$, on each question, $q_i$. This importance degree could be obtained from a set of experts or the staff members of the academic library and it may be different for each academic library.

Once the group of users, $\{e_1, \ldots, e_L\}$, have filled all the questionnaires for a given academic library, the model calculates the quality assessments of the academic library and obtains its weaknesses and strengths using the linguistic aggregation operators LWA and LOWA in the following steps:

- For each question, $q_i$, its global quality assessment of the minimum service level, $MSL_i$, its global quality assessment of the desired service level, $DSL_i$, and its global quality assessment of the perceived performance level, $PPL_i$, are obtained by aggregating the evaluation judgments provided by the group of users on the question, $q_i$, by means of the LWA operator $\Phi$:

$$\begin{aligned} MSL_i &= \Phi_Q((I(e_1, q_i), e_1(q_i^{MSL})), \ldots, ((I(e_L, q_i), e_L(q_i^{MSL}))) \,, \\ DSL_i &= \Phi_Q((I(e_1, q_i), e_1(q_i^{DSL})), \ldots, ((I(e_L, q_i), e_L(q_i^{DSL}))) \,, \qquad (5) \\ PPL_i &= \Phi_Q((I(e_1, q_i), e_1(q_i^{PPL})), \ldots, ((I(e_L, q_i), e_L(q_i^{PPL}))) \,, \end{aligned}$$

where $e_l(q_i^{MSL}) \in S$ is the minimum service level provided by the user $e_l$ on question, $q_i$, $e_l(q_i^{DSL}) \in S$ is the desired service level provided by the user $e_l$ on question, $q_i$, $e_l(q_i^{PPL}) \in S$ is the perceived performance level provided by the user $e_l$ on question, $q_i$, and $I(e_l, q_i) \in S$ is the linguistic importance degree of the user, $e_l$, on question, $q_i$. Therefore, $MSL_i$, $DSL_i$ and $PPL_i$, are the linguistic measures that represents the minimum service level, the desired service level and the perceived performance level, respectively, with respect to question, $q_i$, according to the majority (represented by the fuzzy linguistic quantifier $Q$) of linguistic evaluation judgments provided by the group of users.

Then, gap analysis is done for each item. According to LibQUAL+ model, the minimum and the desired scores establish the boundaries of a zone of tolerance within which the perceived scores should desirably float. The difference between the perceived and minimum scores is called the service adequacy gap, and the difference between the desired and perceived scores is called the service superiority gap. Taking into account these considerations, it is defined two scores which can obtain the strengths and weaknesses of an academic library according to the users' answers:

$$\begin{aligned} SA_i &= J(PPL_i) - J(MSL_i) \,, \\ SS_i &= J(PPL_i) - J(DSL_i) \,, \end{aligned} \qquad (6)$$

where $SA_i$ is the service adequacy score on question, $q_i$, $SS_i$ is the service superiority score on the question, $q_i$, and

$$J : S \to \{0, \ldots, \mathcal{T}\} \mid J(s_i) = i, \ \forall s_i \in S \,. \qquad (7)$$

The $SA$ is calculated by subtracting the minimum score from the perceived score on any given question. It is an indicator of the extent to which academic libraries are meeting the minimum expectations of their users. The $SS$ is calculated by subtracting the desired score from the perceived score on any given question and is an indicator of the extent to which academic libraries are exceeding the desired expectations of their users. Figure 1 depicts the three possible cases for gaps $SA$ and $SS$. The cases when the perceived level of service falls out of the zone of tolerance are denoted as $SA^-$ and $SS^+$. $SA^-$ means that the academic library is not meeting its users minimum expectations, i.e., the perceived score is lower than the minimum one. Likewise, $SS^+$ means that the academic library is exceeding its users desired expectations, i.e., the perceived score is higher than the desired one. Therefore, $SA^-$ can be used to identify academic library functionalities needing improvement, whereas $SS^+$ is an indicator of the extent to which academic library functionalities are exceeding the desired expectations of the users.

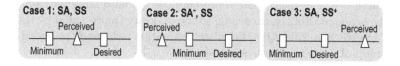

**Fig. 1.** Possible cases for gaps SA and SS

– Finally, for the academic library, the following quality assessments are calculated:

- Its quality assessment on the affect of service dimension, $ASD$, by aggregating the perceived performance level, $PPL_i$, from questions $q_1$ to $q_9$, by means of the LOWA operator $\phi$:

$$ASD = \phi_Q(PPL_1, \ldots, PPL_9), \qquad (8)$$

where $ASD$ is a measure that represents the quality assessment of the affect of service dimension for the academic library, according to the majority (represented by the fuzzy linguistic quantifier $Q$) of linguistic evaluation judgments provided by the group of users about questions $\{q_1, \ldots, q_9\}$.

- Its quality assessment on the library as place dimension, $LPD$, by aggregating the perceived performance level, $PPL_i$, from questions $q_{10}$ to $q_{14}$, by means of the LOWA operator $\phi$:

$$LPD = \phi_Q(PPL_{10}, \ldots, PPL_{14}), \qquad (9)$$

where $LPD$ is a measure that represents the quality assessment of the library as place dimension for the academic library, according to the majority (represented by the fuzzy linguistic quantifier $Q$) of linguistic evaluation judgments provided by the group of users about questions $\{q_{10}, \ldots, q_{14}\}$.

- Its quality assessment on the information control dimension, $ICD$, by aggregating the perceived performance level, $PPL_i$, from questions $q_{15}$ to $q_{22}$, by means of the LOWA operator $\phi$:

$$ICD = \phi_Q(PPL_{15}, \ldots, PPL_{22}), \qquad (10)$$

where $ICD$ is a measure that represents the quality assessment of the information control dimension for the academic library, according to the majority (represented by the fuzzy linguistic quantifier $Q$) of linguistic evaluation judgments provided by the group of users about questions $\{q_{15}, \ldots, q_{22}\}$.

- Its global quality assessment, $r$, by aggregating the perceived performance level, $PPL_i$, from all questions, by means of the LOWA operator $\phi$:

$$r = \phi_Q(PPL_1, \ldots, PPL_{22}), \tag{11}$$

where $r$ is a measure that represents the global quality assessment for the academic library, according to the majority (represented by the fuzzy linguistic quantifier $Q$) of linguistic evaluation judgments provided by the group of users about questions $\{q_1, \ldots, q_{22}\}$.

## 4    Conclusions

In this paper we have presented an extended LibQUAL+ model based on fuzzy linguistic information to obtain both the strengths and the weaknesses of the academic library services according to user satisfaction. Using the ordinal fuzzy linguistic modeling to represent the user's perceptions and taking into account that the users' opinions on the library service levels are not equally important, some drawbacks of the LibQUAL+ model have been overcome. Considerable use has been made of fuzzy set technology to provide the ability to describe the information by using linguistic label in a way that is particularly user friendly. Furthermore, we have applied automatic tools of fuzzy computing with words based on the LOWA and LWA operators to compute quality assessments of academic libraries. In the future, we will extend the concept of quality to other new services of the academic libraries, such as Library 2.0.

**Acknowledgments.** This work has been developed with the financing of the Andalusian Excellence Projects TIC-05299 and TIC-5991, the FEDER funds in FUZZYLING-II Project TIN2010-17876, and "Proyecto de Investigación del Plan de Promoción de la Investigación UNED 2011 (2011V/PUNED/0003)".

## References

1. Bawden, D., Petuchovaite, R., Vilar, P.: Are we effective? How would we know? Approaches to the evaluation of library services in Lithuania, Slovenia and the United Kingdom. New Library World 106, 454–463 (2005)
2. Asemi, A., Kazempour, Z., Rizi, H.A.: Using LibQUAL+TM to improve services to libraries: A report on academic libraries of Iran experience. The Electronic Library 28, 568–579 (2010)
3. Association of Research Libraries: LibQUAL+: charting library service quality. ARL, http://www.libqual.org

4. Cook, C., Heath, F.M.: Users' perception of library service quality: a LibQUAL+ qualitative study. Library Trends 49, 548–584 (2001)
5. Hakala, U., Nygren, U.: Customer satisfaction and the strategic role of university libraries. International Journal of Consumer Studies 34, 204–211 (2010)
6. Jankowska, M.A., Hertel, K., Young, N.J.: Improving library service quality to graduate students: LibQual+(TM) survey results in a practical setting. Portal-Libraries and the Academy 6, 59–77 (2006)
7. López-Gijón, J., Ávila-Fernández, B., Pérez, I.J., Herrera-Viedma, E.: The quality of biomedical academic libraries according to the users. Profesional de la Información 19, 255–259 (2010)
8. Thompson, B., Kyrillidou, M., Cook, C.: Library users' service desires: A LibQUAL+ study. Library Quarterly 78, 1–18 (2008)
9. Likert, R.: A technique for the measurement of attitudes. Archives of Psychology 22, 1–55 (1932)
10. Tsaur, S.H., Chang, T.Y., Yen, C.H.: The evaluation of airline service quality by fuzzy MCDM. Tourism Management 23, 107–115 (2002)
11. Zadeh, L.A.: Fuzzy sets. Information and Control 8, 338–353 (1965)
12. Herrera, F., Herrera-Viedma, E., Verdegay, J.L.: Direct approach processes in group decision making using linguistic OWA operators. Fuzzy Sets and Systems 79, 175–190 (1996)
13. Herrera, F., Herrera-Viedma, E.: Aggregation operators for linguistic weighted information. IEEE Transactions on Systems, Man and Cybernetics, Part A: Systems and Humans 27, 646–656 (1997)
14. Parasuraman, A., Zeithaml, V.A., Berry, L.L.: A multiple-item scale for measuring consumer perceptions of service quality. Journal of Retailing 64, 12–40 (1988)
15. Zeithaml, V.A., Berry, L.L., Parasuraman, A.: Delivering Quality Services - Balancing Customer Perceptions and Expectations. The Free Press, New York (1990)
16. Thompson, B., Cook, C., Heath, F.: The LibQUAL+ gap measurement model: the bad, the ugly, and the good of gap measurement. Peformance Measurement and Metrics 1, 165–178 (2000)
17. Zadeh, L.A.: The concept of a linguistic variable and its applications to approximate reasoning. Part I. Information Sciences 8, 199–249 (1975)
18. Zadeh, L.A.: The concept of a linguistic variable and its applications to approximate reasoning. Part II. Information Sciences 8, 301–357 (1975)
19. Zadeh, L.A.: The concept of a linguistic variable and its applications to approximate reasoning. Part III. Information Sciences 9, 43–80 (1975)
20. Yager, R.R.: On ordered weighted averaging aggregation operators in multicriteria decision making. IEEE Transactions on Systems, Man and Cybernetics 18, 183–190 (1988)
21. Kacprzyk, J.: Group decision making with a fuzzy linguistic majority. Fuzzy Sets Systems 18, 105–118 (1986)
22. Zadeh, L.A.: A computational approach to fuzzy quantifiers in natural languages. Computer & Mathematics with Applications 9, 149–184 (1983)

# Measure of Inconsistency
# for the Potential Method

Lavoslav Čaklović*

Department of Mathematics, University of Zagreb, Croatia
caklovic@math.hr

**Abstract.** The inconsistency of the decision maker's preferences may be measured as a number of violations of the transitivity rule. If the intensity of the preference is available, then the incosistency may be measured by measuring the inconsistency of each cycle of the preference graph. In the Potential Method, this may be accomplished by mesuring an angle (degree) between the preference flow and the column space of the incidence matrix.

In this article a random study is performed to determine the upper bound for admissible inconsistency. The degree distribution is recognized as the Gumbel distribution and the upper bound for admissible inconsistency measure is defined as a $p$-quantile ($p = 0.05$) of that distribution.

**Keywords:** decision making, preference graph, inconsistency measure, condition of order preservation, randomization.

*Subject Classifications*: 62C25, 90B50, 91B06.

## 1 Introduction

This paper is about the inconsistency in the decision maker's input data when it is in the form of the preferences obtained from pairwise comparisons. Inconsistency measure is a useful information which shows a degree of non-transitivity in the decision maker's preferences. The high inconsistency measure may suggests reconsidering the input again and again if necessary. For the Eigenvalue Method (EVM), proposed by Saaty [10], the *consistency index* CI is defined as

$$\mathrm{CI}(A) = \frac{\lambda_{\max}(A) - n}{n - 1},$$

where $A$ is the positive reciprocal matrix of order $n$ and $\lambda_{\max}(A)$ is the Perron root of $A$. It is well-known that $\mathrm{CI} \geq 0$ and $\mathrm{CI}(A) = 0 \iff A$ is consistent. A positive reciprocal matrix $A$ is of *admissible inconsistency* if

$$\mathrm{CI}(A) \leq 0.1 \times \mathrm{MRI}(n)$$

---

* The author thanks to V. Šego and to the referee for valuable comments and suggestions.

V. Torra et al. (Eds.): MDAI 2012, LNAI 7647, pp. 102–114, 2012.

where MRI($n$) is the mean of the random CI. The random index study in AHP context was performed by several authors, from Crawford and Williams [4] to Alonso and Lamata [1]. A nice overview of the results is given in Alonso and Lamata [1, Table 1, p.449].

The Potential Method (PM) (Čaklović [5]) uses a *preference graph* to capture the results of pairwise comparisons. A *preference flow* $\mathcal{F}$ is the non-negative function defined on the set of arcs which captures the intensity of the preferences. An example of the preference graph in a voting procedure was considered by Condorcet [3]. He defined a *social preference* flow as

$$\mathcal{F}_C(u, v) := N(u, v) - N(v, u) \tag{1}$$

where $N(u, v)$ denotes the number of voters choosing $u$ over $v$. We say that $u$ is *socially preferred* to $v$ if $\mathcal{F}_C(u, v) \geq 0$.

In the graph representation of the preferences, inconsistency is closely related to non-transitivity and may be defined even for incomplete graphs, which is not so straightforward for AHP. In simple words, the flow $\mathcal{F}$ is *consistent* if it is consistent along each cycle $c$, i.e. if the sum $\mathcal{F}_c$ of the algebraic components of the flow $\mathcal{F}$ along each cycle $c$ is equal to zero, see Definition 1 and Theorem 1. In

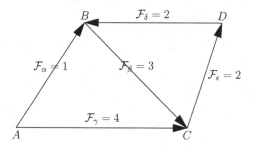

**Fig. 1.** An example of the inconsistent flow. The sum of the flow components along the cycle $CDBC$ is equal to $2 + 2 + 3 = 7$. The flow is consistent along the cycle $ABCA$.

Figure 1, the flow $\mathcal{F}$ is consistent along the cycle $ABCA$ and inconsistent along the cycle $CDBC$ because the sum $\mathcal{F}_c$ along this cycle is equal to $2 + 2 + 3 = 7$. Intuitively, the inconsistency measure of the flow may be defined as the sum $\sum_c \mathcal{F}_c$ over all independent cycles $c$ in the preference graph divided by the 2-norm $\|\mathcal{F}\|_2$. The exact definition is slightly different: this is the angle between the flow $\mathcal{F}$ and the vector space of all consistent flows, see Definition 3. Please, note that the notion of the flow consistency, as defined here, is stronger than pure transitivity. This motivates the search for the upper bound of the admissible inconsistency of the given flow which is done in Section 4.

Saaty's inconsistency [10, Saaty] and flow inconsistency are closely related. There is a theorem which states that a positive reciprocal matrix $A = (a_{ij})$ is consistent if and only if

$$a_{ij}a_{jk} = a_{ik}, \quad i, j, k = 1, \ldots, n \tag{2}$$

Taking the logarithm of this relation, one recognizes the inconsistency condition *4* from Theorem 1 for the flow

$$\mathcal{F}_{(i,j)} := \log(a_{ij}). \tag{3}$$

In the stochastic preference approach [6, French, p. 101] the author introduces a notion of the stochastic preference $p_{ab}$ as a probability of choosing $a$ when offered a choice between $a$ and $b$. Then, it is easy to show that if the stochastic preference satisfies the consistency condition

$$\frac{p_{ab}}{p_{ba}} \cdot \frac{p_{bc}}{p_{cb}} = \frac{p_{ac}}{p_{ca}} \tag{4}$$

for all $a, b, c \in V$ then, it generates a weak preference order on the set of alternatives $V$. If we define a *stochastic flow* $\mathcal{F}$ by

$$\mathcal{F}_{(b,c)} := \log \frac{p_{bc}}{p_{cb}}, \tag{5}$$

then, the stochastic preference is consistent if and only if the flow $\mathcal{F}$ is consistent[1].

Another kind of inconsistency may be considered after the ranking procedure is over. This is the *number of violations of Condition of Order Preservation* (#VCOP) introduced in Costa-Vansnick [2]. This number indicates how far from the measurable value function is the calculated potential. The precise definition of COP is given in (15). The aim of this article is to investigate the correspondence of #VCOP and the flow inconsistency. This is done by performing a random study inspired by the random index study for EVM.

The paper is organized as follows. In Section 2 we introduce the basic notation, develop the idea of consistency and give some other equivalent conditions of consistency.

Section 3 describes the connection of PM with the Geometric Mean, the Ordinal value function and the Stochastic preference model. Some elementary facts about the PM as the social preference are mentioned.

In Section 4 we perform a randomization procedure to determine the admissible level of the flow inconsistency. The Analytic Hierarchy Process (AHP), more precisely the Eigenvalue Method (EVM), serves as a model. It is shown that the empirical distribution of the inconsistency measure DEG may be modeled as the Gumbel distribution. The upper bound for admissible inconsistency is defined as the 0.05-quantile of the theoretical distribution. The randomization is performed for complete graphs only because of the possibility to make a comparison with AHP.

In Section 5 the Condition of Order Preservation (COP) is considered and the number of violations #VCOP is calculated for the random graph and the random reciprocal matrix. It is shown that the *consistency index* (CI) is not correlated with #VCOP while the correlation between the random degree DEG and #VCOP is very good.

---

[1] Moreover, it may be shown that the corresponding potential $X$ from formula (7) is a measurable value function which is true for every consistent flow.

## 2    Preference Graph

### 2.1    Consistent Preference Flow

A *preference graph* is a digraph $\mathcal{G} = (V, \mathcal{A})$ where $V$ is the set of nodes and $\mathcal{A}$ is the set of arcs of $\mathcal{G}$. We say that the node $a$ is *more preferred* than node $b$, in notation $a \succ b$, if there is an arc $(a, b)$ outgoing from $b$ and in-going to $a$. A *preference flow* is a non-negative real function $\mathcal{F}$ defined on the set of arcs. The value $\mathcal{F}_\alpha$ on the arc $\alpha$ is the intensity of the preference on some scale[2]. For the arc $\alpha = (a, b)$, $\mathcal{F}_\alpha = 0$ means that the decision maker is indifferent to the pair $\{a, b\}$. In that case the orientation of the arc may be arbitrary.

The *incidence matrix* $A = (a_{\alpha, v})$ of the graph is defined as the $m \times n$ matrix, $m = \operatorname{Card} \mathcal{A}, n = \operatorname{Card} V$, where

$$a_{\alpha, v} = \begin{cases} -1, \text{ if the arc } \alpha \text{ leaves the node } v \\ \phantom{-}1, \text{ if the arc } \alpha \text{ enters the node } v \\ \phantom{-}0, \text{ otherwise.} \end{cases}$$

It is more convenient to write $a_{ij}$ where $i$ is the index of $i$-th arc and $j$ is the index of $j$-th node. The vector space $\mathbb{R}^m$ is called the *arc space* and the vector space $\mathbb{R}^n$ is called the *vertex space*. The incidence matrix[3] generates an orthogonal decomposition

$$N(A^\tau) \oplus R(A) = \mathbb{R}^m \tag{6}$$

where $R(A)$ is the column space of the matrix $A$ and $N(A^\tau)$ is the null-space of the matrix $A^\tau$. $N(A^\tau)$ is called the *cycle space* because it is generated by all cycles of the graph.

For example, the incidence matrix of the preference graph in Fig. 1 is given in Table 1 (left). The arcs $\alpha, \beta, \gamma, \delta, \epsilon$ form the basis of the ars space. In the last column are the components of the preference flow. Please, note that the cycles $c_1$ and $c_2$ (the basis of the cycle space) are orthogonal to the columns of the incidence matrix according to (6). The columns of the incidence matrix span the space $R(A)$ of the consistent flows.

**Table 1.** Incidence matrix of the preference graph from Table 1 and the cycle space

| arcs$_m$ | A | B | C | D | $\mathcal{F}$ | arcs$_m$ | $c_1$ | $c_2$ |
|----------|-----|-----|-----|-----|---|----------|-----|-----|
| $\alpha$ | -1 | 1 | 0 | 0 | 1 | $\alpha$ | 1 | 0 |
| $\beta$ | 0 | -1 | 1 | 0 | 3 | $\beta$ | 1 | 1 |
| $\gamma$ | -1 | 0 | 1 | 0 | 4 | $\gamma$ | -1 | 0 |
| $\delta$ | 0 | 1 | 0 | -1 | 2 | $\delta$ | 0 | 1 |
| $\epsilon$ | 0 | 0 | -1 | 1 | 2 | $\epsilon$ | 0 | 1 |

Header spanning: nodes$_n$ (A B C D), flow ($\mathcal{F}$), cycle space ($c_1$ $c_2$).

---

[2] For subjective pairwise comparisons the scale is $\{0, 1, 2, 3, 4\}$.
[3] And matrix in general.

**Definition 1.** *A preference flow $\mathcal{F}$ is consistent if there is no component of the flow in the cycle-space.*

The following theorem is evident (the proof is left to the reader).

**Theorem 1.** *The following statements are equivalent:*
1. *$\mathcal{F}$ is consistent.*
2. *$\mathcal{F}$ is a linear combination of the columns of the incidence matrix $A$.*
3. *There exists $X \in \mathbb{R}^n$ such that $AX = \mathcal{F}$.*
4. *The scalar product $y^\tau \mathcal{F} = 0$ for each cycle $y$, i.e. $\mathcal{F}$ is orthogonal to the cycle space.*

We may test the consistency of the given flow $\mathcal{F}$ by solving the equation

$$AX = \mathcal{F}. \tag{7}$$

**Definition 2.** *A solution of the equation $AX = \mathcal{F}$, if it exists, is called the potential of $\mathcal{F}$.*

Evidently, $X$ is not unique because the constant column $\mathbb{1}^\tau = [\, 1 \; \cdots \; 1 \,]^\tau$ is an element of the kernel $N(A)$. For the consistent flow the equation (7) may be rewritten as

$$\mathcal{F}_\alpha = X(a) - X(b), \;\; \forall \alpha = (a, b) \in \mathcal{A} \tag{8}$$

which means that $X$ is a measurable value function i.e. it measures the preference on the interval scale. For the consistent flow it is easy to find a potential $X$ using a spanning tree of the preference graph (if it is connected). The details are left to the reader.

PM calculates the weights of the nodes in the following way. If $X$ denotes the potential of the flow, then the weights $w$ are obtained using the formula

$$w = \frac{a^X}{\|a^X\|_1} \tag{9}$$

where $\| \cdot \|_1$ represents $l_1$-norm. The exponential function $X \mapsto a^X$ is defined by the components and $a > 1$ is a positive constant. Currently, we use the value $a = 2$ but the user may precise some other value. The arguments for such definition is that the flow $\mathcal{F}$, and the potential $X$, are the logarithms of the data on the ratio scale and we should go back on that scale by exponential function.

## 2.2  Potential of the Inconsistent Preference Flow

In practice, a decision maker, while performing pairwise comparisons, does not give the flow which is necessarily consistent. The best approximation of that flow by the column space of the incidence matrix may be calculated in this situation. The *approximative potential* or *potential $X$* is a solution of the Laplace equation

$$A^\tau AX = A^\tau \mathcal{F}, \qquad \sum_{v \in V} X(v) = 0 \tag{10}$$

where the second requirement is for uniqueness.

Because of the linearity of the equations it is evident that $X$ is invariant on the multiplication of $\mathcal{F}$ by a positive number, i.e. $X(\alpha\mathcal{F}) = \alpha X(\mathcal{F})$, $\alpha > 0$. In other words $\mathcal{F}$ is measured on the ratio scale.

It is easy to prove that for the complete flow (the proof is left to the reader):

$$X(v) = \frac{1}{n}\left( \sum_{\alpha\in\text{In}(v)} \mathcal{F}_\alpha - \sum_{\alpha\in\text{Out}(v)} \mathcal{F}_\alpha \right), \tag{11}$$

where $\text{Out}(v)$ and $\text{In}(v)$ denote the set of all outgoing and in-going arcs for $v$. Formula (11) may be simplified by introducing the *flow matrix* $F$

$$F_{ij} = \begin{cases} \mathcal{F}_{(i,j)} \text{ if } (i,j) \in \mathcal{A}, \\ -\mathcal{F}_{(j,i)} \text{ if } (j,i) \in \mathcal{A}, \end{cases}$$

with the convention $F_{ii} = 0$. The matrix $F$ is anti symmetric and the potential $X$, defined by (11), is the arithmetic mean of the columns of $F$, i.e.

$$x_i = \frac{1}{n}\sum_{j=1}^{n} F_{ij}, \ i = 1,\dots,n. \tag{12}$$

**Definition 3.** *Measure of inconsistency of the flow $\mathcal{F}$, in notation* $\text{DEG}(\mathcal{F})$, *is defined as the angle between $\mathcal{F}$ and the column space of the incidence matrix.*

Evidently, $\text{DEG}(\mathcal{F})$ is the angle between $\mathcal{F}$ and $AX$, where $AX$ is the consistent approximation of $\mathcal{F}$ and $\text{DEG}(\mathcal{F}) = 0$ if and only if $\mathcal{F}$ is consistent. In case when $\text{DEG}(\mathcal{F}) = \pi/2$ then, there is no transitivity at all in the preference graph.

## 3    Potential and Other Methods

### 3.1    Potential and Geometric Mean

In AHP the results of pairwise comparisons are measured on the ratio scale and stored in a positive reciprocal matrix $A$. The logarithm of $A$, taken by components,

$$F_{ij} = \log_a a_{ij}, \quad a > 0$$

is an anti-symmetric matrix $F$ which is the flow matrix of some flow $\mathcal{F}$. The potential $X$ of $\mathcal{F}$ may be expressed in terms of the matrix $A$, using the formula (12), as

$$x_i = \frac{1}{n}\sum_j F_{ij} = \frac{1}{n}\sum_j \log_a a_{ij} = \log_a \left( \prod_j a_{ij} \right)^{\frac{1}{n}},$$

and the weight $w_i$, using (9), may be written as the row geometric mean

$$w_i = \left( \prod_j a_{ij} \right)^{\frac{1}{n}}, \ i = 1,\dots,n.$$

## 3.2  Potential as Ordinal Value Function

Suppose, for the moment, that $\mathcal{F}$ is an uni-modular flow, i.e. $\mathcal{F}_\alpha \in \{0,1\}, \forall \alpha \in \mathcal{A}$. In that case, we may define the relation

$$u \succeq_{\mathcal{F}} v \Leftrightarrow \mathcal{F}_{(u,v)} \geq 0.$$

If $\succeq_{\mathcal{F}}$ is a weak preference relation, then the potential $X$ is an ordinal value function, i.e.

$$\mathcal{F}_{(a,b)} \geq 0 \Leftrightarrow X(a) - X(b) \geq 0.$$

The proof may be found in Čaklović [5].

## 3.3  Potential of the Stochastic Flow

For the complete stochastic flow defined by the formula (4) we may calculate the potential $X$ using the formula (12) and formula (5).

$$X(a) = \frac{1}{n} \sum_{b \neq a} F_{ab} = \log \left( \prod_{b \neq a} \frac{p_{ab}}{p_{ba}} \right)^{\frac{1}{n}}$$

and the weight of the node $a$ is, by formula (9),

$$w_a = \left( \prod_{b \neq a} \frac{p_{ab}}{p_{ba}} \right)^{\frac{1}{n}}.$$

## 3.4  Potential as the Social Preference

Let us give a few comments about the social preference and the PM. The starting point is the Condorcet flow $\mathcal{F}_C$ defined by (1). There are two possibilities how PM may be used for ranking the candidates. One of them is *direct PM ranking*, and another one is *indirect PM ranking*. The first one calculates the potential $X$ of the Condorcet flow $\mathcal{F}_C$, and another one calculates the potential $X^u$ of the unimodular flow $\mathcal{F}_C^u$, where the uni-modular flow of a given flow is obtained by taking the sign of the given intensity. A candidate with the maximal $X^u$ value we call the PM-winner.

It is easy to prove that the Condorcet winner[4] is the PM-winner. This may not be true if the social ranking is taken to be direct PM-ranking. It can be also proved the PM winner is in the *minimal domination set*, and that the indirect PM ranking is *clone independent*. The exhaustive list of the social choice properties of PM is under the reconstruction.

---

[4] The Condorcet winner is defined as the winner in all pairwise confrontations with other candidates.

## 4    Admissible Inconsistency

In this section we shall determine the distribution of the inconsistency measure of the random flow (Definition 3). For $n \geq 4$ this distribution is recognized as the Gumbel distribution

$$e(x) = \frac{e^{-e^{\frac{-x+\alpha}{\beta}} + \frac{-x+\alpha}{\beta}}}{\beta}, \tag{13}$$

which parameters depend[5] upon the number of nodes in the graph, see Table 2. For instance, if the randomization is made as a log-normal perturbation of the random consistent flow defined in formula (14), the inconsistency measure is the Gumbel Distribution $E(\alpha = 17.61, \beta = 7.03)$ (Figure 4).

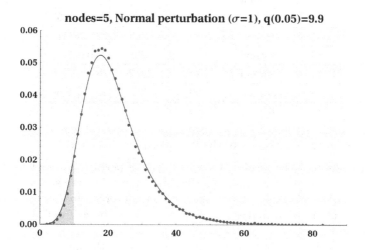

**Fig. 2.** The simulated distribution of the inconsistency measure (dots). 0.05-quantile (9.9) is taken as the upper bound for admissible inconsistency ($10^5$ simulations).

### 4.1    Randomization

A random index study in the AHP context was performed by several authors, from Crawford and Williams [4] to Alonso and Lamata [1]. An overview of the results is given in [1, Table 1, p. 449].

The randomization of the preferences may be designed, generally speaking, as: random *perturbation* (of the consistent situation) and random *distribution*. We performed the following randomizations:

1. normal perturbation of the consistent flow (reciprocal matrix),

---

[5] We also performed a uniform perturbation of the consistent flow and the results are slightly different.

**Table 2.** Quantiles of random degree as a function of the nodes number ($10^5$ simulations)

| number of nodes | perturbation ($\sigma = 1$) | The Gumbel Distribution $E(\alpha, \beta)$ | | | |
|---|---|---|---|---|---|
| | | 0.05-quantile | | Parameters | |
| | | from data | theoretical | $\alpha$ | $\beta$ |
| 4 | normal | 6 | 5.3 | 15.01 | 8.83 |
| 5 | normal | 10 | 9.9 | 17.61 | 7.03 |
| 6 | normal | 13 | 12.7 | 19.18 | 5.91 |
| 7 | normal | 15 | 14.7 | 20.24 | 5.07 |
| 8 | normal | 16 | 16.1 | 21.03 | 4.47 |
| 9 | normal | 17 | 17.2 | 21.64 | 4.02 |
| 10 | normal | 18 | 18.0 | 22.06 | 3.67 |
| 11 | normal | 18 | 18.8 | 22.49 | 3.38 |
| 12 | normal | 19 | 19.3 | 22.77 | 3.16 |
| 13 | normal | 19 | 19.8 | 23.04 | 2.96 |
| 14 | normal | 20 | 20.2 | 23.28 | 2.79 |
| 15 | normal | 20 | 20.6 | 23.47 | 2.66 |

2. uniform perturbation of the consistent flow (reciprocal matrix),
3. unrestricted randomization of the flow (reciprocal matrix).

We present here only the results of the first type of perturbation. The results of the uniform perturbation are just slightly different and the unrestricted randomization generates highly inconsistent flows with the average greater than $50°$. We believe that the decision maker's preferences in real life are well described by the first process.

A random positive reciprocal matrix is obtained as the normal perturbation of the random consistent reciprocal matrix with elements

$$a_{ij} = w_{ij} * \exp(N(0, \sigma)) \tag{14}$$

where $\sigma = 1$ and $w_{ij} := int(rand(1\text{-}9))^\alpha$ $(i < j)$, is the random choice from the set $\{1, 2, 3, 4, 5, 6, 7, 8, 9\}$, powered by $\alpha$ which is the random choice from $\{-1, 1\}$, and $w_{ij} := w_{ji}^{-1}$, for $j < i$, and $w_{ii} = 1$, $\forall i = 1, \ldots, n$.

The random consistent flow is made by random choice of the orientation of the arc and by random choice of the flow value in the set $\{0, 1, 2, 3, 4\}$. Normal perturbation has a standard deviation $\sigma = 1$. The randomization procedure was performed by Perl and data analysis was done by R.

In the AHP context our results are exactly the same as in Noble [9].

## 4.2   Admissible Inconsistency

It seems reasonable to determine the upper bound for admissible inconsistency as a $p$-quantile of the theoretical random degree distribution. Those values for

$p = 0.05$ are given in Table 2 in the column *theoretical*. The quantiles of the generated data are given in the column *from data*.

If the number of vertices in the preference graph is $n = 3$ the distribution is not the Gumbel distribution. The reason may be in the severe restriction on the stochastic flow values, i.e $\mathcal{F}_\alpha \in \{0, 1, 2, 3, 4\}$. The randomization in this case should be recalculated in a slightly different way, perhaps with less restrictions.

## 5    Condition of Order Preservation

We say that the potential $X$ satisfies the Condition of Order Preservation (COP) if

$$\mathcal{F}_{(i,j)} > \mathcal{F}_{(k,l)} \implies X_i - X_j > X_k - X_l. \tag{15}$$

In contrast to the measure of inconsistency $\mathrm{DEG}(\mathcal{F})$, which is an 'a priori' inconsistency measure, the number of violations of COP may be regarded as an 'a posteriori' measure of inconsistency which shows 'how far' the calculated potential $X$ is from the measurable value function.

For a reciprocal positive matrix $A$, we say that COP is satisfied if

$$(a_{ij} > 1 \,\&\, a_{kl} > 1 \,\&\, a_{ij} > a_{kl}) \implies \frac{w_i}{w_j} > \frac{w_k}{w_l},$$

where $w$ is the Perron eigenvector of $A$.

In this section we present the results of a statistical comparison of the number of violations of the COP (#VCOP) between EVM and PM. For this purpose we performed $10^4$ simulations of $5 \times 5$ positive reciprocal matrix. For each randomly generated reciprocal matrix we calculate its consistency index CI and #VCOP(EVM) for EVM. Then, we calculate the measure of inconsistency $\mathrm{DEG}(\mathcal{F})$ of the flow $\mathcal{F}$ defined by formula (3), and #VCOP(PM) generated by PM. The correlation matrix of the random vector (DEG, #VCOP(PM), CI, #VCOP(EVM)) is given in Table 3. The correlation between CI and #VCOP equals 0.460128 while the correlation between DEG and #VCOP equals 0.811266 which suggests that DEG may better predict the #VCOP than CI (in average). We do not impose the zero value of #VCOP as a standard, we just want to say that #VCOP gives some new information about the inconsistency from the metric topology's point of view.

**Table 3.** Correlation matrix of the random vector (DEG, #VCOP(PM), CI, #VCOP(EVM))

|  | DEG | #VCOP(PM) | CI | #VCOP(EVM) |
|---|---|---|---|---|
| DEG | 1. | 0.811266 | 0.55142 | 0.817288 |
| #VCOP(PM) |  | 1. | 0.449875 | 0.950875 |
| CI |  |  | 1. | 0.460128 |
| #VCOP(EVM) |  |  |  | 1. |

## 5.1   Post Festum

The consistency ratio has been criticized because it allows contradictory judgments in matrices (Bana e Costa and Vansnick [2]) or rejects reasonable matrices (Karapetrovic and Rosenbloom [7]). Several authors (Wang-Chin-Luo [11], Korhonen [8]) argued that the implicit information about priority judgments in the AHP matrix should be taken into account and that the above criticism is not justified. That implicit information, according to them, is of the form $a_{ij}a_{jk}$ which is the element of $A^2$. But $A^2$ is 'more consistent'[6] than $A$ regarding the iterative procedure $w_n = A^n e/\|A^n e\|, n \in \mathbb{N}$ of obtaining the Perron vector. The consistency of $A$ should not be measured by the 'consistency' of $A^2$. It seems that the criticism of the criticism is not well-founded either.

During the randomization process we found an $4 \times 4$ AHP matrix with $\lambda_{max} = 4.107$ and the consistency ratio CR $= 0.04$, while its inconsistency degree is DEG $= 73.278$. According to Table 2, the upper bound for admissible inconsistency is 5.3 degrees. Here is the matrix:

$$A = \begin{pmatrix} 1. & 1.024 & 0.852 & 1.521 \\ 0.976 & 1. & 1.41 & 0.719 \\ 1.174 & 0.709 & 1. & 1.197 \\ 0.658 & 1.391 & 0.835 & 1. \end{pmatrix}$$

The preference graph associated with this matrix is given in Figure 3.

The reader who is more familiar with graphs may immediately conclude from Figure 3 that the inconsistency degree of the flow is high. First, a spanning tree should be chosen. In our example, the maximal spanning tree of the graph is drawn (solid line), together with the corresponding chords (dashed lines) which generate the base in the cycle space. Here we have 3 basic cycles, one for each chord. For example, the chord $2 \to 4$ generates the cycle $2 \to 4 \to 3 \to 1 \to 2$, and the sum of the flow components along this cycle equals $0.476 + 0.259 - 0.231 - 0.034 = 0.47$. This number is written in the parentheses beside the flow component ($\Sigma$-value). This value is also the scalar product of the flow with this cycle.

According to the decomposition (6), the arc space may be decomposed as the orthogonal sum of the cycle space and the range space of the incidence matrix $M$ which is given bellow. To be a bit more precise let us fix the canonical base in the arc space: $(2 \to 1, 1 \to 3, 4 \to 3, 3 \to 2, 2 \to 4, 4 \to 1)$. The first 3 elements of the base are the arcs of the spanning tree, the rest are the chords. The cycle space is generated by the first three columns of the matrix $B$

---

[6] It is meaningless to speak about the consistency index of $A^2$ because it is not reciprocal and its Perron root is generaly smaller that the dimension of $A$. But its columns are closer to the Perron eigenvector than those of $A$ and from this point of view we may say that it is more consistent than $A$.

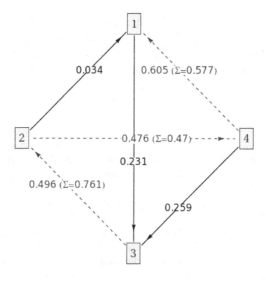

**Fig. 3.** The flow obtained from the matrix $A$

$$
M = \begin{pmatrix}
1 & -1 & 0 & 0 \\
-1 & 0 & 1 & 0 \\
0 & 0 & 1 & -1 \\
0 & 1 & -1 & 0 \\
0 & -1 & 0 & 1 \\
1 & 0 & 0 & -1
\end{pmatrix}
\qquad
B = \begin{pmatrix}
1 & -1 & 0 & 1 & -1 & 0 \\
1 & -1 & 1 & -1 & 0 & 0 \\
0 & 1 & -1 & 0 & 0 & -1 \\
1 & 0 & 0 & 0 & 1 & 0 \\
0 & 1 & 0 & 0 & -1 & 1 \\
0 & 0 & 1 & 1 & 0 & -1
\end{pmatrix}
$$

and the other three columns form the base of the column space of the incidence matrix. For example, the first column of $B$ is the cycle $3 \to 2 \to 1 \to 3$
determined by the chord $3 \to 2$. $\Sigma$-values are the scalar products of the flow

$$\mathcal{F} = (0.034, 0.231, 0.259, 0.496, 0.476, 0.605)$$

with the first 3 columns of the matrix $B$. Because of the high $\Sigma$-values it is
obvious that the flow inconsistency is also high. The precise calculation gives
the inconsistency of 73.278 degrees.

## 6   Conclusion

This paper explains the properties of the Potential Method and the randomization procedure for obtaining the upper bound for admissible inconsistency of the
input data. The upper bound is determined as a 0.5-quantile of the theoretical
distribution if DEG which is recognized as the Gumbel distribution (Table 2).

A comparison with the Eigenvalue Method is given, regarding the correlation of the inconsistency measure and the number of violations of the Condition of Order Preservation (COP). The Potential Method and the Eigenvalue Method are 'equally good' from the point of view of the number of violations (#VCOP) of the COP. On the other side, the inconsistency measure of PM correlates better with #VCOP than the consistency index of AHP (Table 3).

# References

1. Alonso, J.A., Lamata, M.T.: Consistency in the analytic hierarchy process: a new approach. International Journal of Uncertainty, Fuzziness and Knowledge- Based Systems 14(4), 445–459 (2006)
2. Banae Costa, C.A., Vansnick, J.C.: A critical analysis of the eigenvalue method used to derive priorities in AHP. EJOR 187(3), 1422–1428 (2008)
3. Condorcet, M.: Essai sur l'application de l'analyse à la probabilité des décisions rendues à la pluralité des voix. Imprimerie Royale, Paris (1785)
4. Crawford, G., Williams, C.: A note on the analysis of subjective judgement matrices. Journal of Mathematical Psychology 29, 387–405 (1985)
5. Čaklović, L.: Stochastic preference and group decision. Metodološki zvezki (Advances in Methodology and Statistics) 2(1), 205–212 (2005)
6. French, S.: Decision theory - An introduction to the mathematics of rationality. Ellis Horwood, Chichester (1998)
7. Karapetrović, S., Rosenbloom, E.S.: A quality control approach to consistency paradoxes in AHP. EJOR 119(3), 704–718 (1999)
8. Korhonen, P.: Settings "condition of order preservation" requirements for the priority vector estimate in AHP is not justified (2008) Working paper
9. Noble, E.E., Sanchez, P.P.: A note on the information content of a consistent pairwise comparison judgment matrix of an AHP decision maker. Theory and Decision 34(2), 99–108 (1993)
10. Saaty, T.L.: Fundamentals of the Analytic Hierarchy Process. RWS Publications, 4922 Ellsworth Avenue (2000)
11. Wang, Y.M., Chin, K.S., Luo, Y.: Aggregation of direct and indirect judgments in pairwise comparison matrices with a re-examination of the criticisms by Bana e Costa and Vansnick. Information Sciences 179(3), 329–337 (2009)

# Hierarchical Bipolar Sugeno Integral Can Be Represented as Hierarchical Bipolar Choquet Integral

Katsushige Fujimoto[1] and Michio Sugeno[2]

[1] Fukushima University, 1 Kanayagawa Fukushima, 906-1296, Japan
fujimoto@sss.fukushima-u.ac.jp
[2] European centre for Soft Computing, Gonzalo Gutiérrez Quirós S/N 33600 Mieres, Spain
michio.sugeno@softcomputing.es

**Abstract.** Several types of hierarchical fuzzy integral models in decision theory have been proposed and been investigated by many researchers. Some of them aim to simplify the models. Others aim to build a high modeling capability. Recently, the notion of *hierarchical bipolar Sugeno integral* has been proposed by Sugeno and Nakama in order to represent/model almost all *admissible* preference orderings. It is known that the ordinary Sugeno integral can be represented as a hierarchical Choquet integral. However, it has never been known whether the hierarchical bipolar Sugeno integral can be represented as some types of Choquet integrals, or not. This paper will show that the hierarchical bipolar Sugeno integral can be represented as a hierarchical bipolar Choquet integral.

**Keywords:** hierarchical Sugeno integral, bi-cooperative game, Choquet integral.

## 1 Introduction

The Choquet and Sugeno integrals are one of the most important integrals with respect to fuzzy measures. In decision theory, the Choquet integral model is well-known and used as a *cardinal* preference model and the Sugeno integral an *ordinal* one. Since 1995, hierarchical Choquet integrals have been studied by many researchers (e.g., Sugeno, Fujimoto, and Murofushi [12,15], Mesiar, Vivona, and Benvenuti [1,9]), Narukawa and Torra [14,17] ...). Recently, the notion of hierarchical bipolar Sugeno integral, i.e., hierarchical Sugeno integral with respect to bi-capacities, has been proposed by Sugeno and Nakama [16,13] in order to model/represent almost all *admissible*, in a natural/common sense, preference orderings. While, Narukawa and Torra [14] investigated relations with the Choquet and Sugeno integrals. They have proved that the Sugeno integral can be represented as a (2-step) hierarchical Choquet integral (with a constant). However, it has never been known whether the bipolar Sugeno integral can be represented as some types of the Choquet integrals, or not. In this paper, we shall show that the hierarchical bipolar Sugeno integral can be represented as a hierarchical bipolar Choquet integral.

## 2 Preliminaries

Throughout this paper we use the following notations and conventions. Let $N$ be a non empty finite set, $N := \{1, \cdots, n\}$, $U_m$ and $B_m$ a finite set of integers, $U_m := \{0, 1, \cdots, m\}$

V. Torra et al. (Eds.): MDAI 2012, LNAI 7647, pp. 115–126, 2012.

and $B_m = \{-m, \cdots, -1, 0, 1, \cdots, m\}$. For a function $f : N \to U_m$ (resp., $f : N \to B_m$), we denote $\sigma_f$ a permutation on $N$ satisfying that

$$f(\sigma_f(1)) \leq \cdots \leq f(\sigma_f(n)) \quad (\text{resp.}, |f(\sigma_f(1))| \leq \cdots \leq |f(\sigma_f(n))|)$$

with convention $f(\sigma_f(0)) = 0$, and $A_{\sigma_f(k)}$, $A^+_{\sigma_f(k)}$, and $A^-_{\sigma_f(k)}$ the subset of $N$ such as

$$A_{\sigma_f(k)} := \{\sigma_f(k), \cdots, \sigma_f(n)\},$$
$$A^+_{\sigma_f(k)} := A_{\sigma_f(k)} \cap \{i \in N \mid f(i) \geq 0\}, \quad \text{and} \quad A^-_{\sigma_f(k)} := A_{\sigma_f(k)} \setminus A^+_{\sigma_f(k)}$$

for any $k \in N$, respectively. we denote $\mathcal{P}(N) := 2^N = \{S \mid S \subseteq N\}$ and $Q(N) := 3^N = \{(A_1, A_2) \mid A_1, A_2 \in \mathcal{P}(N), A_1 \cap A_2 = \emptyset\}$. Binary operators $\vee, \wedge$ on $U_m$ and $\curlyvee, \curlywedge$ on $B_m$ is defined as follows:    For any $a, b \in U_m$ and $c, d \in B_m$,

$$a \vee b := \max\{a, b\} \quad \text{and} \quad a \wedge b := \min\{a, b\},$$
$$c \curlyvee d := sign(c + d) \cdot (|c| \vee |d|) \quad \text{and} \quad c \curlywedge d := sign(c \cdot d) \cdot (|c| \wedge |d|),$$

where $sign(c) := -1$ if $c < 0$, $:= 0$ if $c = 0$, and $:= 1$ if $c > 0$. These operators $\curlyvee$ and $\curlywedge$ have been introduced by Grabisch [4,5] in order to define the Sugeno integral with respect to a bipolar scale. Moreover, $\bigsqcup : B_m \times \cdots \times B_m \to B_m$ is defined as follows:

$$\bigsqcup_{i=1}^{n} b_i := \min_{i \in \{1, \cdots, n\}} b_i \curlyvee \max_{i \in \{1, \cdots, n\}} b_i.$$

## 2.1    Fuzzy Measures and Bi-Capacities

**Definition 1 (fuzzy measure [13]).** A function $\mu : \mathcal{P}(N) \to U_m$ is a *fuzzy measure* (or capacity) on $\mathcal{P}(N)$ to $U_m$ if it satisfies the following two conditions:
   (i)   $\mu(\emptyset) = 0$,
   (ii)   $E, F \in \mathcal{P}(N), E \subseteq F \Rightarrow \mu(E) \leq \mu(F)$.
$\mu$ is said to be *normalized* if $\mu$ satisfies that $\mu(N) = \max U_m = m$.

**Definition 2 (bi-capacity [6]).** When equipped with the following order $\sqsubseteq$: for arbitrary $(A_1, A_2), (B_1, B_2) \in Q(N)$,

$$(A_1, A_2) \sqsubseteq (B_1, B_2) \quad \text{iff} \quad A_1 \subseteq B_1, A_2 \supseteq B_2,$$

a function $v : Q(N) \to B_m$ is a *bi-capacity* on $Q(N)$ to $B_m$ if it satisfies the following two conditions:
   (i)   $v(\emptyset, \emptyset) = 0$,
   (ii)   $A, B \in Q(N), A \sqsubseteq B \Rightarrow v(A) \leq v(B)$.

If $v$ satisfies that $v(N, \emptyset) = \max B_m = m$ and $v(\emptyset, N) = \min B_m = -m$, $v$ is said to be *normalized*. If $v$ satisfies the condition (i), $v$ is said to be a *bi-cooperative game* [2].

**Definition 3 (the Möbius transform [6]).** To any bi-cooperative game $v : Q(N) \to B_m$, another function (bi-cooperative game) $m^v : Q(N) \to B_l$ for some integer $l$ can be associated by

$$v(A_1, A_2) = \sum_{(B_1, B_2) \sqsubseteq (A_1, A_2)} m^v(B_1, B_2) \quad \forall (A_1, A_2) \in Q(N).$$

This correspondence proves to be one-to-one, since conversely, for $(A_1, A_2) \in Q(N)$,

$$m^v(A_1, A_2) = \sum_{\substack{B_1 \subseteq A_1 \\ A_2 \subseteq B_2 \subseteq A_1^c}} (-1)^{|A_1 \setminus B_1| + |B_2 \setminus A_2|} v(B_1, B_2).$$

**Definition 4 (support, null set).** A subset $S \subseteq N$ is said to be a *support* of $N$ with respect to a bi-cooperative game $v$ on $Q(N)$ if

$$v(A, B) = v(A \cap S, B \cap S) \quad \forall (A, B) \in Q(N).$$

So, $N \setminus S$ is said to be a *null set*.

**Definition 5 (relative bi-cooperative game).** For a bi-cooperative game $v$ on $Q(N)$ and a subset $S$ on $N$, the bi-cooperative game $w$ on $Q(S)$ is said to be the *relative bi-cooperative game* of $v$ on $Q(S)$ if

$$w(A, B) = v(A, B) \quad \forall (A, B) \in Q(S).$$

**Proposition 1.** [3] *Let $v$ be a bi-cooperative game on $Q(N)$. The following two conditions are equivalent to each other:*

(i)   $v$ is a bi-capacity,

(ii)   $\displaystyle\sum_{(C_1, C_2) \sqsubseteq (B_1, B_2) \sqsubseteq (A_1, A_2)} m^v(B_1, B_2) \geq 0$

for all $(C_1, C_2) \in Q(N)$ such as $|C_2| = n - 1$ and all $(A_1, A_2) \in Q(N)$.

**Proposition 2.** *For any bi-cooperative game $v$ on $Q(N)$, there exist two bi-capacities $v_1$ and $v_2$ on $Q(N)$ such that*

$$v(A_1, A_2) = v_1(A_1, A_2) - v_2(A_1, A_2) \quad \forall (A_1, A_2) \in Q(N).$$

**Proof.** Let $m^v$ be the Möbius transform of $v$. Then, we define two functions $m_1$ and $m_2$ on $Q(N)$ via $v$ as follows:

$$m_1(A_1, A_2) := \begin{cases} m^v(A_1, A_2) & \text{if } m^v(A_1, A_2) \geq 0 \text{ and } A_2 \neq N. \\ 0 & \text{if } m^v(A_1, A_2) < 0 \text{ and } A_2 \neq N, \end{cases}$$

$$m_2(A_1, A_2) := \begin{cases} -m^v(A_1, A_2) & \text{if } m^v(A_1, A_2) < 0 \text{ and } A_2 \neq N. \\ 0 & \text{if } m^v(A_1, A_2) \geq 0 \text{ and } A_2 \neq N, \end{cases}$$

$$m_1(\emptyset, N) := - \sum_{\substack{(B_1, B_2) \sqsubseteq (\emptyset, \emptyset) \\ B_2 \neq N}} m_1(B_1, B_2), \quad m_2(\emptyset, N) := - \sum_{\substack{(B_1, B_2) \sqsubseteq (\emptyset, \emptyset) \\ B_2 \neq N}} m_2(B_1, B_2).$$

Through these $m_1$ and $m_2$, we define $v_1$ and $v_2$ as

$$v_i(A_1, A_2) := \sum_{(B_1, B_2) \sqsubseteq (A_1, A_2)} m_i(B_1, B_2) \quad \forall i \in \{1, 2\}.$$

Then, it follows from Proposition 1 that both $v_1$ and $v_2$ are bi-capacities. Moreover, for any $(A_1, A_2) \in Q(N)$,

$$
\begin{aligned}
v_1(A_1, A_2) - v_2(A_1, A_2) &= \sum_{(B_1, B_2) \sqsubseteq (A_1, A_2)} m_1(B_1, B_2) - \sum_{(B_1, B_2) \sqsubseteq (A_1, A_2)} m_2(B_1, B_2) \\
&= \sum_{(B_1, B_2) \sqsubseteq (A_1, A_2)} m^v(B_1, B_2) = v(A_1, A_2),
\end{aligned}
$$

from the definition of the Möbius transform (See, Definition 3).    □

## 2.2 The Sugeno Integral and the Choquet Integral

Let $\mathfrak{F}(N, U_m)$ be the set of all fuzzy measures on $\mathcal{P}(N)$ to $U_m$, $\mathfrak{B}(N, B_m)$ the set of all bi-capacities on $Q(N)$ to $B_m$, and $\mathfrak{B}'(N, B_m)$ the set of all bi-cooperative games on $Q(N)$ to $B_m$.

**Definition 6 (the Sugeno and Choquet integral).** The *Sugeno* (resp., *Choquet*) integral, $S_\mu(f)$ (resp., $C_\mu(f)$), of a function $f : N \rightarrow U_m$ with respect to $\mu \in \mathfrak{F}(N, U_m)$ is given by

$$
S_\mu(f) := \bigvee_{k=1}^{n} \left[ f(\sigma_f(k)) \wedge \mu(A_{\sigma_f(k)}) \right], \tag{1}
$$

$$
C_\mu(f) := \sum_{k=1}^{n} \left( f(\sigma_f(k)) - f(\sigma_f(k-1)) \right) \cdot \mu(A_{\sigma_f(k)}). \tag{2}
$$

**Definition 7 (the bipolar Sugeno and Choquet integral [4,6]).** The *Sugeno* (resp., *Choquet*) integral, $BS_v(f)$ (resp., $BC_v(f)$), of a function $f : N \rightarrow B_m$ with respect to $v \in \mathfrak{B}'(N, B_m)$ is given by

$$
BS_v(f) := \bigsqcup_{k=1}^{n} \left[ |f(\sigma_f(k))| \wedge v(A^+_{\sigma_f(k)}, A^-_{\sigma_f(k)}) \right], \tag{3}
$$

$$
BC_v(f) := \sum_{k=1}^{n} \left( |f(\sigma_f(k))| - |f(\sigma_f(k-1))| \right) \cdot v(A^+_{\sigma_f(k)}, A^-_{\sigma_f(k)}). \tag{4}
$$

**Proposition 3.** *Suppose that $S \subseteq N$ is a support of $N$ with respect to $v \in \mathfrak{B}'(N, B_m)$ and that $w$ is the relative bi-cooperative game of $v$ on $Q(S)$. Then for any $f : N \rightarrow B_m$,*

$$
BC_v(f) = BC_w(f|_S)
$$

*where $f|_S$ is the function on $S$ such that $f|_S(i) = f(i) \; \forall i \in S$.*

**Proof.** It suffice to prove the case where supports $S$ are represented as

$$
S = N \setminus \{k\}
$$

for some null $k \in N$ with respect to $v$. When we denote $g(i)$ instead of $f|_S(i)$, we have a permutation $\sigma_g$ on $S = N \setminus \{k\}$ such as

$$
\sigma_g(i) = \begin{cases} \sigma_f(i) & \text{if } i < \sigma_f^{-1}(k), \\ \sigma_f(i+1) & \text{if } i > \sigma_f^{-1}(k). \end{cases}
$$

Then, it is easy to verify, from the fact that $\{k\}$ is a null set, that

$$v(A^+_{\sigma_f(\sigma_f^{-1}(k))}, A^-_{\sigma_f(\sigma_f^{-1}(k))}) = v(A^+_{\sigma_f(\sigma_f^{-1}(k)+1)}, A^-_{\sigma_f(\sigma_f^{-1}(k)+1)}).$$

Under above notations with convention $(A^+_{\sigma_f(n+1)}, A^-_{\sigma_f(n+1)}) = (A^+_{\sigma_g(n)}, A^-_{\sigma_g(n)}) = (\emptyset, \emptyset)$,

$$
\begin{aligned}
BC_v(f) &= \sum_{i=1}^{n} |f(\sigma_f(i))| \left[ v(A^+_{\sigma_f(i)}, A^-_{\sigma_f(i)}) - v(A^+_{\sigma_f(i+1)}, A^-_{\sigma_f(i+1)}) \right] \\
&= \sum_{i<\sigma_f^{-1}(k)-1} |f(\sigma_f(i))| \left[ v(A^+_{\sigma_f(i)}, A^-_{\sigma_f(i)}) - v(A^+_{\sigma_f(i+1)}, A^-_{\sigma_f(i+1)}) \right] \\
&\quad + |f(\sigma_f(\sigma_f^{-1}(k) - 1))| \left[ v(A^+_{\sigma_f(\sigma_f^{-1}(k)-1)}, A^-_{\sigma_f(\sigma_f^{-1}(k)-1)}) - v(A^+_{\sigma_f(\sigma_f^{-1}(k))}, A^-_{\sigma_f(\sigma_f^{-1}(k))}) \right] \\
&\quad + |f(\sigma_f(\sigma_f^{-1}(k)))| \left[ v(A^+_{\sigma_f(k)}, A^-_{\sigma_f(k)}) - v(A^+_{\sigma_f(\sigma_f^{-1}(k)+1)}, A^-_{\sigma_f(\sigma_f^{-1}(k)+1)}) \right] \\
&\quad + \sum_{i>\sigma_f^{-1}(k)} |f(\sigma_f(i))| \left[ v(A^+_{\sigma_f(i)}, A^-_{\sigma_f(i)}) - v(A^+_{\sigma_f(i+1)}, A^-_{\sigma_f(i+1)}) \right] \\
&= \sum_{i<\sigma_f^{-1}(k)-1} |f(\sigma_f(i))| \left[ v(A^+_{\sigma_f(i)}, A^-_{\sigma_f(i)}) - v(A^+_{\sigma_f(i+1)}, A^-_{\sigma_f(i+1)}) \right] \\
&\quad + |f(\sigma_f(\sigma_f^{-1}(k) - 1))| \left[ v(A^+_{\sigma_f(\sigma_f^{-1}(k)-1)}, A^-_{\sigma_f(\sigma_f^{-1}(k)-1)}) - v(A^+_{\sigma_f(\sigma_f^{-1}(k)+1)}, A^-_{\sigma_f(\sigma_f^{-1}(k)+1)}) \right] \\
&\quad + \sum_{i>\sigma_f^{-1}(k)} |f(\sigma_f(i))| \left[ v(A^+_{\sigma_f(i)}, A^-_{\sigma_f(i)}) - v(A^+_{\sigma_f(i+1)}, A^-_{\sigma_f(i+1)}) \right] \\
&= \sum_{i<\sigma_f^{-1}(k)-1} |g(\sigma_g(i))| \left[ v(A^+_{\sigma_g(i)}, A^-_{\sigma_g(i)}) - v(A^+_{\sigma_g(i+1)}, A^-_{\sigma_g(i+1)}) \right] \\
&\quad + |g(\sigma(\sigma_f^{-1}(k) - 1))| \left[ v(A^+_{\sigma_g(\sigma_f^{-1}(k)-1)}, A^-_{\sigma_g(\sigma_f^{-1}(k)-1)}) - v(A^+_{\sigma_g(\sigma_f^{-1}(k)+1)}, A^-_{\sigma_g(\sigma_f^{-1}(k)+1)}) \right] \\
&\quad + \sum_{n-1\geq i\geq\sigma_f^{-1}(k)} |g(\sigma_g(i))| \left[ v(A^+_{\sigma_g(i)}, A^-_{\sigma_g(i)}) - v(A^+_{\sigma_g(i+1)}, A^-_{\sigma_g(i+1)}) \right] \\
&= \sum_{i=1}^{n-1} |g(\sigma_g(i))| \left[ w(A^+_{\sigma_g(i)}, A^-_{\sigma_g(i)}) - w(A^+_{\sigma_g(i+1)}, A^-_{\sigma_g(i+1)}) \right] = BC_w(g) = BC_w(f|_S). \quad \square
\end{aligned}
$$

**Definition 8 (hierarchical bipolar Sugeno and Choquet integral [13]).** Let $L$ be a positive integer and $\{N_j\}_{j\in\{1,\cdots,L+1\}}$ a family of non empty sets with $N_1 = N$, $N_{L+1} = \{1\}$, and $N_j := \{1, \cdots, n_j\}$ for an integer $n_j$ for all $j \in \{1, \cdots, L+1\} \setminus \{1, L+1\}$. For each $j \in \{1, \cdots, L\}$, let $\{v_i^j\}_{i\in N_{j+1}}$ be a set of bi-cooperative games on $Q(N_j)$ to $B_{m_j}$, i.e., $v_i^j \in \mathfrak{B}'(N_j, B_{m_j})$ for any $i \in N_{j+1}$. Then,

$$\mathfrak{H}^L := (\{N_j\}_{j\in\{1,\cdots,L+1\}}, \{v_i^j\}_{i\in N_{j+1}, j\in\{1,\cdots,L\}})$$

is called a *system of L-step hierarchical bi-cooperative games*. Then, the L-step hierarchical bipolar Sugeno (resp., Choquet) integral, $HBS_{\mathfrak{H}^L}(f)$ (resp., $HBC_{\mathfrak{H}^L}(f)$ ), of a function $f : N \to B_m$ with respect to $\mathfrak{H}^L$ is given by the following procedures (e.g., see Fig. 1.):

   (i)   $F_1(i) := f(i)$  $\forall i \in N(= N_1)$,
   (ii)  $F_j(i) := BS_{v_i^{j-1}}(F_{j-1})$  (resp., $F_j(i) := BC_{v_i^{j-1}}(F_{j-1})$)
$\forall i \in N_j,$  $j \in \{1, \cdots, L+1\} \setminus \{1, L+1\}$.
   (iii)  $HBS_{\mathfrak{H}^L}(f) := BS_{v_1^L}(F_L)$  (resp., $HBC_{\mathfrak{H}^L}(f) := BC_{v_1^L}(F_L)$).

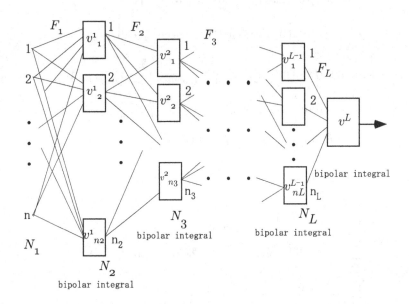

**Fig. 1.** Hierarchical bipolar Sugeno and/or Choquet integral

## 3   Lemmas

### 3.1   Binary Operators ∧, ∨, ⅄, ⅄ and the Bipolar Choquet Integral

**Lemma 1.** *Let $N := \{1, 2\}$. For any $a, b \in B_m$, $a > 0$, we define a function $f : N \to B_m$ as $f(1) = a$ and $f(2) = b$. Then*

$$a \wedge b = BC_v(f),$$

*where*

$$v(A, B) = \begin{cases} 1 & \text{if } A = N, \\ 0 & \text{if } A = \emptyset \text{ or } A \neq N, B = \emptyset, \\ -1 & \text{otherwise.} \end{cases}$$

**Proof.** In the case that $b > a$, i.e., $a \wedge b = a$, $BC_v(f) = a \cdot v(N, \emptyset) + (b - a)v(2, \emptyset) = a$.
   In the case that $a \geq b \geq 0$, i.e., $a \wedge b = b$, $BC_v(f) = b \cdot v(N, \emptyset) + (a - b)v(1, \emptyset) = b$.
   In the case that $-a \leq b \leq 0$, i.e., $a \wedge b = b$, $BC_v(f) = -b \cdot v(1, 2) + (a + b)v(1, \emptyset) = b$.
   In the case that $b \leq -a$, i.e., $a \wedge b = -a$, $BC_v(f) = a \cdot v(1, 2) + (-b - a)v(\emptyset, 2) = -a$.□

**Lemma 2.** *Let* $N := \{1, 2\}$. *For any* $a, b \in B_m$, *we define a function* $f : N \to B_m$ *as* $f(1) = a$ *and* $f(2) = b$. *Then* $a \vee b$ *can be represented as a hierarchical bipolar Choquet integral of* $f$ *as follows:*

(i)   Define a set of non empty set $N_1$ and $N_2$ as $N_1 = N$ and $N_2 = \{1, 2, 3\}$.
(ii)   Define a set of bi-cooperative games $\{v_j^1\}_{j \in N_2}$ in $\mathfrak{B}'(N, B_{2m})$ and $v^2$ in $\mathfrak{B}'(N, B_1)$ as

$$v_1^1(A, B) = \begin{cases} 0 & \text{if } A = \emptyset, \\ 1 & otherwise, \end{cases} \quad v_2^1(A, B) = \begin{cases} 0 & \text{if } B = \emptyset, \\ -1 & otherwise, \end{cases}$$

$$v_3^1(A, B) = (|A| - |B|) \cdot m, \quad v^2(A, B) = \begin{cases} 1 & \text{if } A = \{1, 3\}, \\ -1 & \text{if } B = \{2, 3\}, \\ 0 & otherwise, \end{cases}$$

respectively.

(iii)   Define the function $F_2 : N_2 \to B_m$ as $F_2(i) := BC_{v_i^1}(f) \ \forall i \in N_2$.

*Then, we have that* $a \vee b = BC_{v^2}(F_2)$.

**Proof.** Let $f^+(i) := \max\{f(i), 0\}$ and $f^-(i) := \min\{f(i), 0\}$, for $i \in N_1$.

**Claim 1** : $F_2(1) = \max_{i \in N} f^+(i)$.
  Now we will prove the claim 1. In the case that $b \geq a \geq 0$, i.e., $\max f^+ = b$,
    $F_2(1) = BC_{v_1^1}(f) = (b - a) \cdot v_1^1(2, \emptyset) + a \cdot v_1^1(12, \emptyset) = b$.
  In the case that $a > b \geq 0$, i.e., $\max f^+ = a$,
    $F_2(1) = BC_{v_1^1}(f) = (a - b) \cdot v_1^1(1, \emptyset) + b \cdot v_1^1(12, \emptyset) = a$.
  In the case that $a > -b \geq 0$, i.e., $\max f^+ = a$,
    $F_2(1) = BC_{v_1^1}(f) = (a + b) \cdot v_1^1(1, \emptyset) + (-b) \cdot v_1^1(1, 2) = a$.
  In the case that $-b \geq a \geq 0$, i.e., $\max f^+ = a$,
    $F_2(1) = BC_{v_1^1}(f) = (-b - a) \cdot v_1^1(\emptyset, 2) + a \cdot v_1^1(1, 2) = a$.
  In the case that $b \geq -a \geq 0$, i.e., $\max f^+ = b$,
    $F_2(1) = BC_{v_1^1}(f) = (b + a) \cdot v_1^1(2, \emptyset) + (-a) \cdot v_1^1(2, 1) = b$.
  In the case that $-a > b \geq 0$, i.e., $\max f^+ = b$,
    $F_2(1) = BC_{v_1^1}(f) = (-a - b) \cdot v_1^1(\emptyset, 1) + b \cdot v_1^1(2, 1) = b$.
  In the case that $-a > -b \geq 0$, i.e., $\max f^+ = 0$,
    $F_2(1) = BC_{v_1^1}(f) = (-a + b) \cdot v_1^1(\emptyset, 1) + (-b) \cdot v_1^1(\emptyset, 12) = 0$.
  In the case that $-b > -a \geq 0$, i.e., $\max f^+ = 0$,
    $F_2(1) = BC_{v_1^1}(f) = (-b + a) \cdot v_1^1(\emptyset, 2) + a \cdot v_1^1(\emptyset, 12) = 0$.

**Claim 2** : $F_2(2) = \min_{i \in N} f^-(i)$.
  This claim can be verified similarly to the proof of Claim 1.

**Claim 3** : $F_2(3) = (a + b) \cdot m$.
  This claim can also be verified similarly to the proof of Claim 1.

Here, we consider the following three cases:

$$|F_2(1)| > |F_2(2)|, \quad |F_2(1)| < |F_2(2)|, \quad and \quad |F_2(1)| = |F_2(2)|.$$

In the case that $|F_2(1)| > |F_2(2)|$,
i.e., $|\max f^+| > |\min f^-|$ and $F_2(3) \geq F_2(1) > -F_2(2) \geq 0$. Then,

$$BC_{v^2}(F_2) = (F_2(3) - F_2(1)) \cdot v^2(3, \emptyset) + (F_2(1) + F_2(2)) \cdot v^2(13, \emptyset) - F_2(2) \cdot v^2(13, 2)$$
$$= F_2(1) = \max f^+ = a \vee b, \quad \text{since } \max f^+ \geq 0 \text{ and } |\max f^+| > |\min f^-|.$$

In the case that $|F_2(1)| < |F_2(2)|$,
i.e., $|\max f^+| < |\min f^-|$ and $-F_2(3) \geq -F_2(2) > F_2(1) \geq 0$. Then,

$$BC_{v^2}(F_2) = (-F_2(3) + F_2(2)) \cdot v^2(\emptyset, 3) + (-F_2(2) - F_2(1)) \cdot v^2(\emptyset, 23) + F_2(1) \cdot v^2(1, 23)$$
$$= F_2(2) = \min f^- = a \vee b, \quad \text{since } \min f^- \leq 0 \text{ and } |\max f^+| < |\min f^-|.$$

In the case that $|F_2(1)| = |F_2(2)|$,
i.e., $|\max f^+| = |\min f^-|$ and $F_2(1) = -F_2(2) \geq F_2(3) = 0$. Then,

$$BC_{v^2}(F_2) = F_2(1) \cdot v^2(1, 2) = 0 = a \vee b, \quad \text{since } a = \max f^+ = -\min f^- = -b. \quad \square$$

**Lemma 3.** *There exist bi-cooperative games $v$ and $w \in \mathcal{B}'(N, B_m)$ such that, for any function $f : N \to B_m$,*

$$\max_{i \in N} f^+(i) = BC_v(f) \quad and \quad \min_{i \in N} f^-(i) = BC_w(f),$$

*where $f^+(i) = \max\{f(i), 0\}$ and $f^-(i) = \min\{f(i), 0\}$ for $i \in N$.*

**Proof.** We put $v$ and $w$ as

$$v(A, B) = \begin{cases} 1 & \text{if } A \neq \emptyset, \\ 0 & \text{otherwise,} \end{cases} \quad \text{and} \quad w(A, B) = \begin{cases} -1 & \text{if } B \neq \emptyset, \\ 0 & \text{otherwise.} \end{cases}$$

Then, it is easy to verify that

$$\max_{i \in N} f^+(i) = BC_v(f), \quad \min_{i \in N} f^-(i) = BC_w(f). \quad \square$$

### 3.2 Ordinary Bipolar Sugeno Integral and Hierarchical Bipolar Choquet Integral

**Lemma 4.** *There exists a system of 4-step hierarchical bi-cooperative games $\mathfrak{S}^4$ such that, for any function $f : N \to B_m$, $\bigsqcup_{i \in N} f(i)$ can be represented as a hierarchical bipolar Choquet integral of $f$ with respect to $\mathfrak{S}^4$, i.e.,*

$$\bigsqcup_{i \in N} f(i) = HBC_{\mathfrak{S}^4}(f).$$

**Proof.** Let $N_1 = N$, $N_2 = \{1, \cdots, 2n\}$, and $N_3 = \{1, 2\}$. Put $\{v_j^1\}_{j \in N_2}$ as

$$v_j^1(A, B) = \begin{cases} 1 & \text{if } A \ni j, \\ 0 & \text{otherwise} \end{cases} \text{ if } j \leq n, \quad v_j^1(A, B) = \begin{cases} -1 & \text{if } B \ni j - n, \\ 0 & \text{otherwise} \end{cases} \text{ if } j > n$$

and put $\{v_j^2\}_{j \in N_3}$ as

$$v_1^2(A, B) = \begin{cases} 1 & \text{if } A \neq \emptyset, \\ 0 & \text{otherwise,} \end{cases} \quad \text{and} \quad v_2^2(A, B) = \begin{cases} -1 & \text{if } B \neq \emptyset, \\ 0 & \text{otherwise.} \end{cases}$$

Here, we define a function $F_2$ on $N_2$ as $F_2(j) := BC_{v_j^1}(f)$. Then

$$F_2(j) = \begin{cases} f^+(j) & \text{if } j \le n, \\ f^-(j-n) & \text{if } j > n. \end{cases}$$

Next, define $F_3$ on $N_3$ as $F_3(j) = BC_{v_j^2}(F_2)$. Then, it follows from Lemma 3 that

$$F_3(1) = \max_{i \in N} f^+ \quad \text{and} \quad F_3(2) = \min_{i \in N} f^-.$$

Thus,

$$\bigsqcup_{i \in N} f(i) = F_3(1) \vee F_3(2) \quad \text{since} \quad \bigsqcup_{i \in N} f(i) = \max_{i \in N} f^+ \vee \min_{i \in N} f^-.$$

Then, it follows from Lemma 2 that $\bigsqcup_{i \in N} f(i)$ can be represented as a 4-step bipolar Choquet integral.    □

**Lemma 5.** *Let* $N := \{1, \cdots, n\}$, *and* $v \in \mathfrak{B}(N, B_m)$. *Then, for any* $k \in N$, *there exists a system of 4-step bi-cooperative games,* $\mathfrak{H}_k^4$, *such that*

$$v(A_{\sigma_f(k)}^+, A_{\sigma_f(k)}^-) = HBC_{\mathfrak{H}_k^4}(g^f)$$

*for any function* $f : N \to B_m$, *where* $g^f$ *is the function on* $N \cup \{0\}$ *defined by*

$$g^f(i) = \begin{cases} f(i) & \text{if } i \in N, \\ 1 & \text{if } i = 0. \end{cases}$$

**Proof.** We consider a 4-step bipolar Choquet integral of $g^f$. Let $N_1 := N \cup \{0\} = \{0, \cdots, n\}$, $N_2 := \{0, \cdots, 2n\}$, $N_3 := \{0, \cdots, n\}$, and $N_4 := \{1, \cdots, n\}$. Here, we use the notation $v_i^{sign}$, on the domain considered, to denote, for any $p \in N$,

$$v_p^{sign}(A, B) := \begin{cases} 1 & \text{if } A \ni p, \\ -1 & \text{if } B \ni p, \\ 0 & \text{otherwise.} \end{cases}$$

We put a set of bi-cooperative games $\{v_i^1\}_{i \in N_2}$ as follows:

$$v_i^1(A, B) := v_i^{sign}(A, B) \qquad\qquad \text{if } i \le n,$$

$$v_i^1(A, B) := \begin{cases} 1 & \text{if } |A \cup B| > n + 1 - k \text{ and } A \ni i - n, \\ -1 & \text{if } |A \cup B| > n + 1 - k \text{ and } B \ni i - n, \qquad \text{if } i > n. \\ 0 & \text{otherwise,} \end{cases}$$

Here, we define a function $F_2 : N_2 \to B_m$ on $N_2$ as

$$F_2(i) := BC_{v_i^1}(g^f) \quad \forall i \in N_2.$$

Then,

$$F_2(i) = \begin{cases} g^f(i) & \text{if } i \le n, \\ g^f(i-n) \wedge |g^f(\sigma_{g^f}(k-1))| & \text{if } i > n. \end{cases}$$

Next, we would define $F_3 : N_3 \to B_m$ as

$$
F_3(i) := \begin{cases} F_2(0), \ i.e., \ 1 & \text{if } i = 0, \\ F_2(i) - F_2(n+i), \ i.e., \ g^f(i) - |g^f(\sigma_{g^f}(k-1))| & \text{if } f(i) \geq 0 \text{ and } \sigma_f(i) \geq \sigma_f(k), \\ F_2(i) - F_2(n+i), \ i.e., \ g^f(i) + |g^f(\sigma_{g^f}(k-1))| & \text{if } f(i) < 0 \text{ and } \sigma_f(i) \geq \sigma_f(k), \\ F_2(i) - F_2(n+i), \ i.e., \ 0 & \text{otherwise,} \end{cases}
$$

via some bipolar Choquet integral. To do this, $\{v_i^2\}_{i \in N_3}$ should be $v_0^2 := v_0^{sign}$ and

$$
v_i^2(A, B) = \begin{cases} 1 & \text{if } A \ni i \text{ and } A \not\ni n+i, \\ -1 & \text{if } B \ni i \text{ and } B \not\ni n+i, \\ 0 & \text{otherwise,} \end{cases}
$$

for $i \in N_2 \setminus \{0\}$. Next, we would define $F_4 : N \to B_m$ as $F_4(i) := F_3(0) \wedge F_3(i)$, i.e.,

$$
F_4(i) = \begin{cases} 1 & \text{if } g^f(i) \geq 0 \text{ and } \sigma_{g^f}(i) \geq \sigma_{g^f}(k), \quad \text{i.e, if } f(i) \geq 0 \text{ and } \sigma_f(i) \geq \sigma_f(k), \\ -1 & \text{if } g^f(i) < 0 \text{ and } \sigma_{g^f}(i) \geq \sigma_{g^f}(k), \quad \text{i.e, if } f(i) < 0 \text{ and } \sigma_f(i) \geq \sigma_f(k), \\ 0 & \text{otherwise,} \end{cases}
$$

via some bipolar Choquet integral. It is easy, from Lemma 1, to demonstrate the fact. Finally, we have that

$$
BC_v(F_4) = v(A^+_{\sigma_f(k)}, A^-_{\sigma_f(k)}).
$$

$\square$

**Lemma 6.** *Let $N := \{1, \cdots, n\}$, For any $k \in N$, there exists a bi-cooperative game $v_k \in \mathcal{B}'(N, B_m)$ such that $|f(\sigma_f(k))| = BC_{v_k}(f)$ for any function $f : N \to B_m$.*

**Proof.** We have that $|f(\sigma(k))| = BC_{v^k}(f)$ for any $f : N \to B_m$ via

$$
v^k(A, B) := \begin{cases} 1 & \text{if } |A \cup B| > n - k, \\ 0 & \text{otherwise.} \end{cases}
$$

$\square$

The next lemma is obtained immediately from Lemmas 1, 5, and 6.

**Lemma 7.** *Let $N := \{1, \cdots, n\}$, and $v \in \mathcal{B}(N, B_m)$. Then, for any $k \in N$, there exists a system of 5-step bi-cooperative games, $\mathfrak{H}_k^5$, such that*

$$
|f(\sigma_f(k))| \wedge v(A^+_{\sigma_f(k)}, A^-_{\sigma_f(k)}) = HBC_{\mathfrak{H}_k^5}(g^f)
$$

*for any function $f : N \to B_m$, where $g^f$ is the function on $N \cup \{0\}$ defined by*

$$
g^f(i) = \begin{cases} f(i) & \text{if } i \in N, \\ 1 & \text{if } i = 0. \end{cases}
$$

The next lemma, obtained immediately from Lemmas 4 and 7, shows that the bipolar Sugeno integral can be represented as a hierarchical bipolar Choquet integral.

**Lemma 8.** *Let $N := \{1, \cdots, n\}$, and $v \in \mathfrak{B}(N, B_m)$. Then, there exists a system of 9-step bi-cooperative games, $\mathfrak{H}_k^9$, such that*

$$\bigsqcup_{k \in N} \left[ |f(\sigma_f(k))| \wedge v(A^+_{\sigma_f(k)}, A^-_{\sigma_f(k)}) \right] = HBC_{\mathfrak{H}_k^9}(g^f),$$

*i.e.,*

$$BS_v(f) = HBC_{\mathfrak{H}_k^9}(g^f)$$

*for any function $f : N \to B_m$, where $g^f$ is the function on $N \cup \{0\}$ defined by $f^g(i) := f(i)$ if $i \in N$ and $:= 1$ if $i = 0$.*

## 4   Theorem

**Theorem 1.** *Let $N := \{1, \cdots, n\}$ and $\mathfrak{H}$ a system of hierarchical bi-cooperative games. Then, there exists another system of hierarchical bi-cooperative games, $\mathfrak{H}'$, such that*

$$HBS_{\mathfrak{H}}(f) = HBC_{\mathfrak{H}'}(g^f)$$

*for any function $f : N \to B_m$, where $g^f$ is the function on $N \cup \{0\}$ defined by $f^g(i) := f(i)$ if $i \in N$ and $:= 0$ if $i = 0$.*

That is, the hierarchical bipolar Sugeno integral, of any functions with respect to any systems of hierarchical bi-cooperative games, can be represented as a corresponding hierarchical bipolar Choquet integral.

**Proof.** It is easy to verify from the fact that the hierarchical bipolar Sugeno integral is represented as a hierarchical combination of bipolar Sugeno integrals and that any bipolar Sugeno integral can be represented as a hierarchical bipolar Choquet integral.

$\square$

## 5   Concluding Remarks

In this paper, we show that the hierarchical bipolar Sugeno integral of any function $f$ can be represented as a hierarchical bipolar Choquet integral of $g^f$ which is obtained by extending the domain $N$ of $f$ to $N \cup \{0\}$ (i.e., $g^f(i) = f(i)$ if $i \in N$ and $= 1$ if $i = 0$). The Choquet integral of this $g^f$ is essentially the same as *the Choquet integral with constant*, introduced by Narukawa and Torra [14], of $f$. Moreover, Torra and Narukawa [17] have demonstrated that the bipolar Choquet integral can be represented as a hierarchical CPT-type Choquet integral. That is, the hierarchical bipolar Sugeno integral can be represented as a hierarchical Choquet integral without using bi-capacities (bi-cooperative games).

**Acknowledgment.** This research was partially supported by the Ministry of Education, Culture, Sports, Science and Technology-Japan, Grant-in-Aid for Scientific Research (C), 22510134, 2010-2013. We express deep gratitude for support from all over the world to our Fukushima.

# References

1. Benvenuti, P., Mesiar, R.: A note on Sugeno and Choquet integrals. In: Proc. of IPMU 2000, Madrid, Spain, pp. 582–585 (2000)
2. Bilbao, J.M., Fernández, J.R., Jiménez, N., López, J.J.: A survey of bicooperative games. In: Pareto Optimality, Game Theory And Equilibria, pp. 187–216. Springer, New York (2008)
3. Fujimoto, K., Murofushi, T.: Some characterizations of $k$-monotonicity through the bipolar Möbius transform in bi-capacities. Journal of Advanced Computational Intelligence and Intelligent informatics 9(5), 484–495 (2005)
4. Grabisch, M.: The symmetric Sugeno integral. Fuzzy Sets and Systems 139, 473–490 (2003)
5. Grabisch, M.: The Möbius function on symmetric ordered structures and its application to capacities on finite sets. Discrete Mathematics 287(1-3), 17–34 (2004)
6. Grabisch, M., Labreuche, C.: Bi-capacities. Part I: definition, Möbius transform and interaction. Fuzzy Sets and Systems 151, 211–236 (2005)
7. Grabisch, M., Labreuche, C.: Bi-capacities. Part II: the Choquet integral. Fuzzy Sets and Systems 151, 237–259 (2005)
8. Grabisch, M., Labreuche, C.: A decade of application of the Choquet and Sugeno integrals in multicriteria decision aid. 4OR 6, 1–44 (2008)
9. Mesiar, R., Vivona, D.: Two-step integral with respect to fuzzy measure. Tetra Mt. Math. Publ. 16, 359–368 (1999)
10. Murofushi, T., Narukawa, Y.: A characterization of multi-step discrete Choquet integral. In: Proc. of Sixth International Conference on Fuzzy Sets Theory and Its Applications, Liptovský Ján, Slovakia, pp. 94–94 (2002)
11. Murofushi, T., Narukawa, Y.: A characterization of multi-level discrete Choquet integral over a finite set. In: Proc. of Seventh Workshop on Heart and Mind, pp. 33–36 (2002) (in Japanese)
12. Murofushi, T., Sugeno, M., Fujimoto, K.: Separated hierarchical decomposition of the Choquet integral. International Journal of Uncertainty, Fuzziness, and Knowledge-Based Systems 5(5), 563–585 (1997)
13. Nakama, T., Sugeno, M.: Admissibility of preferences and modeling capability of fuzzy integrals. In: Proc. of 2012 IEEE World Congress on Computational Intelligence, Brisbane, Australia (2012)
14. Narukawa, Y., Torra, V.: Twofold integral and multi-step Choquet integral. Kybernetika 40(1), 39–50 (2004)
15. Sugeno, M., Fujimoto, K., Murofushi, T.: A hierarchical decomposition of Choquet integral model. International Journal of Uncertainty. Fuzziness and Knowledge-Based Systems 3(1), 1–15 (1995)
16. Sugeno, M.: Ordinal Preference Models Based on S-Integrals and Their Verification. In: Li, S., Wang, X., Okazaki, Y., Kawabe, J., Murofushi, T., Guan, L. (eds.) Nonlinear Mathematics for Uncertainty and its Applications. AISC, vol. 100, pp. 1–18. Springer, Heidelberg (2011)
17. Torra, V., Narukawa, Y.: On the meta-knowledge Choquet integral and related models. International Journal of Intelligent Systems 20, 1017–1036 (2005)

# Qualitative Integrals and Desintegrals – Towards a Logical View

Didier Dubois[1], Henri Prade[1], and Agnès Rico[2]

[1] IRIT, CNRS Université de Toulouse, France
[2] ERIC, Université de Lyon, France
{dubois,prade}@irit.fr, agnes.rico@univ-lyon1.fr

**Abstract.** This paper presents several variants of Sugeno integral, and in particular the idea of (qualitative) desintegrals, a dual of integrals. When evaluating an item, desintegrals are maximal if no defects at all are present, while integrals are maximal if all advantages are sufficiently present. This idea leads to a bipolar representation of preferences, by means of a pair made of an integral and a desintegral, whose possibilistic logic counterparts are outlined (in the case where criteria are binary).

**Keywords:** Sugeno integral, possibilistic logic, bipolar representation.

## 1 Introduction

In multi-criteria decision making, Sugeno integrals are commonly used as qualitative aggregation functions for evaluating objects on the basis of several criteria [10]. A Sugeno integral delivers a score between the minimum and the maximum of the partial ratings. The definition of Sugeno integral is based on a capacity (or fuzzy measure) which represents the importance of the sets of criteria. But the importance of the criteria can be exploited in different ways when aggregating partial evaluations. Especially, variants of Sugeno integral can be defined when the evaluation scale is taken as a Heyting algebra using an operator named the residuum.

When Sugeno integral is used, the criteria are considered positive: global evaluation increases with the partial ratings. If we consider negative criteria, then the global evaluation increases when the partial ratings decrease. In such a context it is possible to define other variants of Sugeno integral we call *desintegrals*. With these new kinds of negative aggregation functions, the better a criterion is satisfied the worse is the global evaluation. Besides, Sugeno integral can be encoded as a possibilistic logic base [8]. This paper partially extends this result to the desintegrals when the partial evaluations are binary values, which comes down to a logical encoding of monotonic set functions.

In order to illustrate our motivations, let us first present an example of how we can intuitively use a pair of specifications that can be represented by an integral and a desintegral, as we shall see later, for modeling preferences. We work in the framework of possibilistic logic with symbolic weights as in [11].

V. Torra et al. (Eds.): MDAI 2012, LNAI 7647, pp. 127–138, 2012.

Let $a, b, c$ and $d$ be four properties and $S$ be a scale, supposedly discrete and totally ordered with a greatest element denoted by 1 and a least element denoted by 0. To illustrate the "negative properties" side, associated with the idea of a desintegral, let us assume the following: if the properties $a$ and $b$ are satisfied then the global evaluation should remain below a certain level $\theta < 1$ and if the property $c$ is satisfied, the evaluation should be less than $\lambda < 1$ with $\lambda > \theta$ (where $\lambda, \theta \in S$). For the "positive properties", modeled by an integral, let us assume that if the property $d$ is satisfied the global evaluation should be greater than $\rho < 1$ and that if the properties $a$ and $d$ are satisfied the evaluation must be greater than $\eta$ with $\eta > \rho$ ($\eta, \rho \in S$). As can be seen on the table in Figure 1, we obtain two symbolic distributions corresponding respectively to an upper bound of the evaluation $x$ of the negative aspects and to a lower bound of the evaluation $y$ of the positive aspects, in different situations. Note that if we impose a single evaluation rather than two, i.e; $x = y$, a consistency condition would be needed: $\mu \leq \theta$. But in the following, we consider the two types of evaluation separately.

| a | b | c | d | negative aspects | positive aspects |
|---|---|---|---|---|---|
| 1 | 1 | 1 | 1 | $x \leq \theta$ | $\eta \leq y$ |
| 1 | 1 | 1 | 0 | $x \leq \theta$ | $0 \leq y$ |
| 1 | 1 | 0 | 1 | $x \leq \theta$ | $\eta \leq y$ |
| 1 | 1 | 0 | 0 | $x \leq \theta$ | $0 \leq y$ |
| 1 | 0 | 1 | 1 | $x \leq \lambda$ | $\eta \leq y$ |
| 1 | 0 | 1 | 0 | $x \leq \lambda$ | $0 \leq y$ |
| 1 | 0 | 0 | 1 | $x \leq 1$ | $\eta \leq y$ |
| 1 | 0 | 0 | 0 | $x \leq 1$ | $0 \leq y$ |
| 0 | 1 | 1 | 1 | $x \leq \lambda$ | $\rho \leq y$ |
| 0 | 1 | 1 | 0 | $x \leq \lambda$ | $0 \leq y$ |
| 0 | 1 | 0 | 1 | $x \leq 1$ | $\rho \leq y$ |
| 0 | 1 | 0 | 0 | $x \leq 1$ | $0 \leq y$ |
| 0 | 0 | 1 | 1 | $x \leq \lambda$ | $\rho \leq y$ |
| 0 | 0 | 1 | 0 | $x \leq \lambda$ | $0 \leq y$ |
| 0 | 0 | 0 | 1 | $x \leq 1$ | $\rho \leq y$ |
| 0 | 0 | 0 | 0 | $x \leq 1$ | $0 \leq y$ |

**Fig. 1.** Symbolic distributions corresponding to the idea of desintegral (upper bound) and to the idea of integral (lower bound)

This article deals with the bipolar representation of preferences (see [4] for an overview on the representation of preferences) in a qualitative framework by extending the notion of Sugeno integral. Section 3 presents two types of integrals and two types of desintegrals, and establishes relations between them when the criteria are binary. Section 4 studies the logical counterpart of integrals and desintegrals in this case. Before concluding, Section 5 briefly discusses the idea of using a pair of integral / desintegral to describe acceptable objects. But first we present a reminder of possibilistic logic and we introduce the algebraic framework necessary for the evaluations.

## 2    Framework and Notations

This section provides a reminder on possibilistic logic, and introduces the algebraic framework required for the evaluations of objects using qualitative integrals and desintegrals whose definitions are presented in the next section.

### 2.1    Possibilistic Logic

Let $B^N = \{(\varphi_j, \alpha_j) \mid j = 1, \ldots, m\}$ be a possibilistic logic base where $\varphi_j$ is a propositional logic formula and $\alpha_j \in \mathcal{L} \subseteq [0,1]$ is a priority level [7]. The logical conjunctions and disjunctions are denoted by $\wedge$ and $\vee$. Each formula $(\varphi_j, \alpha_j)$ means that $N(\varphi_j) \geq \alpha_j$, where $N$ is a necessity measure, i.e., a set function satisfying the property $N(\varphi \wedge \psi) = \min(N(\varphi), N(\psi))$. A necessity measure is associated to a possibility distribution $\pi$ as follows:
$N(\varphi) = \min_{\omega \notin M(\varphi)}(1 - \pi(\omega)) = 1 - \Pi(\neg\varphi)$, where $\Pi$ is the possibility measure associated to $N$ and $M(\varphi)$ is the set of models induced by the underlying propositional language for which $\varphi$ is true.

The base $B^N$ is associated to the least informative possibility distribution induced by the constraints $N(\varphi_j) \geq \alpha_j$, namely, $\pi_B^N(\omega) = \min_{j=1,\ldots,m} \pi_{(\varphi_j, \alpha_j)}(\omega)$ on the set of interpretations, where $\pi_{(\varphi_j, \alpha_j)}(\omega) = 1$ if $\omega \in M(\varphi_j)$, and $\pi_{(\varphi_j, \alpha_j)}(\omega) = 1 - \alpha_j$ if $\omega \notin M(\varphi_j)$. An interpretation $\omega$ is all the more possible as it does not violate any formula $\varphi_j$ having a higher priority level $\alpha_j$. Hence, this possibility distribution is expressed as a min-max combination:

$$\pi_B^N(\omega) = \min_{j=1,\ldots,m} \max(1 - \alpha_j, I_{M(\varphi_j)}(\omega))$$

where $I_{M(\varphi_j)}$ is the characteristic function of $M(\varphi_j)$. So, if $\omega \notin M(\varphi_j)$, $\pi_B^N(\omega) \leq 1 - \alpha_j$, and if $\omega \in \bigcap_{j \in J} M(\neg\varphi_j)$, $\pi_B^N(\omega) \leq \min_{j \in J}(1 - \alpha_j)$. It is a description "from above" of $\pi_B^N$. A possibilistic base $B^N$ can be transformed in a base where the formulas $\varphi_i$ are clauses (without altering the distribution $\pi_B^N$). We can still see $B^N$ as a conjunction of weighted clauses, i.e., as an extension of the conjunctive normal form.

A dual representation in possibilistic logic is based on guaranteed possibility measures. Hence a logical formula is a pair $[\psi, \beta]$, interpreted as the constraint $\Delta(\psi) \geq \beta$, where $\Delta$ is a guaranteed (anti-)possibility measure characterized by $\Delta(\phi \vee \psi) = \min(\Delta(\phi), \Delta(\psi))$ and $\Delta(\emptyset) = 1$. In such a context, a base $B^\Delta = \{[\psi_i, \beta_i] \mid i = 1, \ldots, n\}$ is associated to the distribution

$$\pi_B^\Delta(\omega) = \max_{i=1,\ldots,n} \pi_{[\psi_i, \beta_i]}(\omega)$$

with $\pi_{[\psi_i, \beta_i]}(\omega) = \beta_i$ if $\omega \in M(\psi_i)$ and $\pi_{[\psi_i, \beta_i]}(\omega) = 0$ otherwise. If $\omega \in M(\psi_i)$, $\pi_B^\Delta(\omega) \geq \beta_i$, and if $\omega \in \bigcup_{i \in I} M(\psi_i)$, $\pi_B^\Delta(\omega) \geq \max_{i \in I} \beta_i$. So this base is a description "from below" of $\pi_B^\Delta$. A dual possibilistic base $B^\Delta$ can always be transformed in a base in which the formulas $\psi_j$ are conjunctions of literals (cubes) without altering $\pi_B^\Delta$. So $B^N$ can be seen as a weighted combination of cubes, i.e, as an extension of the disjunctive normal form.

A possibilistic logic base $B^\Delta$ expressed in terms of a guaranteed possibility measure can always be rewritten equivalently in terms of a standard possibilistic logic base $B^N$ based on necessity measures [2,1] and conversely with the equality $\pi_B^N = \pi_B^\Delta$. This transformation is similar to a description from below of $\pi_B^N$. Let us note that

- if $\omega \in M(\varphi_j)$, $\pi_B^N(\omega) \geq \min_{k \neq j}(1 - \alpha_k)$,
- and more generally si $\omega \in \bigcap_{j \in J} M(\varphi_j)$, $\pi_B^N(\omega) \geq \min_{k \notin J}(1 - \alpha_k)$,

so $\pi_B^N$ can be rewritten in a max-min form (equivalent to the previous one):

$$\pi_B^N(\omega) = \max_{J \subseteq \{1,\ldots,m\}} \min(\min_{k \notin J}(1 - \alpha_k), I_{M(\wedge_{j \in J}\varphi_j)}(\omega))$$

where $I_{M(\wedge_{j \in J}\varphi_j)}(\omega) = \min_{j \in J} I_{M(\varphi_j)}(\omega)$, and $\min_\emptyset X = 1$. This transformation corresponds to writing the min-max expression of $\pi_B^N$ as a max-min expression by applying the distributivity of min to max. The base obtained is

$$B^\Delta = \{[\wedge_{j \in J}\varphi_j, \min_{k \notin J}(1 - \alpha_k)], J \subseteq \{1, \ldots, m\}\}.$$

Note that this procedure generalizes the transformation of a conjunctive normal form into a disjunctive normal form to the gradual case.

In the following, for convenience, we use a possibilistic logic encoding of the type $\Delta(\psi) \geq \beta$.

## 2.2  Algebraic Framework

We consider a set of criteria $\mathcal{C} = \{1, \cdots, n\}$. Objects are evaluated using these criteria. The evaluation scale, $\mathcal{L}$, associated to each criterion is totally ordered. It may be finite or be the interval $[0,1]$. Then an object is represented by its evaluation on the different criteria, i.e., by $f = (f_1, \cdots, f_n) \in \mathcal{L}^n$. Moreover we consider $\mathcal{L}$ as a Heyting algebra i.e as a complete residuated lattice with a greatest element denoted by 1 and a least element denoted by 0. More precisely, $< \mathcal{L}, \wedge, \vee, \rightarrow, 0, 1 >$ is a complete lattice: $< \mathcal{L}, \wedge, 1 >$ is a commutative monoid (i.e $\wedge$ is associative, commutative and for all $a \in \mathcal{L}$, $a \wedge 1 = a$). The operator denoted by $\rightarrow$ will be the Gödel implication defined by $a \rightarrow b = 1$ if $a \leq b$ and $b$ otherwise.

In the following, we consider positive criteria and negative criteria. In the latter case, 0 will be a good evaluation, 1 will be a bad evaluation and the scale will be said decreasing (the scale is increasing in the case of positive criteria). To handle the directionality of the scale, we also need an operation that reverses the scale. This operation (a decreasing involution) defined on $\mathcal{L}$ is denoted by $1-$. We can then define integrals and desintegrals. A particular integral is Sugeno integral for which the possibilistic logic counterpart has been studied recently [8].

# 3    Qualitative Integrals and Desintegrals

Now we introduce two qualitative integrals and two qualitative desintegrals. See [9] for a more comprehensive framework. This should not be confused with a proposal made in [3] where the integrals considered are generalizations of Sugeno integral on a De Morgan-like algebra, and where the idea of desintegrals does not appear, neither the use of a residuated structure, nor the concern for a weighted logic counterpart.

## 3.1    Qualitative Integrals and Increasing Scale

In this part the criteria are evaluated on an increasing scale and the global evaluation is also on an increasing scale. An importance factor $\pi_i$ is assigned to each criterion $i$. It is all the higher as the criterion $i$ is important. We assume that $\vee_{i=1,...,n}\pi_i = 1$. We view an object as described by a mapping $f : \mathcal{C} \to \mathcal{L}$.

In a loose aggregation of type max-priority, $\vee_{i=1,...,n}f_i \wedge \pi_i$, $\pi_i$ is the maximum possible global score due to the only criterion $i$. Indeed, we obtain $\pi_i$ if $f_i = 1$ and $f_j = 0$ if $j \neq i$. A criterion is all the more important as it can contribute to a higher global evaluation. A demanding aggregation is of the min-priority form $\wedge_{i=1,...,n}f_i \vee (1 - \pi_i)$, where we consider $1 - \pi_i$ as the minimum possible global evaluation solely due to criterion $i$ (we obtain $1 - \pi_i$ if $f_i = 0$ and $f_j = 1$ if $j \neq i$). A criterion is all the more important as it can lead to a lower global evaluation. In this setting, the importance factors act as saturation levels.

We can generalize importance factors from individual criteria to sets thereof by means of a capacity $\mu : 2^{\mathcal{C}} \to \mathcal{L}$: $\mu(A)$ is the importance of set $A$. $\mu$ is an increasing set function such that $\mu(\emptyset) = 0$ and $\mu(\mathcal{C}) = 1$.

An important class of aggregation functions, used in a qualitative framework is the so-called Sugeno integral:

$$\oint_{\mu} (f) = \vee_{A \subseteq \mathcal{C}} \ \mu(A) \wedge (\wedge_{i \in A} f_i). \tag{1}$$

This integral generalizes the prioritized max and min respectively obtained if $\mu$ is a possibility measure or a necessity measure in (1). We can check that $\oint_{\mu}(I_A) = \mu(A)$ where $I_A$ is the characteristic function of $A$.

Another viewpoint is to consider that the importance factor $\pi_i$ acts as follows on the evaluations of an object $f$: If $f_i \geq \pi_i$ then the evaluation becomes 1 and it becomes $1 - \pi_i$ otherwise. Therefore, in this case the evaluation scale of the criterion $i$ is reduced to $\{1 - \pi_i, 1\}$. If $f_i$ is greater than $\pi_i$, we consider that the criterion $i$ is satisfied. If $\pi_i$ is high and $f_i$ is less than $\pi_i$, then the value of $f_i$ is drastically reduced to $1 - \pi_i$. Conversely, if $\pi_i$ is small, and $f_i$ is less than $\pi_i$, then $f_i$ is upgraded to $1 - \pi_i$. In this case, importance factors correspond to tolerance levels.

In such a context, partial evaluations can be aggregated using the minimum: $\wedge_{i=1,...,n}(1 - f_i) \to (1 - \pi_i)$. More generally, if the groups of criteria are weighted, we obtain the following integral with respect to a capacity $\mu$:

$$\oint_{\mu}^{\Uparrow} (f) = \wedge_{A \subseteq C} (\wedge_{i \in A}(1 - f_i)) \to \mu(\overline{A}) \tag{2}$$

We can check that $\oint_{\mu}^{\Uparrow}(I_B) = \wedge_{A \subseteq \overline{B}} \mu(\overline{A}) = \mu(B)$.

## 3.2   Qualitative Desintegrals and Decreasing Scale

In this part, the evaluation scale for each criterion is decreasing, i.e., 0 is better than 1, but the scale for the global evaluation is increasing. In this case the aggregation functions must be decreasing and the capacities are replaced by decreasing set functions $\nu$ such that $\nu(\emptyset) = 1$ and $\nu(C) = 0$, called anti-measures. $\nu(A)$ is the level of tolerance of $A$: the greater $\nu(A)$, the less important is $A$.

A first desintegral is obtained by a saturation effect on a reversed scale:

$$\oint_{\nu}^{\textit{t}} (f) = \vee_{A \subseteq C} \nu(\overline{A}) \wedge (\wedge_{i \in A}(1 - f_i)) \tag{3}$$

where $\nu$ is an anti-measure. We can check that $\oint_{\nu}^{\textit{t}}(I_A) = \nu(A)$. We recognize the definition of Sugeno integral $\oint_{\nu(\cdot)}(1-f)$. Note that $\oint_{\nu}^{\textit{t}}(f) = 1$ if there exists a non important subset of criteria (because completely tolerant) and the evaluations of $f$ with respect to the other criteria are equal to 0.

Moreover, we can verify that $\oint_{\nu}^{\textit{t}}(f) = \wedge_{i=1}^{n}(1-f_i) \vee t_i$ if the anti-measure $\nu$ is a guaranteed possibility, i.e., $\nu(A) = \wedge_{i \in A} t_i =_{def} \Delta_T(A)$ where $t_i$ is the tolerance of criterion $i$ (the greater is $t_i$ the more tolerant is the criterion $i$). Moreover, $\oint_{\nu}^{\textit{t}}(f) = \vee_{i=1}^{n}(1-t_i) \wedge (1-f_i)$ if $\nu(A) = \vee_{i \in \overline{A}}(1-t_i) =_{def} \nabla_t(A) = 1 - \Delta_T(\overline{A})$. This is the counterpart of the fact that the Sugeno integral gives the max or min with priority when the measure is a possibility measure or a necessity measure.

The other viewpoint is to consider that if $f_i > t_i$ then the local evaluation is bad and $f_i$ becomes $t_i$. Otherwise the local evaluation is good and $f_i$ becomes 1. This corresponds to the use of the Gödel implication and the global evaluation $\wedge_{i=1,\dots,n} f_i \to t_i$ which is generalized by the following desintegral:

$$\oint_{\nu}^{\Downarrow} (f) = \wedge_{A \subseteq C}(\wedge_{i \in A} f_i) \to \nu(A) \tag{4}$$

where $\nu$ is an anti-measure (with the convention $\wedge_{i \in \emptyset} f_i = 0$). We can check that $\oint_{\nu}^{\Downarrow}(I_A) = \nu(A)$. Note that $\oint_{\nu}^{\Downarrow}(f) = 1$ if for each subset of criteria, at least one criterion has an evaluation lower than the tolerance of this subset.

In the following the values of the function $f$ are in $\{0, 1\}$ and the values of the set functions are in $[0, 1]$. In this context, the criteria are binary and $f_i = 1$ (resp. $f_i = 0$) can be encoded as a proposition $f_i$ (resp. $\neg f_i$) expressing that the criterion $i$ is satisfied (resp. not satisfied).

## 3.3   Relation between Integrals and Desintegrals

If the criteria are represented on a binary scale then there exist some links between integrals and desintegrals.

**Proposition 1.** *If the values of the functions $f$ are in $\{0,1\}$ then $f$ is a characteristic function $I_A$ and we can check that*

- *for any capacity $\mu$, $\oint_\mu (I_A) = \oint_\mu^\Uparrow (I_A) = \mu(A)$;*
- *for any anti-measure $\nu$, $\oint_\nu^{\downarrow} (I_A) = \oint_\nu^{\Downarrow} (I_A) = \nu(A)$.*

These relations are not true in the general case because if $f$ is such that $f_i < 1, \forall i$, then $\oint_\mu^\Uparrow (f) = 0$ since $\wedge_{i \in \mathcal{C}}(1-f_i) \to \mu(\overline{\mathcal{C}}) = \alpha \to 0 = 0$ with $\alpha > 0$. But generally we have $\oint_\mu (f) \neq 0$. Similarly, if $f_i > 0, \forall i$, $\wedge_{i \in \mathcal{C}} f_i \to \nu(\mathcal{C}) = \beta \to 0 = 0$ with $\beta > 0$. Hence $\oint_\nu^{\Downarrow} (f) = 0$, but generally $\oint_\nu^{\downarrow} (f) \neq 0$.

**Example 1.** *We consider two criteria denoted by $a$ and $b$, the function $f$ is such that $f(a) = 0.5$ and $f(b) = 0.6$.*

*If $\mu$ is the capacity $\mu(a) = 0.4$, $\mu(b) = 0.5$ then $\oint_\mu (f) = \vee(0.5, 0.6 \wedge 0.5) = 0.5$ and $\oint_\mu^\Uparrow (f) = \wedge(0.5 \to 0.5, 0.4 \to 0.4, 0.4 \to 0) = 0$.*

*If $\nu$ is an anti-measure $\nu(a) = \nu(b) = 1$ then $\oint_\nu^{\downarrow} (f) = \vee(0.5, 0.4, 0.4) = 0.5$ and $\oint_\nu^{\Downarrow} (f) = \wedge(0.5 \to 1, 0.6 \to 1, 0.5 \to 0) = 0$.*

**Proposition 2.** *If $\nu$ is an anti-measure, then there exists a capacity $\mu$ such that for all $f$ (Boolean or not) $1 - \oint_\mu (f) = \oint_\nu^{\downarrow} (f)$ and conversely.*

**Proof.** $1 - \oint_\nu^{\downarrow} (f) = 1 - \vee_{A \subseteq \mathcal{C}} \nu(\overline{A}) \wedge \wedge_{i \in A}(1 - f_i) = \wedge_{A \subseteq \mathcal{C}} (1 - \nu(\overline{A})) \vee \vee_{i \in A} f_i = \vee_{A \subseteq \mathcal{C}} 1 - \nu(A) \wedge \wedge_{i \in A} f_i = \oint_{1-\nu}(f)$ because of a Sugeno integral property [12].

So, Sugeno integral is the complement to 1 of its dual desintegral by replacing $\mu$ by $\nu = 1 - \mu$. This result is general for any function $f$. In the Boolean case, the two propositions are summarized in

$$\oint_\mu (I_A) = \oint_\mu^\Uparrow (I_A) = \mu(A) = 1 - \oint_{1-\mu}^{\downarrow} (I_A) = 1 - \oint_{1-\mu}^{\Downarrow} (I_A).$$

The relation $\oint_\mu^\Uparrow (f) = 1 - \oint_{1-\mu}^{\Downarrow}(f)$ is not true in the general case:

**Example 2.** *Let $\mathcal{C} = \{a, b\}$, $f(a) = 1$, $f(b) = 0.8$, $\mu(a) = 0$ $\mu(b) = 0.5$.*

$1 - \oint_{1-\mu}^{\Downarrow}(f) = \vee_{A \subseteq \mathcal{C}}(1 - ((\wedge_{i \in A} f_i) \to (1 - \mu(A)))) = \max(0, 0.5, 1) = 1$, and $\oint_\mu^\Uparrow (f) = \wedge_{A \subseteq \mathcal{C}} (\wedge_{i \in A}(1 - f_i)) \to \mu(\overline{A}) = \min(0 \to 0.5, 0.2 \to 0, 0 \to 0) = \min(1, 0, 1) = 0$.

## 4    Qualitative Integrals as Possibilistic Bases

This section recalls the method presented in [8] for interpreting a Sugeno integral as a possibilistic base when $f$ is a Boolean function. Next we will study the case of a desintegral. More precisely, this section presents logical representations for capacities and anti-measures.

### 4.1   Logical Framework for Sugeno Integral in a Boolean Context

The Sugeno integral $\oint_\mu$ is used to classify objects $f$ according to their evaluation $\oint_\mu(f)$. A possibilistic logic framework can be constructed as follows.

Looking at each criterion $i$ as a predicate $P_i$, $P_i(f_i)$ indicates that the evaluation with respect to the criterion $i$ is $f_i \in L$. Hence the object $f = (f_1, ..., f_n)$ is represented by the logic Boolean formula $P_1(f_1) \wedge \cdots \wedge P_n(f_n)$. Boolean criteria are assumed so we can simplify $P_i(1)$ into $f_i$ and $P_i(0)$ into $\neg f_i$ according to whether criterion $i$ is satisfied or not. If $f = I_A$, then $P_1(f_1) \wedge \cdots \wedge P_n(f_n) = \wedge_{i \in A} f_i \bigwedge \wedge_{i \notin A} \neg f_i$. So any object is encoded as an interpretation of the language induced by the variables associated to the criteria.

In this context, a logical formula corresponds to a set of objects that must satisfy (or not) some criteria. The global evaluation $\oint_\mu(I_A)$ can be seen as a degree of guaranteed possibility of $\wedge_{i \in A} f_i \bigwedge \wedge_{i \notin A} \neg f_i$ (view from below) or as a degree of standard possibility (view from above).

**Example 3.** *We consider three criteria or properties $a, b, c$. Some objects are evaluated with respect to these criteria. The evaluation scale for each criterion is $\{0, 1\}$, an object $f$ is represented by the characteristic function of the subset of criteria (or properties) that it satisfies.*

*Capacities (and anti-measures) are valued in $\mathcal{L} = \{1, 1 - \beta, \beta, 0\}$ with $1 > 1 - \beta > \beta > 0$.*

*We consider the capacity $\mu$ defined by: $\mu(a) = \mu(b) = \beta$, $\mu(c) = 1 - \beta$, $\mu(\{a, b\}) = \beta$, $\mu(\{a, c\}) = 1$, $\mu(\{b, c\}) = 1 - \beta$, $\mu(\{a, b, c\}) = 1$.*

*In this context if we consider $f = I_{\{a,b\}}$ then $\oint_\mu(f) = \mu(\{a, b\})$. As the capacity $\mu$ is increasing, if the properties $a$ and $b$ are satisfied by an object $g$ then $\oint_\mu(g) \geq \mu(\{a, b\})$. This inequality corresponds to a $\Delta$ possibilistic base on language generated by $\{a, b, c\}$, as we shall see.*

### 4.2   Construction of a Possibilistic Base Associated to a Capacity

This section presents the possibilistic base associated to a Sugeno integral in the particular case of binary values i.e, a capacity. The general case was presented in [8]. The following property characterizes the set of objects $f$ solutions of the inequality $\oint_\mu(f) \geq \gamma$, $\gamma \in \mathcal{L}$ when the criteria are not Boolean.

**Proposition 3.** $\{f | \oint_\mu(f) \geq \gamma\} = \{f | \exists A \text{ s.t. } \mu(A) \geq \gamma \text{ and } \forall i \in A, f_i \geq \gamma\}$.

**Proof.** $\oint_\mu(f) = \vee_{A \subseteq C} \mu(A) \wedge \wedge_{i \in A} f_i \geq \gamma$ iff $\exists A$, $\mu(A) \wedge \wedge_{i \in A} f_i \geq \gamma$ i.e., $\exists A$ such that $\mu(A) \geq \gamma$ and $\forall i \in A$ $f_i \geq \gamma$.

Particularly, the non-trivial case is $\gamma > 0$, so if the evaluation scale is Boolean we must suppose $f_i = 1$ in the previous proposition and therefore $\{f | \oint_\mu(f) \geq \gamma\} = \{I_B | \exists A \text{ s.t.} \mu(A) \geq \gamma \text{ and } A \subseteq B\}$. The monotony of $\mu$ and the fact that the set $\mathcal{F}_\mu^\gamma = \{A, \mu(A) \geq \gamma > 0\}$ is closed under inclusion entails that $\mathcal{F}_\mu^\gamma$ has least elements $A_k^\gamma, k = 1, \ldots, p^\gamma$ such that $\mu(A) \geq \gamma \iff \exists k, A_k^\gamma \subseteq A$. In logical terms, the constraint $\mu(A) \geq \gamma$ can be represented with the base

$$B_\gamma^\Delta = \{[\wedge_{i \in A_k^\gamma} f_i, \gamma], k = 1, \ldots, p^\gamma\}.$$

Hence it is obvious that $\oint_\mu(f) \geq \gamma$, with $f = I_A$ if and only if $\wedge_{i \in A} f_i \wedge \wedge_{i \notin A} \neg f_i \models \vee_{k=1,\ldots,p^\gamma} \wedge_{i \in A_k^\gamma} f_i$ which can be written with a $\Delta$ possibilistic base $B_\gamma^\Delta \vdash [f, \gamma]$. So the capacity $\mu$ can be represented by a $\Delta$ possibilistic base: $B^\Delta = \cup_{\gamma \in \mu(2^C)} B_\gamma^\Delta$.

**Example 4.** *We consider the capacity $\mu$ of the previous example. Let us lay bare:*

- *objects $f$ such that $\oint_\mu(f) \geq \beta$: we find $\{A \subseteq C | \mu(A) \geq \beta\} = 2^C \setminus \{\emptyset\}$. So $f = I_A$ with $a \in A$, or $b \in A$ or $c \in A$. These three vectors correspond to the formulas $[a, \beta]$, $[b, \beta]$ and $[c, \beta]$ respectively.*
- *objects $f$ such that $\oint_\mu(f) \geq 1 - \beta$. We find $\{A \subseteq C | \mu(A) \geq 1 - \beta\} = \{\{c\}, \{a, c\}, \{b, c\}, \{a, b, c\}\} = \{A, c \in A\}$. So $f = I_A$ with $c \in A$. which corresponds to the formula $[c, 1 - \beta]$.*
- *objects $f$ such that $\oint_\mu(f) \geq 1$. We find $\{A \subseteq P | \mu(A) \geq 1\} = \{\{a, c\}, \{a, b, c\}\}$. So $f = I_A$ with $\{a, c\} \subseteq A$ which corresponds to the formula $[a \wedge c, 1]$.*
- *According to the definition all objects $f$ are such that $\oint_\mu(f) \geq 0$. We have $\{A \subseteq | \mu(A) \geq 0\} = 2^C = \{A : \emptyset \subseteq A\}$, which entails no $\Delta$-formula because $[\bot, 0]$ is a tautology.*

*The associated possibilistic $\Delta$ base is $K_\mu^\Delta = \{[a, \beta], [b, \beta], [c, 1 - \beta], [a \wedge c, 1]\}$.*

## 4.3   Qualitative Desintegrals as Possibilistic Bases

In this part we consider $\nu$ an anti-measure. Similarly as the positive case, each object $f = (f_1, \cdots, f_n)$ represents a logical interpretation $f_1^{\epsilon_1} \wedge \cdots \wedge f_n^{\epsilon_n}$ where $f_i^{\epsilon_i} \in \{f_i, \neg f_i\}$; so $\oint_\nu^\Downarrow(f) = \oint_\nu^\ell(f) = \nu(A)$ with $f = I_A$. But in this context it is $1 - \oint_\nu^\ell$ which is viewed as the possibilistic degree corresponding to the formula. So the inequality $1 - \oint_\nu^\ell \leq \gamma$, i.e $\oint_\nu^\ell \geq 1 - \gamma$, is linked with a $N$ possibilistic base; and the inequality $\oint_\nu^\ell \leq 1 - \gamma$, is associated to a $\Delta$ possibilistic base. We are going to use the relationship between the desintegral $\oint_\nu^\ell$ and Sugeno integral to come back to the $\Delta$ possibilistic base of a Sugeno integral. In practice in the Boolean case, $\oint_\nu^\ell(f) \leq 1 - \gamma$ means $\nu(A) \leq 1 - \gamma$ i.e $\mu(A) = 1 - \nu(A) \geq \gamma$. Formally we come back to the previous case, but the meaning is totally different. Indeed $f_i = 1$ means that the object is bad for the criterion $i$ (it is a defect) and $\nu(A)$ shows how $f = I_A$ is good. So $\mu(A) = 1 - \nu(A)$ measures how much the object $f$ is bad with respect to the evaluations of its defects. We evaluate the unattractiveness of an object with respect to its defects, in the same way as we evaluate its attractiveness with a Sugeno integral.

**Example 5.** *Let us come back to the example 2 but with an anti-measure. For example we consider $\nu = 1 - \mu$ where $\mu$ is the capacity of the example 2:*
$\nu(a) = \nu(b) = 1 - \beta$, $\nu(c) = \beta$, $\nu(\{a, b\}) = 1 - \beta$, $\nu(\{a, c\}) = 0$, $\nu(\{b, c\}) = \beta$, $\nu(\{a, b, c\}) = 0$.

*In such a context, if we consider $f = I_{\{a,b\}}$, $\oint_\nu^\ell(f) = \nu(\{a, b\})$. As the desintegral is decreasing, if the properties $a$ and $b$ are satisfied for an object $g$ we have $1 - \oint_\nu^\ell(g) \geq 1 - \nu(\{a, b\})$. This inequality corresponds to a $\Delta$ possibilistic base.*

We use the proposition 2 : $1 - \oint_{\nu}^{\xi} = \oint_{1-\nu}$. As $\nu$ is defined by $1 - \mu$ we recover $\oint_{\mu}$ and the possibilistic base associated to the Sugeno integral of example 3.

**Example 6.** We consider the anti-measure $\nu$ defined by $\nu(a) = 1 - \beta$, $\nu(b) = 1$, $\nu(c) = 1 - \beta$, $\nu(\{a,b\}) = 1 - \beta$, $\nu(\{a,c\}) = 1 - \beta$, $\nu(\{b,c\}) = \beta$, $\nu(\{a,b,c\}) = 0$. We have $1 - \oint_{\nu}^{\xi} = \oint_{\mu'}$ with $\mu'(a) = \beta$, $\mu'(b) = 0$, $\mu'(c) = \beta$, $\mu'(\{a,b\}) = \beta$, $\mu'(\{a,c\}) = \beta$, $\mu'(\{b,c\}) = 1 - \beta$, $\mu'(\{a,b,c\}) = 1$ and define the following sets:
$\{A|\mu'(A) \geq \beta\} = \{\{a\}, \{c\}, \{a,b\}, \{a,c\}, \{b,c\}, \{a,b,c\}\}$ which corresponds to the $\Delta$-formulas $[a, \beta]$ and $[c, \beta]$.
$\{A|\mu'(A) \geq 1 - \beta\} = \{\{b,c\}, \{a,b,c\}\}$ which corresponds to the $\Delta$-formula $[b \wedge c, 1 - \beta]$.
$\{A|\mu'(A) \geq 1\} = \{\{a,b,c\}\}$ which corresponds to the $\Delta$-formulas $[a \wedge b \wedge c, 1]$.
So the $\Delta$ possibilistic base associated to $\oint_{\nu}^{\xi}$ is $K_{\nu}^{\Delta} = \{[a, \beta], [c, \beta], [b \wedge c, 1 - \beta], [a \wedge b \wedge c, 1]\}$.

| | a | b | c | d | $\oint_{\nu}^{\xi}(I_S)$ | $\oint_{\mu}(I_S)$ |
|---|---|---|---|---|---|---|
| 1 | 1 | 1 | 1 | 1 | $\theta$ | $\eta$ |
| 2 | 1 | 1 | 1 | 0 | $\theta$ | 0 |
| 3 | 1 | 1 | 0 | 1 | $\theta$ | $\eta$ |
| 4 | 1 | 1 | 0 | 0 | $\theta$ | 0 |
| 5 | 1 | 0 | 1 | 1 | $\lambda$ | $\eta$ |
| 6 | 1 | 0 | 1 | 0 | $\lambda$ | 0 |
| 7 | 1 | 0 | 0 | 1 | 1 | $\eta$ |
| 8 | 1 | 0 | 0 | 0 | 1 | 0 |
| 9 | 0 | 1 | 1 | 1 | $\lambda$ | $\rho$ |
| 10 | 0 | 1 | 1 | 0 | $\lambda$ | 0 |
| 11 | 0 | 1 | 0 | 1 | 1 | $\rho$ |
| 12 | 0 | 1 | 0 | 0 | 1 | 0 |
| 13 | 0 | 0 | 1 | 1 | $\lambda$ | $\rho$ |
| 14 | 0 | 0 | 1 | 0 | $\lambda$ | 0 |
| 15 | 0 | 0 | 0 | 1 | 1 | $\rho$ |
| 16 | 0 | 0 | 0 | 0 | 1 | 0 |

**Fig. 2.** Values of the desintegral and the integral for the different subsets of possible satisfied properties for an object

## 5    Towards the Description of Acceptable Objects

A previous work [13] has already proposed to use a pair of evaluations made of a Sugeno integral and a Sugeno integral reversed by complementation to 1 (which corresponds to a desintegral), in order to describe acceptable objects in terms of properties they must have and of properties that they must avoid. In that work, all properties were assumed to be partitioned between these two categories for a given object. In the general case, as suggested by the example presented in the introduction, the fact that a property is desirable or undesirable depends on the context of other satisfied properties. In this example the property $a$ is undesirable if $b$ is satisfied but the property $d$, which is good by itself, is desirable if the property $a$ is satisfied. Let us now encode the example of the introduction in terms of an integral and a desintegral.

First let us consider the positive aspects. An object is satisfactory to degree $\rho$ (at least) if the property $d$ is satisfied, and to degree $\eta > \rho$ (at least) if the properties $a$ and $d$ are satisfied. So we consider the fuzzy measure[1] $\mu$ such that $\mu(\{d\}) = \rho$ and $\mu(\{a,d\}) = \eta$. Moreover, $\forall A \neq C, A \supseteq \{a,d\}, \mu(A) = \eta$ and $\forall A \supseteq \{d\}, A \not\supseteq \{a\}, \mu(a) = \rho$ and $\mu(B) = 0$ otherwise. The value of the integral $\oint_\mu(I_S)$ is given in Figure 2 for the different possible subsets of properties satisfied by an object. The larger $\oint_\mu(I_S)$, the more satisfactory the object characterized by $S$. Now let us consider the negative aspects. The object is only satisfactory to a degree at most $\lambda$ if the property $c$ is satisfied, and satisfactory to a degree at most $\theta < \lambda$ if the properties $a$ and $b$ are satisfied. This leads us to consider an anti-measure[2] $\nu$ such that $\nu(\{c\}) = \lambda$ and $\nu(\{a,b\}) = \theta$. Moreover, $\forall A \neq \emptyset, A \subseteq \{a,b\}, \nu(A) = \theta$, and $\nu(B) = 0$ otherwise. The value of the desintegral $\oint_\nu^i(I_S)$ is given in Figure 2 for the different possible subsets of properties satisfied by an object. The less $\oint_\nu^i(I_S)$, the less satisfactory the object characterized by $S$. An acceptable object needs to be fully satisfactory w.r.t. its negative aspects, i.e. $\oint_\nu^i(I_S) = 1$, which means that the object should have no potential defect, and to be as satisfactory as possible w.r.t. its positive aspects, i.e. with $\oint_\mu(I_S)$ maximal. In our example, we can verify in Figure 2 that the objects that satisfy $a$ and $d$, but not $b$ or $c$ are the most acceptable, and that the objects which satisfy $b$ and $d$ and not $a$ nor $c$ or $d$ and not $a$ nor $b$ nor $c$ are slightly less acceptable. There is no other acceptable objects. The values of $1 - \oint_\nu^i(I_S)$ indicate how much an object should be rejected. Note that some objects (in this example, there are 5 cases (lines 1, 3, 5, 9 and 13 of the table in Figure 2) are both in some respects satisfactory and in other respects unsatisfactory. It comes close to the problem of choosing objects described by binary properties from the pros and cons of arguments [5].

## 6    Conclusion

We have given the definitions and some properties of qualitative integrals and desintegrals which extend the classical Sugeno integral. We have also studied the particular case of binary properties and we have proposed a logical possibilistic view for these aggregation functions. The general case of gradual properties remains to be studied. Generally speaking, we may think of a level-cut based approach, and we may also take lessons of the logical approach to qualitative decision under uncertainty [6] where two logical bases are used for preferences and knowledge, making a classical parallel between multiple criteria decision and

---

[1] We give the smallest one.
[2] We give the greatest one.

decision under uncertainty. A first attempt at providing a logical representation in the general case has been presented in [8] for Sugeno integrals only, and another type of representation of positive and negative synergy between properties, which is not bipolar (for example you want an object which satisfies $a$ or $b$, but not $a$ and $b$) has been also indicated in [8]. The relation between the two types of representation is a topic for further research.

# References

1. Benferhat, S., Dubois, D., Kaci, S., Prade, H.: Modeling positive and negative information in possibility theory. Int. J. of Intellig. Syst. 23(10), 1094–1118 (2008)
2. Benferhat, S., Kaci, S.: Logical representation and fusion of prioritized information based on guaranteed possibility measures: Application to the distance-based merging of classical bases. Artificial Intelligence 148, 291–333 (2003)
3. de Cooman, G., de Baets, B.: Implicator and coimplicator integrals. In: Proc. 6th Inter. Conf. on Processing and Management of Uncertainty in Knowledge-Based Systems (IPMU 1996), Granada, July 1-5, vol. III, pp. 1433–1438 (1996)
4. Domshlak, C., Hüllermeier, E., Kaci, S., Prade, H.: Preferences in AI: An overview. Artif. Intell. 175(7-8), 1037–1052 (2011)
5. Dubois, D., Fargier, H., Bonnefon, J.F.: On the comparison of decisions having positive and negative features. J. of Artif. Intellig. Res. 32, 385–417 (2008)
6. Dubois, D., Le Berre, D., Prade, H., Sabbadin, R.: Using possibilistic logic for modeling qualitative decision: ATMS-based algorithms. Fund. Inform. 37, 1–30 (1999)
7. Dubois, D., Prade, H.: Possibilistic logic: a retrospective and prospective view. Fuzzy Sets and Systems 144, 3–23 (2004)
8. Dubois, D., Prade, H., Rico, A.: A Possibilistic Logic View of Sugeno Integrals. In: Melo-Pinto, P., Couto, P., Serôdio, C., Fodor, J., De Baets, B. (eds.) Eurofuse 2011. AISC, vol. 107, pp. 19–30. Springer, Heidelberg (2011)
9. Dubois, D., Prade, H., Rico, A.: Qualitative Integrals and Desintegrals: How to Handle Positive and Negative Scales in Evaluation. In: Greco, S., Bouchon-Meunier, B., Coletti, G., Fedrizzi, M., Matarazzo, B., Yager, R.R. (eds.) IPMU 2012, Part III. CCIS, vol. 299, pp. 306–316. Springer, Heidelberg (2012)
10. Grabisch, M., Labreuche, C.: A decade of application of the Choquet and Sugeno integrals in multi-criteria decision aid. Annals of Oper. Res. 175, 247–286 (2010)
11. Hadjali, A., Kaci, S., Prade, H.: Database preference queries - A possibilistic logic approach with symbolic priorities. Ann. Math. in AI (2012)
12. Marichal, J.-L.: Aggregation Operations for Multicriteria Decision Aid. Ph.D.Thesis, University of Liège, Belgium (1998)
13. Prade, H., Rico, A.: Describing acceptable objects by means of Sugeno integrals. In: Proc. 2nd Inter. IEEE Conf. on Soft Computing and Pattern Recognition (SoCPaR 2010), Cergy-Pontoise, pp. 6–11 (December 2010)

# A Quantile Approach to Integration with Respect to Non-additive Measures

Marta Cardin

Department of Economics,
University Cà Foscari of Venice
Cannaregio 873, 30121 Venice, Italy
mcardin@unive.it

**Abstract.** The aim of this paper is to introduce some classes of aggregation functionals when the evaluation scale is a complete lattice.

We focus on the notion of quantile of a lattice-valued function which have several properties of its real-valued counterpart and we study a class of aggregation functionals that generalizes Sugeno integrals to the setting of complete lattices. Then we introduce in the real-valued case some classes of aggregation functionals that extend Choquet and Sugeno integrals by considering a multiple quantile model generalizing the approach proposed in [3].

**Keywords:** Completely distributive lattice, quantile, Sugeno integral, Choquet integral.

## 1 Introduction

Aggregation operators are an important mathematical tool for the combination of several inputs in a single outcome that is used in many applied fields and in particular in the area of artificial intelligence for decision making(see [9] for a general background). Real-valued non-additive measures and their associated integrals are widely used aggregation operators. There are many situations where inputs to be aggregated are qualitative and numerical values are used by convenience. Moreover sometimes we need to evaluate objects with a scale that is not totally ordered. As the aim of this paper is to generalize some well known aggregation functionals in a purely ordinal context. In this case only maximum and minimum are used for aggregation of different inputs. So we study aggregation functionals based on a complete lattices and we consider in particular the class of completely distributive lattices. The quantile is a generalization of the concept of median and it play an important role in statistical and economic literature. We study quantile in a ordinal framework and and we consider an axiomatic representation of quantiles as in [6] and [5].

The structure of the paper is as follows. To make this work self-contained in Section 2 we briefly mention some basic concepts on lattices theory and we provide the necessary definitions. Section 3 is devoted to lattice-valued measures and lattice-valued integrals. Finally in Section 4 we introduce some classes of generalized integrals based on a multiple quantiles model.

V. Torra et al. (Eds.): MDAI 2012, LNAI 7647, pp. 139–148, 2012.

## 2   Notations and Definitions

To introduce our general framework we will need some algebraic preliminaries. Much of this terminology is well known and for further background in lattice theory we refer the reader to Davey and Priestley [7] or Grätzer [10].

A *lattice* is an algebraic structure $\langle L; \wedge, \vee \rangle$ where $L$ is a nonempty set, called *universe*, and where $\wedge$ and $\vee$ are two binary operations, called *meet* and *join*, respectively, which satisfy the following axioms:

**(i)** (idempotency) for every $a \in L$, $a \vee a = a \wedge a = a$;
**(ii)** (commutativity) for every $a, b \in L$, $a \vee b = b \vee a$ and $a \wedge b = b \wedge a$;
**(iii)** (associativity) for every $a, b, c \in L$, $a \vee (b \vee c) = (a \vee b) \vee c$ and $a \wedge (b \wedge c) = (a \wedge b) \wedge c$;
**(iv)** (absorption): for every $a, b \in L$, $a \wedge (a \vee b) = a$ and $a \vee (a \wedge b) = a$.

Every lattice $L$ constitutes a partially ordered set endowed with the partial order $\leq$ such that for every $x, y \in L$, write $x \leqslant y$ if $x \wedge y = x$ or, equivalently, if $x \vee y = y$. If for every $a, b \in L$, we have $a \leqslant b$ or $b \leqslant a$, then $L$ is said to be a *chain*. A lattice $L$ is said to be *bounded* if it has a least and a greatest element, denoted by 0 and 1, respectively.

A lattice $L$ is said to be *distributive*, if for every $a, b, c \in L$,

$$a \vee (b \wedge c) = (a \vee b) \wedge (a \vee c) \quad \text{or, equivalently,} \quad a \wedge (b \vee c) = (a \wedge b) \vee (a \wedge c).$$

Clearly, every chain is distributive. A lattice $L$ is said to be *complete* if for every $S \subseteq L$, its supremum $\bigwedge S := \bigwedge_{x \in S} x$ and infimum $\bigvee S := \bigvee_{x \in S} x$ exist. Clearly, every complete lattice is necessarily bounded.

A complete lattice $L$ is said to be *completely distributive* is the following more stringent distributive law holds

$$\bigwedge_{i \in I} \left( \bigvee_{j \in J} x_{ij} \right) = \bigvee_{f \in J^I} \left( \bigwedge_{i \in I} x_{if(i)} \right),$$

for every doubly indexed subset $\{x_{ij} : i \in I, j \in J\}$ of $L$. Note that every complete chain (in particular, the extended real line and each product of complete chains) is completely distributive. Moreover, complete distributivity reduces to distributivity in the case of finite lattices. Throughout this paper, $A$ denotes an arbitrary nonempty set and $L$ a lattice. The set $L^A$ of all functions from $A$ to $L$ constitutes a lattice under the operations $\wedge$ and $\vee$ defined pointwise, i.e.,

$$(f \wedge g)(x) = f(x) \wedge g(x) \quad \text{and} \quad (f \vee g)(x) = f(x) \vee g(x) \quad \text{for every } f, g \in L^A.$$

In particular, for any lattice $L$, the cartesian product $L^n$ also constitutes a lattice by defining the lattice operations componentwise. Observe that if $L$ is bounded (distributive), then $L^A$ is also bounded (resp. distributive). We denote by $\mathbf{0}$ and $\mathbf{1}$ the least and the greatest elements, respectively, of $L^A$. Likewise, for each $c \in L$, we denote by $\mathbf{c}$ the constant $c$ map in $L^A$. Moreover or each $X \subset A$, we denote by **X** the *characteristic function* of $X$ in $L^A$ defined by $\mathbf{X}(x) = 1$ if $x \in X$ and $\mathbf{X}(x) = 0$ if $x \notin X$.

The following notion extends that of homomorphism between lattices. A map $\gamma\colon L \to L$, where $L$ is a complete lattice, is said to be *continuous* if it preserves arbitrary meets and and arbitrary joins, i.e., for every $S \subseteq L$,

$$\gamma\left(\bigwedge S\right) = \bigwedge \gamma(S) \quad \text{and} \quad \gamma\left(\bigvee S\right) = \bigvee \gamma(S).$$

The term continuous is justified by the following fact (see [11]): if $\gamma\colon L \to L$ is continuous, then it is continuous with respect to the Lawson topology on $L$.

# 3    Lattice-Valued Measures and Lattice-Valued Integrals

The following definitions are natural extensions of the well known concepts of real -valued non-additive (or fuzzy) measures and their associated integrals.

We follow the approach proposed by Greco in [13] and more recently by Ban and Fechete in [2] for lattice-valued measures and integrals and we refer to [16] and [17] for the standard case. Let $(A, \mathcal{A})$ be a measurable space and $L$ a a bounded lattice. A *non-additive measure* on $A$ with values in $L$ is a function $m\colon \mathcal{A} \to L$ such that $m(\emptyset) = 0$, $m(A) = 1$ and $m(X) \le m(Y)$ whenever $X \subseteq Y$. A function $f\colon A \to L$ is said to be *measurable* if the sets $\{x : f(x) \leqslant a\}$ and $\{x : f(x) \geqslant a\}$ are elements of $\mathcal{A}$ for every $a \in L$. We will use $\{f \geqslant x\}$ to indicate the weak upper level set $\{t \in L : f(t) \geqslant x\}$.

We denote by $\mathcal{M}$ the set of all fuzzy measures on A with values in $L$ and by $\mathcal{F}$ the set of the measurable functions $f\colon A \to L$. Following [15] and [17] we give the following definition.

**Definition 1.** *A mapping $I\colon \mathcal{F} \times \mathcal{M} \to L$ will be called a* lattice-valued integral *if the following properties are satisfied:*

*(i) for every $c \in L$ and $m \in \mathcal{M}$ , $I(\mathbf{c}, m) = c$;*
*(ii) for each $m_1, m_2, \in \mathcal{M}$ with $m_1 \le m_2$ and $f_1, f_2, \in \mathcal{F}$ with $f_1 \le f_2$ we have $I(m_1, f_1) \le I(m_2, f_2)$ .*

This general definition has to be completed by a variety of additional properties. In some cases we consider a lattice-valued integral as a function defined in $\mathcal{F}$.

In order to obtain the additivity of the integral it is useful the concept of comonotonic functions. The concept of comonotonicity emerges quite naturally in many different fields such as aggregation theory, decision theory, finance and actuarial sciences (see [22] )and comonotonicity was already used under different names by many authors. We refer to Denneberg [8] for the definition as well as for different characterizations of comonotonicity. In [22] several multivariate extensions of the classical definition have been studied. In this paper we propose a generalization of the notion of comonotonicity to the case of lattice-valued functions.

If A is a non empty set and $L$ is a lattice two function $f, g: A \to L$ are said to be *comonotone* if for every $x \in L$

$$\text{either} \quad \{f \geqslant x\} \supseteq \{g \geqslant x\} \quad \text{or} \quad \{g \geqslant x\} \supseteq \{f \geqslant x\}$$

We consider now some of the properties that a lattice-valued integral $I: \mathcal{F} \to L$ may or may not satisfy:

**(i)** (homogeneity) $I(f \wedge \mathbf{c}) = I(f) \wedge c$ for every $c \in L$ and for every $f \in \mathcal{F}$;

**(ii)** (invariance): $I(\gamma \circ f) = (\gamma \circ I)(f)$ for every continuous mapping $\gamma: L \to L$ and for every $f \in \mathcal{F}$;

**(iii)** (comonotone maxitivity): $I(f \wedge g) = I(f) \wedge I(g)$ if $f, g$ are comonotone elements of $\mathcal{F}$.

It is easy to prove that an invariant integral is homogeneous.

## 4    Quantiles and Sugeno Integrals in Complete Lattices

Here we provide a definition and characterization of quantiles for lattice-valued operators. In this section we assume that $L$ is a completely distributive lattice and that $\mathcal{A} = 2^A$.

**Definition 2.** *If $\alpha$ is an element of $L$ the lattice-valued quantile of level $\alpha$ is the functional $Q_\alpha: \mathcal{F} \times \mathcal{M} \to L$ such that*

$$Q_\alpha(f, m) = \bigvee \{x : m(\{f \geqslant x\}) \geqslant \alpha\}.$$

It can be proved that this definition extends the well known definition of quantile for real-valued functions(see [5]).

Say that a collection of sets $\mathcal{U} \subseteq 2^A$ is an *upper set* if $X \in \mathcal{U}$ and $X \subset Y$ implies that $Y \in \mathcal{U}$. Then we can prove the following result.

**Proposition 1.** *A lattice-valued integral $I: \mathcal{F} \to L$ is a lattice-valued quantile with respect to a non-additive measure $m: \mathcal{A} \to L$ if and only if there exists a upper set $\mathcal{U}$ such that*

$$I(f) = \bigvee_{X \in \mathcal{U}} \bigwedge_{x \in X} f(x)$$

*or if and only if there exists a upper set $\mathcal{U}$ such that*

$$I(f) = \bigwedge_{X \in \mathcal{U}} \bigvee_{x \in X} f(x)$$

*Proof.* If $I$ is a lattice-valued quantile we can consider the upper set $\mathcal{U} = \{X \in 2^A : m(X) \geq \alpha\}$ and we can get

$$I(f) = \bigvee_{X \in \mathcal{U}} \bigwedge_{x \in X} f(x).$$

By Theorem 5 in [4] we have that

$$\bigvee_{X \in \mathcal{U}} \bigwedge_{x \in X} f(x) = \bigwedge_{X \in \mathcal{V}} \bigvee_{x \in X} f(x)$$

where $\mathcal{V} = \{Y \subseteq A : X \cap Y \neq \emptyset$ for every $X \in \mathcal{U}\}$. It is easy to prove that $\mathcal{V}$ is an upper set being $\mathcal{U}$ an upper set.

Conversely given the upper set $\mathcal{U}$ if we define a non-additive measure $m : \mathcal{A} \to L$ such that $m(X) = 1$ if $X \in \mathcal{U}$ and $m(X) = 0$ if $X \notin \mathcal{U}$ $I(f)$ is a a lattice-valued quantile with respect to the non-additive measure $m$.

We can immediately prove that a lattice-valued quantile $Q_\alpha : \mathcal{F} \times \mathcal{M} \to L$ is an integral.    The following proposition characterizes quantiles as functionals $Q_\alpha : \mathcal{F} \to L$ in completely distributive lattices.

**Proposition 2.** *A lattice-valued quantile is an invariant and comonotone maxitive functional such that for every $X \subseteq A$ either $Q_\alpha(\mathbf{X}) = \mathbf{1}$ or $Q_\alpha(\mathbf{X}) = \mathbf{0}$. If $I : \mathcal{F} \times \mathcal{M} \to L$ is an integral such that either $I(\mathbf{X}) = 1$ or $I(\mathbf{X}) = 0$ then $I$ is a lattice-valued quantile if and only if $I$ is invariant.*

*Proof.* The result follows easily from Theorem 3 in [5], we have only to prove that a lattice-valued quantiles is comonotone maxitive.

If $f, g : A \to L$ are two comonotone functions, then $\{(f \vee g) \geqslant x\} = \{f \geqslant x\}) \cup \{g \geqslant x\}$ is equal to $\{f \geqslant x\}$ or to $\{g \geqslant x\}$. Then we can prove that $m(\{(f \vee g) \geqslant x\}) = m(\{(f \geqslant x\}) \vee m(\{(g \geqslant x\})$. Hence it follows that $Q_\alpha(f \vee g) = \bigvee\{x : m(\{(f \vee g) \geqslant x\}) \geqslant \alpha\} = Q_\alpha(f) \vee Q_\alpha(g)$.

We are interested in a class of integral functionals defined on a complete lattice. Following the approach in [13] we consider the functionals $S_l, S_u$ defined by :

$$S_l(m, f) = \bigvee_{x \in L} (x \wedge m(\{f \geqslant x\})) \quad \text{and}$$

$$S_u(m, f) = \bigwedge_{x \in L} (x \vee m(\{x : f(x)) \nleq x\})) .$$

If $L$ is a completely distributive lattice $S = S_l = S_u$ and the functional $S$ extends Sugeno integral to an ordinal framework and so is called the *lattice-valued Sugeno integral* of $f$ with respect to $m$. The following proposition provides an axiomatic representation of this functional.

**Proposition 3.** *A lattice-valued integral $I : \mathcal{F} \times \mathcal{M} \to L$ is a lattice-valued Sugeno integral that is*

$$I(m, F) = \bigvee_{x \in L} (x \wedge m(\{f \geqslant x\}))$$

*if and only if it is homogeneous and comonotone maxitive.*

*Proof.* It is straightforward to show that a lattice-valued Sugeno integral is monotone and homogeneous. Using the properties of comonotone functions as in the proof of Proposition 2 we can prove that the functional $S(f)$ is comonotone maxitive. Then we consider a comonotone maxitive and homogeneous integral $I = I(f)$. In [13] it is proved that this functional is a Sugeno integral if when $f, g: A \to L$ are two functions such that $f \leq g$ and $a \in L$ we have that

$$I(f \vee (a \wedge g)) = I(f) \vee (a \wedge I(g)).$$

If $a \geq x$ then $\{f \geqslant x\} \supseteq \{g \wedge a \geqslant x\}$ and if $a \not\geq x$ we have that $\{g \wedge a \geqslant x\} = \emptyset$ and then the functions $f, g \wedge a$ are comonotone. Hence we have $I(f \vee (a \wedge g)) = I(f) \vee I(a \wedge g)$ and then since $I$ is homogeneous and $I(a \wedge g)) = a \wedge I(g)$ an so the claim is proved.

Here we characterize quantiles as a subclass of the Sugeno integrals.

**Proposition 4.** *A lattice-valued Sugeno integral $S: \mathcal{F} \to L$ is a lattice-valued quantile if and only if there exists a $\{0, 1\}$-valued non-additive measure $m \in \mathcal{M}$ such that $I(f) = S(f, m)$.*

*Proof.* If $S$ is a lattice-valued quantile of level $\alpha$ with respect to the non additive measure $m \in \mathcal{M}$ can consider the non-additive measure $m^* \in \mathcal{M}$ such that $m^*(X) = 1$ if $m^*(X) \geq \alpha$ and $m^*(X) = 0$ otherwise. Conversely if $S$ is lattice-valued quantile of level $\alpha$ with respect to the $\{0, 1\}$-valued non-additive measure $m \in \mathcal{M}$, $S$ is a a lattice-valued quantile of level $\alpha$ for every $\alpha \neq 0$.

We can also prove that the subclass of quantiles generates the class of Sugeno integrals.

**Proposition 5.** *If $S: \mathcal{F} \times \mathcal{M} \to L$ is a lattice-valued Sugeno integral then for every $f \in \mathcal{F}$ and for every $m \in \mathcal{M}$ we have*

$$S(f, m) = \bigvee_{\alpha \in L} (Q_\alpha(f, m) \wedge \alpha)$$

*Proof.* If $f$ is an element of $\mathcal{F}$ and $m$ is an element of $\mathcal{M}$ we have that the set $\{Q_\alpha(f, m) : \alpha \in L\} \subseteq L$ and then $S(f, m) = \bigvee_{x \in L} (x \wedge m(\{f \geqslant x\})) \geq \bigwedge_{\alpha \in L} (Q_\alpha(f, m) \wedge \alpha)$.

If $x \in L$ is such that $m(\{f \geqslant x\}) = \alpha$ then $x \geq Q_\alpha$ hence we can prove that $S(f, m) = \bigvee_{x \in L} (x \wedge m(\{f \geqslant x\})) \leq \bigwedge_{\alpha \in L} (Q_\alpha(f, m) \wedge \alpha)$.

## 5   Real Valued Quantiles and Integrals: Some Extensions

Throughout this section $(A, \mathcal{A})$ be a measurable space (if $A$ is a finite set we usually assume that $\mathcal{A} = 2^A$) and $L$ is the real interval $[0, 1]$. It can be noticed that several type of integrals and in particular the Choquet integral were introduced considering real interval different from $[0, 1]$, but they can be transformed into the $[0, 1]$ framework.

If $\mathcal{M}$ is the class of $[0,1]$-valued non-additive measures defined on $A$, $\mathcal{F}$ is the class of $[0,1]$-valued measurable functions defined on $A$ and $f \in \mathcal{F}$ the Choquet integral is a mapping $C: \mathcal{F} \times \mathcal{M} \to [0,1]$ defined by

$$C(f, m) = \int f \, dm = \int_0^1 m(\{f \geqslant x\}) \, dx.$$

It is well known (see [8] and [9] for example) that Choquet integral is a *comonotone linear* functional i.e. $I(af + bg) = I(af) + I(bg)$ if $f, g$ are comonotone elements of $\mathcal{F}$ and $0 \leq a, b \leq 1$.

Moreover it can be proved that every integral functional $I: \mathcal{F} \times \mathcal{M} \to [0,1]$ that is comonotone linear is a Choquet integral (see [8] or [9]).

The following proposition proves that, as in the case of Sugeno integrals, the quantiles functionals are Choquet integrals and that the subclass of quantiles generates the class of Choquet integrals.

**Proposition 6.** *The Choquet integral $I: \mathcal{F} \to [0,1]$ is a quantile if and only if there exists a $\{0,1\}$-valued non-additive measure $m \in \mathcal{M}$ such that $I(f) = C(f, m)$.*

*If $C: \mathcal{F} \times \mathcal{M} \to L$ is a $[0,1]$-valued Choquet integral then for every $f \in \mathcal{F}$ and for every $m \in \mathcal{M}$ we have*

$$C(f, m) = \int_0^1 Q_\alpha(f, m) \, d\alpha.$$

*Proof.* If $I$ is a lattice-valued quantile of level $\alpha$ with respect to the non additive measure $m \in \mathcal{M}$ we consider the non-additive measure $m^* \in \mathcal{M}$ such that $m^*(X) = 1$ if $m^*(X) \geq \alpha$ and $m^*(X) = 0$ otherwise.

Then it is easy to prove that if $q = Q_\alpha(f, m)$

$$C(f, m^*) = \int f \, dm^* = \int_0^q m^*(\{f \geqslant x\}) \, dx = q.$$

The equality $C(f, m) = \int_0^1 Q_\alpha(f, m) \, d\alpha$ follows directly from proposition 1.4 in [8].

We have considered quantiles with respect to a (possibly non-additive) measure and not necessarily with respect to an endogenous probability. as in the classical case. Now we define quantiles with respect to a family of non-additive measures considering different attitude for low or high input values as in the definition of level-dependent integrals (see [18] and [12] ).

If $(m_t)$ is a family of elements of $\mathcal{M}$ and $t \in [0,1]$ we consider the *generalized quantile* of level $\alpha$, the *generalized Sugeno integral* and the *generalized Choquet integral* of a function $f \in \mathcal{F}$ as follows:

$$GQ_\alpha(f) = \bigvee \{x : m_\alpha(\{f \geqslant x\}) \geqslant \alpha\};$$

$$GS(f) = \bigvee_{\alpha \in L} (Q_\alpha(f, m_\alpha) \wedge \alpha) \quad \text{and}$$

$$GC(f) = \int_0^1 Q_\alpha(f, m_\alpha) \, d\alpha.$$

The proposed generalized integrals satisfy some minimal properties.

**Proposition 7.** *The generalized Sugeno integral and Choquet integral are monotone functionals. The generalized Sugeno integral is comonotone maxitive while the generalized Choquet integral is comonotone additive.*

*Proof.* Let us check comonotone maxitivity (additivity) of generalized Sugeno (Choquet) integral. If $f, g \in \mathcal{F}$ are comonotone functions then for every $\alpha \in$ then $GQ_\alpha(f \vee g) = GQ_\alpha(f) \vee GQ_\alpha(g)$ and $GQ_\alpha(f + g) = GQ_\alpha(f) + GQ_\alpha(g)$ and then the two properties are easily proved.

The integrals with respect to non-additive measures proved useful in many areas of decision theory. As it is well known in decision under uncertainty it leads to a generalization of expected utility. In this framework the aggregation functionals defined above introduces a a multiple quantiles model for a decision-making process under ambiguity in which a decision-maker is supposed to consider a rank among outcomes. Following the approach in [3] the generalized functionals are able to represent asymmetric attitude on extreme events (unexpected gains or unusual losses) and a rational prudence on ordinary events.

## 6    Application to Citation Analysis

Assessment of the quality of research has become increasingly necessary in recent years and many different indicators have been studied. We consider the approach in which the quality of a a research output is measured by citation analysis. Among the numerous bibliometric indices that have been used to evaluate the scientific production of a researcher or a scientific journal, a very popular index is the $h$-index which take into account the quality of the output of a scientist represented by the number of citations per paper and the impact represented by the number of paper ([14]). This index is relatively recent but the scientific community has shown a considerable interest for this indicator. The $h$-index of a researcher is the maximum number $h$ of papers of the considered scientist having at least $h$ citations each. The $h$-index is a particular case of Sugeno integral (see [20]) and obviously it is an aggregation operator.

Many recent papers generalize this approach and use aggregation functions in the analysis of citation data (see [21]). . Prospect Theory in an am We introduce a generalization of this index to quantify an individual's scientific work, considering in particular excellent papers.

Let $\mathbb{I} = [0, +\infty)$ denote the interval of nonnegative real numbers. We consider also a finite index set $N = \{1, \ldots, n\}$ and a non-additive measure $m: 2^N \to \mathbb{I}$. Let an author's output be characterized by a set of $N$ publications and, for each publication, the number of citations of that paper. We take here the number of citations as given and we consider the number of citations of a paper as a measure of the paper's quality. We represent a researcher by a function $f: N \to \mathbb{I}$ where

$f(i)$ is the number of citations of the $i$th publication and we assume $f(i) = 0$ if the considered author has less than $i$ publications.

An *impact index* is an aggregation function $I: \mathbb{I}^N \to \mathbb{I}$.

If the measure $m$ is the counting measure the impact index defined by

$$F(f) = S(m, f) = \int_0^1 Q_\alpha(f, m)\, d\alpha$$

is the h-index. The $h$-index identifies the most cited papers and it is insensitive to low cited papers. However this index don't identify researcher that have a moderate level of production but a very high impact. In some cases, to have large $h$-index is to have many good papers while a scientist with few papers with a high number of citations per paper in general has not a high $h$-index. For a discussion of this and other weakness of the $h$-index we refer to [1]. Now where are considered also a number of $h$-type indices proposed in the literature.

If $(m_t)$ is a family of non-additive measures on $N$ and $t \in \mathbb{I}$ we may consider the impact index defined by

$$GC(f) = \int_0^{+\infty} Q_\alpha(f, m_\alpha)\, d\alpha \quad \text{where}$$

$$Q_\alpha(f) = \bigvee_{\alpha \in \mathbb{I}} \{x : m_\alpha(\{f \geqslant x\}) \geqslant \alpha\}.$$

In this case we obtain a more flexible index which takes into account the degree of importance of a given level of citations. The proposed indices can be considered as generalized $h$-type indices. For a more detailed discussion on a number of generalized $h$-type indices proposed in the literature see the paper [1].

## 7   Concluding Remarks

We introduced a unified qualitative framework for studying non-additive measures and integration theory based on the notion of quantile.

The focus has been on aggregation functionals defined on lattices. In particular we have introduced integral-based aggregation functionals defined on completely distributive lattices.

For real-valued functions we introduce some functionals that generalize Sugeno and Choquet integrals and a further research direction is that of an axiomatic characterization of the considered aggregation functionals.

It is important to note that the proposed definition of generalized Sugeno integral can be easily extended to an ordinal framework. We have shown with an example that the proposed generalized integrals can be applied in real problems.

## References

1. Alonso, S., Cabrerizo, F.J., Herrera-Viedmac, E., Herrera, F.H.: H-index: A Review Focused in its Variants, Computation and Standardization for Different Scientific Fields. Journal of Informetrics 3, 273–1289 (2009)

2. Ban, A., Fechete, I.: Componentwise decomposition of some lattice-valued fuzzy integrals. Information Sciences 177, 1430–1440 (2007)
3. Basili, M., Chateauneuf, A.: Extreme events and entropy: a multiple quantile utility model. International Journal of Approximate Reasoning 52, 1095–1102 (2011)
4. Cardin, M., Couceiro, M.: Invariant functionals on completely distributive lattices. Fuzzy Sets and Systems 167, 45–56 (2011)
5. Cardin, M., Couceiro, M.: An Ordinal Approach to Risk Measurement. In: Perna, C., Sibillo, M. (eds.) Mathematical and Statistical Methods for Actuarial Science and Finance, pp. 79–86. Springer (2012)
6. Chambers, C.: Ordinal Aggregation and Quantiles. Journal of Economic Theory 137, 416–443 (2007)
7. Davey, B.A., Priestley, H.A.: Introduction to Lattices and Order. Cambridge University Press, New York (2002)
8. Denneberg, D.: Non-additive measure and integral. Kluwer Academic Publisher, Dordrecht (1994)
9. Grabisch, M., Marichal, J.L., Mesiar, R., Pap, E.: Aggregation Functions. In: Encyclopedia of Mathematics and its Applications. Cambridge University Press, Cambridge (2009)
10. Grätzer, G.: General Lattice Theory. Birkhäuser, Berlin (2003)
11. Gierz, G., Hofmann, K.H., Keimel, K., Lawson, J.D., Mislove, M.D., Scott, D.S.: Continuous Lattices and Domains. Cambridge University Press, Cambridge (2003)
12. Giove, S., Greco, S., Matarazzo, B.: The Choquet Integral with respect to level dependent capacity. Fuzzy Sets and Systems 175, 1–35 (2011)
13. Greco, G.H.: Fuzzy Integral and Fuzzy Measures with Their Values in Complete Lattices. Journal of Mathematical Analysis and Applications 126, 594–603 (1987)
14. Hirsch, J.E.: An index to quantify an individivual's scientific research output. Proceedings of the National Academy of Sciences of USA 102, 16569–16572 (2005)
15. Klement, E.P., Mesiar, R., Pap, E.: A Universal Integral. In: Proceedings EUSFLAT 2007, Ostrava, vol. I, pp. 253–256 (2007)
16. Mesiar, R.: Fuzzy measures and integrals. Fuzzy Sets and Systems 156, 365–370 (2005)
17. Mesiar, R.: Fuzzy integrals and linearity. International Journal of Approximate Reasoning 4, 352–358 (2008)
18. Mesiar, R., Mesiarová Zemánková, A., Ahmad, K.: Level-dependent Sugeno Integral. IEE Transactions on Fuzzy Systems 171, 167–172 (2009)
19. Murofushi, T., Sugeno, M.: Some quantities represented by the Choquet integral. Fuzzy Sets and Systems 5, 229–235 (1993)
20. Narukawa, Y., Torra, V.: The h-Index and the Number of Citations: Two Fuzzy Integrals. IEE Transactions on Fuzzy Systems 16, 795–797 (2008)
21. Narukawa, Y., Torra, V.: Multidimensional generalized fuzzy integral. Fuzzy Sets and Systems 160, 802–815 (2009)
22. Puccetti, G., Scarsini, M.: Multivariate comonotonicity. Journal of Multivariate Analysis 101, 291–304 (2010)

# Analysis of On-Line Social Networks Represented as Graphs – Extraction of an Approximation of Community Structure Using Sampling

Néstor Martínez Arqué[1] and David F. Nettleton[1,2]

[1] Dept. Information Technology and Communications,
Universitat Pompeu Fabra, Barcelona, Spain
{nestor.martinez,david.nettleton}@upf.edu
[2] IIIA-CSIC, Bellaterra, Spain

**Abstract.** In this paper we benchmark two distinct algorithms for extracting community structure from social networks represented as graphs, considering how we can representatively sample an OSN graph while maintaining its community structure. We also evaluate the extraction algorithms' optimum value (modularity) for the number of communities using five well-known benchmarking datasets, two of which represent real online OSN data. Also we consider the assignment of the filtering and sampling criteria for each dataset. We find that the extraction algorithms work well for finding the major communities in the original and the sampled datasets. The quality of the results is measured using an NMI (Normalized Mutual Information) type metric to identify the grade of correspondence between the communities generated from the original data and those generated from the sampled data. We find that a representative sampling is possible which preserves the key community structures of an OSN graph, significantly reducing computational cost and also making the resulting graph structure easier to visualize. Finally, comparing the communities generated by each algorithm, we identify the grade of correspondence.

**Keywords:** Data mining, social networks.

## 1 Introduction

Finding structure in ad-hoc networks without any a priori knowledge about the expected result is a complex task. With the advent of online social networks (OSNs), the study of how to extract a vision of the network in terms of 'communities' has become an active field. By 'communities' we understand the 'sociological' interpretation in which individuals (humans beings) interact socially in some way (by email, using some online application, collaborating in some endeavour such as the writing of scientific papers, or forming some other social group such as a club, association, and so on). We can also extent our definition of individuals to include the study of the behaviour and social interactions of living beings in general (such as Dolphins, Simians and so on).

The results of extracting a community structure are highly dependent on the statistical and topological characteristics of the graph dataset, such as the average degree,

V. Torra et al. (Eds.): MDAI 2012, LNAI 7647, pp. 149–160, 2012.
© Springer-Verlag Berlin Heidelberg 2012

clustering coefficient and level of fragmentation. It may also be dependent on the community extraction algorithm used, many of which are stochastic and non-deterministic. Other problems include the large volume of data in many OSN logs, the presence of 'noise' or 'unreliable' and outdated links in the graph, and highly fragmented graphs.

In the present study we apply two distinct community extraction algorithms [1,2] to five structurally distinct datasets, and compare the results. The extraction algorithms represent an optimization process based on an entropy type metric (modularity). For the three largest datasets we have also applied a filtering processing to reduce the number of nodes tested, while maintaining the key community structure information. This implies a very considerable saving in computational cost of processing by the community search algorithm. In this paper we describe how a filter based on degree and/or clustering coefficient enables us to extract the core parts of the communities in a complex graph. This filtering also significantly improves the results of visualization, using, for example, the Gephi software tool (http://gephi.org/), avoiding the typical "hairball" [3] appearance of many high data volume social networks.

The structure of the paper is as follows: in Section 2 we present the state of the art and related work; in Section 3 we present our approach for filtering and sampling the data; in Section 4 we define the datasets used and the experimental setup; in Section 5 we present the empirical tests and results for the community extraction, comparing the two algorithms and sampling Vs. using the complete dataset; finally in Section 6 we give the conclusions.

## 2     State of the Art and Related Work

The following briefly reviews the related work and key authors in the field of OSN graph processing, community detection and OSN graph sampling.

The study of community structure in social networks has been of interest for many years as a multidisciplinary field [4, 5]. More recently, with the advent of online social networks (Facebook, Twitter, etc.) research in this area has been given a great impulse due to the availability of (some) of this online data for analysis, by authors such as [1, 6, 7, 8, 9], and which deal specifically with the mining of social networks as graphs [10, 11]. In this paper we benchmark two community structure extraction algorithms: Newman[1], which we have implemented in Python NetworkX and **(ii)** the Louvain method[2] using the default version available in the Gephi graph processing software.

Newman's algorithm[1] focuses on how to extract a community structure from social network graph data. Two main approaches are defined: (i) the identification of groups around a prototypic nucleus defined in terms of the 'most central' edges, an adjacency matrix being used as the basis to calculate the weights; (ii) identification of groups by their boundaries, using the least central edges (frontiers). This metric is also referred to as "edge betweenness", and is based on Freeman's "betweenness centrality measure" [5]. The algorithm is as follows: (a) calculate the betweenness for all edges in the graph; (b) remove the edge with the highest betweenness; (c) recalculate betweennesses for all edges affected by the removal; (d) repeat from step (b) until no edges remain. Newman's fast algorithm [12] is used for calculating betweenness.

Newman's algorithm[1] extracts the communities by successively dividing the graph into components, using a metric to quantify the quality of the community partitions 'on the fly'. The value calculated by the quality metric for a given community is called the modularity. For a graph divided into $k$ communities, a symmetrical matrix $e$ of order $k2$ is defined whose elements $e_{ij}$ are the subset of edges from the total graph which connect the nodes of communities i and j.

The modularity metric is defined as the fraction of edges in the graph which connect vertices in the same community, minus the expected value of the same number of edges in the graph with the same community partitions but with random connections between their respective nodes. If the number of intra-community edges shows no improvement on the expected value, then the modularity would be Q=0. On the other hand, Q approaches a maximum value of 1 when the community structure is strong. According to [1], the usual empirical range for Q is between 0.3 and 0.7.

The Louvain method[2] can be considered an optimization of Newman's method, in terms of computational cost. Firstly, it looks for smaller communities by optimizing modularity locally. As a second step, it aggregates nodes of the same community and builds a new network whose nodes are the communities. These two steps are repeated iteratively until the modularity value is maximized. The optimization consists of evaluating the modularity gain, which is done by performing a local calculation of the change in modularity for a given community, caused by moving each node from it to an adjacent community. With each iteration the number of nodes to test quickly reduces (due to the aggregation of the corresponding nodes), and the computational cost is reduced in the same order.

With respect to the evaluation of community detection algorithms, Lancichinetti and Fortunato in [13] carried out an exhaustive benchmarking of 12 different methods, including the Louvain method[2] and Newman's method[1]. However, they used synthetic datasets for their tests and did not evaluate sampling of the networks. In the current work we have used real datasets and considered sampling. In [13], partition comparison between methods used the fraction of correctly identified nodes measure (NMI- Normalized Mutual Information). Lancichinetti also benchmarked a second measure, called 'LFR', which also takes into account degree power law distributions and community size. In our current work we have used Girvan and Newman's benchmark [14], using only the top N communities for evaluation, chosen by studying the size distributions.

Sampling is a key aspect of processing large graph datasets, when it becomes increasingly difficult to process the graph as a whole due to memory and/or time constraints. Sampling is related to, but not the same as filtering. Filtering eliminates records from the complete dataset according to some criteria, for example, "remove all nodes with degree equal to one". Sampling, on the other hand, tries to maintain the statistical distributions and properties of the original dataset. For example, if 10% of the nodes have degree = 1 in the complete graph, in the sample the same would be true. Chakrabarti, in [15], compares two sampling methods: *(i)* a full graph data collection and *(ii)* the Snowball method. The latter is implemented by taking well connected seed nodes and growing a graph around them. However the authors confirm the general consensus in the literature that although 'snowballing' is an adequate technique for graph sampling, it tends to miss out isolated individuals. In order to solve this problem, the authors propose a random or probabilistically weighted selection of seeds. However, for community sampling, we propose that this bias is advantageous because we are interested in identifying key hubs and highly connected neighbors, as opposed to the more isolated regions and nodes of the graph. A key consideration in sampling is the choice of the initial starting nodes (or 'seeds') for extracting the sample. Another consideration is how to measure the 'quality' of the derived sample. These two aspects are studied in [16, 17].

## 3     Our Filtering/Sampling Approach

In this section, with reference to Figures 1a and 1b, we will see how we apply a two step process, consisting of filtering followed by sampling, in order to obtain a subset of a complete graph consisting of three communities. We emphasize that we have defined a process which is customized for extracting community structure. Thus we make emphasis on identifying hub nodes, high density regions, and their neighbors, rather than on an equitative percentage of all types of nodes. Hub nodes are identified by their degree, and high density regions are identified by the clustering coefficient. Once we have selected the "seed nodes" based on their degree or clustering coefficient, then we apply a sampling at 1 hop to obtain all the neighbors of each "seed node". Again, instead of applying a proportional number of nodes based on their distribution of the complete dataset, we let the search be biased to nodes with a high degree or a high clustering coefficient. In Fig. 1 we see a schematic representation of the filtering and sampling process.

Now we will see how we would process a simple graph consisting of 3 communities. In Fig. 2 we see the assigning of seeds (encircled nodes) using the 92.5 percentile of the degree *(a)* and clustering coefficient *(b)* values, respectively, and then including all the seed's neighbors (indicated by rectangles) at one hop. This means that the degree of the seed nodes (Fig. 2a) will be in the top 7.5% of the degree distribution for the complete graph. Likewise, the clustering coefficient of the seed nodes (Fig. 2b) will be in the top 7.5 of the distribution of the clustering coefficient for the complete graph.

We see that a very good coverage is obtained of the three communities in this graph, without having to expand the inclusion of nodes (in a "snowball" fashion) to 2 or more hops. This is because the regions we are interested in, the community cores, will be generally made up of a lattice of high degree and/or highly interlinked nodes. Therefore, selecting precisely these nodes as the seeds and including their neighbors will cover a high percentage of the core component of the major communities. In the empirical section we see how this result applies for much more complex and fragmented networks, and how we decide when to use the degree as the filter, or the clustering coefficient.

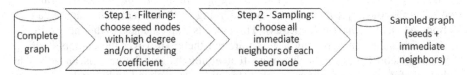

**Fig. 1.** Schematic representation of the node filtering and selection process

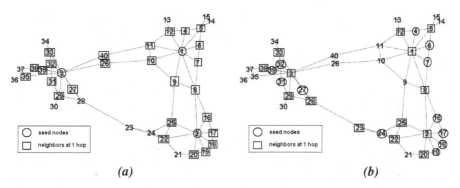

**Fig. 2.** (a) Selection of seed nodes using 92,5 percentile of degree values; (b) Selection of seed nodes using 92,5 percentile of clustering coefficient values frequency

## 4    Datasets Used and Experimental Setup

In this section we briefly describe the procedure we have followed, the datasets used and their basic statistics. We also give the details of the sampling for the three large datasets, and their statistics after sampling. We use five different benchmark datasets: Karate [4], Dolphins [18], ArXiv GrQc-General Relativity and Quantum Cosmology [19], Enron [7] and Facebook [9]. In Table 1 we see a summary of the basic graph statistics for each test dataset. For community extraction we apply Newman's algorithm[1] (which we implemented in Python NetworkX) and the Louvain method[2] (standard version available in Gephi) to the datasets.

### 4.1    Values for Filtering and Sampling

With reference to Table 2, we see the summary for the sampling methods, per dataset. The filter and the value were chosen by different trial and error tests in order to obtain the desired overall percentage, which is the sum of the seed nodes plus all the neighbors of each of these. We used the recommendations of [20] as a guideline for the approximate optimum percentage of the complete dataset, which Ahn stated as being 25% for the degree as filter, and 20% for the clustering coefficient as filter. However, we found that the real sample size depends on the dataset and the distributions of the degree and clustering coefficient values.

**Table 1.** Summary of graph statistics for the five original datasets

|  | Karate | Dolphins | GrQc | Enron | Facebook |
|---|---|---|---|---|---|
| **#Nodes** | 34 | 62 | 5242 | 10630 | 31720 |
| **#Edges** | 78 | 159 | 14496 | 164837 | 80592 |
| **Avg. degree** | 4.59 | 5.13 | 5.530 | 31.013 | 5.081 |
| **Clust. coef.** | 0.57 | 0.26 | 0.529 | 0.383 | 0.079 |
| **Avg. path length** | 2.408 | 3.356 | 6.049 | 3.160 | 6.432 |
| **Diameter** | 5 | 8 | 17 | 20 | 9 |

**Table 2.** Summary of sampling criteria/methods for sampled datasets

|  | Filter | Value | Resulting sample size | Sample |
|---|---|---|---|---|
| **ArXiv-GrQc** | Degree | ≥30 | 17.91% | All neighbors |
| **Enron** | Clustering coef. | =1 | 20.83% | All neighbors |
| **Facebook** | Clustering coef. | ≥0.5 | 10.75% | All neighbors |

**Table 3.** Summary of graph statistics for the three largest sampled datasets

|  | GrQc | Enron | Facebook |
|---|---|---|---|
| **#Nodes** | 939 | 2218 | 3410 |
| **#Edges** | 5715 | 14912 | 6561 |
| **Avg. degree** | 12.17 | 12.315 | 3.848 |
| **Clust. coef.** | 0.698 | 0.761 | 0.632 |
| **Avg. path length** | 4.51 | 3.143 | 8.388 |
| **Diameter** | 10 | 7 | 27 |

In Table 3 we summarize the basic graph statistics for each test dataset, after sampling. As we mentioned in Sec. 3, we are interested in extracting the strong community structure of the graph and therefore the fact that the before/after statistics are distinct is not an issue, which is caused mainly by the omission of isolated and low connectivity areas of the graph. One negative aspect would be the possible loss of some of the bridge nodes between communities, as commented in the previous section.

# 5    Empirical Tests and Results

In Section 5.1 we first document the results of applying Newman's method to the sampled datasets; then in Section 5.2 we compare the results with those of the literature; in Section 5.3 we apply the Louvain method and compare the communities extracted from the original datasets to those extracted from the sampled datasets, using an NMI type metric for node assignments to communities; finally, in Section 5.4 we compare the communities extracted by Newman's method and the Louvain method using the same NMI metric.   N.B. *In the following text we will now refer throughout to Newman's method as NG and the Louvain method as LV.*

## 5.1    Evaluation of Newman's (NG) Method with the Sampled Datasets

For the **ArXiv-GrQc** dataset[19] and with reference to Fig. 3 (GrQc) and Table 4, the optimum modularity was obtained at Q=0.777, produced at iteration 56 and which partitioned the sampled version of the dataset in 57 communities. As can be seen in Fig. 3 (GrQc), the modularity value rises rapidly to a global maximum, which it maintains during approx. 100 iterations and then decays smoothly. With reference to Fig. 4a, the greatest community (lowest part of the Figure), is formed by 16.29% of the total nodes. The next two communities represent 10% of the total nodes.

For the **Enron** dataset, with reference to Fig. 3 (Enron) and Table 4, the optimum modularity was found at iteration 865, corresponding to Q=0.42, and dividing the dataset into 869 communities, of which the biggest represented 66.95% of the total nodes. We see from Fig. 2 that the modularity ascends rapidly during the first 70 iterations, and then keeps increasing with a much shallower gradient until it reaches the optimum, after which it begins to decay significantly. The version we finally used for the sampled Enron dataset was that generated an early iteration (51), which partitioned the dataset into 56 communities with a modularity close to the optimum obtained later at iteration 864.

For the **Facebook** dataset, with reference to Fig. 3 (Facebook) and Table 4, the optimum modularity was found at iteration 40, with Q=0.87 (a relatively high value with respect to the other datasets), resulting in a total of 190 communities (Fig. 4b). This value was obtained as a consequence of the low clustering coefficient in the dataset. As can be seen in Fig. 3 (Facebook), the optimum value is found relatively early on in the process, followed by a linear decay from that point onwards.

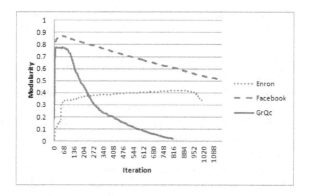

**Fig. 3.** Evolution of the modularity Q for the three largest sampled datasets

**Table 4.** Summary of community structure processing statistics for five test datasets using the NG Method

|  | It. | Q | C | Original or Sampled |
|---|---|---|---|---|
| Karate | 4 | 0.494 | 5 | O |
| Dolphins | 5 | 0.591 | 6 | O |
| GrQc | 56 | 0.777 | 57 | S |
| Enron | 865 | 0.421 | 869 | S |
| Enron Early* | 51 | 0.325 | 56 | S |
| Facebook | 40 | 0.870 | 190 | S |

It.=number of iterations, Q=modularity, C=number of communities, *Early termination with a semi-optimal Q.

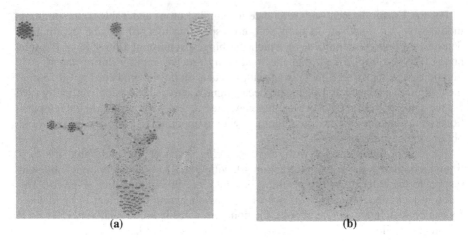

**Fig. 4.** **NG Method:** (a) Visualization of the principal communities extracted for the sampled arXiv-GrQc dataset; (b) Visualization for the Facebook dataset, with a partitioning corresponding to 190 communities obtained at iteration 40 (both using 'Force Atlas' visualization metric)

## 5.2    Discussion of Results with Reference to the Literature

In the following we reference the literature in order to obtain some idea of what we could consider the correct number of communities in each of the three largest datasets. For the arXiv-GrQc dataset, Xie[21] reported 499 communities using a "Label Propagation Algorithm" and 605 communities using a "Clique Propagation Algorithm". In the case of the Enron dataset, Shetty[7] defined the dataset as consisting of 151 Enron employees, and 5 key (hub) persons, however the graph is generated by emails sent between these persons, including "cc" and "re:" mailings to external emails, which greatly increases the number of entities in the dataset, and therefore nodes in the graph. Finally, for the Facebook dataset, there are no definitive values for the number of clusters in the literature, however Viswanath in [9] reported a high fragmentation into small communities, and Leskovec in [19] also reported a relatively high fragmentation of communities in other online social networks similar to Facebook. However the fragmentation of this particular dataset is also probably influenced by the measure of interaction (writes to wall) used to define links between users.

**Table 5.** Summary of community Q and C values for the three largest graphs using LV on the sampled and original versions of the datasets

|          | Original | | Sampled | |
|----------|----------|------|----------|------|
|          | Q        | C    | Q        | C    |
| **GrQc**     | 0.856    | 390  | 0.789    | 11   |
| **Enron**    | 0.491    | 43   | 0.560    | 68   |
| **Facebook** | 0.681    | 1105 | 0.519    | 33   |

Q=modularity, C=number of communities.

## 5.3    Communities Extracted using the LV Method from Sampled Datasets Vs. Communities Extracted using the LV Method from Original Datasets

In the following Section we have used the LV Method (rather than the NG Method) to compare the communities in the sampled and original datasets, given the very large computational cost of NG to process the larger (original) graphs, and because LV allows us to obtain an adequate benchmark for the datasets which is comparable to the results in the literature.

Table 5 shows a summary of the Q (modularity) and C (number of communities) values generated for the original and the sampled datasets. For the original datasets we see a high fragmentation of small communities in the arXiv-GrQc and Facebook datasets, which in the sampled datasets is greatly reduced so as to include only the most significant communities. Curiously, for the Enron dataset more communities are found in the sampled dataset than in the original dataset (68 with respect to 43), however for the former (sampled dataset) the number of communities it finds (value of 68 in Table 5) is similar to that of NG (value of 56 in Table 4).

**NMI (Normalized Mutual Information).** With reference to Table 6, we now compare the results of the community labeling by counting the number of nodes which are assigned in each community, and the number of nodes which are assigned to the same corresponding community, in the sampled and original datasets. This represents a NMI (Normalized Mutual Information) type metric [13] in which the nodes are labeled by the community of the original dataset, following the method we commented in Section 2. Table 6 summarizes the "purity" of the correspondences, in which a "purity" of 100% would mean that all the nodes which were assigned to communities $C_1...C_N$ in the sampled dataset were also assigned to communities $C_1...C_N$ in the original dataset.

The matching is made more difficult given that the LV Method (and the NG Method) are stochastic and non-deterministic. This means that each execution may produce slightly different results (nodes are assigned to different communities), although we assume the extraction of the most important communities will be similar. Also, the labels assigned to the communities (community 1, 2, etc.) may vary. Hence, in order to realize a comparison, we must first find the majority matching of each community label in the first execution with each community label in the second, in order to establish the correspondence. We chose the top N communities (in general N=10) by studying the size distribution.

In column B of Table 6 we see the difference for the sampled dataset with itself (for two different executions of the algorithm), and in column C we see the difference for the original dataset with itself. Hence, if we consider the correspondence of the

**Table 6.** Comparison of correspondence (NMI) of community assignments between original datasets and sampled datasets (using the LV Method)

|  | NMI orig. Vs. sampled (A) | NMI sampled Vs. sampled (B) | NMI orig. Vs. orig. (C) | Net loss (C - A) |
|---|---|---|---|---|
| **GrQc** | 0.66559 | 0.82544 | 0.77301 | 0.10742 |
| **Enron** | 0.69069 | 0.86903 | 0.82012 | 0.12943 |
| **Facebook** | 0.58996 | 0.73249 | 0.69215 | 0.10219 |

original dataset with itself as the baseline (column C), then the net precision loss (last column of Table 6) will be the difference between the baseline and the correspondence between the communities of the original dataset with those of the sampled dataset (column A).

We observe from the final column of Table 6 that the precision loss is between 10% and 13%, depending on the dataset, and the average correspondence is between 58% and 70% (column A). The NMI of the sampled datasets (column B) represents a significant improvement with respect to the original datasets (column C). Finally, we inspected the correspondence between communities in the original dataset and those in the sampled dataset. We found that the most "pure" communities and the most "impure" in general remained the same, that is, communities with a high relative correspondence remained so and those with a low relative correspondence also remained so.

### 5.4    Communities Extracted by LV Method vs. Communities Extracted by NG Method

**In Terms of 'Q' (Modularity Value) and 'C' (Number of Communities Created):** We first compare the methods referring to Table 4 (NG, columns 3 and 4) and Table 5 (LV, 2 rightmost columns). In the case of the sampled version of the GrQc dataset, the number of communities extracted by LV (11) was different from that of NG (57). In terms of Q (modularity), both methods gave the same value (0.77 Vs. 0.79). For the sampled version of the Enron dataset, for NG we took the early cutoff version. The number of communities found was similar (56 for NG Vs 68 for LV), however the Q value was significantly lower for NG (0.32 for NG Vs. 0.56 for LV).

Finally, Facebook, gave the biggest difference in terms of C (190 communities for NG Vs. 33 for LV) and Q (0.87 for NG Vs. 0.52 for LV). We propose that a key factor in this result the lack of an identifiable cut-off point for NG, and the high fragmentation of communities in the Facebook dataset. In general, we can conclude that NG and LV may give distinct results in terms of the number of communities and modularity values.

**In Terms of NMI (Normalized Mutual Information):** In Table 7 we compare the assignments of nodes between the top $N$ communities $\{C_{LV}\}$ extracted by LV and those extracted by NG $\{C_{NN}\}$, for the sampled data. We note that for column A we have used the $N$ largest communities $\{C_{LV}\}$ created by LV, by number of nodes, then we find the percentage of corresponding nodes of the principal corresponding communities $\{C_{NN}\}$ of NG.

**Table 7.** Normalized Mutual Information (NMI) comparison of correspondence of node assignments to communities: LV Method Vs NG Method for sampled data

|  | NMI LV Vs. NG (A) | NMI NG Vs. LV (B) | NMI orig. Vs. orig. (C) | Net loss C - Avg. (A, B) |
|---|---|---|---|---|
| GrQc | 0.69116 | 0.87243 | 0.77301 | -0.00878 |
| Enron | 0.31313 | 0.68796 | 0.82012 | 0.31958 |
| Enron early | 0.83437 | 0.44320 | 0.82012 | 0.18133 |
| Facebook | 0.62056 | 0.54551 | 0.69215 | 0.10911 |

Contrastingly, in column B we first identify the top $N$ communities $\{C_{NN}\}$ created by NG by number of nodes, then we find the percentage of corresponding nodes of the principal communities of LV $\{C_B\}$. In column C we define the same baseline used in Table 6, which is the NMI of the communities of the same dataset for two different executions of LV. Finally, in the final column we take the difference between the baseline and the average of columns A and B. From Table 7 we observe that the correspondence, in terms of node assignments, between the two methods is dataset dependent, with Enron (maximum number of iterations) having the least similarity (0.31) and GrQc having the greatest similarity (-0.01). We have also considered the 'early cutoff' version of applying NG to the Enron data, given that it produced a much smaller number of communities. As can be seen, there is a significant improvement with respect to the version which was allowed to run much longer (0.18 Vs. 0.31).

In conclusion with respect to the comparison of the methods, the empirical tests and results show there is a significant difference between the assignment of the nodes between methods.

## 6     Conclusions

We have benchmarked five statistically and topologically distinct datasets, applying two community structure elicitation algorithms and sampling on the three biggest datasets. The sampling is designed to maintain the overall community structure by choosing "hub" type nodes and high density regions, based on degree and clustering coefficient. The results indicate that it is possible to identify the principal communities for large complex datasets, using this type of sampling. The sampling method maintains the key facets of the community structure of a dataset, while reducing significantly (80 to 90%) the dataset size. We have also established, due to the stochastic nature of the algorithms, that a significant difference is found in the assignment of nodes to communities between different executions and methods. However, by inspection of the communities, we observe that the overall structure is consistent.

**Acknowledgments.** This research is partially supported by the Spanish MEC, (project HIPERGRAPH TIN2009-14560-C03-01).

## References

1. Newman, M.E.J., Girvan, M.: Finding and evaluating community structure in networks. Phys. Rev. E 69, 26113 (2004)
2. Blondel, V.D., Guillaume, J.L., Lambiotte, R., Lefebure, E.: Fast unfolding of communities in large networks. In: J. of Stat. Mech.: Theory and Experiment (10), P1000 (2008)
3. Traud, A.L., Kelsic, E.D., Mucha, P.J., Porter, M.A.: Comparing Community Structure to Characteristics in Online Collegiate Social Networks. SIAM Review 53(3), 526–543 (2011)
4. Zachary, W.W.: An Information Flow Model for Conflict and Fission in Small Groups. Journal of Anthropological Research 33, 452–473 (1977)
5. Freeman, L.C.: A Set of Measures of Centrality Based on Betweenness. Sociometry 40(1), 35–41 (1977)

6. Newman, M.E.J.: Modularity and community structure in networks. PNAS 103(23), 8577–8582 (2006)
7. Shetty, J., Adibi, J.: Discovering Important Nodes through Graph Entropy - The Case of Enron Email Database. In: Proc. 3rd Int. W. on Link Discovery, pp. 74–81 (2005)
8. Mislove, A., Marcon, M., Gummadi, K.P., Druschel, P., Bhattarcharjee, B.: Measurement and analysis of online social networks. In: Proc. 7th ACM SIGCOMM Conference on Internet Measurement, IMC 2007, pp. 29–42 (2007)
9. Viswanath, B., Mislove, A., Cha, M., Gummadi, K.P.: On the Evolution of User Interaction in Facebook. In: Proc. 2nd ACM Workshop on Online Social Networks, WOSN 2009, Barcelona, Spain, pp. 37–42 (2009)
10. Kleinberg, J.M.: Challenges in mining social network data: processes, privacy, and paradoxes. In: Proc. 13th Int. Conf. on K. Disc. & Data Mining (KDD 2007), pp. 4–5 (2007)
11. Kumar, R., Novak, J., Tomkins, A.: Structure and Evolution of Online Social Networks. In: Link Mining: Models, Algorithms, and Applications, Part 4, pp. 337-357. Springer (2010)
12. Newman, M.E.J.: Scientific collaboration networks. I. Network construction and fundamental results. Phys. Rev. E 64(1), 016131 (2001)
13. Lancichinetti, A., Fortunato, S.: Community Detection Algorithms: a comparative analysis. Physical Review E 80, 056117 (2009)
14. Girvan, M., Newman, M.E.J.: Community structure in social and biological networks. Proc. Natl. Acad. Sci. USA 99, 7821–7826 (2002)
15. Chakrabarti, D., Faloutsos, C.: Graph mining: Laws, generators, and algorithms. ACM Computing Surveys 38(1) (March 2006)
16. Bartz, K., Blitzstein, J., Liu, J.: Graphs, Bridges and Snowballs: Monte Carlo Maximum Likelihood for Exponential Random Graph Models. Presentation, January 8 (2009)
17. Lee, S.H., Kim, P.J., Jeong, H.: Statistical properties of sampled networks. Phys. Rev. E 73, 016102 (2006)
18. Lusseau, D., Schneider, K., Boisseau, O.J., Haase, P., Slooten, E., Dawson, S.M.: The bottlenose dolphin community of Doubtful Sound features a large proportion of long-lasting associations. Behavioral Ecology and Sociobiology 54(4), 396–405 (2003)
19. Leskovec, J., Lang, K.J., Dasgupta, A., Mahoney, M.W.: Community Structure in Large Networks: Natural Cluster Sizes and the Absence of Large Well-Defined Clusters. Internet Mathematics 6(1), 29–123 (2009)
20. Ahn, Y.Y., Han, S., Kwak, H., Moon, S., Jeong, H.: Analysis of topological characteristics of huge online social networking services. In: Proc. 6th Int. Conf. WWW, pp. 835–844 (2007)
21. Xie, J., Szymanski, B.K., Liu, X.: SLPA: Uncovering Overlapping Communities in Social Networks via a Speaker-listener Interaction Dynamic Process. Cornell University Library (2011), http://arxiv.org arXiv:1109.5720v3

# Using Profiling Techniques to Protect the User's Privacy in Twitter

Alexandre Viejo, David Sánchez, and Jordi Castellà-Roca

Departament d'Enginyeria Informàtica i Matemàtiques,
UNESCO Chair in Data Privacy, Universitat Rovira i Virgili
Av. Països Catalans 26, E-43007 Tarragona, Spain
{alexandre.viejo,david.sanchez,jordi.castella}@urv.cat

**Abstract.** The emergence of microblogging-based social networks shows how important it is for common people to share information worldwide. In this environment, Twitter has set it apart from the rest of competitors. Users publish text messages containing opinions and information about a wide range of topics, including personal ones. Previous works have shown that these publications can be analyzed to extract useful information for the society but also to characterize the users who generate them and, hence, to build personal profiles. This latter situation poses a serious threat to users' privacy. In this paper, we present a new privacy-preserving scheme that distorts the real user profile in front of automatic profiling systems applied to Twitter. This is done while keeping user publications intact in order to interfere the least with her followers. The method has been tested using Twitter publications gathered from renowned users, showing that it effectively obfuscates users' profiles.

**Keywords:** Microblogging, Noise addition, Privacy, Profiling, Twitter.

## 1 Introduction

Twitter is a very popular online social network and microblogging system. Basically, this social tool allows registered users to share short text-based posts (named *tweets*) up to 140 characters with anybody else on the Internet. Nowadays, the company behind this social network claims to have 100 million active users who generate 230 million tweets on average per day [1]. This fact proves two points: (i) how sharing information has become a real necessity for common people; and (ii) the relevance of Twitter in the present day.

Nevertheless, tweets generally contain personal information [2] and this fact motivates the existence of systems that analyze those publications and build *user profiles*. This implies that profiling data such as user preferences can be linked with her identity. This may invite malicious attacks from the cyberspace (*e.g.*; personalized spamming, phishing, etc) and even from the real world (*e.g.*, stalking) [4].

Several profiling mechanisms that gather information from Twitter (and other Internet services) can be found in the literature [5–10]. In general, all these tools

V. Torra et al. (Eds.): MDAI 2012, LNAI 7647, pp. 161–172, 2012.

generate *user profiles* that contain the interests of the users. In order to achieve that, they generally use a knowledge base (*e.g.,* Wikipedia [11], news repositories, the Web) to semantically interpret message contents and the number of term occurrences/co-occurrences to calculate the weight of each topic of interest (*e.g.,* sports, technology, health, etc), which is found in the tweets of a certain user. As an example of this process, the authors in [8] present a knowledge-based framework that builds user profiles from text messages shared in social platforms. Researchers [9, 10] have also noted the difficulty of analyzing textual contents in Twitter due to the use of short text-based messages. Tweets are, quite commonly, ungrammatical and noisy (due to hashtags, re-tweets, URLs). Hence, due to the difficulty of applying exhaustive syntactical analyses, Twitter profiling systems [7–10] usually analyse the whole user tweets as a bag-of-words.

In order to prevent privacy problems that are inherent to the existence of *profilers* and *user profiles*, it is convenient to design privacy preserving tools that allow users to protect themselves from profilers. Nevertheless, it is worth to mention that privacy-preserving methods based on restricting the visibility of the user-generated content compromise as well the capability of users to gain attention from others. Indeed, since this is one of the main motivations of using Twitter, this straightforward approach may not be widely adopted. Therefore, a successful privacy-preserving scheme should protect the privacy without limiting the visibility of user-generated data.

### 1.1   Contribution and Plan of This Paper

In this paper we propose a new scheme designed to prevent automatic text-based data extraction techniques from profiling users (*i.e.,* to discover dominant profile categories) of microblogging services. The basic idea is to introduce a set of fake messages within the user account while maintaining users' messages intact. We have used Twitter to test our proposal but the results achieved can be applied to any other social platform which works with text-based messages.

Section 2 discusses previous work aiming at providing privacy to microblogging services and introduces the basis of our approach. Section 3 presents our privacy-preserving method, detailing how it distorts the user profile (in order to hide it) as new messages are published. Section 4 evaluates the proposal, applying it to several well-differentiated Twitter users. The final section presents the conclusions and some lines of future research.

## 2   Previous Work

In the literature there is an important lack of mechanisms that try to preserve the privacy of the users of Twitter and similar platforms.

In [12], the authors present a Firefox extension that allows users to specify which data or activity need to be kept private. Sensitive data is substituted with fake one, while the real data is stored in a third party server that can be only accessed by the allowed users. Note that this solution requires a centralized infrastructure which must be honest and always available. Clearly, this requirement is a very strong shortcoming of this proposal.

This behaviour avoids profiling but it also jeopardizes the capability of the users to gain attention from others. As explained previously getting attention (which in turn provides publicity, vanity and ego gratification) is the main motivation for Twitter users (and other similar platforms) [3].

In order to avoid this major problem, we propose a method which aims to distort user profiles while keeping their original messages intact. In order to do so, we studied techniques proposed in the research field of *unstructured text anonymization* which aim to hide original information while not completely removing/transforming it. In [13], authors explain three different methods to deal with the anonymization of textual documents: (i) *Named entity generalization*: sensitive entities can be generalized (*e.g.*, iPhone → cell phone) to achieve some degree of privacy while preserving some of their semantic meaning in the document [14]; (ii) *Entity swapping*: this method is based on swapping relatively similar entities [15] between documents of the same set, or within the same document depending on the concrete case; and (iii) *Entity noise addition*: this method introduces new entities to user documents that can help to hide the original information.

The main problem of the first method is that it may introduce significant loss of information in user messages derived from the degree of generalization introduced. The second technique is designed to work with documents of an adequate length and which are properly structured. Clearly, tweets do not hold these requirements. The last method is the more suitable for short-length documents and, hence, it might be applied to Twitter and similar microblogging services. The main shortcoming is that, if noise is added within user posts, it may generate uncomfortable publications like the *Entity swapping* technique. Nevertheless, it might be successful if it is implemented allowing readers to distinguish between legitimate and distorted publications, which is precisely the goal of our method.

## 3   Our Proposal

The proposed method follows two main steps: *user profiling* and *semantic noise addition*. In the first one, our method uses similar techniques as those proposed by profiler systems to characterize the user profile according to her published tweets. In the second one, it assesses which are the dominant and dominated categories in her profile and which should be the contents of the fake tweets to be added in the user account (*i.e.*, distortion) to achieve a balanced profile.

It is important to note that fake tweets are constructed as a concatenation of terms which lack the semantically-coherent discourse that a human reader would expect. In this manner, human readers could easily discern between user tweets and fake ones while general approaches for profiling users would be unable to do it. Obviously, it is always possible to design an ad-hoc solution trained to discern between user messages and fake ones. Nevertheless, developing an ad-hoc system is costly and, in turn, it might be defeated by modifying certain aspects of the way in which fake messages are generated.

In the following two sections, our method is formalized and described in detail.

### 3.1   User Profiling

Let us consider a user $u$ that starts posting a first tweet $t_1$ in her Twitter account. Let us also consider that the user profile $P$ is defined (like in related works [5] and [6]) as a set of well-defined categories $C = \{c_1, \ldots, c_k\}$ ($e.g.$, science, health, society, sports, etc), for which their relative weight ($v_i$) should be computed according to the amount of evidences found, for each category, in user tweets. Hence, after the $i$-th tweet $t_i$, the user profile ($P_i$) will be characterized according to the set of weighted profile categories obtained from analyzing all tweets from $t_1$ to $t_i$: $P_i = \{< c_1, v_1 >, \ldots, < c_k, v_k >\}$.

At the beginning of its execution, our system has no idea about the initial profile of the user $P_0$. Therefore, the profile is initialized as $P_0 = \{< c_1, 0 > , \ldots, < c_k, 0 >\}$. For each new user tweet $t_i$ published by $u$, the system progressively computes and updates $P_i$, mimicking the common way to build user profiles of related works.

First, since tweets are slightly-grammatical short texts which are difficult to analyze [8, 10] syntactically and semantically, we opted, as done in some related works [8, 9], to implement a shallow linguistic parsing which, instead of trying to analyze well-formed sentences, focuses on extracting pieces of text with semantic content that can contribute to characterize the user profile: $noun$-$phrases$ ($NPs$).

NPs are built around a noun whose semantics can be refined by adding new nouns or adjectives ($e.g.$, $iPhone \rightarrow new\ iPhone$). Each NP either refers to a generic concept ($e.g.$, $water\ sports$) or it can be considered a proper noun that is an instance of a concept ($e.g.$, $iPhone$ is an instance of a $cell$-$phone$).

Accordingly, **the first step** of the profiling process consist in extracting NPs from user tweets. To achieve that, we rely on several commonly-used natural language processing tools[1]: $sentence\ detection,\ tokenization,\ part$-$of$-$speech\ tagging$ and $syntactic\ parsing$ ($i.e.$, chunking). As a result of this analysis, prepositional or nominal phrases are detected. From these, only NPs (which are those with semantic content) are extracted [16]. It is worth to mention that this process also removes superfluous elements like misspelled words, abbreviations, emoticons, etc.

As output of a tweet $t$, the set $M = \{< NP_1, w_1 >, \ldots, < NP_p, w_p >\}$ is obtained, in which $NP_i$ is each Noun Phrase extracted from $t$ and $w_i$ is its number of appearances.

The **second step** of the profiling process consists, as done in related works [8, 9], in semantically analyzing extracted NPs in order to classify them, if possible, according to the defined categories $C$. By doing this, the number of NPs corresponding to each category can be evaluated to characterize the user profile in a later step.

---

[1] OpenNLP Maxent Package: http://maxent.sourceforge.net/about.html

To enable this classification from a semantic point of view (*i.e.*, to associate each NP to its conceptual abstraction), we rely on a predefined knowledge base [17]. This knowledge base can be a taxonomy, folksonomy or a more formal ontology [20] as long as it offers a structured conceptualization of one or several knowledge domains expressed by, at least taxonomic relationships. In order to improve the recall of the semantic analysis and due to the proliferation of proper nouns in user tweets, a large base that potentially includes up-to-date Named Entities (NEs) [18, 19] is desired.

In our work, we rely on the *Open Directory Project (ODP)* hierarchy of categories because it is the largest, most comprehensive human-edited directory of the Web, constructed and maintained by a vast community of volunteer editors [21]. The purpose of ODP is to list and categorize web sites. Manually created categories are taxonomically structured and populated with related web resources. Nowadays, it classifies almost 5 million web sites in more than 1 million categories (considering also up-to-date Named Entities).

In order to semantically classify NPs in $M$, the system queries each $NP_i$ to ODP. If found, ODP returns the most likely hierarchy $H_i$ of categories ($H_i = h_{i,1} \rightarrow \ldots \rightarrow h_{i,l}$) to which $NP_i$ belongs. For example, if the system queries the NP *"iPhone"*, ODP returns: *iPhone $\rightarrow$ Smartphones $\rightarrow$ Handhelds $\rightarrow$ Systems $\rightarrow$ Computers*.

If $NP_i$ is not found in ODP, the system tries with simpler forms of the NP by removing adjectives/nouns starting from the one most on the left (*e.g.*, *"new iPhone" $\rightarrow$ "iPhone"*) to improve the recall while maintaining the core semantics. The fact that NPs incorporate qualifiers is quite common in texts, but these are hardly covered in knowledge structures which try to model them in a generic way.

The **third and final step** of the profiling process applied to the first tweet $t_1$ consists in updating the user profile $P_1$ according to the categories to which extracted NPs belong. Concretely, for each $NP_i$, ODP has retrieved a hierarchy $H_i$ of categories, hence, the system checks if any of the profile categories ($c_j$ in $C$) is included in $H_i$. In the affirmative case, the system states that $NP_i$ is a taxonomical specialization of $c_j$ (*i.e.*, $NP_i$ *is-a* $c_j$) and it adds the contribution of $NP_i$ to $c_j$ by adding the amount of occurrences of $NP_i$ ($w_i$) to the category weight (*i.e.*, $v_j$ of $c_j$). As more NPs found to be a taxonomical specialization of $c_j$ are considered, the weight $v_j$ of $c_j$ is incremented accordingly, as follows:

$$v_j = v_j + \sum_{\forall NP_i\ is\text{-}a\ c_j} w_i \qquad (1)$$

Once all NPs are considered, the user profile $P_1$ corresponding to the first tweet $t_1$ is defined by a ranked list of categories, according to their weights: $P_1 = \{< c_1, v_1 >, \ldots, < c_k, v_k >\}$, where $v_j$ states the sum of contributions according to the number of term occurrences/co-occurrences related to each particular profile category $c_j$.

## 3.2  Semantic Noise Addition

After the publication of $t_1$ and the characterization of the user profile $P_1$, the objective of our system is to introduce additional terms in the user account as a new fake tweet $ft$ that will balance the user profile towards a uniformly distributed one (according to the considered categories), while maintaining the original tweets unaltered. In this manner, dominant profile categories will become indistinguishable for a profiling system.

In the **first step** of the semantic noise addition process, the system uses the user profile $P_1$ constructed after the publication of $t_1$ to analyze the set of weighted categories and selects the one with the maximum weight (*i.e.*, $< c_{MAX}, v_{MAX} >= argmax_{\forall <c_i, v_i> \in P_1}(v_i)$). Then, for the rest of the categories $c_j$ in $P_1$, it computes the difference with respect to the maximum one ($\Delta(< c_j, v_j >; P_1)$), as follows:

$$\Delta(< c_j, v_j >; P_1) = v_{MAX} - v_j \qquad (2)$$

This difference quantifies the number of term occurrences/co-occurrences that are needed to balance each non-maximal category $c_j$ with respect to the dominant one $c_{MAX}$.

In the **second step** of the semantic noise addition process, for all non-maximal categories, and starting from the $c_j$ for which its $\Delta$ is the largest (*i.e.*, the one with the least dominance in the user profile), the system randomly retrieves $\Delta(< c_j, v_j >; P_1)$ terms from the ODP hierarchy under the corresponding category $c_j$.

In the **third and final step** of this process, retrieved subcategories for all non-maximal categories are then put together in the form of a new fake tweet $ft_1$ to be published after the user tweet $t_1$. This represents the semantically correlated noise added to balance the user profile.

Note that, due to limitations imposed by Twitter regarding message lengths (a maximum of 140 characters), the number of terms to be added in order to fake tweets $ft$ should fulfill this restriction. Hence, even though a certain number of terms should be added to obtain a -theoretic- perfectly balance user profile, in practice, that number could be lower to fulfill Twitter restriction. The fact that a lower amount of fake terms are allowed to be added will cause a slower balancing of the user profile, as it will be shown in the evaluation section. As a general rule, considering that the average length of terms in ODP is 8 characters, up to 15 terms (counting separator whitespaces) could be fitted, in average, in each $ft$.

Also note that, since fake tweets are raw lists of terms of different domains put together without a narrative thread, human readers would easily distinguish them from those created by the original user. On the contrary, an automatic profiling based on term distribution is assumed to fail when characterizing the user profile, due to the added *semantic noise*.

The **whole compound process** (profiling+noise addition) is iteratively executed as new tweets are posted by the user. Concretely, for the $i$-th legitimate tweet, $t_i$, the profile characterization $P_i$ will reflect the aggregation of all previous ones (*i.e.*, both legitimate, from $t_1$ to $t_i$, and fake ones, from $ft_1$ to $ft_i - 1$). Note that each new tweet (both legitimate or fake) contributes to update category weights, increasing previous values according to new extracted category terms. As a result of $P_i$ profiling, the system will create a new fake tweet $ft_i$ that balances the characterization of $P_i$. As new fake tweets are added, the user profile will tend to balance, while the system adapts its behavior (*i.e.*, semantic noise addition) to the new user messages. Reaction time for profile balancing will depend on the amount of noise required to be added (that would depend on the homogeneity of user messages according to the computed profile) and on the maximum number of terms allowed to be published (according to the Twitter length limitation of messages). The fact that the system dynamically re-computes the user profile after each new tweet, enables our proposal to adapt to changes in user preferences or topics of interest, considering also the past history.

## 4   Evaluation

In this section, we evaluate the performance of the proposed system in balancing and, hence, hiding Twitter user profiles.

As evaluation data, we took eight well-differentiated Twitter users, whose profiles should correspond to eight root categories in the ODP hierarchy, as shown in Table 1. To select individual users, we used the *WhoToFollow* [22] search engine provided by Twitter. It provides a list of the most relevant Twitter users according to a specific topic. For each profile category (taken from ODP), we took the most relevant Twitter user as indicated by *WhoToFollow* for the corresponding topic. These are also shown in Table 1. For each user, we took the 100 most recently published tweets as evaluation data.

To numerically quantify the degree of balancing, $\theta$, of a user profile $P$ after each published tweet (both legitimate or fake), the proposed system sums the differences $\Delta$ in the number of occurrence $v_j$ for each category $c_j$ in $P$ with respect to the maximum one, $c_{MAX}$ (see Section 3.2). Then, the result is normalized by the total number of occurrences needed to balance a profile with respect to $c_{MAX}$ in the worst case (*i.e.*, when the contribution of the other non-maximal categories is zero). The normalizing factor corresponds to the product of the number of non-maximal categories in $P$, this is $|P| - 1$, by the number of occurrences of $c_{MAX}$, that is $v_{MAX}$.

$$\theta(P) = \frac{\sum_{\forall <c_j, v_j> \in P} (\Delta(< c_j, v_j >; P))}{v_{MAX} \times (|P| - 1)} \tag{3}$$

The numerical interval of $\theta$ goes from 0 to 1, where 0 means a perfectly balanced profile (*i.e.*, zero difference between all non-maximal categories with respect to the most dominant one) and 1 means maximal difference (*i.e.*, the contribution of all categories except the maximum one is zero).

**Table 1.** ODP Categories, corresponding WhoToFollow topics, most relevant Twitter users for each one and user description (last accessed: January 22th, 2012)

| ODP category | WhoToFollow topic | Twitter user | User description |
|---|---|---|---|
| Arts | Arts and Design | @johnmaeda | President of the Rhode Island School of Design |
| Business | Business | @businessinsider | The latest business news and analysis |
| Computers | Technology | @guardiantech | News from the Guardian tech team |
| Health | Health | @CDC_eHealth | Center for Disease Control, USA |
| Science | Science | @ReutersScience | Science by Reuters.com |
| Shopping | Fashion | @glamour_fashion | Glamour magazine's fashion team |
| Society | News | @nytimes | The New York Times |
| Sports | Sports | @espn | Sports news |

Due to the lack of related works which propose method of profile distortion methods for Twitter (and microblogging services in general), we compare our method with the original data (*i.e.,* no fake tweets are added) and with a naive distortion method which adds a fixed amount of *random noise* (*i.e.,* a number of random terms taken from ODP) per each fake tweet. In this last case, the semantics associated to user tweets are not considered in the construction of fake tweets.

Finally, in order to quantify the influence of the amount of noise added per fake tweet on our method, we have fixed several upper bounds. In the most favorable setting (*high semantic noise*), we allowed up to 15 terms to be added per fake tweet, which corresponds, in average, to the maximum amount allowed by Twitter (see Section 3.2). In the intermediate situation (*medium semantic noise*), we allowed up to 7 terms per fake tweet. The most constrained scenario (*low semantic noise*) only allowed up to 3 terms per fake tweet. Note that, for the *random* approach, the amount of added noise is constant and fixed to *high* (*i.e.,* 15 terms per fake tweet).

Figure 1 shows profile balancing results according to the number of user tweets analyzed (up to 100) for the different methods and scenarios. Note that the horizontal scale quantifies the number of analyzed *user tweets* including also, in the case of the *semantic* and *random noise* addition methods, the corresponding fake ones.

Several conclusions can be drawn from the analysis of the graphs. First, as expected, the addition of semantic noise results in the best profile balancing (*i.e.,* it is closer to zero) because the distribution of category terms at the $i$-th tweet tends to be uniformly distributed.

The amount of semantic noise added per tweet directly influences the results (even though in some case more than in others). When the maximum amount of noise (*i.e.* up to 15 terms per fake tweet) is allowed, user profiles are rapidly

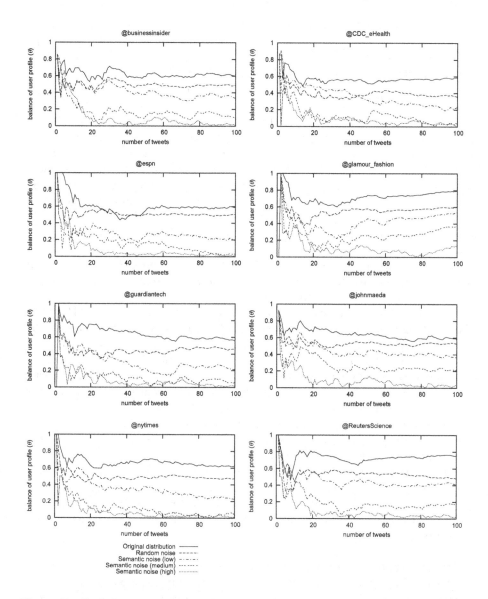

**Fig. 1.** Profile balancing results for the different methods and scenarios for the eight Twitter users

balanced with $\theta$ values below 0.1. Only around 20 user tweets (and, hence, 20 additional fake ones) are needed to achieve that figure. On the contrary, when restricting the amount of semantic noise to the minimum (*i.e.*, up to 3 terms per tweet), an ideal profile balancing (near to zero) can be hardly achieved. Even though more than 100 tweets could be considered, it seems unlikely to achieve a balance due to the curve shape tend to stabilize horizontally around

$\theta$ values between 0.2 and 0.4. This indicates that, in this case, the amount of allowed noise is not enough to achieve a good balance, especially considering that there are *eight* profile categories that should be balanced. The case in which a medium amount of semantic noise is allowed (*i.e.*, up to 7 terms, which is closer to the number of profile categories), presents a more variable behavior. For some users (like *@espn*, covering topics of Arts, Sports, Society and Shopping, *@CDC_eHealth* covering topics of Health, Society and Business, or *@ny_times*, posting about Society, Business, Arts, Computers and Shopping), the system was able to catch the lower figures obtained by the noisiest setting, despite a smoother decreasing shape. These cases correspond to users who post more heterogeneous tweets (covering several profile categories) and, hence, requiring from less noise (terms from other categories) to achieve an equilibrium. For other users (*e.g.*, *@glamour_fashion*, *@johnmaeda*) with profiles more focused towards a dominant category (*e.g.*, Arts, in these cases), profile balancing requires more noise to achieve the balance.

In the graphs, it can also be observed how spikes present in the original distribution (*e.g.*, around the 30th tweet for *@businessinsider*, 50th for *@espen* or 65th for *@guardiantech*), which represent a notorious change in the accumulated user profile, are also reflected in the balanced profiles even more clearly. However, the fact that user profiles are re-computed after each new user tweet allows the system to dynamically adapt its behavior (*i.e.*, categories of added noise) to eventual or long-term changes in user preferences or interests.

In any case, when comparing the semantic setting with the random scenario, it can be observed that, even though the random approach always introduces the maximum amount of noise (*i.e.*, 15 terms per fake tweet), it poorly balances the user profile. In fact, assuming that random noise is uniformly distributed according to profile categories, it would hardly lower the provided figures if the user maintains her preferences throughout time.

Finally, more spikes can be observed in the 1-20 tweets zone when analyzing curves of semantic scenarios. This corresponds to the zone in which the user profile is being characterized, and the dominant category/ies may change from one tweet to the next. In consequence, noise categories may also vary from one fake tweet to the next, producing more pronounced spikes in the profile balancing. As stated in related works [9, 10], individual tweets are too short and ungrammatical to enable an accurate profile characterization. From the analysis of the results, it can be concluded that, at least, 10 tweets (and preferably 20) are required to obtain a stable profile (and hence, a more accurate and coherent noise addition), even though the concrete number may vary from one user to another, according to the topical coherency of her tweets and the homogeneity of her profile.

## 5   Conclusion

In this paper, we have proposed a new system that prevents text-based profilers from characterizing the dominant profile categories of the users of microblogging services like Twitter. Our scheme generates and publishes fake tweets together with legitimate ones. These fake publications are constructed according to two basic principles: (i) They contain specially tailored terms, introducing a semantic distortion in the user profile in order to hide user characteristics (*i.e.*, dominant profile categories) in front of automatic profiling methods; and (ii) They are formed by a concatenation of terms, which leads to a lack of semantically-coherent discourse that allows human readers to easily discern between user tweets and fake ones, while preventing automatic profilers who analyze tweets according to term distribution to discover dominant topics/categories.

The evaluation results obtained show that: (i) the proposed system effectively balances user profiles in front of profilers based on term distribution; (ii) it achieves that balance with a quite limited number of publications (between 10 and 20 tweets are enough to obtain a stable profile); and (iii) it dynamically adapts its behaviour to eventual or long-term changes in user preferences or interests.

Regarding future work, it would be interesting to evaluate the use of the presented approach by regular users in their daily duties in order to confirm the qualities showed by the simulations and also to rate the usability of the system and its level of intrusiveness in a real situation.

**Disclaimer and Acknowledgments.** Authors are solely responsible for the views expressed in this paper, which do not necessarily reflect the position of UNESCO nor commit that organization. This work was partly supported by the Spanish Ministry of Science and Innovation (through projects eAEGIS TSI2007-65406-C03-01, CO-PRIVACY TIN2011-27076-C03-01, ARES-CONSOLIDER IN-GENIO 2010 CSD2007-00004 and Audit Transparency Voting Process IPT-430000-2010-31), by the Spanish Ministry of Industry, Commerce and Tourism (through projects eVerification2 TSI-020100-2011-39 and SeCloud TSI-020302-2010-153) and by the Government of Catalonia (under grant 2009 SGR 1135).

# References

1. McMillan, G.: Twitter Reveals Active User Number, How Many Actually Say Something. Time - Techland (September 2011)
2. Consumer Reports National Research Center, Annual State of the Net Survey, Consumer Reports 75(6) (2010)
3. Rui, H., Whinston, A.: Information or attention? An empirical study of user contribution on Twitter. Information Systems and E-Business Management, 1–16 (2011)
4. Zhang, C., Sun, J., Zhu, X., Fang, Y.: Privacy and Security for Online Social Networks: Challenges and Opportunities. IEEE Network 24(4), 13–18 (2010)
5. TweetPsych (2011), http://tweetpsych.com
6. Peerindex (2011), http://www.peerindex.com
7. Ebner, M., Mühlburger, H., Schaffert, S., Schiefner, M., Reinhardt, W., Wheeler, S.: Getting Granular on Twitter: Tweets from a Conference and Their Limited Usefulness for Non-participants. Key Competencies in the Knowledge Society 324, 102–113 (2010)

8. Zoltan, K., Johann, S.: Semantic Analysis of Microposts for Efficient People to People Interactions. In: Proc. of the Roedunet International Conference – RoEduNet 2011, pp. 1–4 (2011)

9. Michelson, M., Macskassy, S.A.: Discovering Users' Topics of Interest on Twitter: a First Look. In: Proc. of the Fourth Workshop on Analytics for Noisy Unstructured Text Data (2010)

10. Abel, F., Gao, Q., Houben, G.-J., Tao, K.: Semantic Enrichment of Twitter Posts for User Profile Construction on the Social Web. In: Antoniou, G., Grobelnik, M., Simperl, E., Parsia, B., Plexousakis, D., De Leenheer, P., Pan, J. (eds.) ESWC 2011, Part II. LNCS, vol. 6644, pp. 375–389. Springer, Heidelberg (2011)

11. DBpedia (2011), http://dbpedia.org/

12. Luo, W., Xie, Q., Hengartner, U.: Facecloak: an Architecture for User Privacy on Social Networking Sites. In: Proc. of the 2009 International Conference on Computational Science and Engineering, pp. 26–33 (2009)

13. Abril, D., Navarro-Arribas, G., Torra, V.: On the Declassification of Confidential Documents. In: Torra, V., Narakawa, Y., Yin, J., Long, J. (eds.) MDAI 2011. LNCS, vol. 6820, pp. 235–246. Springer, Heidelberg (2011)

14. Martínez, S., Sánchez, D., Valls, A.: Semantic adaptive microaggregation of categorical microdata. Computers & Security 31(5), 653–672 (2012)

15. Sánchez, D., Batet, M.: Semantic similarity estimation in the biomedical domain: An ontology-based information-theoretic perspective. Journal of Biomedical Informatics 44(5), 749–759 (2011)

16. Sánchez, D.: A methodology to learn ontological attributes from the Web. Data and Knowledge Engineering 69(6), 573–597 (2010)

17. Martínez, S., Sánchez, D., Valls, A., Batet, M.: Privacy protection of textual attributes through a semantic-based masking method. Information Fusion 13(4), 304–314 (2012)

18. Sánchez, D., Moreno, A.: Pattern-based automatic taxonomy learning from the Web. AI Communications 21(1), 27–48 (2008)

19. Sánchez, D., Isern, D., Millan, M.: Content Annotation for the Semantic Web: an Automatic Web-based Approach. Knowledge and Information Systems 27(3), 393–418 (2011)

20. Guarino, N.: Formal Ontology in Information Systems. In: Proc. of the 1st International Conference on Formal Ontology in Information Systems, FOIS 1998, pp. 3–15 (1998)

21. Open Directory Project (2012), http://www.dmoz.org/docs/en/about.html

22. Twitter - WhoToFollow (2012), http://twitter.com/#!/who_to_follow

# Detecting Sensitive Information from Textual Documents: An Information-Theoretic Approach

David Sánchez, Montserrat Batet, and Alexandre Viejo

Departament d'Enginyeria Informàtica i Matemàtiques,
UNESCO Chair in Data Privacy, Universitat Rovira i Virgili
Av. Països Catalans 26, E-43007 Tarragona, Spain
{david.sanchez,montserrat.batet,alexandre.viejo}@urv.cat

**Abstract.** Whenever a document containing sensitive information needs to be made public, privacy-preserving measures should be implemented. Document sanitization aims at detecting sensitive pieces of information in text, which are removed or hidden prior publication. Even though methods detecting sensitive structured information like e-mails, dates or social security numbers, or domain specific data like disease names have been developed, the sanitization of raw textual data has been scarcely addressed. In this paper, we present a general-purpose method to automatically detect sensitive information from textual documents in a domain-independent way. Relying on the Information Theory and a corpus as large as the Web, it assess the degree of sensitiveness of terms according to the amount of information they provide. Preliminary results show that our method significantly improves the detection recall in comparison with approaches based on trained classifiers.

**Keywords:** Privacy, Document sanitization, Information Theory.

## 1 Introduction

In the context of the Information Society, many documents are needed to be made public every day [1]. Since some of these documents may contain confidential information about private entities, measures should be taken prior their publication to avoid revealing sensitive data or disclosing individuals' identities.

Document sanitization precisely pursuits the removal of sensitive information from text (which can yield to revealing private information/identities of the entities referred in the document) so that it may be distributed to a broader audience.

In the past, sanitization has been usually tackled manually by governments and companies. Standard guidelines [2] detailing the correct procedures to ensure irreversible suppression or distortion of sensitive parts in physical and electronic documents have been proposed. In the medical context, the *Health Insurance Portability and Accountability Act (HIPAA)* [3] states safe harbor rules about the kind of personally identifiable information which should be removed in medical documents prior allowing their publication.

V. Torra et al. (Eds.): MDAI 2012, LNAI 7647, pp. 173–184, 2012.
© Springer-Verlag Berlin Heidelberg 2012

However, manual sanitization is expensive, time-consuming [4], prone to disclosure risks [5] and does not scale as the volume of data increases [6]. Considering the amount of digital textual information made available daily (*e.g.,* the US Department of Energy's OpenNet initiative [7] requires sanitizing millions of documents yearly), one can realize of the need of automatic text sanitization methods. This need is manifested in initiatives like the DARPA's request for new technologies to support the declassification of confidential documents [8] or the creation of the Consortium for Healthcare Informatics Research (CHIR) [9], which aims at building new methods and tools for de-identification of medical data in order to utilize them for research and operational purposes.

To tackle this problem, semi-automatic applications assisting the sanitization process have been developed, focusing on structured sensitive data like email addresses, dates, telephone numbers or credit card/social security numbers. Commercial applications like Adobe Acrobat Professional [10] incorporate patterns that are able to recognize this kind of data thanks to its regular structure. However, they leave the detection of sensitive textual data (like names, locations or descriptive assertions) to a human expert. In fact, the sanitization of this kind of free text data (which is the most usually available one) is specially challenging due to its unbounded and unstructured nature [11].

In this paper, we tackle the problem of automatic detection of sensitive text for sanitization purposes. Relying on the foundations of the Information Theory, we mathematically formulate what we consider *sensitive information* and how it can be applied to detect potentially sensitive textual entities. Our method has been compared to other general-purpose approaches relying on trained classifiers, showing that it is able to improve the recall detection while offering a more general and less constrained solution.

The rest of the paper is organized as follows. Section 2 describes related works focusing on detecting sentitive terms in textual documents. Section 3 presents our method, discussing its theoretical premises and formalizing its design. Section 4 details preliminary experiments carried out with highly identifying biographical sketches, showing promising results regarding the detection recall. The final section depicts the conclusions and presents some lines of future research.

## 2   Related Work

Among the unsupervised sanitization methods available, one of the first approaches that can be found is the Scrub system [12]. It finds and replaces patterns of identifying information such as Social Security number, medical terms, age, date, etc. Similar schemes that focus on removing sensitive terms from medical records [13,14] use very specific patterns designed according to the HIPAA "Safe Harbor" rules that mention 18 data elements that must be removed from clinical data in order to anonymize it [3]. Examples of those sensitive elements are: names, dates, medical record numbers, biometric identifiers, full face photographs, etc.

The authors in [6] present a scheme that detects sensitive elements using a database of entities (persons, products, diseases, etc.) instead of patterns. Each entity in this database is associated with a set of terms related to the the entity; this set is the context of the entity that should be hidden (*e.g.,* the context of a person entity could include her name, birth date, etc).

The method proposed in [11] focuses on domain-independent unstructured documents. Authors propose the use of named entity recognition techniques to identify the entities of the documents that require protection. It is worth to mention that this proposal assumes that named entities (such as person and organization names and locations) are always sensitive data and, hence, they should be sanitized.

The authors in [5] present a semi-automatic tool build into Microsoft Word that suggests to the user the entities that should be anonymized. Regarding the entity detection process, this work focuses on documents directly linked to certain companies (*i.e.,* documents to be sanitized describe certain companies/organizations or their activities). The data to be detected is divided into two categories: (i) *Client Identifying Information:* this information includes any words and phrases that reveal what company the document pertains to; and (ii) *Personally Identifying Information:* this includes any person names, location names, phone numbers, etc. Similarly to [11], authors uses the Stanford Named Entity Recognizer [15] to automatically recognize people, organizations and locations. Additionally, specific patterns are used to detect social security numbers or telephone numbers. Regarding the Client Identifying Information, a Naive Bayes classifier is implemented to recognize it.

# 3    A General Purpose Method to Detect Sensitive Terms in Textual Documents

Our method pursuits to automatically detect sensitive pieces of text in a general and unconstrained way, so that it can be applied to heterogeneous documents (both regarding its structure and knowledge domain), and to any kind of textual term (instead of predefined types or lists). To do so, we first discuss the notion of sensitive information and how it can be detected.

*Sensitive information* regards to pieces of text that can either reveal the identity of a private entity or refer to confidential information. To discover sensitive information, problem-specific related works rely on predefined lists of sensitive words [6] or use machine learning methods (like trained classifiers [5] or pattern-matching techniques [14]) aimed at detecting specific types of information. The former can provide accurate results, but lists have to be manually compiled (which is costly and time-consuming) for specific problems (which lacks generality); the latter methods manually train/design classifiers/patterns to detect domain specific sensitive data (like PHIs in the medical context [14,9] or organizational data [5]), which can be hardly generalized.

On the other hand, general purpose methods [11] usually associate the discovery of sensitive data to the detection of generic Named Entities (NEs). Due to their specificity and the fact that they represent individuals rather than concepts, NEs are likely to reveal private information. NEs can be accurately detected in an automatic manner, either using patterns [16,17] or trained classifiers [15]. However, they are hampered by several problems. First, some detected NEs could refer to very general entities (*e.g.*, continents), which are not needed to be sanitized and whose removal would result in unnecessary information loss. On the other hand, some words or combinations of words, which may be omitted since they are not NEs, could refer to very concrete concepts (*e.g.*, rare diseases, concrete employments), which are likely reveal confidential or identifiable information. Moreover, most generic NE recognition packages only detect a limited amount of NE types, usually *persons*, *locations* and *organizations* [11,15]. Finally, they are language-dependent, since the NE recognition accuracy depends on the availability of training data, which is expressed in a concrete language. These problems negatively affect the detection recall, which is crucial to avoid disclosure risk.

To overcome these problems, we base the text sanitization on a more general notion of *sensitive information*. In our approach sensitive terms are those that, due to their specificity, provide *more information* than common terms. Hence, the key-point to detect them is to quantify *how much information* each textual term provides, sanitizing those that provide *too much information* (according to a sanitization criteria).

To quantify the amount of information provided by a textual term, we rely on the information theory and the notion of Information Content (IC).

### 3.1   Information Content Estimation

The Information Content (IC) of a term measures the amount of information provided by the given term when appearing in a context (*e.g.*, a document). Specific terms (*e.g.*, *pancreatic cancer*) provide more IC than those more general ones (*e.g.*, *disease*). Formally, the IC of a term $t$ is computed as the inverse of the probability of encountering $t$ in a corpus ($p(t)$). In this way, infrequent concepts obtain a higher IC than more common ones.

$$IC(t) = -\log_2 p(t) \tag{1}$$

Classical methods [18] used tagged textual data as corpora, so that term frequencies can be computed unambiguously. The use of this kind of corpus provided accurate results in the past, when applied to general terms [18] at the cost of manually compiling and tagging it. However, the limited coverage and relative small size of used corpora resulted in data sparseness problems (*i.e.*, the fact that not enough data is available to extract reliable conclusions from their analysis) when computing the IC of concrete terms (*e.g.*, rare diseases), NEs (*e.g.*, names) or recently minted/trending terms (*e.g.*, netbook, tablet) [19,20]. Considering that document sanitization focuses precisely on concrete (*i.e.*, highly informative) terms, a wider corpus covering them would be desirable to obtain robust IC values.

When looking for a general-purpose corpus covering as much terms as possible, the Web stands out. Its main advantages are its free and direct access and its wide coverage of almost any possible up-to-date term. In fact, it has been argued that the Web is so large and heterogeneous that represents the true current distribution of terms at a social scale [21]. Since IC calculus relies on term distribution to compute probabilities, the characteristics of the Web makes it specially convenient [19].

The main problem of computing term appearances in the Web is that the analysis of such an enormous repository is impracticable. However, the availability of Web Information Retrieval tools (IRs) like Web Search Engines (WSEs) can help in this purpose. WSEs directly provide web-scale page counts (stating term appearances) for a given query. Many authors [22,19,23] have used these page counts to compute term probabilities in the Web. Hence, by estimating term probabilities at a social/Web scale one can compute, in an unsupervised and domain-independent manner, their IC.

Taking into consideration the Web size, its high coverage for any kind of terms (including concrete ones and NEs) and the possibility of obtaining web-scale term distribution measures in an immediate way, in this work, we quantify the IC of a potentially sensitive term $t$ found in a document $d$ to be sanitized, as follows:

$$IC_{web}(t) = -\log_2 p_{web}(t) = -\log_2 \frac{page\_counts(t)}{total\_webs} \qquad (2)$$

where $page\_counts(t)$ is the number provided by a WSE when querying $t$ and $total\_webs$ quantifies the total amount of web sites indexed by the search engine. (e.g., around 3.5 billions in Bing[1]).

To avoid the need of on-line querying that, in addition to overhead the process, may disclose sensitive words, one can use databases of some WSEs that can be stored and queried off-line [24,25].

## 3.2   Extracting Terms from Textual Documents

In this section, we detail how sensitive terms $t$ are extracted from a document $d$ to be sanitized. Given an input text like the one shown in Figure 1, sensitive data is such corresponding to concrete concepts (e.g., pancreatic cancer) or individual names (e.g., Peter Greenow) that reveal too much information. These are referred in text by means of nouns or, more generally, noun phrases (NPs). Hence, the detection of sensitive terms focuses on NPs found in the input document.

To detect NPs, we rely on several natural language processing tools [26], which perform (i) sentence detection, (ii) tokenization (i.e., word detection, including contraction separation), (iii) part-of-speech tagging (POS) of individual tokens and (iv) syntactic parsing of POS tagged tokens, so that they are put together according to their role, obtaining verbal (VPs), prepositional (PPs) or nominal phrases (NPs). From these, NPs are considered (see an example of the output

---

[1] http://www.worldwidewebsize.com/ [last accessed: May 8th, 2012]

> Peter Greenow, from Syracuse, United States, suffers from pancreatic cancer. He was given treatment in the Community General Hospital for his condition by an oncologist.

**Fig. 1.** Sample text of a document to sanitize

of this analysis in Figure 2). As discussed in the previous section, the *amount of information* NPs provide (IC) will be quantified by querying them in a WSE and using eq. 2.

> [*NP* Peter Greenow] , from [*NP* Syracuse], [*NP* United States], suffers from [*NP* pancreatic cancer]. [*NP* He] was given [*NP* treatment] in [*NP* the Community General Hospital] for [*NP* his condition] by [*NP* an oncologist].

**Fig. 2.** Noun Phrases (NP) detected in sample text

In order to focus the IC-based analysis on the information provided by the conceptualization to which each NP refers, we also remove *stop words*. Stop words configure a finite list of domain independent terms like determinants, prepositions or adverbs which can be removed from NPs without altering their conceptualizations (*e.g.,* an oncologist → oncologist). The motivation of removing stop words is to avoid their influence in the computation of IC values by means of web queries. For example, in a WSE like Bing[2], the query *"an oncologist"* results in a page count (654.000) an order of magnitude lower than the query *"oncologist"* (5.870.000), even though both refer to the same concept and, hence, both should provide the same *amount of information.*

Note that, even though both natural language processing tools and stop words are language-dependent, both are available for many languages, including English, Spanish, Portuguese, German or Danish [26].

### 3.3   Detecting Sensitive Terms

The final step consists on assessing which of the NPs provide *too much information* according to their computed IC; these will be considered as *sensitive.*

As discussed at the beginning of the section, NE-based methods assume that NEs *always* provide too much information. From an information theoretic perspective, this is a rough criteria that may result in unnecessarily sanitizing very general terms (*e.g., "United States"* results in 1.300 million pages in Bing, obtaining a very low IC); at the same time, more informative terms are omitted because they are not NEs (*e.g.,* the concept "pancreatic cancer" results in around 6,5 million page counts in Bing, which provides a comparatively much higher IC).

---

[2] http://www.bing.com/ [accessed: May 8th, 2012]

Relying on the notion of IC, our proposal enables a more comprehensive and adaptable sanitization, which considers as *sensitive* those NPs whose IC (computed using eq. 2) is higher or equal than a given value $\beta$, which acts as the *detection threshold*. This value, which is also expressed in terms of IC, represents the *degree of informativeness* above which terms are considered to reveal *too much information*. $\beta$ can be defined in an intuitive way by associating it to the IC of the most general feature that should remain hidden in the sanitized document.

Formally, any $NP_i$ in $d$ (*i.e.*, the document to be sanitized) whose IC is higher or equal than $\beta$ will be considered as sensitive:

$$Sensitive\_NPs = \{NP_i \in d | IC_{web}(NP_i) >= \beta\} \qquad (3)$$

For example, if we would like to sanitize the text shown in Figure 1 so that a potential attacker cannot discover that *Peter Greenow* has *cancer* (and any other more concrete information, like his name and detailed census data), we can specify the detection threshold as $\beta = IC_{web}(cancer)$. In this manner, any reference to *cancer* or any other more concrete (*i.e.*, more informative) term like *pancreatic cancer* or *Community General Hospital* will be considered as sensitive. Table 1 shows the detection results, presenting sensitive terms (according to the specified threshold) in **bold**. One can realize that some concrete concepts that are not NEs (*i.e.*, oncologist, pancreatic cancer) have been appropriately tagged as sensitive, whereas very general NEs (*i.e.*, United States) have not. Compared to NE-based methods [5,11], the former case minimizes the disclosure risk, whereas the later case contributes to retain the sanitized document's utility.

**Table 1.** Detected Noun phrases (NP) with their corresponding *page_counts* (from Bing) and $IC_{web}$. Words in (brackets) are stop words that are not considered in the IC calculus. **Bold** rows correspond to sensitive terms according to the detection threshold (*i.e.*, $IC_{web}(cancer) = 2.7$, given that *page_counts(cancer)* = 536.000.000). The last column states which ones are Named Entities (NE).

| NP | page_counts | $IC_{web}$ | NE? |
|---|---|---|---|
| **Peter Greenow** | **21** | **27.3** | **Yes** |
| **Syracuse** | **68.000.000** | **5.7** | **Yes** |
| United States | 1.300.000.000 | 1.4 | Yes |
| **pancreatic cancer** | **6.550.000** | **9.1** | **No** |
| (He) | Not Considered | N/C | No |
| treatment | 616.000.000 | 2.5 | No |
| **(the) Community General Hospital** | **146.000** | **14.5** | **Yes** |
| (his) condition | 702.000.000 | 2.3 | No |
| **(an) oncologist** | **7.200.000** | **8.9** | **No** |

## 4   Experiments

In this section, some preliminary results are presented, showing the accuracy of the detection when applying our method to highly sensitive textual documents. Since most general-purpose related works [5,11] rely on the detection of NEs to sanitize text, our method has been compared against the state-of the art *Stanford Named Entity Recognizer* [15], which is able to detect and classify NEs as *persons, locations* or *organizations*. Both approaches have been evaluated against the criterion of two human experts stating which pieces of text could reveal too information about the described entity.

To test our method in a realistic setting, we use *real* raw texts containing highly sensitive information. In particular, we used biographical sketches describing *actors/actresses* taken from English Wikipedia articles. Wikipedia descriptions of concrete entities usually contain an high amount of potentially identifiable information, which makes the detection of sensitive information a challenging task. Two types of actors have been selected: three American actors (*Sylvester Stallone, Arnold Schwarzenegger* and *Audrey Hepburn*), so that most terms and NEs appearing in text would be expressed in English (easing the detection for English-trained NE recognizers), and three Spanish (but well-known) actors (*Antonio Banderas, Javier Bardem* and *Jordi Mollà*) for which, even though their descriptions are written in English, could include NEs expressed with non-translatable Spanish words or localisms. In this manner, we can also compare the degree of language-dependency of our method against NE recognizers based on English-trained classifiers.

To evaluate the results obtained by both methods, we requested two human experts to select and agree on which terms (*i.e.,* words or NPs, including NEs) reveal too much information, considering that it is desired to hide the fact that the described entities are *actors*. Hereinafter, we will refer to the set of sensitive terms selected by the human experts as $Human\_Sensitive\_NPs$. Coherently with our method's design, we set $\beta = IC_{web}(actor)$, so that any term providing more information than the term *actor* will be detected as *sensitive*. To compute the IC of terms Bing Web Search Engine have been used, fixing the total amount of indexed web sites in 3.5 billions [3]. The detection performance is quantified by means of *precision, recall* and *F-measure*.

*Precision* (eq. 4) is calculated as the ratio between the number of automatically detected sensitive terms ($Sensitive\_NPs$) that have been also selected by the human experts ($Human\_Sensitive\_NPs$), and the total amount of automatically detected terms (*i.e.,* $|Sensitive\_NPs|$). The higher the precision, the lower the amount of incorrectly detected sensitive terms.

$$Precision = \frac{|Sensitive\_NPs \cap Human\_Sensitive\_NPs|}{|Sensitive\_NPs|} \cdot 100 \qquad (4)$$

*Recall* (eq. 5) is calculated as the ratio between the number of terms in $Sensitive\_NPs$ that also belong to $Human\_Sensitive\_NPs$, and the total

---

[3] http://www.worldwidewebsize.com/ [last accessed: May 8th, 2012]

amount of terms in $Human\_Sensitive\_NPs$. Recall indicates the number of detected sensitive terms. The higher the recall, the less the disclosure risk, because a lower amount of non-detected sensitive terms would remain in the text.

$$Recall = \frac{|Sensitive\_NPs \cap Human\_Sensitive\_NPs|}{|Human\_Sensitive\_NPs|} \cdot 100 \qquad (5)$$

Finally, the *F-measure* (eq. 6) quantifies the harmonic mean of recall and precision, summarizing the accuracy of the detection stage:

$$F\text{-}measure = \frac{2 \cdot Recall \cdot Precision}{Recall + Precision} \qquad (6)$$

The obtained values for *precision*, *recall* and *F-measure* for our method (*IC*) and for the method based on NE detection (*NE*) are listed in Table 2.

**Table 2.** Precision, Recall and F-measure for evaluated entities and methods

|  |  | Sylvester Stallone | Arnold Schwarz. | Audrey Hepburn | Antonio Banderas | Javier Bardem | Jordi Mollà |
|---|---|---|---|---|---|---|---|
| Precision | NE | 100% | 100% | 87.10% | 100% | 100% | 75% |
|  | IC | 82.35% | 72% | 81.13% | 94.44% | 100% | 83.33% |
| Recall | NE | 46.67% | 47.37% | 58.69% | 52.94% | 33.33% | 57.14% |
|  | IC | 93.33% | 94.74% | 93.48% | 100% | 100% | 95.24% |
| F-measure | NE | 63.64% | 64.28% | 70.13% | 69.23% | 50% | 64.86% |
|  | IC | 87.5% | 81.82% | 86.87% | 97.14% | 100% | 88.89% |

Analyzing *precision*, we realize that, in most cases, the NE-detection method provided better results than ours. Since precision mainly depends on the number of false positives, this states that our method tends to select too much terms as sensitive. A reason for this is the fact that our method detected, in some cases, syntactically complex NPs as sensitive terms, even though they may refer to general (non-revealing) concepts. Complex NPs are those composed by several words and using complex syntactic constructions that, when queried in a Web Search Engine, tend to provide a relatively low page count, giving the impression of a high IC. The fact that the page count depends on the lexico-syntactical construction of queried terms is caused by the strict terminological matching implemented by Web search engines in which our method relies. On the other hand, since the NE-based method obtained a perfect precision in most cases, this suggest that most (but no all) NEs are sensitive. A worth-noting case is *Jordi Mollà*, in which the NE-based method provided a lower precision than ours. In this case, the NE-detection package tagged general entities like *United States* or *Spain* (since they represent a *location*) which were not considered as sensitive by human experts due to their generality. Our method, on the contrary, relying on the low IC these term provide, behave inversely, achieving a higher precision and retaining more information.

*Recall* represents a more important dimension in the context of document sanitization, since a low recall implies that a number of terms considered as sensitive will appear in the sanitized document. In this case, recall figures for NE-based methods are significantly lower than ours. In fact, when our method was able to stay in the 95-100% range in most cases, the NE-based method resulted in recall values around 50%. On the one hand, this shows that not only NEs appearing in text are sensitive, but also NPs (*e.g. film and fashion icon of the twentieth century*, referred to *Audrey Hepburn*) referring to concrete concepts. On the other hand, NE-based methods are limited by the scope of the trained classifiers. The fact that only certain NE types (*locations, persons* and *organizations*, in this case) are detected, resulted in the omission of an amount sensitive NEs like *movie titles*. Moreover, the worst results were obtained for an Spanish actor (*Javier Bardem*) due to the presence of Spanish localisms and Spanish movie titles, which are difficult to detect for an English-trained classifier. It is worth mentioning that several of these omissions were highly revealing, resulting in an instant disclosure (*e.g. Rocky* for *Sylvester Stallone* or *Governator* for *Arnold Schwarzenegger*). This shows the limitations of classifiers based on training data: they base the recognition on the fact that the entity or a similar one has been previously tagged. When aiming at designing a general-purpose method, training data may be not enough when dealing with specific entities, or they may be outdated with regards to recently minted entities. This is, however, the most common sanitization scenario. In comparison, our method bases the detection on the fact that few evidences are found in the Web. This is a more desirable behavior because sensitive data is detected when it is very likely to act as an identifier. The reliance on the lack of evidences rather than on the presence of them also avoids being affected by the data sparseness that characterizes manual training/knowledge-based models [19]. Moreover, on the contrary to tagged corpora, the Web offers up-to-date results and covers almost any possible domain [19].

As a result of the significant differences between methods' recalls (*i.e.,* disclosure risk), when comparing them according to their global accuracy (*i.e., F-measure*), our method surpass NE-based ones in all cases.

## 5    Conclusions and Future Work

In this paper, an automatic method to detect sensitive information in text documents is presented. The method's generality is given by the theoretical foundations of the Information Theory and a corpus as general/global as the Web. As a result, it can be applied to heterogeneous textual data (and not only NEs [5,11]) in a domain-independent fashion.

As future work, we plan to tackle the limitations observed in the IC calculus regarding the too strict query matching applied by Web search engines. In this case, different lexico-syntactical forms of the same terms can be queried and page count results can be aggregated to obtain a more general (and accurate) estimation of their informativeness.

Moreover, it is worth noting that even though most sanitization models propose removing those terms that are detected as potentially sensitive [6,14,13], this is not the most desirable strategy. Since the purpose of document sanitization is to provide a privacy-preserved but still useful version of the input document to the audience, a systematic removal of sensitive terms may hamper the document's utility. In fact, since semantics are the mean to interpret and extract conclusions from the analysis of textual data, the retention of text semantics is crucial to maintain the utility of documents [27,28]. To tackle this problem, recent methods [5,11] propose replacing sensitive information by generalized versions (*e.g., "iPhone"* → *"cell phone"*) instead of removing it. In this manner, the document still retains a degree of semantics (and hence, a level of utility) while revealing less information. To enable term generalizations, a knowledge base (KB) modeling the taxonomical structure of sanitized terms is needed. We plan to use general-purpose KBs to provide more accurate sanitizations, exploiting them from an information theoretic perspective.

**Disclaimer and Acknowledgments.** Authors are solely responsible for the views expressed in this paper, which do not necessarily reflect the position of UNESCO nor commit that organization. This work was partly supported by the Spanish Ministry of Science and Innovation (through projects eAEGIS TSI2007-65406-C03-01, CO-PRIVACY TIN2011-27076-C03-01, ARES-CONSOLIDER INGENIO 2010 CSD2007-00004 and Audit Transparency Voting Process IPT-430000-2010-31), by the Spanish Ministry of Industry, Commerce and Tourism (through projects eVerification2 TSI-020100-2011-39 and SeCloud TSI-020302-2010-153) and by the Government of Catalonia (under grant 2009 SGR 1135).

# References

1. U.S. Department of Justice: U.S. freedom of information act (FOIA) (2012)
2. Nat. Security Agency: Redacting with confidence: How to safely publish sanitized reports converted from word to pdf. Technical Report I333-015R-2005 (2005)
3. Department of Health and Human Services, Office of the Secretary: The health insurance portability and accountability act of 1996. Technical Report Federal Register 65 FR 82462 (2000)
4. Dorr, D.A., Phillips, W.F., Phansalkar, S., Sims, S.A., Hurdle, J.F.: Assessing the difficulty and time cost of de-identification in clinical narratives. Methods of Information in Medicine 45(3), 246–252 (2006)
5. Cumby, C., Ghan, R.: A machine learning based system for semi-automatically redacting documents. In: Proceedings of the 23rd Innovative Applications of Artificial Intelligence Conference, pp. 1628–1635 (2011)
6. Chakaravarthy, V.T., Gupta, H., Roy, P., Mohania, M.: Efficient techniques for document sanitization. In: Proceedings of the ACM Conference on Information and Knowledge Management, CIKM 2008, pp. 843–852 (2008)
7. U.S. Department of Energy: Department of energy researches use of advanced computing for document declassification (2012)
8. DARPA: New technologies to support declassification. Request for Information (RFI) Defense Advanced Research Projects Agency. DARPA-SN-10-73 (2010)

9. Meystre, S.M., Friedlin, F.J., South, B.R., Shen, S., Samore, M.H.: Automatic de-identification of textual documents in the electronic health record: a review of recent research. BMC Medical Research Methodology 10(70) (2010)

10. National Security Agency: Redaction of pdf files using Adobe Acrobat Professional X (2011)

11. Abril, D., Navarro-Arribas, G., Torra, V.: On the Declassification of Confidential Documents. In: Torra, V., Narakawa, Y., Yin, J., Long, J. (eds.) MDAI 2011. LNCS, vol. 6820, pp. 235–246. Springer, Heidelberg (2011)

12. Sweeney, L.: Replacing personally-identifying information in medical records, the scrub system. In: Proceedings of the 1996 American Medical Informatics Association Annual Symposium, pp. 333–337 (1996)

13. Tveit, A., Edsberg, O., Rost, T.B., Faxvaag, A., Nytro, O., Nordgard, M.T., Ranang, M.T., Grimsmo, A.: Anonymization of general practioner medical records. In: Proceedings of the Second HelsIT Conference (2004)

14. Douglass, M.M., Cliffford, G.D., Reisner, A., Long, W.J., Moody, G.B., Mark, R.G.: De-identification algorithm for free-text nursing notes. In: Proceedings of Computers in Cardiology 2005, pp. 331–334 (2005)

15. Finkel, J.R., Grenager, T., Manning, C.: Incorporating non-local information into information extraction systems by gibbs sampling. In: Proceedings of the 43rd Annual Meeting of the Association for Computational Linguistics, pp. 363–370 (2005)

16. Sánchez, D., Isern, D.: Automatic extraction of acronym definitions from the web. Applied Intelligence 34(2), 311–327 (2011)

17. Sánchez, D., Isern, D., Millan, M.: Content annotation for the semantic web: an automatic web-based approach. Knowledge and Information Systems 27(3), 393–418 (2011)

18. Resnik, P.: Using information content to evalutate semantic similarity in a taxonomy. In: Proceedings of 14th International Joint Conference on Artificial Intelligence, pp. 448–453 (1995)

19. Sánchez, D., Batet, M., Valls, A., Gibert, K.: Ontology-driven web-based semantic similarity. Journal of Intelligent Information Systems 35(3), 383–413 (2010)

20. Sánchez, D., Batet, M., Isern, D.: Ontology-based information content computation. Knowledge-based Systems 24(2), 297–303 (2011)

21. Cilibrasi, R.L., Vitanyi, P.M.B.: The google similarity distance. IEEE Transactions on Knowledge and Data Engineering 19(3), 370–383 (2006)

22. Turney, P.D.: Mining the Web for Synonyms: PMI-IR versus LSA on TOEFL. In: Flach, P.A., De Raedt, L. (eds.) ECML 2001. LNCS (LNAI), vol. 2167, pp. 491–502. Springer, Heidelberg (2001)

23. Sánchez, D.: A methodology to learn ontological attributes from the web. Data and Knowledge Engineering 69(6), 573–597 (2010)

24. Cafarella, M.J., Etzioni, O.: A search engine for natural language applications. In: Proceedings of the 14th International Conference on WWW, pp. 442–452 (2005)

25. Open Directory Project: ODP (2012)

26. Apache Software Foundation: OpenNLP (2012)

27. Martínez, S., Sánchez, D., Valls, A., Batet, M.: Privacy protection of textual attributes through a semantic-based masking method. Information Fusion 13(4), 304–314 (2012)

28. Martínez, S., Sánchez, D., Valls, A.: Semantic adaptive microaggregation of categorical microdata. Computers and Security 31(5), 653–672 (2012)

# A Study of Anomaly Detection in Data from Urban Sensor Networks

Christoffer Brax[1] and Anders Dahlbom[2]

[1] Training Systems & Information Fusion,
Business Area Electronic Defence Systems,
Saab AB, Skövde, Sweden
christoffer.brax@saabgroup.com
[2] Informatics Research Centre, University of Skövde,
P.O. Box 408, SE-541 28 Skövde, Sweden
anders.dahlbom@his.se

**Abstract.** In many sensor systems used in urban environments, the amount of data produced can be vast. To aid operators of such systems, high-level information fusion can be used for automatically analyzing the surveillance information. In this paper an anomaly detection approach for finding areas with traffic patterns that deviate from what is considered normal is evaluated. The use of such approaches could help operators in identifying areas with an increased risk for ambushes or improvised explosive devices (IEDs).

**Keywords:** Anomaly detection, decision support, traffic flow analysis.

## 1 Introduction

In both peace keeping and peace enforcing missions, task forces mainly operate in asymmetric conflict environments where the situation most often can be described as being in a grey area between peace and war. Threats are usually camouflaged and hiding within the population and the regular activities of everyday life and the warfare is often carried out using terrorism, sabotage, IEDs, smuggling operations, etc. Task force security has become an increasingly important issue during missions in these environments, and new requirements are put on the technological support that is needed. It is not sufficient to detect the presence of an object in order to determine the threat that it might constitute.

A common tactic in asymmetric conflicts is the use of various forms of ambush attacks against the least defended elements followed by subsequent rapid movement away from the area of the attack [1]. In this way, attackers are only exposed for a very limited amount of time. Moreover, attackers carefully avoid open confrontation with larger and more heavily equipped forces.

In order to successfully identify these types of threats, suspicious object behaviors need to be detected and connected to imminent attacks or the preparation for them. This puts requirements on detailed information about detected objects along with robust processing of data received over time.

V. Torra et al. (Eds.): MDAI 2012, LNAI 7647, pp. 185–196, 2012.

Modern sensor systems are not designed to accomplish such surveillance tasks and it is left to human operators to detect and analyze these types of situations. Presented to an operator, all individual vehicle tracks in a city would however become unmanageable since it is difficult to maintain focus on more than a few tracks simultaneously. The operator, having limited cognitive ability [2], will also have trouble finding small changes in a situation. This might pose a problem in situations that extend over a long period of time with only small incremental changes. Additionally, there is a risk of operators experiencing information overload, which in the end can lead to poor threat detection. Only in exceptional cases can sufficient situation awareness in the ground theatre be achieved through conventional detect and track methods.

To address the situation awareness problem in urban environments, one has to either drastically limit the area of surveillance, or one has to look for objects that in some way deviate from a large background of similar objects (e.g. behavior, position). Anomaly detection is an interesting approach which pursuits the second alternative. The purpose of an anomaly detection function is to assist an operator by analyzing situation data to filter out important parts and give early warnings when suspicious behaviors are detected. When the function detects an anomaly it must be able to characterize it in such a way that the operator can easily understand the anomaly and make an informed assessment as to whether to monitor the object, take preventive action or to classify it as irrelevant.

Much of the focus in previous anomaly detection approaches in the surveillance domain has been on tracking and analyzing single objects to find objects that behave anomalous [3,4,5,6]. While this is important, it requires high quality tracking of the surveyed objects, something that is not always available in crowded urban environments.

This paper proposes a Gaussian anomaly detector which, in contrast to modeling the behavior of single objects, focuses on modeling the collective behavior of objects. This is carried out by constructing a model of normalcy based on the average flow and speed of objects in relation to geographical areas. Measuring the average flow and speed of objects does not require the use of advanced sensor systems with high quality tracking, and it is thus easier to carry out in urban areas. The proposed anomaly detector is evaluated using data from a simulation platform that simulates traffic in an urban area.

## 2   Anomaly Detection

One of the first fields to use anomaly detection was IT security where anomaly detection was used to build self-learning intrusion detection systems capable of detecting previously unknown viruses, trojans and break-in attacks [7]. Today it is also used in military and civilian surveillance systems. The concept of anomaly detection is, however, somewhat vague and there is no clear definition of anomaly detection is or even what constitutes an anomaly. Some argue that an anomaly is something that is known beforehand (i.e. can be described by a domain expert) but which seldom occurs [1]. Others argue that an anomaly is something unknown that has not been seen before [6]. Anomalous objects are also referred to variously as outliers, novelties or deviations [8]. In this paper, we adopt the definition of anomaly detection and anomalies suggested by Tan et al.:

> "Anomaly detection is the task of identifying observations whose characteristics are significantly different from the rest of the data. Such observations are known as anomalies or outliers." [9]

A large variety of anomaly detection techniques have been suggested. Many of these have been specifically developed and tailored for specific problems and application domains, while others are more generic. Furthermore, anomaly detection has been the topic of a number of excellent surveys and review articles, see e.g. [10,11,8,12,7]. These articles mainly address methods which model data based on their statistical properties and use this information to investigate if the new incoming data originates from the same distribution or not.

During the past decade many different approaches for anomaly detection have been investigated in the surveillance domain. In [3] self-organizing maps are used together with Gaussian mixture models for detecting anomalous vessel traffic. This type of approach is also used in combination with interactive visualization in [13]. In [6] the focus is also on detecting anomalous vessel activity, however, they employ semantic networks composed of connected spiking neurons that are laid out in a grid over an area of interest. This allows for some degree of temporal information to be modeled. Other work focusing on detecting anomalous vessel behavior include the use of Bayesian networks [14,15], kernel density estimation and conformal prediction [5,16] and trajectory clustering [17]. Besides the maritime domain, work has also been carried out on anomaly detection based on e.g. video data. In [18], trajectory clustering is used for detecting anomalous traffic behavior and in [4] abstract state space modeling combined with Gaussian models is used for detecting anomalous behaviors in public areas.

Anomaly detection approaches in the surveillance domain have mostly focused on tracking and analyzing single objects to find objects that behave anomalous. While this is important, it requires high quality tracking of the surveyed objects, something that is not always available in crowded urban environments.

## 2.1   Anomaly Detection in Urban Environments

Urban environments are characterized by numerous closely spaced targets moving in rather confined spaces. In such crowded scenarios, the origin of observations is often highly ambiguous, meaning that it is a complex task to associate observations to new or existing tracks. In some situations, it is even impossible to determine the origin of the observation, no matter how advanced the tracking algorithm is. For this type of operational environment, additional functionality must be incorporated into the technical systems in order to analyze situation data and support the human operator.

It is thus interesting to look at techniques that do not rely on accurate tracks and advanced information about individual objects, but which instead make use of the collective behavior of objects as expressed by uncorrelated observations.

# 3    Gaussian Anomaly Detector

This paper investigates the use of a simple Gaussian normalcy modeling scheme modeling the collective behavior of objects in an urban setting using average speed and flow of objects. An assumption is that the collective behavior in an area where a threat is located is inclined to change. Moreover, road blocks and similar changes to trafficabillity will also have an effect on movement patterns. The idea is that threats possibly can be detected by finding changes in the traffic around them. It might however not be sufficient to construct a model over the complete area of interest, but rather to construct models for subspaces laid out in e.g. a grid, similar to [6]. In urban settings there also exists contextual information that can be exploited, e.g. maps of road networks.

The detection performance might however vary depending on the type of measurements that are used and on the type and degree of subdivision that is used. These are important factors to evaluate for deciding what kind of sensors that are appropriate and if contextual data such as road networks can be used for improving performance. Three important questions have thus been identified: (1) should we divide the area of interest using a simple grid or using additional contextual road segment information, (2) should we measure the average flow or the average speed of objects, and (3) how does the degree of subdivision affect the detection performance?

An anomaly detection system consisting of five components has been constructed for addressing these questions (figure 1 shows a schematic structure).

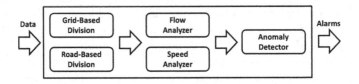

**Fig. 1.** Components in the anomaly detection system

The first two components handle the division of objects into subsets based on a grid or a road network. The next two components analyze the flow and speed of objects in each subset. The last component is the anomaly detector. The components at each step can be connected to any of the components at the next step, in order to easily evaluate various configurations of the system.

## 3.1    Geographical Division

The *Grid-Based Division* component creates a grid over the area of interest. The grid is defined by the upper right and lower left coordinates and the number of rows and columns. Each grid cell is represented by a geographical zone. When a new object is received by the system it will be assigned to the geographical zone that contains the coordinates of the object. The use of a rectangular grid may result in e.g. roads being split into multiple grid cells and cells with few data points due to low coherence with actual geography.

The *Road-Based Division* component uses GIS data to create a number of geographical zones based on the segments in the road network. The GIS data defines a number of waypoints and information about how these are connected. The component assumes a 10 meter wide road between every pair of connected waypoints. When a new object is received by the system, it is assigned to one of the zones containing the position of the object. Note that there might be overlapping zones where road segments meet. This should not significantly impact the results of the experiments since the same zone is always chosen when there are multiple overlapping zones.

## 3.2   Flow and Speed Analayzers

The *Flow Analyzer* component measures the flow in an area by counting the average number of objects in the area over time. The analyzer handles multiple areas at the same time and these can be supplied by any of the two subdivision abilities. The flow analyzer could in a live system be implemented using some form of tracking or tripwire sensor.

The *Speed Analyzer* component measures the average speed of objects present in an area. Similarly to the flow analyzer, the areas can be supplied by either of the division abilities. The average speed is measured over all the objects in the area. In a live system, the speed could easily be measured using a doppler radar and it would not require any tracking.

## 3.3   Anomaly Detector

The Anomaly Detector component is responsible for the actual detection of anomalies. It can run in two modes: learning and detection. In learning mode, the component receives flow or speed statistics based on grid or road division. These statistics are saved and the mean and standard deviations are constantly updated for each geographic zone (grid cell or road segment).

In detection mode, a previously learned normal model is loaded and used to classify new data as normal or anomalous. This is carried out by inspecting if the present mean value is within $n$ standard deviations ($\sigma$) of the mean in the previously learned model. If the new mean lay beyond this range, an alarm is sent to the presentation system. In this context an alarm consists of time, reference to geographical area and the actual deviation. In some cases, the standard deviation for a geographic zone is very small and therefore a parameter called minimum deviation ($\sigma_{min}$) is also defined. This threshold value defines a minimum required distance between the present mean and the model mean, for sending an alarm.

The last feature of the anomaly detector component is a sliding window that allows the detector to set which of the new pieces of data the detector should use when calculating the mean. The available options are: all data, last minute, last five minutes and last fifteen minutes. Setting the sliding window value to a low value, e.g. last minute, will increase the reactivity of the detector but it will also make it more sensitive to variations in the input data.

## 4  Experimental System

In order to investigate the questions put forth in section 3 an experimental system has been constructed. This system has been built by integrating a number of existing software components and by extending their functionality to support evaluation of the proposed anomaly detection system described in section 3. Figure 2 shows the architecture of the experimental system.

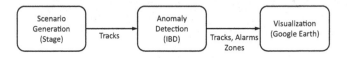

**Fig. 2.** High-level architecture of the experimental system

As can be seen, data sets are generated in real-time using Stage[1]. Individual tracks of simulated objects are fed into the Anomaly Detection System (implemented on the Intelligent Behavior Detector (IBD) platform [19]). The IBD routes the tracks to Google Earth[2] for visualization while at the same time using them internally for anomaly detection. The result of the anomaly detection are alarms that also are sent to Google Earth together with information about geographic zones that are used by the anomaly detector. To show information in Google Earth overlaid on the map, we use kml files[3]. The information in the kml file is regularly updated to reflect the current situation and it is generated from the current tracks, alarms and zones defined in the IBD.

### 4.1  Experimental Setup

A number of different experiments have been carried out to evaluate the usefulness of data-driven anomaly detection for detecting threats such as roadblocks and IEDs. Two simulated scenarios have been used for evaluation, where the first consist of normal traffic from an area of interest. The output from this scenario represents the training dataset that is used to train the anomaly detection algorithms. The second scenario is similar to the first scenario, but with a number of roadblocks/IEDs added to it. Due to the definition of roadblock, the generated traffic will automatically avoid these areas and instead use alternative routes to reach their corresponding destinations. The output from the second scenario has been used for evaluating different settings of the anomaly detection algorithm.

---

[1] More information about Stage can be found at http://www.presagis.com/products_services/products/modeling-simulation/simulation/stage/

[2] Information about Google Earth can be found at http://www.google.com/earth/

[3] Keyhole Markup Language, http://code.google.com/intl/sv-SE/apis/kml/

## 4.2   Simulated Road-Block Scenario

To create the two versions of the previously described scenario, it has been identified that Stage needs to be extended to fulfill the requirements in table 1.

**Table 1.** Requirements for creating simulated road-block scenarios

| |
|---|
| 1. It should be possible to generate traffic that follows a road network. |
| 2. It should be possible to dynamically spawn vehicles at multiple locations. |
| 3. Vehicles should be able to take different routes though the road network to simulate different driving behaviors. |
| 4. It should be possible to set the spawn interval for each spawn location. |
| 5. It should be possible to set the parameters of spawned vehicles such as speed, type and initial heading. |
| 6. It should be possible to turn the spawning on and off to simulate an uneven flow of new vehicles. |
| 7. It should be possible to alter the road network and add roadblocks where traffic cannot pass. |
| 8. It should be possible to extract the ground truth from the simulation as well as objects detected by simulated sensors. |

The requirements are fulfilled by creating three new entities in Stage: (1) a spawner entity, (2) a spawn controller entity and (3) a vehicle entity. The spawner entity is responsible for spawning new vehicles and it can be configured to spawn vehicles at different intervals. The spawn controller entity is responsible for turning the spawner entity on and off at certain intervals, to simulate an uneven flow of traffic. The vehicle entity represents individual objects in the simulation. Each spawned vehicle is given a mission that it starts to implement as soon as it is spawned. The mission tells the entity how to behave. A number of different missions are defined such as *Go to beach*, *Exit the area east*, *Go to shopping centre* and *Go to beach and then to a specific parking lot*.

Figure 3 shows the area of interest and it also illustrates a single road block that has been placed outside a shopping mall in the evaluation scenario.

## 4.3   Experimental Process

The process that has been used for evaluation includes 7 steps:

1. Start the simulation with the normal flow scenario. Wait for the system to reach a stable state (usually five minutes).
2. Start the Anomaly Detector in learning mode and let it run for one hour.
3. Turn off the Anomaly Detector and save the normal model.
4. Turn off the simulation.
5. Start the simulation with the scenario including a number of flow interrupting elements. Wait for the simulation to reach a stable state.
6. Start the Anomaly Detector in detection mode. Wait five minutes to let the statistics settle and start measuring which anomalies are found.
7. Stop the anomaly detector and the simulation. Evaluate the results.

**Fig. 3.** The area-of-interest used in the simulation. The figure shows spawners and spawn controllers as blue triangles and the location of the road-block as a red circle.

A stable state means that there is a constant flow of traffic on the roads. When the simulation starts there is no traffic on the roads and vehicles begin to spawn at the spawning locations. After a while, when enough vehicles have been removed after reaching their destinations with new vehicles simultaneously spawning, the simulation will behave in a stable manner, i.e. the roughly the same number of vehicles are present in the simulation at any time.

# 5    Results

The first set of experiments is based on the ground truth from the simulation, i.e. the correct position for all objects at all times. This is not possible in real-world scenarios where real sensors must be used. Therefore, a simple sensor was implemented in the Stage tool for the second set of experiments. The sensor corresponds to the Saab SIRS 1600 short range radar sensor [20] that has a detection range of 1600 meters and a field-of-view of about 15 degrees. The SIRS 1600 can detect and track objects such as humans, cars, buses and trucks.

## 5.1    Results Using Ground-Truth Data

The default parameters in the experiments were $n = 2$ and $\sigma_{min} = 0$. In some experiments, the parameters have been altered in order to find any anomalies.

A total of eight experiments have been carried out in order to answer the questions in section 3. Each experiment has been carried out using the process described in section 4.3. Table 2 shows the results from the experiments.

**Table 2.** Results from experiments with ground truth data

| Experiment | Division | Analyzed parameter | Result |
|---|---|---|---|
| 1 | Road | Speed | Works very well, no false alarms in normal data or during evaluation. An example of the alarms can be found in figure 4. |
| 2 | Road | Flow | Works very well, no false alarms in normal data or during evaluation. |
| 3 | Grid (10x10) | Speed | Hard to find the anomalies with the default thresholds. With $\sigma_{min} = 0.2$ some anomalies are found but the output are intermittent. |
| 4 | Grid (10x10) | Flow | No anomalies found. The road block is located on the border between two grid cells. |
| 5 | Grid (5x5) | Speed | No anomalies found. With $\sigma = 1$ and $\sigma_{min} = 0.1$ some anomalies are found as well as false alarms. |
| 6 | Grid (5x5) | Flow | No anomalies found. With $\sigma = 1$ and $\sigma_{min} = 0.1$ some anomalies are found as well as false alarms. |
| 7 | Grid (25x25) | Speed | Anomalies are found. Some intermittent false alarms. |
| 8 | Grid (25x25) | Flow | Anomalies are found. Some intermittent false alarms. |

*Should we divide the area of interest using a simple grid or using additional contextual road segment information?*

Based on the results it is more efficient to use road segment information for subdivision, compared to using a simple grid. The optimal size of grid cells is however not obvious although the finest grid (25x25) yielded the best results in the experiments. The results from using the grid approach also resulted in unstable statistics in some cells. The normal variations were sometimes higher than two standard deviations which resulted in false alarms when feeding normal data into the anomaly detector. The road segment approach is therefore preferred if such data is available; otherwise the grid based approach can be used.

**Fig. 4.** Alarms from experiment 1

*Should we measure the average flow or the average speed of objects?*

In the experiments, the normal model based on both approaches performed similar. An advantage of using flow is however that the flows can be measured with simpler sensors.

*How does the degree of subdivision affect the detection performance?*

The experiments show that having too coarse a grid will decrease the ability to detect anomalies. The finest grid resulted in the most detected anomalies and the fewest false alarms.

## 5.2    Results Using Simple Sensors

Both the grid cell and the road based division of the geographic area were evaluated with the flow and speed analyzers. It was however decided that the grid cell approach should only be evaluated using the grid parameters that gave the best results, i.e. 25x25 grid cells. Each experiment has been carried out using the process described in section 4.3. The results are presented in table 3.

**Table 3.** Results from experiments using simple sensors

| Experiment | Division | Analyzed parameter | Result |
|---|---|---|---|
| 9 | Road | Speed | Works very well, three anomalies detected in the vicinity of the road block. |
| 10 | Road | Flow | Works very well, two anomalies detected in the vicinity of the road block. |
| 11 | Grid (25x25) | Speed | Six anomalies are found. Four in the vicinity of the road block and one in each of the east and south entrances to the road network. |
| 12 | Grid (25x25) | Flow | Two anomalies are found. One near the road block and one at the west entrance to the road network. |

The conclusion of the experiments with simple sensors is that the anomaly detection work almost as well as with ground truth data. This is however very dependent on the placement of the sensors and the number of sensors used. In the experiments, seven sensors were used to cover the most important roads in the area of interest. With fewer sensors, the performance would not be as good.

## 6    Conclusions

The detection of threats in urban asymmetric conflict environments, e.g. ambushes and IEDs, has become an increasingly important objective for increased task force security. Urban environments are characterized by numerous moving objects in crowded areas, making it difficult to only rely on detect and track methods. In this paper a Gaussian Anomaly Detector has been suggested for generating early warnings that can be used to assist human operators in the detection of threats in such environments. The proposed anomaly detector focuses on modeling the collective behavior of objects of interest through the use of average speed and flow in relation to small geographic areas.

The initial experiments that have been carried out using a simulated urban threat scenario have investigated (1) different ways of dividing the area of interest, i.e. square grid cells or based on road-segments, and (2) if the average speed or flow was the best measure for modeling the normal behavior of a set of vehicles. The evaluation shows that, using the proposed anomaly detector, there is only small difference in performance between measuring speed and flow. It also shows that the use of contextual map information to divide the area based on road segments, yields more stable performance than using a grid cell approach. A conclusion is that road network information should be used if it is available; otherwise, acceptable performance can be achieved using grid cells.

An advantage of using an anomaly detector that operates on average speed or flow information is that it puts lower requirements on sensor systems and their tracking performance. It is not critical to have perfect tracking of all objects at all times; instead, it is sufficient to be able to measure the number of objects or the average speed of objects. This can be achieved using simple sensors.

Although it has been shown that the proposed anomaly detector can be used for detecting anomalies in a simulated urban scenario, more research and development is needed to achieve an operational system. The normalcy modeling scheme should be extended to handle more contextual information and to better capture variations in the data. Moreover, it is not enough to evaluate the feasibility using only simulated data. Data from a real sensor network deployed in an urban area should be collected and used for evaluation.

**Acknowledgments.** This work was supported by the European Defence Agency under the Defence R&T Joint Investment Programme on Force Protection; Contract No. A-0828-RT-GC Data Fusion in Urban Sensor Networks D-FUSE. The research has also been supported by the Infofusion Research Program (University of Skövde, Sweden) in partnership with Saab AB and the Swedish Knowledge Foundation under grant 2010/0320 (UMIF).

# References

1. Ayling, S., Benchoam, D.: New antennas for new battlefields. In: Proceedings of the Military Communications and Information Systems Conference, Canberra, Australia, November 9-11 (2010)
2. Nilsson, M., van Laere, J., Ziemke, T., Edlund, J.: Extracting rules from expert operators to support situation awareness in maritime surveillance. In: Proceedings of the 11th International Conference on Information Fusion, Cologne, Germany, June 30 - July 3 (2008)
3. Kraiman, J.B., Arouh, S.L., Webb, M.L.: Automated anomaly detection processor. SPIE, vol. 4716, pp. 128–137 (2002)
4. Brax, C., Niklasson, L., Smedberg, M.: Finding behavioural anomalies in public areas using video surveillance data. In: Proceedings of the 11th International Conference on Information Fusion (Fusion 2008), Cologne, Germany, June 30 - July 3, pp. 1655–1662 (2008)

5. Laxhammar, R., Falkman, G., Sviestins, E.: Anomaly detection in sea traffic - a comparison of the gaussian mixture model and the kernel density estimator. In: Information Fusion, pp. 756–763 (2009)
6. Bomberger, N., Rhodes, B., Seibert, M., Waxman, A.: Associative learning of vessel motion patterns for maritime situation awareness. In: 2006 9th International Conference on Information Fusion, pp. 1–8 (2006)
7. Chandola, V., Banerjee, A., Kumar, V.: Anomaly detection: A survey. ACM Comput. Surv. 41(3) (2009)
8. Patcha, A., Park, J.: An overview of anomaly detection techniques: Existing solutions and latest technological trends. Computer Networks 51(12), 3448–3470 (2007)
9. Tan, P.N., Steinbach, M., Vipin, K.: Introduction to Data Mining. Pearson Education, Inc., Boston (2006)
10. Portnoy, L., Eskin, E., Stolfo, S.: Intrusion detection with unlabeled data using clustering. In: Proceedings of ACM CSS Workshop on Data Mining Applied to Security (DMSA 2001), pp. 5–8 (2001)
11. Markou, M., Singh, S.: Novelty detection: a review-part 1: statistical approaches. Signal Processing 83(12), 2481–2497 (2003)
12. García-Teodoro, P., Díaz-Verdejo, J., Maciá-Fernández, G., Vázquez, E.: Anomaly-based network intrusion detection: Techniques, systems and challenges. Computers and Security 28(1-2), 18–28 (2009)
13. Riveiro, M., Falkman, G., Ziemke, T.: Improving maritime anomaly detection and situation awareness through interactive visualization. In: Proceedings of the 11th International Conference on Information Fusion (Fusion 2008), Cologne, Germany, June 30 - July 3, pp. 47–54 (2008)
14. Johansson, F., Falkman, G.: Detection of vessel anomalies - a bayesian network approach. In: International Conference on Intelligent Sensors, Sensor Networks and Information Processing, pp. 395–400 (2007)
15. Mascaro, S., Korb, K.B., Nicholson, A.E.: Learning abnormal vessel behaviour from ais data with bayesian networks at two time scales. Technical report, TR 2010 / Bayesian Intelligence (2010)
16. Laxhammar, R., Falkman, G.: Conformal prediction for distribution-independent anomaly detection in streaming vessel data. In: Proceedings of the First International Workshop on Novel Data Stream Pattern Mining Techniques (StreamKDD 2010), Washington D.C., USA, July 25, pp. 47–55 (2010)
17. Dahlbom, A., Niklasson, L.: Trajectory clustering for coastal surveillance. In: Proceedings of the 10th International Conference on Information Fusion (Fusion 2007), Québec, Canada, July 9-12 (2007)
18. Snidaro, L., Piciarelli, C., Foresti, G.L.: Fusion of trajectory clusters for situation assessment. In: Proceedings of the 9th International Conference on Information Fusion (Fusion 2006), Florence, Italy, July 10-13 (2006)
19. Saab AB: Intelligent behaviour detector. Internet, http://www.saabgroup.com/Global/Documents%20and%20Images/Civil%20Security/Maritime%20Transportation%20and%20Port%20Security/IBD/ibd-ver1-090505a.pdf (accessed January 28, 2012)
20. Saab AB: Sirs radar sensor. Internet, http://www.saabgroup.com/Global/Documents%20and%20Images/Civil%20Security/Land%20Transport%20and%20Urban%20Security/SIRS%20200ITS/SIRS200ITS_downloads_Product%20sheet.pdf (accessed December 28, 2011)

# Comparing Random-Based
# and *k*-Anonymity-Based Algorithms
# for Graph Anonymization

Jordi Casas-Roma, Jordi Herrera-Joancomartí, and Vicenç Torra

[1] Department of Computer Engineering, Multimedia and Telecomunications (EIMT),
Universitat Oberta de Catalunya (UOC), Barcelona
jcasasr@uoc.edu
[2] Department of Information and Communications Engineering (DEIC),
Universitat Autònoma de Barcelona (UAB), Bellaterra
jherrera@deic.uab.cat
[3] Artificial Intelligence Research Institute (IIIA), Spanish National Research Council
(CSIC), Bellaterra
vtorra@iiia.csic.es

**Abstract.** Recently, several anonymization algorithms have appeared for privacy preservation on graphs. Some of them are based on randomization techniques and on *k*-anonymity concepts. We can use both of them to obtain an anonymized graph with a given *k*-anonymity value. In this paper we compare algorithms based on both techniques in order to obtain an anonymized graph with a desired *k*-anonymity value. We want to analyze the complexity of these methods to generate anonymized graphs and the quality of the resulting graphs.

**Keywords:** Privacy, Anonimization, Social networks, Graphs, k-Anonymity.

## 1 Introduction

Currently, the data mining processes require large amounts of data, which often contain personal and private information of users and individuals. Although basic processes are performed on data anonymization, such as removing names or other key identifiers, remaining information can still be sensitive, and useful for an attacker to re-identify users and individuals. E.g., birthday and ZIP codes might be enough to re-identify individuals [1]. To solve this problem, methods that perform introduction of noise in the original data have been developed in order to hinder the subsequent processes of re-identification.

In this paper we will discuss anonymization techniques applied to graph formatted data. One of the most well known data that can be represented as graphs are social networks. Social networks are very interesting for their analysis by scientists and companies, nevertheless any release to a third party for their analysis requires the application of a protection procedure.

There are multiple methods for privacy preservation in graphs. One of the most used are random-based methods, which modify graphs at random to

V. Torra et al. (Eds.): MDAI 2012, LNAI 7647, pp. 197–209, 2012.
© Springer-Verlag Berlin Heidelberg 2012

hinder re-identification processes. Other methods are based on the concept of $k$-anonymity [1]. These methods are more complex than random-based. In this paper we ask ourselves if we can get the same results using a random algorithm and a algorithm to select $k$-anonymous graphs.

This paper is organized as follows. In Section 2, we review different anonymization methods for graphs' privacy preservation. Section 3 presents our experimental framework, including anonymization algorithms, graph and re-identification risk assessment and data sets used in our experiments. In Section 4, we show the experiments and discuss the results. Finally, in Section 5, we discuss conclusions and future work.

## 2    State of the Art

Anonymization methods depend on the type of data they are intended to work with. In this paper, we will work with simple, undirected and unlabelled graphs. Because these graphs have no attributes or labels in the edges, information is only in the structure of the graph itself and, due to this, the adversary can use information about the structure of the network to attack the privacy. However, since all of the information is contained in it, we want to preserve the structure of the graph.

### 2.1    Random-Based Methods

One widely adopted strategy of graph modification approaches are randomization methods. Randomization methods are based on adding random noise in original data. There are two basic approaches to work with graph data: (1) *Rand Add/Del*: randomly add and delete the same number of edges from the original graph (this strategy keeps the number of edges) and (2) *Rand Switch*: exchange edges between pairs of nodes (this strategy keeps the number of edges and the degree of all nodes).

Hay et al. [2] proposed a method to anonymize unlabelled graphs. This method is called *Random Perturbation* and is based on two phases: first, $m$ edges are randomly removed from the graph and then false $m$ edges are randomly added. The set of vertices is not changed and the number of edges is preserved in the anonymized graph.

Ying et al. [3] proposed a variation of *Rand Add/Del* method, called *Rand Add/Del-B*. This method implements modifications (by adding and removing edges) on the nodes at high risk of re-identification, not at random over the entire set of nodes. The authors expect to introduce fewer perturbations (with better utility preservation) to achieve the same privacy protection.

### 2.2    $k$-Anonymity-Based Methods

Another strategy widely adopted for privacy-preserving is based on the concept of $k$-anonymity. This concept was introduced by Sweeney [1] for the privacy preservation on relational data. Formally, the $k$-anonymity model is defined as: let $RT(A_1, \ldots, A_n)$ be a table and $QI_{RT}$ be the quasi-identifier associated with it. $RT$ is said to satisfy $k$-anonymity if and only if each sequence of values in $RT[QI_{RT}]$ appears with at least $k$ occurrences in $RT[QI_{RT}]$. The $k$-anonymity model indicates that an attacker can not distinguish between different $k$ records although he manages to find a group of quasi-identifiers. Therefore, the attacker can not re-identify an individual with a probability greater than $\frac{1}{k}$.

Different concepts can be used to apply the $k$-anonymity model on graphs. A widely option is to use the node degree as a quasi-identifier [4]. It is called $k$-degree anonymity. We assume that the attacker knows the degree of some nodes. If the attacker identifies a single node with the same degree in the anonymized graph, then he has re-identified this node. $K$-anonymity methods are based on modifying the graph structure (by adding and removing edges) to ensure that all nodes satisfy the $k$-anonymity properties for the degrees of all the nodes. In other words, the main objective is that all nodes have at least $k - 1$ other nodes sharing the same degree.

### 2.3 Graph Assessment

Several measures and metrics have been used to quantify network structure in graph formatted data. Usually, the authors compare the values obtained by the original data and the anonymized data in order to quantify the noise introduced by the anonymization process.

Hay et al. [2] proposed five structural properties from graph theory for quantifying network structures. For each node, the authors evaluate closeness centrality (average shortest path from the node to every other node), betweenness centrality (proportion of all shortest paths through the node) and path length distribution (computed from the shortest path between each pair of nodes). For the graph as a whole, they evaluate the degree distribution and the diameter (the maximum shortest path between any two nodes). The objective is to keep these five steps closer to their original values, assuming that it involves little distortion in the anonymized data.

Zou et al. [6] defined a simple method for evaluating information loss on undirected and unlabelled graphs. The method is based on the difference between the original and the anonymized graph edges. Formally, $Cost(G, \widetilde{G}) = (E \cup \widetilde{E}) - (E \cap \widetilde{E})$ where $G(V, E)$ is the original graph, $V$ is the node set, $E$ is the edge set, and $\widetilde{G}(\widetilde{V}, \widetilde{E})$ is the anonymized graph.

### 2.4 Risk Assessment

Re-identification risk in anonymized graph is important to evaluate the quality of any anonymization process. Determining the knowledge of the adversary is the main problem. From the knowledge of the adversary, different methods for assessing the re-identification risk have been developed.

Zhou et al. [13] model the background knowledge of adversaries in various ways: Identifying attributes of nodes, nodes degrees, link relationship, neighbourhoods, embedded sub-graphs and graph metrics. We focus on a knowledge of the adversary based on degree nodes.

Hay et al. [2] [5] proposed a method, called *Vertex Refinement Queries*, to model the knowledge of the adversary. This class of queries, with increasing attack power, models the local neighbourhood structure of a node in the network. The weakest knowledge query, $\mathcal{H}_0(v_j)$, simply returns the label of the node $v_j$. The queries are successively more descriptive: $\mathcal{H}_1(v_j)$ returns the degree of $v_j$, $\mathcal{H}_2(v_j)$ returns the list of each neighbours' degree, and so on. The queries can be defined iteratively, where $\mathcal{H}_i(v_j)$ returns the multi-set of values which are the result of evaluating $\mathcal{H}_{i-1}$ on the set of nodes adjacent to $v_j$:

$$\mathcal{H}_i(v_j) = \{\mathcal{H}_{i-1}(v_1), \mathcal{H}_{i-1}(v_2), \dots, \mathcal{H}_{i-1}(v_m)\} \tag{1}$$

where $v_1, v_2, \dots, v_m$ are the nodes adjacent to $v_j$.

A candidate set for a query $\mathcal{H}_i$ is a set of all nodes with the same value of $\mathcal{H}_i$. Therefore, the cardinality of a candidate set for $\mathcal{H}_i$ is the number of indistinguishable nodes in $G$ under $\mathcal{H}_i$. Note that if the cardinality of the smallest candidate set under $\mathcal{H}_1$ is $k$, the probability of re-identification is $\frac{1}{k}$. Hence, the $k$-degree anonymity value for $G$ is $k$.

## 3    Experimental Set Up

Our main objective is to compare random-based and $k$-anonymity-based algorithms for privacy preservation on graphs. If we want to anonymize a graph to a specific value of $k$-anonymity, then we should ask ourselves, what kind of method is the best to achieve this purpose. I.e., we want to compare random-based and $k$-anonymity-based methods to anonymize graphs to a specific value of $k$-anonymity.

Random-based methods modify the structure of the graph, so they can modify the value of $k$-degree anonymity. But we can not specifically control the desired value. Therefore, if we want to get an anonymized graph with a specific value of $k$-anonymity, we must generate multiple anonymized graphs until we find one with the desired $k$-anonymity value.

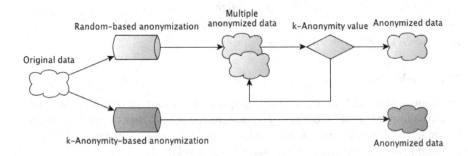

**Fig. 1.** Experimental framework

To conduct this experiment, we choose three graph formatted datasets, two anonymization algorithms and several quality measures. Figure 1 shows our experimental framework. First, we anonymize the graphs data sets (details are shown in Section 3.1) using two anonymization algorithms (Section 3.2). Then, we evaluate original and anonymized data using measures for quantifying network structures (Section 3.3). And finally, we use risk assessment measures (Section 3.4) to assess the improvement in privacy-preserving on anonymized data.

### 3.1   Data Sets

Three different data sets are used in our experiments. Table 1 shows a summary of the data sets' main features. The data sets considered are the following ones:

- Zachary's Karate Club [7] is a graph widely used in the literature. The graph shows the relationships among 34 members of a karate club.
- American College Football [8] is a graph of American football games between Division IA colleges during regular season Fall 2000.
- Jazz Musicians [9] is a graph of jazz musicians and their relationships.

**Table 1.** Data sets properties

| Data set | Nodes | Edges | Average degree | Average distance | Diameter |
|----------|-------|-------|----------------|------------------|----------|
| Zachary's Karate Club | 34 | 78 | 4.588 | 2.408 | 5 |
| American College Football | 115 | 613 | 10.661 | 2.508 | 4 |
| Jazz Musicians | 198 | 2,742 | 27.697 | 2.235 | 6 |

### 3.2   Anonymization Methods

We choose the following random-based and $k$-anonymity-based anonymization algorithms for our experiments.

#### Random-Based Algorithm

Among all existing random-based anonymization algorithms, we use **Random Perturbation** (RP) [2]. This algorithm removes and adds the same number of edges from the original graph, by keeping the total number of edges in the graph.

As Figure 1 describes, we perform multiple anonymizations using RP algorithm. The total number of anonymized graphs depends on each data set and will be specified in each experiment. For each $k$-anonymity value we want to achieve, we execute the RP algorithm iteratively, until we get a graph with the $k$-anonymity value. The process starts with an anonymization percentage of 1%.

If a graph with the desired $k$-anonymity value is generated, it is the solution and anonymization process finishes. After 100 iterations, and if a graph with the desired $k$-anonymity value is not found, the anonymization percentage is increased at 1%. This process is repeated until the anonymization percentage reaches a limit of 50%, when the process stops without a solution. Therefore, 5,000 randomly anonymized graphs are generated before RP finishes without solution.

### $k$-Anonymity-Based Algorithm

We use the **Genetic Graph Anonymization** (GGA) [10] for obtaining an anonymized graph which preserves $k$-anonymity on the degree. The approach, based on genetic algorithms, can be described in terms of the following two steps.

1. Given the degree sequence of the nodes of $G(V, E)$, $d = \{d_1, \cdots, d_n\}$, we construct a new sequence $\widetilde{d}$ that is $k$-degree anonymous and minimizes the distance between the two sequences.
2. We construct a new graph $\widetilde{G}(\widetilde{V}, \widetilde{E})$ with degree sequence $\widetilde{d}$ in which $\widetilde{V} = V$ and $\widetilde{E} \cap E \approx E$.

Our proposal for the first step of the anonymization algorithm uses genetic algorithms. These algorithms use the mutation process and the fitness function defined as follows.

**Mutation Process.** Add one to an element of the degree sequence and subtract one from another element. This transaction represents a change in one of the nodes of an edge. For example, if we modify node $v_1$ to $v_2$ on the edge $e_{0,1} = (v_0, v_1)$ we get the edge $e_{0,2} = (v_0, v_2)$. This node change is represented in the degree sequence as subtracting one to the degree value of node $v_1$ and adding one to the degree value of node $v_2$. Note that our genetic algorithm does not use the recombination of pairs of parents, since this process systematically breach the rule that preserves the number of edges in the graph, and therefore generate no valid candidates.

**Fitness Function.** The fitness function, which evaluates candidates, is computed from three parameters: (1) the current $k$-anonymity value, where the objective is to achieve a $k$-anonymity value greater than or equal to the desired value. (2) The distance between the anonymized and the original degree sequence, computed by Equation 2, where the objective is to minimize this value. And (3), the number of nodes that do not meet the desired value of $k$-anonymity, which will decrease until it reaches 0, when we get the desired $k$-anonymity value.

$$D(d, \widetilde{d}) = \sum_{i=0}^{n} |\widetilde{d}_i - d_i| \qquad (2)$$

where $n = |V|$.

In the second step we make the necessary changes in the original graph to obtain the anonymized graph. The changes that have occurred in the degree sequence indicate the nodes that should change its degree. I.e., indicate the edges to be modified.

## 3.3   Graph Assessment

We use different measures for quantifying network structure. These measures and metrics are used to compare both the original and the anonymized data in order to quantify the level of perturbation introduced in the anonymized data by the anonymization process. These measures and metrics evaluate some key graph properties.

In the rest of this section we review the measures used. In the definitions we use $G(V, E)$ and $\widetilde{G}(V, \widetilde{E})$ to indicate the original and the anonymized graphs, $n = \mid V \mid$ to denote the number of nodes, and $d_{ij}$ to denote the length of the shortest geodesic path from node $v_i$ to $v_j$.

The first one is **average distance**. It is defined as the average of the distances between each pair of nodes in the graph. It measures the minimum average number of edges between any pair of nodes. Formally, it is defined as:

$$AD(G) = \frac{\sum_{i,j} d_{ij}}{\binom{n}{2}} \tag{3}$$

The second is **edge intersection**. It is defined as the intersection of the edges set. Formally:

$$EI(G, \widetilde{G}) = \frac{E \cap \widetilde{E}}{\mid E \cup \widetilde{E} \mid} \tag{4}$$

The third is **betweenness centrality**, which measures the fraction of the number of shortest paths that go through each vertex. This measure indicates the centrality of a node based on the flow between other nodes in the graph. A node with a high value indicates that this node is part of many shortest paths in the graph, which will be a key node in the graph structure. This measure is normalized to be in the range [0,1]. Formally, we define the betweenness centrality of a node $v_i$ as:

$$BC(v_i) = \frac{1}{n^2} \sum_{st} \frac{g_{st}^i}{g_{st}} \tag{5}$$

where $g_{st}^i$ is the number of geodesic paths from $s$ to $t$ that pass through $v_i$, and $g_{st}$ is the total number of geodesic paths from $s$ to $t$.

The fourth one is **closeness centrality**, which is defined as the inverse of the average distance to all accessible nodes. It is normalized in the range [0, 1]. Closeness is an inverse measure of centrality in that a larger value indicates a less central node while a smaller value indicates a more central node. Formally, we define the closeness centrality of a node $v_i$ as:

$$CC(v_i) = \frac{n}{\sum_j d_{ij}} \tag{6}$$

The betweenness and closeness centrality lead to different results for the same graph as they focus on different aspects of centrality. As shown above, both compute a value for each node. To compare the original and the protected graph, it is convenient to aggregate these values in a single one. For each of the two measures, we compute an average difference using the root mean square (other average functions [11] could be used here as well) as follows:

$$Diff(G, \hat{G}) = \sqrt{\frac{1}{n}((g_1 - \hat{g}_1)^2 + \ldots + (g_n - \hat{g}_n)^2)} \tag{7}$$

where $g_i$ is either the betweenness centrality or the closeness centrality of node $v_i$.

The number of nodes, edges and average degree are not considered to assess the anonymization process because the methods analysed in this work keep these values constant.

### 3.4    Risk Assessment

As we have discussed above, it is necessary to define the adversary's knowledge to define a method for assessing the re-identification risk. In this paper we assume a knowledge of the adversary based on the degree of the nodes and we use Vertex Refinement Queries of level 1 ($\mathcal{H}_1$) as a re-identification risk measures.

The $\mathcal{H}_1(v_i)$ indicates the degree of node $v_i$ and the candidate set of $\mathcal{H}_1$, $cand_{\mathcal{H}_i}$, is the set of all nodes grouped by their degree. That is, one subset corresponds to all nodes of degree value equal to 1, another to all nodes of degree value equal to 2, and so on. Therefore, the minimum cardinality of the subsets corresponds to the value of $k$-degree anonymity. But $cand_{\mathcal{H}_i}$ also shows interesting information about how re-identification risk is distributed on all nodes of the graph.

$$cand_{\mathcal{H}_1} = \{v_j \in V \mid \mathcal{H}_1(v_i) = \mathcal{H}_1(v_j)\} \tag{8}$$

In our experiments we analyse how the candidate set evolves, so this allows us to see how the graph evolves in terms of re-identification's risk.

## 4    Experimental Results

In this section, we show the results of our experiments. We compare RP and GGA algorithms to anonymize a graph with a specific $k$-anonymity value.

## 4.1   Zachary's Karate Club

The original graph has a $k$-anonymity value equal to 1. RP algorithm achieves anonymized graphs with a $k$-anonymity values equal 2,3 and 4, while GGA achieves graphs with a $k$-anonymity values equal to 2, 4 and 5. So, GGA algorithm gets a higher $k$-anonymity value than RP algorithm. Table 2 shows that RP algorithm is much faster than GGA algorithm.

**Table 2.** Zachary's Karate Club generation time

| Algorithm | $k=2$ | $k=3$ | $k=4$ | $k=5$ |
|-----------|-------|-------|-------|-------|
| RP | 00:01 sec | 00:06 sec | 00:53 sec | - |
| GGA | 00:33 sec | - | 01:38 sec | 02:34 sec |

Average distance is shown in Figure 2a. GGA algorithm achieves better results than RP algorithm for all values of $k$. Note that the $k = 1$ values correspond to the original graph. Figure 2b shows edge intersection between original and anonymized graphs. Also, GGA algorithm achieves better results for all values of $k$. In addition, RP algorithm obtains a very bad result for $k=4$, where edge intersection measure falls to 20%.

The RMS error of the betweenness centrality, Figure 2c, and the RMS error of the closeness centrality, Figure 2d, show similar results on both measures, where GGA introduces less perturbation than RP.

Figures 2e and 2f show the details of the $cand_{\mathcal{H}_1}$ results for RP and GGA algorithms. Nodes with a candidate set of size 1 have been uniquely re-identified (6 nodes, 17.64%, on the original graph). Nodes with a candidate set of size between 2 and 4 are in high risk of re-identification (5 nodes, 14.70%, on the original graph). However, nodes with candidates set between 5 and 10 and greater than 10 are well-protected (23 nodes, 67.64%, on the original graph).

If we compare the results of anonymized graphs with a value of $k=2$, we can see that the GGA algorithm achieves a smaller set of nodes at high risk of re-identification (41.17% RP and 35.29% GGA). If we compare the results with a value of $k=4$, we can see that the GGA algorithm achieves a smaller set of nodes at high risk of re-identification and a bigger set of well-protected nodes.

## 4.2   American College Football

The original graph has a $k$-anonymity value equal to 1. RP algorithm get values of $k$-anonymity of 2, 3, 4, 5 and 6, and GGA algorithm get values of $k$-anonymity of 4 and 10. Table 3 shows generation time for both algorithms. Like in the previous experiment, RP algorithm is faster than GGA algorithm, but GGA algorithm achieves higher values of $k$-anonymity than RP algorithm.

In Figure 3a we can see that GGA algorithm gets much better results on average distance than RP algorithm, especially for $k$-anonymity values greater than 5. Figure 3b shows the same behaviour for edge intersection.

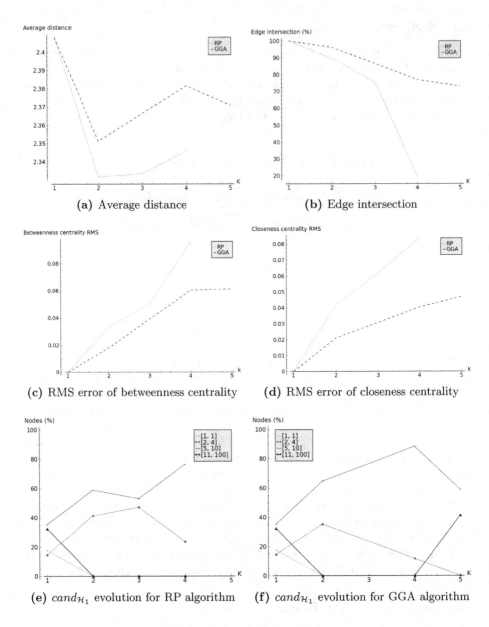

(a) Average distance

(b) Edge intersection

(c) RMS error of betweenness centrality

(d) RMS error of closeness centrality

(e) $cand_{\mathcal{H}_1}$ evolution for RP algorithm

(f) $cand_{\mathcal{H}_1}$ evolution for GGA algorithm

**Fig. 2.** Zachary's Karate Club

**Table 3.** American College Football generation time

| Algorithm | k=2 | k=3 | k=4 | k=5 | k=6 | k=7 | k=8 | k=9 | k=10 |
|---|---|---|---|---|---|---|---|---|---|
| RP | 00:01 sec | 00:01 sec | 00:04 sec | 00:06 sec | 01:21 sec | - | - | - | - |
| GGA | - | - | 00:51 sec | - | - | - | - | - | 02:01 sec |

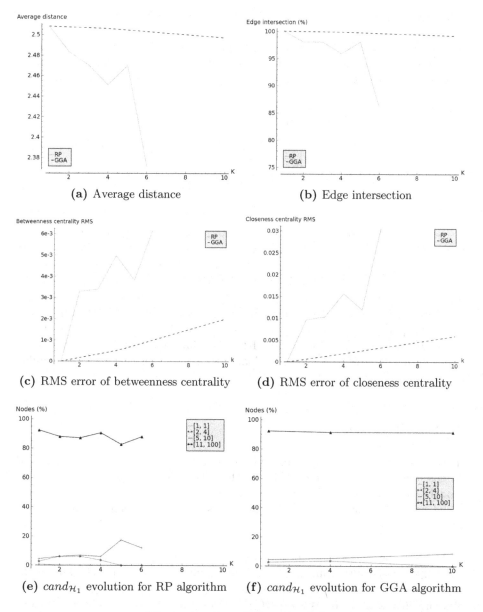

(a) Average distance          (b) Edge intersection

(c) RMS error of betweenness centrality    (d) RMS error of closeness centrality

(e) $cand_{\mathcal{H}_1}$ evolution for RP algorithm    (f) $cand_{\mathcal{H}_1}$ evolution for GGA algorithm

**Fig. 3.** American College Football

The RMS error of the betweenness centrality, Figure 3c, and the RMS error of the closeness centrality, Figure 3d, show that GGA algorithm introduces less perturbation in both measures.

Figures 3e and 3f show the details of the $cand_{\mathcal{H}_1}$ results for RP and GGA. GGA algorithm achieves excellent results at all the range of anonymization. RP algorithm achieves good results too, but fall short of those achieved by GGA.

### 4.3   Jazz Musicians

The original graph has a $k$-anonymity value equal to 1. Both algorithms only achieve graphs with a $k$-anonymity value equal to 2. Table 4 shows generation time for these processes.

**Table 4.** Jazz Musicians generation time

| Algorithm | $k=2$ |
|-----------|-------------|
| RP | 3:14:27 sec |
| GGA | 5:26:51 sec |

Using this data set, average distance decreases smoothly on GGA, like in previous data sets. Edge intersection gets a value of 99.53% on GGA $k = 2$ anonymized graph, while RP gets a value of 33.04% for a graph with the same $k$-anonymity value. This value indicates that RP affects the quality and the usefulness of the anonymized data. The RMS error of the betweenness centrality and the RMS error of the closeness centrality show that GGA algorithm introduces less perturbation in both measures. The details of the $cand_{\mathcal{H}_1}$ results for RP and GGA algorithms sows that RP algorithm increases the well-protected nodes until a value of 52%, while GGA algorithm maintains the data very similar to the initial values.

## 5   Conclusions

In this paper we have reported an experimental study of two anonymization algorithms. One of them is random-based, while the other is based on $k$-anonymity model. We have applied these anonymization algorithms on three real world social networks that have well-documented structures: Zachary's Karate Club network, American College Football teams' network and Jazz Musicians' network.

After seeing the results of the experiments, we can clearly see that $k$-anonymity-based algorithm gets the best results on all data sets. This algorithm, called *GeneticGraphAnonymization* (GGA), achieves a greater degree of anonymity and produces less perturbation on graphs. So, it produces a more useful data and a more protected data. However, GGA algorithm is slower than RP on all data sets.

Many interesting directions for future research have been uncovered by this work. Other graph anonymization methods should be evaluated and compared with our $k$-anonymity-based algorithm. Also, another interesting area is to evaluate different measures for the re-identification risk. There are several measures and it is interesting to compare all of them. Finally, another graph types will be considered, such as weighted [12] or directed graphs.

**Acknowledgments.** This work was partially supported by the Spanish MCYT and the FEDER funds under grants TSI2007-65406-C03 "E-AEGIS", TIN2010-15764 "N-KHRONOUS", CONSOLIDER CSD2007-00004 "ARES", and TIN2011-27076-C03 "CO-PRIVACY".

# References

1. Sweeney, L.: k-anonymity: a model for protecting privacy. Of Uncertainty Fuzziness and Knowledge Based 10(5), 557–570 (2002)
2. Hay, M., Miklau, G., Jensen, D., Weis, P., Srivastava, S.: Anonymizing Social Networks. Science, 1–17 (2007)
3. Ying, X., Pan, K., Wu, X., Guo, L.: Comparisons of randomization and k-degree anonymization schemes for privacy preserving social network publishing. In: The 3rd Workshop on Social Network, pp. 10:1–10:10. ACM, New York (2009)
4. Liu, K.: Towards identity anonymization on graphs. In: Proceedings of the 2008 ACM SIGMOD International, pp. 93–106. ACM, New York (2008)
5. Hay, M., Miklau, G., Jensen, D., Towsley, D., Weis, P.: Resisting structural re-identification in anonymized social networks. Proc. VLDB Endow. 1(1), 102–114 (2008)
6. Zou, L., Chen, L., Ozsu, M.T., A-zsu, M.T.: K-Automorphism: A General Framework For Privacy Preserving Network Publication. In: VLDB 2009: Proceedings of the Thirtieth International Conference on Very Large Data Bases, vol. 2, pp. 946–957. VLDB Endowment, Lyon (2009)
7. Zachary, W.: An information flow model for conflict and fission in small groups. Journal of Anthropological Research 33, 452–473 (1977)
8. Girvan, M., Newman, M.E.J.: Community structure in social and biological networks. Proceedings of the National Academy of Sciences of the United States of America 99(12), 7821–7826 (2002)
9. Gleiser, P., Danon, L.: Adv. Complex Syst. 6, 565 (2003)
10. Casas-Roma, J., Herrera-Joancomartí, J., Torra, V.: Algoritmos genéticos para la anonimización de grafos. In: XII Spanish Meeting on Cryptology and Information Security (RECSI), Donostia-San Sebastián, Spain (2012)
11. Torra, V., Narukawa, Y.: Modeling decisions: information fusion and aggregation operators. Springer (2007)
12. Das, S., Egecioglu, A., Abbadi, A.E.: Anonymizing weighted social network graphs. In: ICDE, pp. 904–907. IEEE (2010)
13. Zhou, B., Pei, J., Luk, W.: A brief survey on anonymization techniques for privacy preserving publishing of social network data. ACM SIGKDD Explorations Newsletter 10(2), 12–22 (2008)

# Heuristic Supervised Approach
# for Record Linkage

Javier Murillo[1], Daniel Abril[2,3], and Vicenç Torra[3]

[1] CIFASIS-CONICET, Universidad Nacional de Rosario, Argentina
[2] Universitat Autònoma de Barcelona (UAB), Barcelona, Spain
[3] Institut d'Investigació en Intel·ligència Artificial (IIIA), Consejo Superior de
Investigaciones Científicas (CSIC), Barcelona, Spain

**Abstract.** Record linkage is a well known technique used to link records
from one database to records from another database which make reference to the same individuals. Although it is usually used in database
integration, it is also used in the data privacy field for the disclosure risk
evaluation of protected datasets. In this paper we compare two different supervised algorithms which rely on distance-based record linkage
techniques, specifically using the Choquet integral's fuzzy integral to
compute the distance between records. The first approach uses a linear
optimization problem which determines the optimal fuzzy measure for
the linkage. While, the second approach is a kind of gradient algorithm
with constraints for the fuzzy measures' identification. We show the advantages and drawbacks of both algorithms and also in which situations
they will work better.

**Keywords:** Fuzzy measure, Choquet integral, Record linkage, Heuristic, Optimization.

## 1 Introduction

Record Linkage is the task of identifying records from different databases (or
data sources in general) that refers to the same entity. This technique was firstly
used for database integration in [14] and further developed in [24,16], and it is
nowadays a popular technique used by statistical agencies, research communities,
and corporations. The main applications of record linkage are database and
datasets integration [1,10,29,30], data cleaning and quality control [5,31]. An
example of the last application is the detection of duplicate records between
several datasets [15]. However, more recently, in the context of data privacy [21],
record linkage has emerged as an important technique to evaluate the disclosure
risk of protected data [26,32]. By identifying the links between the protected
dataset and the original one, we can evaluate the re-identification risk of the
protected data [12].

Among record linkage approaches we have focused on those based on a
distance function between records, that is, it links records by their closeness.
There are previous works [28,4,3] that have considered the use of different

V. Torra et al. (Eds.): MDAI 2012, LNAI 7647, pp. 210–221, 2012.
© Springer-Verlag Berlin Heidelberg 2012

parameterized distances together with a supervised learning approach. This supervised approach relies on an optimization problem which finds the best combination of distance's parameters in order to maximize the number of correct re-identifications. In this paper we compare two different supervised learning approaches relying on distance-based record linkage for data privacy which are based on the Choquet integral [9,27]. Both supervised approaches allow the use of a fuzzy measure to weight the attributes in the datasets. However, one is based on an adaptation of the gradient descent algorithm proposed by Grabisch in [17] and the other is based on a linear optimization problem [2]. That means that the first one will find the parameters of a local minimum in a reasonable time, while the other approach will find the optimal parameters that give the maximum number of re-identifications. The goal of this comparison is to analyse if the Grabisch heuristic method can achieve similar results than the optimization problem. These results are based on the number of correct linkages between the records from two databases, the computational time needed and whether weights are much fitted for the training set, producing overfitting.

The paper is organized as follows. Section 2 introduces the record linkage techniques in the data privacy context. In Section 3 we define both supervised approaches that are compared. Section 4 shows the results of the comparison taking into account all the factors mentioned. Finally, Section 5 concludes the paper and present the future work.

## 2   Record Linkage in Data Privacy

In data privacy, record linkage can be used to re-identify individuals from a protected dataset. It serves as an evaluation of the protection method used by modeling the possible attack to be performed on the protected dataset.

A dataset $X$ can be viewed as a matrix with $N$ rows (*records*) and $V$ columns (*attributes*), where each row refers to a single individual. The attributes in a dataset can be classified, depending on their capability to identify unique individuals, as follows:

- *Identifiers*: attributes that can be used to identify the individual unambiguously. A typical example of identifier is the passport number.
- *Quasi-identifiers*: attributes that are not able to identify a single individual when they are used alone. However, when combining several quasi-identifier attributes, they can unequivocally identify an individual. Among the quasi-identifier attributes, we distinguish between confidential ($X_c$) and non-confidential ($X_{nc}$), depending on the kind of information that they provide. An example of non-confidential quasi-identifier attribute would be the zip code, while a confidential quasi-identifier might be the salary.

Before releasing the data, a protection method $\rho$ is applied, leading to a protected dataset $X'$. This protection method will protect the non-confidential quasi-identifiers $X'_{nc} = \rho(X_{nc})$. However, to ensure the privacy, identifiers are either remover or encrypted. The confidential quasi-identifiers are not modified because they are interesting for third parties. Therefore, the protected dataset, $X' = X'_{nc}||X_c$ can be published and made available. This scenario, which was first used in [12] to compare several protection methods, has also been adopted in other works like [26].

In data privacy, record linkage can be used to re-identify individuals between the protected dataset and a part or the whole original dataset as an evaluator of disclosure risk. There are two main approaches of record linkage. The **Probabilistic record linkage (PRL)** [20] and the **Distance-based record linkage (DBRL)** [25], which links each record $a$ to the *closest* record in $b$, by means of a distance function. Both approaches have been used extensively in the area of data privacy to evaluate the disclosure risk of protected data. Nevertheless, the work in this paper is focused on distance-based record linkage, specifically using the Choquet integral as a distance. This is further described in the next section.

## 2.1   Record Linkage Based on the Choquet Integral

The main point of distance-based record linkage is the definition of a distance. In this paper we consider the parametrization of distance-based record linkage. This distance parameterization allows us to weight data attributes in order to express the importance of the variables in the linkage process.

It is well known that the multiplication of the Euclidean distance by a constant will not change the results of any record linkage algorithm. Due to this, we can express the Euclidean distance used for attribute-standardized data as a weighted mean of the distances for the attributes.

We will use $V_1^X, \ldots, V_n^X$ and $V_1^Y, \ldots, V_n^Y$ to denote the set of variables of file $X$ and $Y$, respectively. Using this notation, we express the values of each variable of a record $a$ in $X$ as $a = (V_1^X(a), \ldots, V_n^X(a))$ and of a record $b$ in $Y$ as $b = (V_1^Y(b), \ldots, V_n^Y(b))$. $\overline{V_i^X}$ corresponds to the mean of the values of variable $V_i^X$.

In a formal way, we redefine the Euclidean distance as follows,

$$d^2(a,b) = \sum_{i=1}^{n} \frac{1}{n} \left( \frac{V_i^X(a) - \overline{V_i^X}}{\sigma(V_i^X)} - \frac{V_i^Y(b) - \overline{V_i^Y}}{\sigma(V_i^Y)} \right)^2$$

In addition, we will refer to each squared term of this distance as,

$$d_i^2(a,b) = \left( \frac{V_i^X(a) - \overline{V^X}_i(a)}{\sigma(V_i^X)} - \frac{V_i^Y(b) - \overline{V^Y}_i(b)}{\sigma(V_i^Y)} \right)^2$$

Using these expressions we can define the squared of the Euclidean distance as follows.

**Definition 1.** *Given two datasets $X$ and $Y$ the square of the Euclidean distance for attribute-standardized data is defined by:*

$$d^2 AM(a,b) = AM(d_1^2(a,b), \ldots, d_n^2(a,b)),$$

*where $AM$ is the arithmetic mean $AM(c_1, \ldots, c_n) = \sum_i c_i/n$.*

In general, any aggregation operator $\mathbb{C}$ [27] might be used in the place of arithmetic mean. So, we can consider the following generic distance.

$$d^2\mathbb{C}(a,b) = \mathbb{C}(d_1^2(a,b),\ldots,d_n^2(a,b))$$

From this definition, it is straightforward to consider weighted versions of the Euclidean distance. In this case we have focused on fuzzy measures of the Choquet integral, these permit us to represent, in the computation of the distance, information like redundancy, complementariness, and interactions among the variables, which are not used in other parametrized distances. Therefore, tools that use fuzzy measures to represent background knowledge permit us to consider variables that, for example, are not independent.

**Definition 2.** *Let $\mu$ be an unconstrained fuzzy measure on the set of variables $V$, i.e. $\mu(\emptyset) = 0$, $\mu(V) = 1$, and $\mu(A) \leq \mu(B)$ when $A \subseteq B$ for $A \subseteq V$, and $B \subseteq V$. Then, the Choquet integral distance is defined as:*

$$d^2 CI_\mu(a,b) = CI_\mu(d_1(a,b)^2,\ldots,d_n(a,b)^2) \tag{1}$$

*where $CI_\mu(c_1,\ldots,c_n) = \sum_{i=1}^{n}(c_{s(i)} - c_{s(i-1)})\mu(A_{s(i)})$, given that $c_{s(i)}$ indicates a permutation of the indexes so that $0 \leq c_{s(1)} \leq \ldots \leq c_{s(i-1)}$, $c_{s(0)} = 0$, and $A_{s(i)} = \{c_{s(i)},\ldots,c_{s(n)}\}$.*

The interest of this variation is that we do not need to assume that all the attributes are equally important in the re-identification. This would be the case if one of the attributes is a key-attribute, e.g. an attribute where $V_i^X = V_i^Y$. In this case, the corresponding weight would be assigned to one, and all the others to zero. Such an approach would lead to 100% of re-identifications. Moreover the interaction of coalitions of variables is taken into account by the fuzzy measure.

## 3   Supervised Learning Approaches for Record Linkage

In this section we describe the two learning processes used on this work. Firstly, we describe the optimization problem approach and then we introduce the heuristic approach, which is based on a gradient descent algorithm. Both approaches take as input a matrix formed by $n+1$ columns ( $n$ attributes + target value) and $m$ rows (each row represent one example). The output of both algorithms are the coefficients of the fuzzy measure that maximizes the number of re-identifications.

### 3.1   Linear Optimization Problem

We start discussing the notation we have used.

Let $X$ represent the original file, and $Y$ the protected file, both with variables $V_1,\ldots,V_n$. Then, $V_k(a_i)$ represents the $k$th variable of the $i$th record. Using this notation, for all $a_i \in X$ we have $a_i = (V_1(a_i),\ldots,V_n(a_i))$ and for all $b_i \in Y$ we have $b_i = (V_1(b_i),\ldots,V_n(b_i))$. For the application of the record linkage algorithm we will consider the sets of values $d(V_k(a_i), V_k(b_j))$ for all pairs of records $a_i \in X$ and $b_j \in Y$.

For the sake of simplicity in the formalization of the process, we presume that each record $a_i$ of $X$ is the protected version of $b_i$ of $Y$. That is, files are aligned.

Then, two records are correctly linked using an aggregation operator $\mathbb{C}$ when the aggregation of the values $d(V_k(a_i), V_k(b_i))$ for all $k$ is smaller than $d(V_k(a_i), V_k(b_j))$ for all $i \neq j$. In optimal conditions this should be true for all records $a_i$.

We have formalized the learning process into an optimization problem with an objective function and some constraints. As the correct linkage will not always be satisfied because of the errors in the data cause by the protection method a relaxation is needed. The relaxation is based on the concept of blocks. We consider a block as the set of equations concerning record $a_i$. Therefore, we define a block as the set of all the distances between one record of the original data and all the records of the protected data. Then, we assign to each block a variable $K_i$. Therefore, we have as many $K_i$ as the number of rows of our original file. Besides, we need for the formalization a constant $C$ that multiplies $K_i$ to overcome the inconsistencies and satisfies the constraint.

The rationale of this approach is as follows. The variable $K_i$ indicates, for each block, if all the corresponding constraints are accomplished ($K_i = 0$) or not ($K_i = 1$). Then, we want to minimize the number of blocks non compliant with the constraints. This way, we can find the best weights that minimize the number of violations, or in other words, we can find the weights that maximize the number of re-identifications between the original and protected data.

Using these variables $K_i$ and the constant $C$, we have that all pairs $i \neq j$ should satisfy

$$\mathbb{C}(d(V_1(a_i), V_1(b_j)), \ldots, d(V_n(a_i), V_n(b_j))) - $$
$$- \mathbb{C}(d(V_1(a_i), V_1(b_i)), \ldots, d(V_n(a_i), V_n(b_i))) + C K_i > 0$$

As $K_i$ is only 0 or 1, we use the constant $C$ as the factor needed to really overcome the constraint. In fact, the constant $C$ expresses the *minimum distance* we require between the correct link and the other incorrect links. The larger it is, the more correct links are distinguished from incorrect links.

Using these constraints and the Choquet integral aggregation operator $d^2 CI_\mu(a, b)$, explained in Definition 2, the minimization problem is defined in a generic form as:

$$Minimize \sum_{i=1}^{N} K_i$$

$Subject\ to:$

$$\sum_{i=1}^{N} \sum_{j=1}^{N} CI_\mu(d(V_1(a_i), V_1(b_j)), \ldots, d(V_n(a_i), V_n(b_j))) -$$
$$- CI_\mu(d(V_1(a_i), V_1(b_i)), \ldots, d(V_n(a_i), V_n(b_i))) + C K_i > 0$$
$$K_i \in \{0, 1\}$$
$$\mu(\emptyset) = 0$$
$$\mu(V) = 1$$
$$\mu(A) \leq \mu(B) \text{ when } A \leq B$$

where $N$ is the number of records, and $n$ the number of variables. This problem is considered a mixed integer linear problem, because it is dealing with integer and real-valued variables in the objective function and in the constraints, respectively. See more details of the implementation and complexity in [2].

## 3.2    Gradient Descent Algorithm

Inspired in HLMS (Heuristic Least Mean Squares), a gradient descent algorithm, introduced by Grabisch in [17], we introduce an new record linkage process relying on it. HLMS takes as input a training dataset P like the following:

$$P = \begin{pmatrix} x_1^1 & \dots & x_i^1 & \dots & x_n^1 & T^1 \\ \vdots & \ddots & \vdots & \ddots & \vdots \\ x_1^z & \dots & x_i^z & \dots & x_n^z & T^z \\ \vdots & \ddots & \vdots & \ddots & \vdots \\ x_1^N & \dots & x_i^N & \dots & x_n^N & T^N \end{pmatrix}$$

where $x_i^j$ is the value of sample $j$ for attribute $i$, and $T^j$ its target value. The algorithm finds the fuzzy measure $\mu$ that minimized the difference $C_\mu(\{x_1^j, x_2^j, ..., x_n^j\}) - T^j \, \forall j$ . The error made in the approximation can be calculated as:

$$E(\mu) = \sum_{j=1}^{N} (C_\mu(\{x_1^j, x_2^j, ..., x_n^j\}) - T^j)^2$$

The formula represents simply the squared difference between the target $T^j$ and the Choquet integral of sample $j$ using $\mu$, summed over all training examples. The direction of steepest descent along the error surface can be found by computing the derivative of E with respect to each component of the vector $\mu$.

$$\nabla E(\mu) \equiv [\frac{\delta E}{\delta \mu_{(1)}}, \frac{\delta E}{\delta \mu_{(2)}}, \dots, \frac{\delta E}{\delta \mu_{(n)}}]$$

Since the gradient specifies the direction of steepest increase of E, the training rule for gradient descent is:

$$\mu_{(i)} \leftarrow \mu_{(i)} - \lambda \nabla E(\mu_{(i)})$$

Here $\lambda$ is a positive constant called the learning rate, which determines the step size in the gradient descent search. The negative sign is present because we want to move the attributes of the aggregation operator in the direction that decreases $E$. The record linkage problem cannot be addressed directly with HLMS since the target value is unknown. To simplify notation let $V_k(a_i) = x_k^i$ and $V_k(b_i) = x'^i_k$. As in the previous approach we have divided the problem in blocks, so a block $D_k$ is now defined as follows:

$$D_k = \begin{pmatrix} (x_1^k - x'^1_1)^2 & \dots & (x_i^k - x'^1_i)^2 & \dots & (x_n^k - x'^1_n)^2 \\ \vdots & \ddots & \vdots & \ddots & \vdots \\ (x_1^k - x'^z_1)^2 & \dots & (x_i^k - x'^z_i)^2 & \dots & (x_n^k - x'^z_n)^2 \\ \vdots & \ddots & \vdots & \ddots & \vdots \\ (x_1^k - x'^N_1)^2 & \dots & (x_i^k - x'^N_i)^2 & \dots & (x_n^k - x'^N_n)^2 \end{pmatrix}$$

The original dataset has $N$ different blocks, one for each row in $X$. The algorithm must find the fuzzy measure $\mu$ that makes for block $k$ that the value of

$$C_\mu(\{(x_1^k - x'^z_1)^2, \ldots, (x_i^k - x'^z_i)^2, \ldots, (x_n^k - x'^z_n)^2\}) \tag{2}$$

to be minimum when $k == z$.

The approach used for each block $k$ is the following:

The fuzzy measure is initialized to the equilibrium state ($\mu_i = \frac{|i|}{n}$). The Choquet integral of each row in $D_k$ is calculated. If the minimum of the Choquet integral is for row k, then proceed with the next block. If the minimum of the Choquet integral is not for row k, calculate the gradient direction that makes the value of the Choquet minimum increases and the gradient of the Choquet integral for row $k$ decreases.

The algorithm for this approach is shown in Algorithm (1).

---

**Algorithm 1.** Description of the heuristic algorithm for record linkage

Let $X$ be the original database and $X'$ the protected one with $N$ samples and $n$ attributes each.

```
———————— Initialization ————————
for  i ∈ P(X) do
    μ_i = |i|/|X|
end for
——————— For each Block ———————
for  i ∈ [1..N] do
    ——————— For each row in X_i ∈ X ———————
    d_j ← (X_i − X'_j)² ∀j ∈ [1..N]
    s = {j|C(d_j) ≤ C(d_i) ∀ j ∈ [1..N]}
    ——————— Update step ———————
    for all j ∈ s do
        Update the fuzzy measure, so that the difference C(d_i) − C(d_j) decreases
    end for
    ——————— Monotonicity check ———————
    Check monotonicity
end for
return  μ
```

---

The algorithm does not guarantee the convergence to a global minimum. Some other minor modifications were done to the algorithm with no significant improvement.

## 4    Results

In this section we have compared both approaches; the heuristic algorithm for record linkage ($HRLA$) and the Choquet integral optimization algorithm ($d^2CI$) over different protected files. This comparison is divided in two parts to tackle the optimization problem. In the first part we have focused on the scores' comparison, in terms of the number of correct linkages and also the required times taken from both approaches. In the second part we have focused on the overfitting problem, testing both approaches with a small set for training and a big set for test.

To do our experiments we have applied different protection methods to an amount of records, randomly selected, from the original file. This file was selected from the Census dataset[8] of the European CASC project [7], which contains 1080 records and 13 variables, and has been extensively used in other works, such as [4,13,22].

To solve the Choquet optimization problem , we used the simplex optimizer algorithm from the IBM ILOG CPLEX tool [19], (version 12.1). The problem is first expressed into the MPS (Mathematical Programming System) format by means of the R statistical software[1] , and then, it was processed with the optimization solver. The $HRLA$ was completely programmed in the R statistical software.

## 4.1   Precision Comparison

The first part of the comparison is made with two different protected files using *Microaggregation* [11], a well-known microdata protection method, which broadly speaking, provides privacy by means of clustering the data into small clusters of size $k$, and then replacing the original data by the centroid of their corresponding clusters. The parameter $k$ determines the protection level: the greater the $k$, the greater the protection and at the same time the greater the information loss.

We have considered two protected files of 400 records, which were protected with two different protection levels.

- $M4 - 28$ : 4 variables, first 2 variables with $k = 2$, and last 2 with $k = 8$.
- $M5 - 38$ : 5 variables, first 3 variables with $k = 3$, and last 2 with $k = 8$.

Note that, we have applied two different protection degrees to different attributes of the same file. The values used range from 2 to 8. This is especially interesting when variables have different sensitivity.

Table 1 shows the percentage of re-identifications and the consumed time in the training step of both presented approaches ($d^2CI$ and $HRLA$). It is clear that both supervised approaches have obtained better results than the arithmetic mean ($d^2AM$). However, if we make a comparison between them, we can see that the $HRLA$ has an error between 2% and 5% respect to the optimum value, achieved by $d^2CI$. Recall that the $HRLA$ is initialized with an equilibrium fuzzy measure. Therefore, in the first iteration the $HRLA$ is at least as good as the Euclidean distance ($d^2AM$). It is worth mention that, since $HRLA$ is an algorithm that finds the local minimum of a function, the results shown in that table correspond to the average of ten runs with the same configuration.

We have also compared training computational times of all the approaches. Table 1 shows that in almost all the situations, the time required by the $HRLA$ to achieve similar results than $d^2CI$ is much lower than the optimization algorithm. However, we have to remember that the time's factor of the $HRLA$ approach could be different depending on the learning rate and the number of iterations which are parameters of the algorithm set up in its initialization.

---

[1] http://www.r-project.org/

**Table 1.** Percentage of re-identifications and computational time

|  | Dataset | $d^2AM$ | $d^2CI$ | $HRLA$ |
|---|---|---|---|---|
| % Re-identifications | $M4\text{-}28$ | 68.50 | 93.75 | 91.75 |
|  | $M5\text{-}38$ | 39.75 | 91.25 | 86.75 |
| Computational Time | $M4\text{-}28$ | - | 30 minutes | 20 minutes |
|  | $M5\text{-}38$ | - | 4 days | 20 minutes |

## 4.2  Overfitting

In the last part of this algorithm comparison, we have evaluated the scenario where an attacker would have a small amount of samples of the original database with its correct linkage between those samples and the samples in the public protected database. Therefore, the attacker is able to find the set of weights that achieve more number of linkages between the known samples (training set) and then, with those obtained weights, he/she is able to try the re-identification between the rest of records (test set) of two datasets in order to discover new confidential information.

In this experiment we have anonymized the whole original file (Census) by means of four different protection methods with several degrees of protection. The selected protection methods are briefly explained below; *RankSwapping* [23], where the values of a variable $V_i$ are ranked in ascending order; then each ranked value is swapped with another ranked value randomly chosen within a restricted range. *AdditiveNoise* [6] which consists of adding Gaussian noise to the original data to get the masked data. If the standard deviation of the original variable is $\sigma$, noise is generated using a $N(0, \rho\sigma)$ distribution. Finally, we have also considered the *JPEG* [18], The idea is to regard a numerical microdata file as an image (with records being rows, variables being columns, and values being pixels) and then use this lossy compression algorithm, and then the compressed image is interpreted as a masked microdata file.

We suppose that the attacker has a prior knowledge, so, a linkage of 200 records between the original and the protected files (labeled training set) could be made. Then, using a supervised approach the set of Choquet integral coefficients are learned to re-identify the rest of records (880 records), i.e., the test set.

Table 2 shows the results of the training and the test steps. Note the lack of training results in the Euclidean distance approach, since it does not require a learning step. Besides, the hyphen indicates that the corresponding computation was not finished, because it needed more than 300 hours. In the training process evaluation we have considered the time need to learn the parameters and the precentage of re-identifications. The minimum consumed times are in bold, most achieved by the $HRLA$, so the optimization problem has needed more than 14 minutes. However, it has achieved the best performance in the training set (9% of improvement at most). With respect to the test step the heuristic algorithm for record linkage has achieved an improment of at most 6% compared with the optimization problem, this is a clear indicator of overfitting. Nevertheless, $HRLA$ has achieved similar re-identification results than $d^2AM$. This is due

**Table 2.** Percentage of re-identifications and time consumed

| Dataset | $d^2AM$ Train | Test | $d^2CI$ Time | Train | Test | HRLA Time | Train | Test |
|---|---|---|---|---|---|---|---|---|
| Rankswap-20 | 14.00 | 2.61 | – | – | – | 14min | 14.50 | 2.73 |
| Rankswap-15 | 24.50 | 9.89 | – | – | – | 14min | 26.00 | 8.98 |
| Rankswap-12 | 43.50 | 17.50 | – | – | – | 14min | 44.50 | 17.73 |
| Rankswap-5 | 94.00 | 78.86 | 4min | 97.5 | 77.61 | 14min | 94.50 | 79.20 |
| Rankswap-4 | 95.50 | 85.23 | 9sec | 100.00 | 80.91 | 14min | 97.00 | 85.11 |
| Mic3-9 | 83.00 | 60.23 | 18min | 89.50 | 57.16 | 14min | 83.00 | 60.11 |
| Mic3-5 | 91.00 | 77.39 | 1.5min | 96.50 | 74.66 | 14min | 93.00 | 76.93 |
| Mic3-8 | 82.50 | 65.00 | 5min | 91.00 | 62.95 | 14min | 83.00 | 65.11 |
| Mic4-4 | 84.50 | 61.48 | 2min | 88.00 | 58.52 | 14min | 84.50 | 61.70 |
| Mic4-8 | 70.00 | 37.27 | 13min | 75.50 | 35.68 | 14min | 70.00 | 37.16 |
| Mic4-5 | 80.00 | 52.50 | 37min | 85.00 | 50.45 | 14min | 80.00 | 52.50 |
| Micz-3 | 0.00 | 0.23 | 3sec | 0.00 | 0.11 | 14min | 0.00 | 0.23 |
| MicMull-3 | 54.50 | 22.50 | 2.5days | 64.50 | 21.70 | 14min | 58.00 | 23.52 |
| Noise-16 | 87.00 | 70.11 | 1days | 92.50 | 67.50 | 14min | 87.00 | 70.11 |
| Noise-12 | 92.00 | 86.59 | 22min | 97.00 | 80.57 | 14min | 93.00 | 86.82 |
| Noise-1 | 100.00 | 100.00 | 4sec | 100.00 | 99.66 | 14min | 100.00 | 100.00 |
| Jpeg-80 | 84.50 | 76.93 | 2.5hours | 94.50 | 73.30 | 15min | 85.50 | 76.48 |
| Jpeg-65 | 58.50 | 40.00 | 15days | 67.00 | 36.59 | 15min | 58.50 | 40.00 |

to the fact that $HRLA$ is initialized with the equilibrated weights and they were slightly changed by this algorithm. Although all the protection processes are different, they mainly rely on the addition of noise to each variable, so a distance function as the Euclidean distance can clearly re-identify some of the records, obviously always depending on the amount of noise added, that is the protection degreed applied for the method.

## 5   Conclusions

In this paper we have introduced an adaptation of the gradient descent algorithm proposed by Grabisch in order to use it as a disclosure risk evaluation in the data privacy context. The use of this heuristic algorithm was motivated on the high computational times required to find the fuzzy measures by another previously presented non-heuristic method which relies on a linear optimization problem. We have evaluated and compared both of them in two different ways.

The first part of the evaluation is focused on a scenario where original and protected files are available, and an evaluation of the protected dataset is performed. This is the worst scenario, where all the information is known, so, a good estimation of the disclosure risk is obtained. This comparison shows that although the linear optimization process ($d^2CI$) guarantees the convergence to the optimal solution, it requires a lot of time, from seconds to hours or even days depending on the level of protection applied, while the time required by the $HRLA$ remains low and stable. Regarding to the results in this comparison we have achieved an error rate from 2% to 5% higher for $HRLA$.

The second part of this work cope with the overfitting problem. In this scenario the results show that when the training dataset is small, the linear optimization problem get better results for training data than $HRLA$, while for test data the results are worst. This suggest that there is an overfitting of the data.

To sum up, if we have an exhaustive disclosure risk evaluation and we have enough computational resources and time it is recommended to use the optimization approach so we will have the optimal weights to analyse the risk and we can also analyse more efficiently if there is some attribute or a set of them that disclose more information than the others. Otherwise, if the resources needed are not available we can use the heuristic approach, that provide a good approximation to the optimal solution.

In view of the results, some additional tasks remains as future work. Firstly, to program the $HRLA$ approach in $C++$ and be able to make a fairer comparison between two compiled approaches. Lastly, to use the fuzzy measures returned by $HRLA$ as a first solution of the linear optimization process, to see if the amount of time required to solve the hard datasets reduces.

**Acknowledgements.** Partial supports by the Spanish MICINN (projects TSI2007-65406-C03-02, ARES-CONSOLIDER INGENIO 2010 CSD2007-00004, TIN2010-45764 and TIN2011-27076-C03-03) and by the EC (FP7/2007-2013) Data without Boundaries (grant agreement number 262608) are acknowledged. Some of the results described in this paper have been obtained using the Centro de Supercomputación de Galicia (CESGA). This partial support is gratefully acknowledged.

# References

1. Statistics Canada. Record linkage at statistics canada (2010),
   http://www.statcan.gc.ca/record-enregistrement/index-eng.htm
2. Abril, D., Navarro-Arribas, G., Torra, V.: Choquet integral for record linkage. Annals of Operations Research, 1–14, 10.1007/s10479-011-0989-x
3. Abril, D., Navarro-Arribas, G., Torra, V.: Supervised learning using mahalanobis distance for record linkage. In: Bernard De Baets, R.M., Troiano, L. (eds.) Proc. of 6th International Summer School on Aggregation Operators - AGOP 2011, pp. 223–228 (2011), Lulu.com
4. Abril, D., Navarro-Arribas, G., Torra, V.: Improving record linkage with supervised learning for disclosure risk assessment. Information Fusion 13(4), 274–284 (2012)
5. Batini, C., Scannapieco, M.: Data Quality: Concepts, Methodologies and Techniques (Data-Centric Systems and Applications). Springer-Verlag New York, Inc. (2006)
6. Brand, R.: Microdata Protection through Noise Addition. In: Domingo-Ferrer, J. (ed.) Inference Control in Statistical Databases. LNCS, vol. 2316, pp. 97–116. Springer, Heidelberg (2002)
7. Brand, R., Domingo-Ferrer, J., Mateo-Sanz, J.: Reference datasets to test and compare sdc methods for protection of numerical microdata. Technical report, European Project IST-2000-25069 CASC (2002)
8. U.S. Census Bureau. Data extraction system
9. Choquet, G.: Theory of capacities. Annales de l'institut Fourier 5, 131–295 (1953)
10. Colledge, M.: Frames and business registers: An overview. Business Survey Methods. Wiley Series in Probability and Statistics (1995)

11. Defays, D., Nanopoulos, P.: Panels of enterprises and confidentiality: The small aggregates method. In: Proc. of the 1992 Symposium on Design and Analysis of Longitudinal Surveys, Statistics, Canada, pp. 195–204 (1993)
12. Domingo-Ferrer, J., Torra, V.: A quantitative comparison of disclosure control methods for microdata. In: Doyle, P., Lane, J., Theeuwes, J., Zayatz, L. (eds.) Confidentiality, Disclosure, and Data Access: Theory and Practical Applications for Statistical Agencies, pp. 111–133. Elsevier (2001)
13. Domingo-Ferrer, J., Torra, V.: Ordinal, continous and heterogeneous anonymity through microaggregation. Data Mining and Knowledge Discovery 11(2), 195–212 (2005)
14. Dunn, H.: Record linkage. American Journal of Public Health 36(12), 1412–1416 (1946)
15. Elmagarmid, A., Ipeirotis, P.G., Verykios, V.S.: Duplicate record detection: A survey. IEEE Transactions on Knowledge and Data Engineering 19(1), 1–16 (2007)
16. Fellegi, I., Sunter, A.: A theory for record linkage. Journal of the American Statistical Association 64(328), 1183–1210 (1969)
17. Grabisch, M.: A new algorithm for identifying fuzzy measures and its application to pattern recognition. In: Fourth IEEE International Conference on Fuzzy Systems, Yokohama, Japan, pp. 145–150 (1995)
18. J. P. E. Group. Standard IS 10918-1 (ITU-T T.81) (2001), http://www.jpeg.org
19. I. IBM ILOG CPLEX. High-performance mathematical programming engine. International Business Machines Corp. (2010)
20. Jaro, M.A.: Advances in record-linkage methodology as applied to matching the 1985 census of tampa, florida. Journal of the American Statistical Association 84(406), 414–420 (1989)
21. Lane, J., Heus, P., Mulcahy, T.: Data access in a cyber world: Making use of cyberinfrastructure. Transactions on Data Privacy 1(1), 2–16 (2008)
22. Laszlo, M., Mukherjee, S.: Minimum spanning tree partitioning algorithm for microaggregation. IEEE Trans. on Knowl. and Data Eng. 17(7), 902–911 (2005)
23. Moore, R.: Controlled data swapping techniques for masking public use microdata sets. U.S. Bureau of the Census (1996) (unpublished manuscript)
24. Newcombe, H.B., Kennedy, J.M., Axford, S.J., James, A.P.: Automatic linkage of vital records. Science 130, 954–959 (1959)
25. Pagliuca, D., Seri, G.: Some results of individual ranking method on the system of enterprise acounts annual survey. Esprit SDC Project, Deliverable MI-3/D2 (1999)
26. Torra, V., Abowd, J.M., Domingo-Ferrer, J.: Using Mahalanobis Distance-Based Record Linkage for Disclosure Risk Assessment. In: Domingo-Ferrer, J., Franconi, L. (eds.) PSD 2006. LNCS, vol. 4302, pp. 233–242. Springer, Heidelberg (2006)
27. Torra, V., Narukawa, Y.: Modeling Decisions: Information Fusion and Aggregation Operators. Springer (2007)
28. Torra, V., Navarro-Arribas, G., Abril, D.: Supervised learning for record linkage through weighted means and owa operators. Control and Cybernetics 39(4), 1011–1026 (2010)
29. USA Government, http://data.gov (2010)
30. UK Government, http://data.gov.uk (2010)
31. Winkler, W.E.: Data cleaning methods. In: Ninth ACM SIGKDD International Conference on Knowledge Discovery and Data Mining (2003)
32. Winkler, W.E.: Re-identification Methods for Masked Microdata. In: Domingo-Ferrer, J., Torra, V. (eds.) PSD 2004. LNCS, vol. 3050, pp. 216–230. Springer, Heidelberg (2004)

# Sampling Attack against Active Learning in Adversarial Environment

Wentao Zhao, Jun Long, Jianping Yin, Zhiping Cai, and Geming Xia

National University of Defense Technology, Changsha, Hunan 410073, China
{wtzhao,junlong,jpyin,zpcai}@nudt.edu.cn, xiageming@126.com

**Abstract.** Active learning has played an important role in many areas because it can reduce human efforts by just selecting most informative instances for training. Nevertheless, active learning is vulnerable in adversarial environments, including intrusion detection or spam filtering. The purpose of this paper was to reveal how active learning can be attacked in such environments. In this paper, three contributions were made: first, we analyzed the sampling vulnerability of active learning; second, we presented a game framework of attack against active learning; third, two sampling attack methods were proposed, including the adding attack and the deleting attack. Experimental results showed that the two proposed sampling attacks degraded sampling efficiency of naive-bayes active learner.

**Keywords:** Active Learning, Adversarial Environment.

## 1 Introduction

Recently, the security of machine learning has received widely attention because a learning-based system may not work well in environments (including intrusion detection, spam filtering and so on) with adversarial opponents which try to disturb the learning process. In such situation, adversarial opponents can launch attacks aiming at the learning process. For example, the attacker can pollute the training set to mislead the trained classifier. In such situation, the learning system will fail to classify instances correctly. For this sake, researchers began to investigate how attackers can destroy the standard supervised learning[1].

As an important style of machine learning, active learning can reduce the labeling efforts by just selecting the most informative instances for labeling to relieve human experts from time consuming, boring labors. Nevertheless, to the best of our knowledge, nobody has studied the security of active learning before. This may be dangerous when people utilize active learning to solve problems.

The process of standard supervised learning includes the training phase and the testing phase whose vulnerability were studied by many researchers[1,2]. But different from standard supervised learning, there is a sampling phase before the training and the testing phase in active learning. Thus, the vulnerability of the sampling phase should also be considered.

V. Torra et al. (Eds.): MDAI 2012, LNAI 7647, pp. 222–233, 2012.

We analyzed the vulnerability of the sampling phase of active learning and found attacks aiming at it.

Our aim is reminding researchers and developers to notice such vulnerability when applying active learning in the adversarial environment.

The rest of the paper can be described as follows: the notations and concepts were introduced in section 2; then, the related work was introduced in section 3; and we analyzed the sampling vulnerabilities of active learning in section 4; then we proposed two sampling attack methods against active learning in section 5; and we showed the experimental results in section 6; finally, we drew the conclusions in section 7.

## 2   Preliminaries

### 2.1   Basics of Supervised Learning

In the supervised learning problem, the learner tries to find the target function by training on a dataset of several instances selected from the instance space.

The instance space $X$ is a nonempty set containing several instances. Each instance $x_i$ is a feature vector $< x_{i1}, x_{i2}, \cdots, x_{im} >$. Let $Y = \{y_1, y_2, \cdots, y_p\}$ be the set of all possible labels.

The target function $f$ is a function $f : X \rightarrow Y$ that maps any $x \in X$ to a member of $Y$. The hypothesis space $H$ is a function set in which we try to find $f$ or a function approximating $f$. The notion $< x, f(x) >$ denotes a labeled instance and $< x, ? >$ denotes an unlabeled instance where $? \in Y$ but is unknown. $L$ denotes the whole set of labeled instances and $U$ denotes the whole set of unlabeled instances. A training set $D^{(train)}$ is a dataset used for training and a testing set $D^{(test)}$ is a dataset used for testing.

A loss function $Loss(s, t)$ is a function to calculate the difference between $s$ and $t$. Generally, $Loss(s, t) = 0$ if $s = t$ and $Loss(s, t)$ equals to a positive value if $s \neq t$.

A standard supervised machine learning process can be divided into 2 phases: the training phase and the testing phase.

In the training phase, the system collects lots of labeled instances from $L$ to generate a training set $D^{(train)}$ and then trains a classifier,

$$h(x) = \arg\min_{f \in H} \sum_{<x,y> \in D^{(train)}} (Loss(y, f(x))) \tag{1}$$

which means the learner tries to find a function which minimizes the misclassification cost given a loss function $Loss$ on $D^{(train)}$.

In the testing phase, a testing set $D^{(test)}$ consisting of numerous new coming unlabeled instances is submitted to the trained classifier $h$ and then $h$ returns all $h(x), x \in D^{(test)}$ to the system.

The whole process of the supervised learning algorithm can be described in algorithm 1.

**Algorithm 1.** the process of supervised learning

**Learning:**
**1.1** Obtain Labeled training set $D^{(train)} \subseteq L$.
**1.2** Learn hypothesis: $h(x) = \arg\min_{f \in H} \sum_{<x,y> \in D^{(train)}} (Loss(y, f(x)))$.
**Evaluation:**
**2.1** Obtain dataset $D^{(test)} \subseteq U$.
**2.2** Compare predictions $f(x)$ with $y$ for each instance $< x, y > \in D^{(test)}$.

## 2.2 Basis of Active Learning

To obtain a classifier with high accuracy, the supervised learner needs a large number of labeled instances. But there are some real world applications in which labeling instances may be time-consuming, tedious, error prone or costly (for example, in text classification problem, labeling documents may be tiresome but collecting a large number of unlabeled is very easy). Thus, those learners can obtain a small number of labeled instances and a lot of unlabeled instances in such tasks. To resolve such problems, the active learning method is proposed.

Active learning includes three phases: the sampling phase, the training phase and the testing phase. In the sampling phase, the active learner chooses the most informative instance from the unlabeled set $U$ and asks human experts to label them. After labeled, these labeled instances are inserted into the training set.

The whole process of active learning can be described in algorithm 2. Initially, a small training set $D^{(train)}$ of labeled instances and an unlabeled set $U$ are available. Then, the active learner trains a base classifier on the training set $D^{(train)}$. After that, the active learner chooses the most informative instance $x$ from $U$ and then labels $x$ by human experts before $< x, t(x) >$ is added into $D^{(train)}$ where $t(x)$ is the label of $x$. Then the active learner retrains the base classifier on updated $D^{(train)}$. The whole process runs repeatedly until the accuracy of the base classifier or iteration times reaches the preset value.

Depending on the criterion used to choose the most informative instances, the current research falls under several categories: uncertainty reduction, expected-error minimization, version space reduction and misclassification priority.

- The uncertainty reduction[3] approach selects the instances on which the current classifier has the least certainty to predict. Many sampling methods apply this strategy[4,5].
- The expected-error minimization approach[6,7] samples the instances that minimize the future expected error rate on the testing set. Such methods expect to achieve the lowest error, but they are computationally expensive.
- The version space reduction method tends to sample the instance which can divide the version space in halves. Query-by-Committee[8,9] is a representative method of this approach that constructs a committee consists of randomly selected hypotheses from the version space and selects the instances on which the disagreement within the committee is the greatest.
- The misclassification priority method prefers to sample instances which are easily be misclassified by current base classifier[10].

**Algorithm 2.** the process of active learning

**Learning:**
**repeat**
    **1.1** Obtain Labeled training set $D^{(train)} \subseteq L$.
    **1.2** Learn hypothesis: $h(x) = \arg\min_{f \in H} \sum_{<x,y> \in D^{(train)}} (Loss(y, f(x)))$.
    **1.3** Query the expert to label a set of instances $Q \subset U$ according to a sampling criteria $c$, and add $Q$ into $D^{(train)}$.
**until** stop condition is satisfied
**Evaluation:**
**2.1** Obtain dataset $D^{(test)}$.
**2.2** Compare predictions $f(x)$ with $y$ for each instance $< x, y > \in D^{(test)}$.

# 3 Related Work

## 3.1 Machine Learning in Adversarial Environments

The adversarial environment denotes an environment where there exists malicious opponents to destroy normal activity of application systems.

Traditional supervised learning methods can be utilized in a wide scope of adversarial environments, including intrusion detection and spam filtering. In such situation, a training set consisting of malicious instances and normal instances can be collected to train a classifier for predicting future instances. Since machine learning can discover hidden patterns in the training set, the trained classifier can be adaptive to detect future unknown malicious instances better than the signature-based detection system in which the signatures of the intrusions are defined by human experts.

In this paper, we focus on the learning-based intrusion detection problem in which machine learning methods have been introduced for several years. For example, Wenke Lee et al. [11] utilized machine learning method to find features relevant to intrusions and proposed an anomaly filter to block such intrusions.

The current intrusion detection techniques include misuse and anomaly detection.

- *Misuse detection.* Attack behaviors are explicitly defined and all events matching these specification are classified as intrusions.
- *Anomaly detection.* A model of the normal events is built and all events deviating the normal model are predicted as intrusions.

There are only few researchers focusing on machine learning for misuse detection, such as the methods proposed by C. Kruegel et al. [12] and Dae-Ki Kang et al. [13].

Currently, anomaly detection is the major application of machine learning techniques in intrusion detection. Related work includes K-Nearest Neighbor Classifier [14], Application-Layer intrusion detection[15], instance-based approaches[16], clustering methods[17], probabilistic learning methods[18] and so on.

## 3.2  Attacks against Machine Learning

The adaptiveness which machine learning brings to such learning based detection systems can also bring vulnerability which can be used by the attackers.

The process of supervised learning method can be divided into the training phase and the testing phase. Each phase can be the target of the attackers.

According to the target phase of attacks, typical attacks against machine learning can be summarized into two categories: training-targeted and testing-targeted.

- *Training-Targeted*: The attackers pollute the training instances to mislead the trained classifier. Such attacks include the red herring attack[2], the correlated outlier attack[2], the allergy attack [19] and so on.
- *Testing-Targeted*: The attackers do not alter the training instances but probe the trained classifier to find the its classification boundary. Thus, the attackers could know the instances which can be misclassified by the classifier and they can submit such instances to the learning based detection system. Typical attacks include the polymorphic blending attack [20], the reverse engineering attack [21], the mimicry attack against "stide" [22] and so on.

Furthermore, Barreno proposed a game-based framework of supervised learning [23], which can be described as algorithm 3:

---

**Algorithm 3.** the adversarial game of machine learning

---

**Defender** Choose a learning algorithm.
**Attacker** Choose the attack algorithms $A^{(train)}$ and $A^{(test)}$.
**Learning:**
**1.1** Obtain the Labeled training set $D^{(train)}$ with contamination from $A^{(train)}$.
**1.2** Learn hypothesis: $f = \arg\min_{f \in H} \sum_{<x,y> \in D^{(train)}} (Loss(y, f(x)))$.
**Evaluation:**
**2.1** Obtain dataset $D^{(test)}$ with contamination from $A^{(test)}$.
**2.2** Compare predictions $f(x)$ with $y$ for each instance $< x, y > \in D^{(test)}$.

---

# 4  Security Analysis of Active Learning in Adversarial Environment

## 4.1  Vulnerability of the Sampling Stage in Active Learning

The process of active learning comprises three parts: the sampling phase, the training phase and the testing phase. There is no need to analyze the vulnerability of the training phase and the testing phase because they are investigated thoroughly in previous work which focus on the security of supervised learning[1].

In the sampling phase, the active learner selects the most informative instances from $U$ for labeling according to the sampling criterion. The attacker can influence such phase to mislead the active learner.

In normal situation, the active learner will select the most informative instance for labeling. But the pool of unlabeled instances is not checked before it is provided for selecting. Therefore, if the unlabeled instances are polluted, the active learner may select some instances which are useless for training a classifier with high accuracy. The attackers can launch attacks aiming at this vulnerability.

## 4.2 Attacker Capabilities and Knowledge

Capabilities of attackers are application-related. In the intrusion detection problem, the attacker can set a filter at any point in the the network, which means the attacker can sniffer, inject, reject or generate network packets before they are submitted to the learning based intrusion detection system. Then the system will construct the labeled set $L$ and the unlabeled set $U$ based on these received packets but the attacker can not influence the following process from that time. Therefore, the capabilities of the attackers can be summarized as follows:

- the attacker knows the distribution of the instance set collected by the learning system but does not know the exact instance set.
- the attacker can add, delete or modify instances before they are collected by the learning system but can not change their labels if the learning system asks human experts to label them.

We can also assume the attackers know the feature space, the process of the active learner and the base classifier.

# 5 Sampling Attacks

## 5.1 The Basic Setting of the Active Learner

In this paper, we focus on pool-based active learning and we set the base classifier as the naive bayes classifier. For simplicity, we assume all the features are independent with other features.

The naive bayes classifier $f$ can be defined as follows:

$$f(x) = \arg\max_{y_j \in Y} P(y_j | x_1, x_2, \ldots, x_m) \tag{2}$$

$$= \arg\max_{y_j \in Y} \frac{P(x_1, x_2, \ldots, x_m | y_j) P(y_j)}{P(x_1, x_2, \ldots, x_m)} \tag{3}$$

$$= \arg\max_{y_j \in Y} P(x_1, x_2, \ldots, x_m | y_j) P(y_j) \tag{4}$$

$$= \arg\max_{y_j \in Y} P(y_j) \prod_i P(x_i | y_j) \tag{5}$$

$P(x_i | y_j)$ and $P(y_j)$ can be calculated on the training set $D^{(train)}$.

And we assume the sampling criterion is uncertainty reduction. The uncertainty of an instance $x$ can be defined as the label distribution entropy of that instance.

$$C(x) = \sum_{y \in Y} (-P(y|x) log P(y|x)) \tag{6}$$

Formula 6 shows the definition of uncertainty where $x$ denotes an instance and $Y$ denotes the label set.

## 5.2    The Adversarial Game of Active Learning

Based on the game framework of supervised learning propose by Barreno [23], we present a game framework of active learning which can be described as algorithm 4.

---

**Algorithm 4.** the adversarial game of active learning

---

**Defender** Choose a learning algorithm.
**Attacker** Choose the attack algorithms $A^{(sample)}$, $A^{(train)}$ and $A^{(test)}$.
**Learning:**
**repeat**
    **1.1** Obtain the training set $D^{(train)} \subseteq L$ with contamination from $A^{(train)}$.
    **1.2** Learn the hypothesis: $f = \arg \min_{f \in H} \sum_{<x,y> \in D^{(train)}} (Loss(y, f(x)))$.
    **1.3** Query the expert to label a set of instances $Q \subset U$ with contamination from $A^{(sample)}$ according to the sampling criteria $c$, and add $Q$ into $L$.
**until** stop condition is satisfied
**Evaluation:**
**2.1** Obtain dataset $D^{(test)}$ with contamination from $A^{test}$.
**2.2** Compare predictions $f(x)$ with $y$ for each instance $< x, y > \in D^{test}$.

---

In the game framework, except the attacks targeting at the training and the testing phases, we show that the attacker can launch the attack algorithm $A^{(sample)}$ to pollute the unlabeled instances.

## 5.3    The Attack Process

The attack process includes two steps:

1. Pre-cluster: Pre-cluster the unlabeled set $U$ using the k-means algorithm[24] to find the areas with high density in the instance space.
2. Pollution: Pollute the unlabeled set $U$ by adding fake clusters or deleting some true clusters.

The idea of such steps is based on the following consideration. According to clustering assumption [25], which says *If two points $x_1$, $x_2$ are close, then so should be the corresponding outputs $y_1$, $y_2$*. That means the classification boundaries go across areas with low density in the instance space. Such assumption makes sense in many real world applications. Thus, if we change the density of some areas by adding or deleting instances, the classification boundaries of the trained classifier would be changed. Therefore, we propose the pollution method by changing the distribution of the unlabeled instances.

For the pollution step, we propose two polluting methods as follows:

- ADDING: The attacker adds some fake clusters to the unlabeled set $U$. Such clusters would be the most uncertain clusters judging by the current base classifier.
- DELETING: The attacker deletes some clusters from $U$ before they are submitted to the learning system.

For ADDING, we take the following steps described in algorithm 5.

---

**Algorithm 5.** ADDING

**repeat**
    **1** Generate an instance set $S$ randomly.
    **1** Select instance $x_n \in S$ where

$$x_n = \arg\max_{x \in S} C_{D^{(train)}}(x) \tag{7}$$

    **2** Generate instance set $\Omega$ containing $l$ instances whose features are light disturbing of instance $x$.
    **3** Add $\Omega$ into $U$.
**until** stop criterion is satisfied

---

In algorithm 5, $C_{D^{(train)}}(x)$ is the label distribution entropy of $x$ predicted by the naive bayes classifier trained on $D^{(train)}$. Then the instance $x_n$ will obtain the largest uncertainty in $S$ and the active learner tends to sample instances in $\Omega$.

For DELETING, we take the following steps described in algorithm 6.

---

**Algorithm 6.** DELETING

**repeat**
    **1** Calculate $C(x) = \sum_{y \in Y}(-P(y|x)logP(y|x))$ where $x \in U$.
    **2** Select instance $x = \arg\max_{x \in U}\sum_{y \in Y}(-P(y|x)logP(y|x))$.
    **3** Select Cluster $Clu$ where $x \in Clu$.
    **4** Delete $Clu$ from $U$
**until** stop criterion is satisfied

---

By both ADDING and DELETING, we can change the distribution of $U$ to influence the sampling process.

# 6    Experimental Results

We conducted a series of experiments to test the sampling attacks proposed in this paper. Each experiment can be described as follows:

- **NBAL**. We tested the active learner without malicious opponents. In this test, the active learner utilize the naive-bayes classifier as the base learner.
- **ADDING**. We tested the active learner under the ADDING attack. The base classifier is also the naive-bayes classifier, but a malicious opponent exists and launches the ADDING attacks presented as algorithm 5.
- **DELETING**. We tested the active learner under the DELETING attack. The base classifier is also the naive-bayes classifier, but a malicious opponent exists and launches the DELETING attacks presented as algorithm 6.

We selected the dataset from the 1999 KDD intrusion detection contests as our testing dataset. The dataset was provided by MIT Lincoln Lab. It was gathered from a local-area network simulating a typical military network environment with a wide variety of intrusions over a period of 9 weeks. The dataset can be downloaded form the UCI KDD repository [26].

In the dataset, each instance $x$ is a vector with 41 features and the label set includes 23 different labels denoting 22 attack types and 1 normal event. For simplification, we transformed the dataset into a 2-class dataset with two labels: "normal" and "malicious". The whole intrusion detection dataset is quite large and we selected 5000 instances from the original dataset randomly as our benchmark dataset.

When testing each method we listed above, the dataset was randomly divided into three parts: the unlabeled set $U$, the labeled set $L$ and the testing set $T$. $T$ contained 800 instances. When the experiment began, $L$ contained only 1 instance selected randomly and $U$ contained the rest instances. Then the tested active learner selected 1 instance in $U$ according to the uncertainty reduction criterion and move it with its true label into $L$ in each iteration. Thus the active learner trained the base classifier on $L$ and recorded the accuracy of the classifier on $T$. In each experiment the active learner sampled 100 times. When we tested ADDING and DELETING, we preclustered $U$ into 22 clusters which correspond to 22 intrusion types (using c-means clustering algorithm).

## 6.1    Results

We take three metrics to evaluate the performance for these active learners. The three metrics are:

**Top Accuracy** the highest accuracy which the tested active learners finally reached after 100 samples.
**Stable Point** the number of samplings when the accuracy of the base classifier was stable which means the accuracy variance of neighboring samplings was below 2%.

**Rising Efficiency** how much accuracy increased when the active learners reached the stable point compared with the accuracy of the initial sampling.

Table 1 showed these metrics of the three tested active learners after 100 samples, where TA denotes Top Accuracy, SP denotes Stable Point and RE denotes Rising Efficiency. In the *Pollution* column, *"m clusters, n instances"* means *"adding m clusters and each cluster included n fake instances"* for ADDING while *"m clusters"* means *"deleting m clusters"* for DELETING. Each recorded data is the average of 10 runs.

**Table 1.** Metrics of the active learners

| Active Learner | Pollution | TA | SP | RE |
|---|---|---|---|---|
| NBAL | | 97.67% | 4 | 31.1% |
| ADDING | 2 clusters, 10 instances | 81.23 % | 5 | 26.77% |
| | 10 clusters, 10 instances | 69.57 % | 7 | 20.19% |
| | 10 clusters, 100 instances | 66.28 % | 8 | 18.46% |
| DELETING | 2 clusters | 82.74 % | 5 | 25.28% |
| | 5 clusters | 75.22 % | 6 | 22.92% |
| | 10 clusters | 63.98 % | 6 | 18.65% |

From Table 1, we found that NBAL can reach the accuracy of 97.67% after 100 samplings while ADDING and DELETING can not reach it. For both ADDING and DELETING, the accuracy dropped while the number of the added or deleted clusters and instances increased. Moreover, the *Stable Point* of these tested active learners did not vary so much. This indicated that these classifiers can stabilize quickly even under heavy attacks with large numbers of clusters and instances added or deleted.

We also recorded the accuracy of the base classifiers at each sampling in these experiments. In figure 1, we showed the learning curves of three active learners: NBAL, ADDING(10 clusters added, 100 instances in each cluster) and DELETING(10 clusters deleted). the vertical axis showed the accuracy of the classifiers and the horizontal axis showed the number of samplings. Each recorded data is the average of 10 runs.

In Figure 1, we can see all the learning curves had a sharp rise at the very beginning and then climbed slowly. The sharp rise corresponded to the first several samplings, which let accuracy of these active learners obtained rapid increase. The learning curve of NBAL reached 96.67%. That means when we utilized NBAL to detect intrusions without any attacks aiming at the sampling phase, there were only 3.33% of network events may be misclassified. But when there are attackers trying to influence the sampling process, the learning curves of the active learners have an obvious decline than the one without attacks. From the learning curves of ADDING and DELETING, we found the accuracy of the base classifiers decreased about 30% at the same sampling times compared to

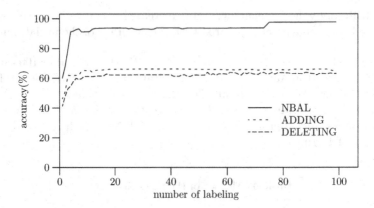

**Fig. 1.** Learning curves of NBAL

NBAL. This means the classifier failed to detect the intrusions. And the learning curve of DELETING is even lower than the one of ADDING.

Thus the results of the conducted experiments showed that the proposed sampling attacks including ADDING and DELETING could weaken the active learners.

## 7    Conclusion

We made several contributions in this paper. First, the vulnerability of active learning was analyzed. Second, a game framework of active learning was proposed. Third, two sampling attacks against active learning were presented.

We would like to pursue the following directions: constructing the framework to evaluate the instance complexities of attack and defence algorithms and designing the detection techniques of the sampling attacks.

**Acknowledgments.** This research was supported by the National Natural Science Foundation of China (No.61105050).

## References

1. Barreno, M., Nelson, B., Sears, R., Joseph, A.D., Tygar, J.D.: Can machine learning be secure? In: ASIACCS, pp. 16–25 (2006)
2. Newsome, J., Karp, B., Song, D.: Paragraph: Thwarting Signature Learning by Training Maliciously. In: Zamboni, D., Kruegel, C. (eds.) RAID 2006. LNCS, vol. 4219, pp. 81–105. Springer, Heidelberg (2006)
3. Lewis, D.D., Gale, W.A.: A sequential algorithm for training text classifiers. In: 17th ACM International Conference on Research and Development in Information Retrieval, pp. 3–12. Springer (1994)
4. Tong, S., Koller, D.: Support vector machine active learning with applications to text classification. Journal of Machine Learning Research 2, 45–66 (2001)
5. Campbell, C., Cristianini, N., Smola, A.: Query learning with large margin classifiers. In: Proc. 17th International Conf. on Machine Learning, Madison, pp. 111–118. Morgan Kaufmann (2000)

6. Cohn, D.A., Ghahramani, Z., Jordan, M.I.: Active learning with statistical models. Journal of Artificial Intelligence Research 4, 129–145 (1996)
7. Roy, N., McCallum, A.: Toward optimal active learning through sampling estimation of error reduction. In: Proc. 18th International Conf. on Machine Learning, pp. 441–448. Morgan Kaufmann, San Francisco (2001)
8. Seung, H.S., Opper, M., Sompolinsky, H.: Query by committee. In: Proceedings of the Fifth Workshop on Computational Learning Theory, San Mateo, CA, pp. 287–294. Morgan Kaufmann (1992)
9. Freund, Y., Seung, H.S., Shamir, E., Tishby, N.: Selective sampling using the query by committee algorithm. Machine Learning 28, 133–168 (1997)
10. Long, J., Yin, J., Zhu, E., Zhao, W.: Active learning with misclassification sampling based on committee. International Journal of Uncertainty, Fuzziness and Knowledge-Based Systems 16(suppl.1), 55–70 (2008)
11. Lee, W., Stolfo, S.J.: A framework for constructing features and models for intrusion detection systems. ACM Trans. Inf. Syst. Secur. 3(4), 227–261 (2000)
12. Kruegel, C., Tóth, T.: Using Decision Trees to Improve Signature-based Intrusion Detection. In: Vigna, G., Kruegel, C., Jonsson, E. (eds.) RAID 2003. LNCS, vol. 2820, pp. 173–191. Springer, Heidelberg (2003)
13. Kang, D.-K., Fuller, D., Honavar, V.: Learning Classifiers for Misuse Detection Using a Bag of System Calls Representation. In: Kantor, P., Muresan, G., Roberts, F., Zeng, D.D., Wang, F.-Y., Chen, H., Merkle, R.C. (eds.) ISI 2005. LNCS, vol. 3495, pp. 511–516. Springer, Heidelberg (2005)
14. Liao, Y.: Machine learning in intrusion detection. PhD thesis, Davis, CA, USA (2005)
15. Rieck, K.: Machine Learning for Application-Layer Intrusion Detection. PhD thesis, Berlin, Germany (2009)
16. Liao, Y., Vemuri, V.R.: Use of k-nearest neighbor classifier for intrusion detection. Computers & Security 21(5), 439–448 (2002)
17. Lazarevic, A., Ertöz, L., Kumar, V., Ozgur, A., Srivastava, J.: A comparative study of anomaly detection schemes in network intrusion detection. In: SDM (2003)
18. Mahoney, M.V., Chan, P.K.: Learning nonstationary models of normal network traffic for detecting novel attacks. In: KDD, pp. 376–385 (2002)
19. Chung, S.P., Mok, A.K.: Collaborative intrusion prevention. In: WETICE, pp. 395–400 (2007)
20. Fogla, P., Lee, W.: Evading network anomaly detection systems: formal reasoning and practical techniques. In: ACM Conference on Computer and Communications Security, pp. 59–68 (2006)
21. Lowd, D., Meek, C.: Adversarial learning. In: KDD, pp. 641–647 (2005)
22. Tan, K.M.C., Killourhy, K.S., Maxion, R.A.: Undermining an Anomaly-Based Intrusion Detection System Using Common Exploits. In: Wespi, A., Vigna, G., Deri, L. (eds.) RAID 2002. LNCS, vol. 2516, pp. 54–73. Springer, Heidelberg (2002)
23. Barreno, M., Nelson, B., Joseph, A.D., Tygar, J.D.: The security of machine learning. Machine Learning 81(2), 121–148 (2010)
24. Lloyd, S.P.: Least squares quantization in pcm. IEEE Transactions on Information Theory 28(2), 129–136 (1982)
25. Zhu, X., Goldberg, A.B.: Introduction to Semi-Supervised Learning. Synthesis Lectures on Artificial Intelligence and Machine Learning. Morgan & Claypool Publishers (2009)
26. Archive, T.U.K.: Kdd cup 1999 data (October 1999)

# Dynamic Credit-Card Fraud Profiling

Marc Damez, Marie-Jeanne Lesot, and Adrien Revault d'Allonnes

LIP6, Université Pierre et Marie Curie-Paris 6, UMR7606
4 place Jussieu, Paris cedex 05, 75252, France
{marc.damez,marie-jeanne.lesot,adrien.revault-d'allonnes}@lip6.fr

**Abstract.** The paper proposes a scalable incremental clustering algorithm to process heterogeneous data-streams, described by both categorical and numeric features, and its application to the domain of credit-card fraud analysis, to establish dynamic frauds profiles. The aim is to identify subgroups of frauds exhibiting similar properties and to study their temporal evolution and, in particular, the emergence of fraudster behaviours. The application to real data corresponding to a one year fraud stream highlights the relevance of the approach that leads to the identification of significant profiles.

**Keywords:** Incremental Clustering, Data-Streams, Heterogeneous Data, Bank Fraud, Credit Card Security.

## 1 Introduction

Credit and debit cards have become ubiquitous modes of payment, which have brought with them the important issue, both for banks and card-holders, of card fraud [1]. The ensuing economic losses have motivated a vast field of study in the machine learning community [2–4], in particular concerned with electronic business, both for supervised –fraud detection– and unsupervised –fraud characterisation– learning tasks.

This paper considers the latter, i.e. the identification of ever changing groups of similar frauds. This problem is challenging in its twofold dynamic nature. First, fraudsters are inventive and continuously adapt to circumvent anti-fraud policies, elaborating new types of frauds. Second, the data are constantly incoming, building a never-ending stream. Finally, the very nature of the heterogeneous data, described by both numeric and categorical attributes, renders part of the classic methods unusable.

In this paper, we propose a methodology to process incoming streams presenting these characteristics and study the derived profiles and the dynamics of their contents. Contrary to most clustering tasks, the method proposed in this paper is less interested in summarising the data into abstract and possibly non-existent prototypes than it is in identifying precisely observable behaviours, those which are most likely to belong to an actual type of fraudster. The proposed method is tested on a real dataset representing one year of frauds

The paper is organised as follows: after outlining related work in Section 2, we present in Section 3 the methodology we propose. Section 4 analyses the profiles

V. Torra et al. (Eds.): MDAI 2012, LNAI 7647, pp. 234–245, 2012.
© Springer-Verlag Berlin Heidelberg 2012

obtained using this methodology on a real dataset corresponding to a one year fraud-stream.

## 2   Related Work

Data representing credit-card frauds combine two characteristics that both require specific clustering algorithms, namely their data-stream and heterogeneous nature. This section outlines algorithms that have been proposed to deal with either data type.

### 2.1   Clustering Data-Streams

**Data-Stream Characteristics:** As opposed to classic data, data-streams are characterised by their production mode: the dataset is incrementally built, which means that not all data are available at once. They usually lead to very large datasets having restrictive characteristics which require specific algorithms to perform data-mining [5].

Indeed, data-streams first demand incremental algorithms that process data progressively, incorporating them as they come in to update the learnt model. Moreover, because of their production mode as well as memory constraints imposed by their quantity, data-streams usually require single-pass algorithms.

Another characteristic of data-streams is their dynamic feature: apart from the data arriving progressively, their underlying distribution generally evolves with time. This is in contradiction with the hypothesis of identically distributed data most classic data-mining algorithms rely on.

**General Principles:** The main existing methods for clustering such data-streams belong to the framework of incremental clustering. First introduced to address the issue of very large datasets, these algorithms decompose the dataset into samples of manageable size. They consist in iteratively processing each sample individually and merging the corresponding partial results into the final partition. In the case of very large datasets, samples are automatically extracted, e.g. randomly drawn so as to fit in memory. In the case of data-streams, where samples are imposed by the time-line, and defined as the set of data becoming available in a given time interval, samples can be seen as data buffers.

Incremental clustering algorithms can be divided according to the way the partial results are merged, this fusion being either progressive or final. Progressive fusion means including, in the clustering step of a given sample, the results from the previous steps. Final fusion is performed at the end, when all samples have been processed. As detailed below, the same distinction can be applied to data-stream clustering algorithms.

**Online Clustering:** Online clustering algorithms are incremental approaches with progressive fusion: the previously seen data are summarised by extracted clusters, possibly weighted by their sizes, and this summary is processed together with the next sample.

This approach has been applied to most classic clustering algorithms, e.g. leading to incremental variants for $k$-means [6], fuzzy $c$-means [7], fuzzy $c$-medoids [8] or DBSCAN [9]. Likewise STREAM [10], which is one of the first algorithms dedicated to data-streams, achieves the same kind of result with memory size limitations and theoretical guarantees on the result quality.

**Two-Level Approaches:** The online approaches are said not to be able to fully adapt to the dynamics of the data, as they do not question previous cluster merges [11]. Therefore, two-level approaches reject progressive fusion and postpone the fusion step until the final partition is required by the user. More precisely, they combine an online part, updating compact representations of the data seen so far, with an offline part, which extracts a final partition from this compact representation. The offline step does not take into account the temporal component: no comparison can be made between two consecutive demands of the user. The two steps are also respectively called micro and macro-clustering.

Some approaches in this framework apply the same clustering algorithm to both steps, e.g. fuzzy $c$-means [12] or fuzzy $c$-medoids [8]. The centres or medoids obtained from each data buffer are in turn clustered.

Other methods perform a preclustering step of a different nature than the final one, e.g. incrementally updating quantities to compute cluster statistics such as cluster average or standard deviation. This approach is exemplified by BIRCH [13], that incrementally builds a compact representation of the dataset, based on structured summaries that optimise memory usage along user-specified requirements. CLUSTREAM [11] generalises the representation, taking into account the temporal dimension in the precluster description. These algorithms try to add a new data point to one of the preclusters. If this fails, then a new cluster is created to represent the data point. To maintain the memory size, either one of the previously identified clusters is deleted, e.g. based on a recentness criterion, or two clusters are merged, e.g. the two most similar.

This principle has also been applied to density-based clustering [14]: it also combines online micro-cluster maintenance with offline generation of the final clusters with a variant of DBSCAN. Two types of clusters are distinguished: core and outlier-clusters, which can become core-clusters if they reach a size threshold. To prevent memory overload, the outlier-clusters are periodically pruned.

The proposed methodology described in the Section 3 is similar to this one, with two main differences, as detailed below: first it relies on partitioning and not density-based clustering. Second, it uses a different representation of the so-called outlier clusters.

### 2.2 Clustering Heterogeneous Data

Heterogeneous data are defined as data described by both numeric and categorical attributes. The presence of categorical attributes rule out the computation of average values and, therefore, the usage of all mean-centred clustering techniques, in particular the very commonly applied $k$-means and its variants.

Two main approaches can be distinguished: first, so-called relational methods which rely on the pairwise dissimilarity matrix (e.g. pairwise distances) and not on vector descriptions of the data. This type of approach includes hierarchical clustering methods, density-based methods [15] as well as relational variants of classic algorithms [16, 17].

On the other hand, medoid-based methods [18, 19] constitute variants of the mean-centered methods that do not define the cluster representative as the average of its members, but as its medoid, that is, the data point that minimises the, possibly weighted, distance to cluster members.

## 2.3   Linearised Fuzzy $c$-Medoids

The linearised fuzzy $c$-medoids algorithm [19], written *l-fcmed* hereafter, is a scalable medoid-based clustering algorithm: it can process data that are both in vast amounts and heterogeneous. Moreover, being a fuzzy variant, it offers properties of robustness and independence from random initialisation. We present it in greater detail here because the proposed method described in Section 3 depends on it.

The algorithm's inputs are $\mathcal{D} = \{x_i \mid i = 1, \ldots, n\}$ the dataset to be clustered, $d$ the metric used to compare data, $c$ the desired number of clusters, $m$ the *fuzzifier* which sets the desired fuzziness and $p$ the size of the neighbourhood in which medoid updates are looked for.

After initialisation of the cluster centres as $c$ data in $\mathcal{D}$ the algorithm alternately updates memberships and cluster centres using the following equations:

$$u_{ir} = \left[ \sum_{s=1}^{c} \left( \frac{d(x_i, v_r)}{d(x_i, v_s)} \right)^{\frac{2}{m-1}} \right]^{-1} \qquad v_r = \operatorname*{argmin}_{k \in N_p(v_r)} \sum_{i=1}^{n} u_{ri}^m d(x_k, x_i) \qquad (1)$$

where $u_{ir}$ denotes the membership degree of $x_i$ to cluster $r$, $v_r$ the cluster centres and $N_p(v_r)$ the neighbourhood of centre $v_r$, which looks to update medoids in their vicinity, alleviating computational costs. The latter is defined as the $p$ data maximising membership to cluster $r$. Both updates are iterated until medoid positions stabilise.

# 3   Proposed Methodology

This section describes the methodology we propose to dynamically cluster a heterogeneous stream, as sketched in Algorithm 1. It belongs to the family of two-level approaches, performing a micro-clustering step based on a partitioning approach.

The micro-clusters, i.e. the cluster information which is updated for each new datum, are defined as cluster medoids. As in [11, 13, 14], we test whether a new data point fits an existing cluster. If the test succeeds, no update is performed; if it fails, instead of creating a new cluster immediately, we add the data point to a buffer $\mathcal{B}$. Cluster creation then only takes place when the buffer reaches a size threshold and is the result of a partitioning algorithm applied to $\mathcal{B}$.

We propose, for this, a variant of the linearised fuzzy c-medoids [19], adding a cluster selection step to increase cluster homogeneity. This step makes it possible to identify atypical data grouped in a set of as yet unassigned data, denoted $\mathcal{U}$: similarly to [14], we distinguish outliers from core-clusters. The main difference here is that the outliers are not clusters of their own but a single set.

The next subsections respectively describe in more detail the test for assigning a point to an existing cluster, called cluster augmentation criterion, and the cluster selection criterion, as well as the global architecture of the algorithm.

### 3.1   Cluster Augmentation Criterion

Cluster augmentation consists in testing whether an incoming point can be assigned to one of the already identified clusters, or whether it should be buffered as a candidate clusterable. This procedure has two immediate advantages. The first is to reduce, for a given volume of data, the number of times the buffer is filled and, thus, the identification of new clusters, making the global process run faster. The second advantage is that it avoids the discovery, at a later stage, of clusters too similar to those already identified: it, indeed, ensures that the next data considered for clustering are adequately separate from the previously selected clusters. It, therefore, suppresses the need for a posterior fusion step, where the results of the buffer clustering are merged to the previously obtained clusters. This, again, alleviates the computation costs of the global method.

We propose to consider that a data point can be assigned to an existing cluster only under the condition that it does not deteriorate its compactness: addition is not aimed at generalising a cluster, rather at processing new data quickly. Therefore, we impose that a point can be added to a cluster if and only if it falls within the mean distance to the medoid at the time the cluster was selected. This can be written formally as:

$$d(x, \nu_C) \leq \frac{1}{|C|} \sum_{y^* \in C} d(y^*, \nu_C) \qquad (2)$$

where $y^*$ denotes any element in the cluster at its creation. This augmentation condition defines a local criterion that adapts to the compactness of each cluster. In particular, for a cluster containing only exact replicas of a data point, the augmentation criterion will exclusively allow the addition of more replicas. Once again, this severe criterion is intended to help the identification of fraudster behaviours, more than it is meant to offer a summarisation of the observed data.

### 3.2   Cluster Selection Criterion: The *l-fcmed-select* Algorithm

Medoid selection is a substep that actually modifies the *l-fcmed* algorithm, leading to the variant we propose, called *l-fcmed-select*. One of its motivations comes from the issue of determining the appropriate number of clusters, $c$: as all partitioning clustering algorithms, *l-fcmed* always produces $c$ clusters, whether $c$ is relevant for the considered data or not. In order to bypass this difficulty,

---

**Algorithm 1.** $\mathcal{DS}$-$l$-$fcmed$-$select$

---

For each new data point $x$

$\triangleright$ Process $x$

**if** $\exists C \in \mathcal{C}$ such that $x$ can be added to $C$ according to Eq. 2 **then**
    Update $\mathcal{C}$: $C \leftarrow C \cup \{x\}$
**else**
    Update $\mathcal{B} \leftarrow \mathcal{B} \cup \{x\}$
**end if**

$\triangleright$ Process $\mathcal{B}$

**if** $|\mathcal{B}| = \tau_\mathcal{B}$ **then**
    Apply $l$-$fcmed$-$select$ to $\mathcal{B} \rightarrow (\mathcal{C}', \mathcal{U}')$
    Update $\mathcal{C} \leftarrow \mathcal{C} \cup \mathcal{C}'$
    Update $\mathcal{U} \leftarrow \mathcal{U} \cup \mathcal{U}'$
**end if**

$\triangleright$ Process $\mathcal{U}$

**if** purging criterion on $\mathcal{U}$ is fulfilled **then**
    Apply $\mathcal{DS}$-$l$-$fcmed$-$select$ recursively to $\mathcal{U}$, treated as a data-stream
**end if**

---

we propose to ask for a 'reasonable' –probably overestimated– number of clusters, and then to select only *some* of the produced clusters.

The proposed selection condition is, again, a compactness criterion: we keep only the clusters of sufficient size that exhibit a very high homogeneity, evaluated as the radius of the cluster. The selection criterion can, thus, be formalised as:

$$|C| > \tau_C \qquad \text{and} \qquad \max_{x \in C} d(x, \nu_C) \leq \xi \qquad (3)$$

where $\tau_C$ is the minimal acceptable size and $\xi$ a compactness threshold.

This algorithm does not return a data partition as some of the data, that assigned to discarded clusters, remain unattributed. More formally, $l$-$fcmed$-$select$ outputs a set of clusters, $\mathcal{C} = \{C_1, ..., C_{c'}\}$, with $c' \leq c$, and a set of unassigned data $\mathcal{U}$. The latter represents atypical cases, or as yet unexplained data, that do not deserve, for the time being, medoids of their own in the clustering solution.

## 3.3 Global Architecture and Parameters

The global architecture of the proposed methodology, called $\mathcal{DS}$-$l$-$fcmed$-$select$, is given in Algorithm 1. The set of clusters $\mathcal{C}$, the buffer $\mathcal{B}$ and the set of unassigned data $\mathcal{U}$ are initially assigned the empty set.

When a new data point then arrives, the algorithm tries to assign it to one of the existing clusters. If this substep fails, the point is added to the buffer $\mathcal{B}$.

Once the buffer reaches a user-defined size threshold, $\tau_\mathcal{B}$, meaning that too many points differ from the previously identified clusters, then $l$-$fcmed$-$select$ is applied anew to the data in the buffer. Previously testing addability ensures that all obtained clusters are distinct enough from the already identified clusters.

The *l-fcmed-select* algorithm imposes handling the set $\mathcal{U}$ of unassigned data. The corresponding data are 'more atypical' than the ones in $\mathcal{B}$, insofar as they have been submitted at least once to clustering, whereas buffered data have only been tested against existing clusters. However, it may be the case that atypical behaviours eventually become less isolated, warranting a cluster of their own.

Therefore, when some purging condition on $\mathcal{U}$ is reached, the data it contains are considered once more for core-clustering. We apply $\mathcal{DS}$-*l-fcmed-select*, processing it as a fictitious data-stream. The purging condition could be linked to the $\mathcal{U}$'s size, yet this induces the risk of running back to back purges, if $\mathcal{U}$'s size does not fall far enough below the threshold. To avoid this, we harness $\mathcal{U}$'s purge to the amount of processed data, with the idea that these regular purges still attest to the aging of the data. A purge, thus, starts every time $\tau_{\mathcal{U}}$ points of the data-stream have been processed and stops as soon as $\mathcal{U}$'s size stops decreasing.

Overall, the proposed method relies on three sets of parameters, that is the *l-fcmed* parameters: the number of clusters $c$, the fuzzifier $m$ and the neighbourhood size for the medoid update $p$; the cluster selection criteria: the minimal acceptable cluster size $\tau_C$ and diameter $\xi$; and the size thresholds: the size at which $\mathcal{B}$ is subjected to *l-fcmed-select*, $\tau_{\mathcal{B}}$ and the rate at which $\mathcal{U}$ is purged, $\tau_{\mathcal{U}}$.

## 4   Experimental Results

### 4.1   Data and Experimental Setup

We applied the proposed methodology to a real fraud-stream covering almost one year (49 weeks) and containing close to a million fraudulent transactions. Each is described by its amount in Euros, a positive real number, as well as categorical attributes, namely the country where it took place and the merchant category code, a general categorisation of transacted products.

The distance $d$ between two transactions $t_1$ and $t_2$, represented as vectors of their features, is defined as $d(t_1, t_2) = 1/q \sum_{i=1}^{q} d_i(t_{1i}, t_{2i})$, where $d_i$ is the distance for attribute $A_i$. Two cases are distinguished: $d_i$ is either $d_i = d_{cat}$, if $A_i$ is categorical, or $d_i = d_{num}$, if it is numeric. Each is defined as follows:

$$d_{cat}(x, y) = \begin{cases} 1 & \text{if } x \neq y \\ 0 & \text{otherwise} \end{cases} \qquad d_{num}(x, y) = \frac{|x - y|}{\max(x, y)}$$

The distance for numeric attributes is thus defined as a relative gap: the assumption being that a difference of 2€ in amount, say, should not have the same impact if the compared amounts are around 5€ or if they are closer to 1 000€.

For the algorithm parameters, we use the following setup: *l-fcmed* is applied with $c = 400$, $m = 2$ and $p = \lfloor \tau_{\mathcal{B}}/c \rfloor = 12$. Cluster selection is based on $\tau_C = 10$ and $\xi = 0.15$. Note that, due to the distance choice, this low value imposes that data assigned to a given cluster all have the same country and the same activity, variability is only (moderately) tolerated for the amount. The size threshold for the buffer is $\tau_{\mathcal{B}} = 5\,000$ and the set of unassigned data $\mathcal{U}$ is purged every time $\tau_{\mathcal{U}} = 50\,000$ transactions have been processed. The experiments were run using a parallel implementation of $\mathcal{DS}$-*l-fcmed-select* on a multicore cluster.

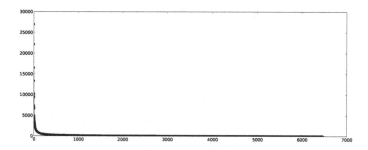

**Fig. 1.** Sizes of the clusters once all data has been processed, in descending order

## 4.2 Experimental Results

With these parameters, the proposed methodology finds a total of 6476 clusters covering a wide range of sizes and time-spans. In the following, we first analyse the global results of the final step, then comment the dynamics of fraud profiles, both globally and at a more detailed level.

**Final Global Analysis:** In order to study the obtained clusters when the whole data-stream has been processed, we focus on cluster sizes. Figure 1 shows the final sizes of the clusters in decreasing order.

It can be observed that, despite a very severe compactness criterion, the largest cluster groups 26917 frauds, highlighting the redundancy of the data, in which there exists a largely dominant type of fraud. More generally, the 25% largest clusters cover 77.7% of all affected frauds. Conversely, the 25% smallest clusters only cover 3.6% of the assigned frauds. This shows that the fraud behaviours include a large number of rather rare fraud procedures, either due to atypical country, activity, amounts or a combination of the above.

Moreover, at the end of the process there remain 21.7% of the whole as unassigned frauds, i.e. too atypical to be clustered according to the criteria imposed on the clustering result.

**Global Dynamic Analysis:** We first analyse the dynamics of fraud profiles globally, examining the distribution of each profile time-span. These lengths are defined as the number of weeks between dates for the first and last frauds assigned to each cluster. We should point out that our analysis of the dynamics is more concerned with the evolution of the contents, the frauds, then it is with the displacement of the clusters, as is usually the case.

Figure 2 shows the repartition of these durations, both in terms of the number of clusters, on the left, as well as the number of frauds they represent, on the right. Two types of clusters co-exist: a first category, representing 46.9% of all clusters, groups long-lasting profiles that nearly span over the whole year, with periods longer than 40 weeks. This comes along with a dominance in terms of data quantities, as they represent 81.6% of the assigned fraudulent transactions.

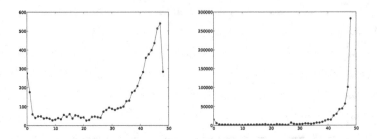

**Fig. 2.** Time-span histograms: (left) number of clusters covering each number of weeks, (right) cumulative size of clusters covering each number of weeks

A second type of profiles can be described as 'flash profiles': these correspond to behaviours that are very precisely dated, disappearing within a single week or, possibly, up to two or three weeks. This category represents 9.1% of the clusters, and still appear to be significant in terms of number of frauds. More precisely, the average size of the 275 clusters observed only during a single week is 57, with sizes ranging from 10 to 3001.

Figure 3 provides another view on the dynamics of credit-card fraud profiles. The full line on top shows the evolution of the number of active clusters each week, where a cluster is called active at a given date if it is created or if at least one transaction is assigned to it at that date. The number of active clusters is globally increasing over the entire period, excepting the final period which is shorter than the other weeks and not augmented by new data. The ratio between new and existing clusters is clarified by the two additional dashed lines, where the one nearer the bottom shows cluster creation per week and the one closer to cluster activity traces the number of augmented clusters in the period. Cluster creations tend to slow down significantly after each peak and the range between consecutive maximum and minimum shrinks as time goes. This is explained by the fact that most relevant profiles are found and that new ones do not appear all the time. Cluster augmentation, for its part, explains most of the activity, as the closeness between plots shows, since the more clusters there are, the more likely a fraud will fit one. It should also be observed that the number of active clusters is always much lower than the total number of clusters.

**Detailed Dynamic Analysis.** In order to offer a more detailed analysis, Figure 4 shows the size evolutions of the 40 largest clusters, where the ranking is based on the clusters' final sizes, as in Figure 1.

Four main types are observed, relative to growth speed: the curves can be linear, corresponding to a constant speed, concave, as in a deceleration behaviour, convex, showing an acceleration, or, as already discussed, flash profiles. We discuss each in turn below.

The constant increase profiles, globally exemplified by ten of the eleven largest clusters, represent fraud behaviours that are observed all year with very little

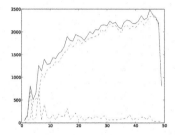

**Fig. 3.** Number of active (full line), augmented (top dashed) and new (bottom dashed) clusters per week

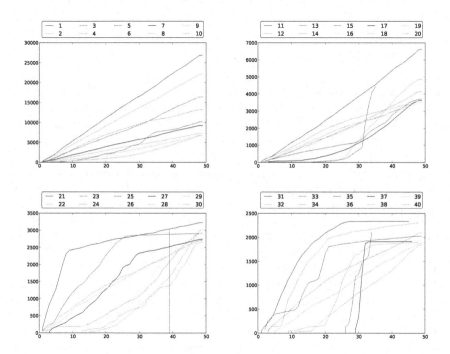

**Fig. 4.** Evolution of the size of the 40 largest clusters, where the cluster id equals its rank in terms of final size

variations in terms of relative distribution, i.e. with an approximate constant number of additional representatives each week. These are the 'fraud basics' or classic behaviours fraudsters know they can rely on. It should be pointed out that the constant increase is correlated to the fact that these clusters are the largest ones.

The second profile type, characterised by concave curves, presents a deceleration or even an abrupt halt in exploitation. This type is illustrated by clusters 21,

25 or 27. It corresponds to obsolete types of frauds on their way out. It would be interesting to check with credit-card fraud experts whether these observations can be related to the introduction of specific anti-fraud policies.

The third type is complementary: it groups convex curves, more precisely a parabolic increase in size or even a sudden augmentation, like cluster 13. It would likewise be relevant to discuss these results with experts to see if these observations correlate with credit-card policy modifications.

Over the given temporal window, some clusters combine a sudden increase and progressive deceleration, as clusters 35 and 38 do. These hybrid profiles introduce some doubt as to whether clusters of the third type are going to slow down at some point.

The last type is the already discussed very short-lived clusters , e.g. cluster 23. It is interesting to note that some of these are indeed large enough to join the 40 largest clusters, in which most are represented over a much longer time-span.

## 5    Conclusion and Future Work

A methodology to process incoming streams of heterogeneous data and study the derived profiles and their dynamics is proposed in this paper, together with its application to a real data set of credit-card frauds. The analysis of the results shows the relevance of the approach, which makes it possible to establish distinct types of fraud profiles depending on their temporal evolution and to identify specific fraudster profiles. The proposed algorithm efficiently processes data streams of heterogeneous nature, in a parallel implementation. It is based on incremental micro-clustering performed by a variant of the *l-fcmed* algorithm, defined as its combination with a cluster selection step guaranteeing highly compact clusters. The processing decomposes the data into three subsets, namely clusters, data to-be-clustered in a buffer and atypical data, remaining unassigned. The purge of the unassigned data, and to a lesser extent the clustering of buffered data, make it possible to manage some data beside the current flow of the data stream. This is particularly helpful in the case of credit-card fraud profiling, where frauds can be declared a long time after the transaction has been recorded.

Ongoing work aims at combining these micro-clustering results with a macro-clustering step, e.g. through a hierarchical method, to get a more synthetic view of the clusters, even if the detailed analysis also provides relevant and meaningful information. Future work includes comparisons with existing algorithms, as well as the discussion with experts in anti-fraud policies to obtain a semantic validation of the observed results. Another perspective is to take more information into account in fraud comparisons, in particular related to transaction sequences.

**Acknowledgements.** This work was supported by the project eFraudBox funded by ANR - CSOSG 2009.

# References

1. Banque de France: Annual Report of the Observatory for Payment Card Security (2011), http://www.banque-france.fr/observatoire/home_gb.html
2. Bolton, R.J., Hand, D.J.: Statistical fraud detection: a review. Statistical Science 17, 235–255 (2002)
3. Phua, C., Lee, V., Smith, K., Gayler, R.: A comprehensive survey of data mining-based fraud detection research. Artificial Intelligence Review (2005)
4. Laleh, N., Azgomi, M.A.: A taxonomy of frauds and fraud detection techniques. Information Systems, Technology and Management Communications in Computer and Information Science 3, 256–267 (2009)
5. Aggarwal, C.C. (ed.): Data Streams: Models and Algorithms. Springer (2007)
6. Farnstrom, F., Lewis, J., Elkan, C.: Scalability for clustering algorithms revisited. SIGKDD Explorations 2, 51–57 (2000)
7. Hore, P., Hall, L., Goldgof, D.: Single pass fuzzy c means. In: Proc. of the IEEE Int. Conf. on Fuzzy Systems, FUZZ-IEEE 2007, pp. 1–7 (2007)
8. Labroche, N.: New incremental fuzzy c medoids clustering algorithms. In: Proc. of the NAFIPS 2010, pp. 145–150 (2010)
9. Ester, M., Kriegel, H.P., Sander, J., Wimmer, M., Xu, X.: Incremental clustering for mining in a data warehousing environment. In: Proc. of the 24th Very Large DataBases Conference, VLDB 1998, pp. 323–333 (1998)
10. O'Callaghan, L., Meyerson, A., Motwani, R., Mishra, N., Guha, S.: Streaming-data algorithms for high-quality clustering. In: Proc. of the 18th Int. Conf. on Data Engineering, ICDE, pp. 685–694 (2002)
11. Aggarwal, C.C., Han, J., Wang, J., Yu, P.S.: A framework for clustering evolving data streams. In: Proc. of Very Large Data Bases, VLDB 2003, pp. 81–92 (2003)
12. Hore, P., Hall, L., Goldgof, D., Cheng, W.: Online fuzzy c means. In: Proc. of NAFIPS 2008, pp. 1–5 (2008)
13. Zhang, T., Ramakrishnan, R., Livny, M.: Birch: an efficient data clustering method for very large databases. In: Proc. of the ACM Int. Conf on Management of Data, SIGMOD 1996, pp. 103–114. ACM Press (1996)
14. Cao, F., Ester, M., Qian, W., Zhou, A.: Density-based clustering over an evolving data stream with noise. In: Proc. of the 6th SIAM Int. Conf. on Data Mining (2006)
15. Sander, J., Ester, M., Kriegel, H.P., Xu, X.: Density-based clustering in spatial databases: the algorithm DBSCAN and its application. Data Mining and Knowledge Discovery 2, 169–194 (1998)
16. Hathaway, R., Bezdek, J.: Nerf c-means: non euclidean relational fuzzy clustering. Pattern Recognition 27, 429–437 (1994)
17. Hathaway, R., Bezdek, J., Davenport, J.: On relational data versions of c-means algorithms. Pattern Recognition Letters 17, 607–612 (1996)
18. Kaufman, L., Rousseeuw, P.: Finding groups in data, an introduction to cluster analysis. John Wiley & Sons, Brussels (1990)
19. Krishnapuram, R., Joshi, A., Nasraoui, O., Yi, L.: Low complexity fuzzy relational clustering algorithms for web mining. IEEE Transactions on Fuzzy Systems 9, 595–607 (2001)

# Representing Fuzzy Logic Programs by Graded Attribute Implications*

Tomas Kuhr and Vilem Vychodil

DAMOL (Data Analysis and Modeling Laboratory)
Dept. Computer Science, Palacky University, Olomouc
17. listopadu 12, CZ–77146 Olomouc, Czech Republic
tomas.kuhr@upol.cz, vychodil@acm.org

**Abstract.** We present a link between two types of logic systems for reasoning with graded if-then rules: the system of fuzzy logic programming (FLP) in sense of Vojtáš and the system of fuzzy attribute logic (FAL) in sense of Belohlavek and Vychodil. We show that each finite theory consisting of formulas of FAL can be represented by a definite program so that the semantic entailment in FAL can be characterized by correct answers for the program. Conversely, we show that for each definite program there is a collection of formulas of FAL so that the correct answers can be represented by the entailment in FAL. Using the link, we can transport results from FAL to FLP and *vice versa* which gives us, e.g., a syntactic characterization of correct answers based on Pavelka-style Armstrong-like axiomatization of FAL.

**Keywords:** logic programming, attribute implications, functional dependencies, ordinal scales, residuated lattices.

## 1 Introduction

This paper contributes to the field of reasoning with graded if-then rules and presents an initial study of a link between two logic systems that have been proposed and studied independently. Namely, we focus on fuzzy logic programming in sense of [17] and fuzzy attribute logic presented in [3]. Both the systems play an important role in artificial intelligence—they can be used for approximate knowledge representation and inference, description of dependencies found in data, representing approximate constraints in relational similarity-based databases, etc. Although the systems are technically different and were developed to serve different purposes, they share common features: (i) they are based on residuated structures of truth degrees, (ii) use truth-functional interpretation of logical connectives, (iii) both the systems can be used to describe if-then dependencies in problem domains when one requires a formal treatment of inexact matches, (iv)

---

* Supported by grant no. P103/11/1456 of the Czech Science Foundation and internal grant of Palacky University no. PrF_2012_029. DAMOL is supported by project reg. no. CZ.1.07/2.3.00/20.0059 of the European Social Fund in the Czech Republic.

V. Torra et al. (Eds.): MDAI 2012, LNAI 7647, pp. 246–257, 2012.
© Springer-Verlag Berlin Heidelberg 2012

models of theories form particular closure systems and semantic entailment (from theories) can be expressed by means of least models.

In this paper, we show that the basic notions of correct answers and semantic entailment that appear in the systems are mutually reducible and allow to transport results from one theory to the other and *vice versa*. In the rest of this section, we outline the form of the rules. Section 2 presents preliminaries and recalls basic notions from FLP and FAL. Further sections are devoted to the reductions in both the directions. We also present a new Pavelka-style Armstrong-like axiomatization of FAL over infinite attribute sets and over arbitrary complete residuated lattices taken as the structure of truth degrees.

*Fuzzy logic programming* [5, 12, 17] is a generalization of the ordinary logic programming [11] in which logic programs consist of facts and complex rules containing a head (an atomic predicate formula) and a tail (a formula composed from atomic predicate formulas using connectives and aggregations interpreted by monotone truth functions) connected by a residuated implication. In addition, each rule (and fact) in a program is assumed to be valid to a degree (i.e., programs are theories in sense of Pavelka's abstract fuzzy logic [8, 15]). As a consequence, fuzzy logic programs are capable of expressing graded dependencies between facts. As an example, we can consider the following rule:

$$suitable(\mathbb{X}) \overset{0.8}{\Leftarrow} \mathsf{w@}\big(near(\mathbb{X}, \mathsf{stadium}) \wedge near(\mathbb{X}, \mathsf{center}), quality(\mathbb{X}), cost(\mathbb{X})\big), \quad (1)$$

which expresses how much suitable is a hotel (variable $\mathbb{X}$) for a sport fan. This rule describes the degree of hotel suitability (atomic formula $suitable(\mathbb{X})$) as weighted average (aggregator $\mathsf{w@}$) of degrees of being conveniently located, having high quality ($quality(\mathbb{X})$) and having low prices ($cost(\mathbb{X})$). The convenience of hotel location is specified here as a conjunction ($\wedge$) of being near to the stadium ($near(\mathbb{X}, \mathsf{stadium})$) and being near to the city center ($near(\mathbb{X}, \mathsf{center})$). The rule is valid in a degree 0.8, that can be understood so that we put "almost full emphasis on the rule".

The basic result of FLP is the completeness which puts in correspondence the declarative and procedural semantics of logic programs [17, Theorem 1 and Theorem 3] represented by correct answers and computed answers.

*Fuzzy attribute logic* [3] was developed primarily for the purpose of describing if-then dependencies that hold in object-attribute relational data where objects are allowed to have attributes to degrees. The formulas of FAL, so-called fuzzy attribute implications (FAIs) can be seen as implications $A \Rightarrow B$ between two graded sets of attributes (features), saying that if an object has all the attributes from $A$ (the antecedent), then it has all the attributes from $B$ (the consequent). The fact that $A$ and $B$ are graded sets (fuzzy sets) allows us to express graded dependencies between attributes. As an example

$$\{^{0.7}/lowAge, \, ^{0.9}/lowMileage\} \Rightarrow \{^{0.6}/highFuelEconomy, \, ^{0.9}/highPrice\} \quad (2)$$

is an attribute implication saying that cars with low age (at least to degree 0.7) and low mileage (at least to 0.9) have also high fuel economy (at least to 0.6) and high price (at least to 0.9).

FAIs have an alternative interpretation as similarity-based functional dependencies in relational databases [2]. For instance

$$\{{}^{0.8}/timeStamp, {}^{1}/creditCardNumber\} \Rightarrow \{{}^{0.9}/geographicalLocation\} \qquad (3)$$

can be seen as a rule saying that if two records (e.g., two tuples in a relational database table) have similar values of the attribute *timeStamp* at least to degree 0.8 and similar values of the attribute *creditCardNumber* to degree 1, then they must have similar values of the attribute *geographicalLocation* at least to degree 0.9. Rules like (3), if interpreted in a database of credit card transaction records (containing information about the transaction time, card number and location of the ATM machine) equipped with graded similarity relations on domains, can help detect a possible credit card misuse—a low degree of satisfaction of the rule means that the same credit card has been used in very different places during a short period of time.

The main results on FAL include syntactico-semantically complete axiomatization with ordinary-style and graded-style (Pavelka style, see [15]) notions of provability and results on descriptions of nonredundant bases of FAIs describing dependencies present in object-attribute data and ranked data tables over domains with similarities [2–4].

## 2  Preliminaries

We first recall basic notions common to both the fuzzy attribute implications and fuzzy logic programming. We then present a short survey of notions from both the theories used in the subsequent reductions.

### 2.1  Adjoint Operations and Residuated Structures

We consider here a complete lattice $\mathbf{L} = \langle L, \wedge, \vee, 0, 1 \rangle$ with $L$ representing a set of degrees (bounded by 0 and 1) and the corresponding lattice order $\leq$. As usual, 0 and 1 are interpreted as degrees representing the (full) falsity and (full) truth, each $0 < a < 1$ is an intermediate degree of truth. In order to express truth functions of general logical connectives, we assume that $\mathbf{L}$ is equipped by a collection of pairs of the form $\langle \otimes, \rightarrow \rangle$ such that $\langle L, \otimes, 1 \rangle$ is a commutative monoid, and $\otimes$ and $\rightarrow$ satisfy the adjointness property (w.r.t. $\mathbf{L}$):

$$a \otimes b \leq c \quad \text{iff} \quad a \leq b \rightarrow c \qquad (4)$$

for any $a, b, c \in L$. As usual, $\otimes$ (called a multiplication) and $\rightarrow$ (called a residuum) serve as truth functions of binary logical connectives "fuzzy conjunction" and "fuzzy implication". The mutual relationship of $\otimes$ and $\rightarrow$ posed by (4) has been derived from a graded counterpart to the classic deduction rule *modus ponens*, see [1, 7, 8]. If $\otimes$ and $\to$ satisfy (4), then $\mathbf{L} = \langle L, \wedge, \vee, \otimes, \to, 0, 1 \rangle$ is called a complete residuated lattice. Note that there are complete lattices that cannot be equipped with adjoint operations. On the other hand, there are complete lattices with multiple possible adjoint operations. Examples of complete residuated lattices include residuated lattices on the real unit interval given by left-continuous t-norms [6, 10], e.g. standard Gödel, Goguen, and Łukasiewicz algebras, see [1, 8] for details. A particular case of $\mathbf{L}$ is the two-element Boolean algebra with $L = \{0, 1\}$ and $\wedge = \otimes$ and $\to$ being the truth functions of the classic conjunction and implication which plays a central role in the classic propositional and predicate logics [13].

We use of the following notions from fuzzy relational systems [1]: An $\mathbf{L}$-set (fuzzy set) $A$ in universe $U$ is a map $A \colon U \to L$, $A(u)$ being interpreted as "the degree to which $u$ belongs to $A$". $L^U$ denotes the collection of all $\mathbf{L}$-sets in $U$. By $\{^a/u\}$ we denote an $\mathbf{L}$-set $A$ in $U$ such that $A(u) = a$ and $A(v) = 0$ for $v \neq u$. An $\mathbf{L}$-set $A \in L^U$ is called crisp if $A(u) \in \{0, 1\}$ for all $u \in U$. The operations with $\mathbf{L}$-sets are defined componentwise. For instance, union of $\mathbf{L}$-sets $A, B \in L^U$ is an $\mathbf{L}$-set $A \cup B$ in $U$ such that $(A \cup B)(u) = A(u) \vee B(u)$ for each $u \in U$, etc. For $a \in L$ and $A \in \mathbf{L}^U$, we define $\mathbf{L}$-sets $a \otimes A$ ($a$-multiple of $A$) and $a \to A$ ($a$-shift of $A$) by $(a \otimes A)(u) = a \otimes A(u)$, $(a \to A)(u) = a \to A(u)$ for all $u \in U$. For $\mathbf{L}$-sets $A, B \in L^U$, we define a subsethood degree of $A$ in $B$:

$$S(A, B) = \bigwedge_{u \in U} \big( A(u) \to B(u) \big), \tag{5}$$

where $\to$ is a residuum. Described verbally, $S(A, B)$ represents the degree to which $A$ is a subset of $B$. In addition, we write $A \subseteq B$ iff $S(A, B) = 1$, i.e. if $A$ is fully included in $B$. Using adjointness, (5) yields that $A \subseteq B$ iff, for each $u \in U$, $A(u) \leq B(u)$.

## 2.2 Fuzzy Attribute Implications

We assume here that $\mathbf{L}$ is a complete residuated lattice with a fixed pair of adjoint operations $\otimes$ and $\to$. Let $Y$ be a nonempty set of *attributes*. A *fuzzy attribute implication* (or, a graded attribute implication, shortly a FAI) is an expression $A \Rightarrow B$, where $A, B \in L^Y$. It is easily seen that (2) represents a FAI with $A \in L^Y$ being an $\mathbf{L}$-set in $Y = \{lowAge, lowMileage, highFuelEconomy, highPrice, \ldots\}$ so that $A(lowAge) = 0.7$, $A(lowMileage) = 0.9$ and analogously for $B$. The intended meaning of $A \Rightarrow B$ is: "if it is (very) true that an object has all attributes from $A$, then it has also all attributes from $B$". Formally, for an $\mathbf{L}$-set $M \in L^Y$ of attributes, we define a degree $\|A \Rightarrow B\|_M \in L$ to which $A \Rightarrow B$ is true in $M$ by

$$\|A \Rightarrow B\|_M = S(A, M)^* \to S(B, M), \tag{6}$$

where $S(\cdots)$ denote subsethood degrees (5), $\to$ is the residuum from $\mathbf{L}$ and $^*$ is an additional unary operation on $L$ satisfying the following conditions: (i) $1^* = 1$, (ii) $a^* \leq a$, (iii) $(a \to b)^* \leq a^* \to b^*$, and (iv) $a^{**} = a^*$ for all $a, b \in L$. An operation $^*$ satisfying (i)–(iv) shall be called a hedge (more precisely, an idempotent truth-stressing hedge) and can be seen as a truth function of a logical connective "very true", see [9]. We use * as a parameter of the interpretation of $A \Rightarrow B$. Namely, if * is set to identity, then $||A \Rightarrow B||_M = 1$ means that $S(A, M) \leq S(B, M)$, i.e. $B$ is contained in $M$ at least to the degree to which $A$ is contained in $M$. On the other hand, if * is defined as a globalization [16]:

$$a^* = \begin{cases} 1, & \text{if } a = 1, \\ 0, & \text{otherwise.} \end{cases} \tag{7}$$

then $||A \Rightarrow B||_M = 1$ means that if $A$ is fully contained in $M$ (i.e., $A \subseteq M$), then $B$ is fully contained in $M$ (i.e., $B \subseteq M$). Thus, two different important ways to interpret $||\cdots||_M = 1$ are obtained from the general definition (6) by different choices of *.

We consider two types of entailment of FAIs: (i) semantic entailment based on satisfaction of FAIs in systems of models, and (ii) syntactic entailment based on the notion of provability. We recall here the semantic entailment (the syntactic entailment will be discussed and extended in Section 4). Recall that $M$ is a *model* of an **L**-set $T$ of FAIs if $T(A \Rightarrow B) \leq ||A \Rightarrow B||_M$ for all $A, B \in L^Y$. Denoting the set of all models of $T$ by $\text{Mod}(T)$, we define a degree $||A \Rightarrow B||_T$ to which $A \Rightarrow B$ *semantically follows from* $T$ as follows:

$$||A \Rightarrow B||_T = \bigwedge_{M \in \text{Mod}(T)} ||A \Rightarrow B||_M. \tag{8}$$

Let us note that $||A \Rightarrow B||_T$ is a general degree from $L$, not necessarily 0 or 1.

*Remark 1.* In [3], we have shown a complete axiomatization of $||\cdots||_T$ using the notion of a degree of provability $|\cdots|_T$ in sense of Pavelka's abstract logic [8, 15]. The result has been proved for arbitrary complete residuated lattice **L** and finite $Y$ using an Armstrong-like axiomatization consisting of four deduction rules, one of them being an infinitary rule [18].

## 2.3   Fuzzy Logic Programming

We recall here the standard notions of (fuzzy) logic programming used in [11, 14, 17] and depart from the standard notation only in cases when it simplifies formulation of the subsequent results. According to [17], we consider a complete lattice **L** with $L$ being the real unit interval with its genuine ordering $\leq$ of real numbers. The approach in [17] uses multiple adjoint operations on **L**. It is even more general in that it allows "conjunctors" (and analogously "disjunctors") with weaker properties than postulated here (commutativity and associativity is not required in general). For simplicity, we do not discuss the issue here (some comments are in Section 5).

We consider programs as particular formulas written in a *language* $\mathcal{L}$ which is given by a finite nonempty set $R$ of *relation symbols* (predicate symbols in terms of LP) and a finite set $F$ of *function symbols* (functors in terms of LP). Each $r \in R$ and $f \in F$ is given its *arity* denoted by $\text{ar}(r)$ and $\text{ar}(f)$, respectively. We assume that $F$ contains at least one symbol for a *constant* (i.e., a function symbol f with $\text{ar}(f) = 0$) and $R$ is nonempty or that $R$ contains at least one *propositional symbol* (i.e., a relation symbol $p$ with $\text{ar}(p) = 0$). Moreover, we assume a denumerable set of *variables*. The variables are denoted by $\mathbb{X}, \mathbb{Y}, \mathbb{X}_i, \ldots$ As usual, *terms* are defined recursively using variables (as the base cases) and function symbols. An *atomic formula* is any expression $r(t_1, \ldots, t_k)$ such that $r \in R$, $\text{ar}(r) = k$, and $t_1, \ldots, t_k$ are terms. Moreover, *formulas* are defined recursively using atomic formulas (as the base cases) and symbols for binary logical connectives $\wedge_1, \wedge_2, \ldots$ (fuzzy conjunctions), $\vee_1, \vee_2, \ldots$ (fuzzy disjunctions), $\Rightarrow_1, \Rightarrow_2, \ldots$ (fuzzy implications), and symbols for *aggregations* $\text{ag}_1, \text{ag}_2, \ldots$ We accept the usual rules on the omission of parentheses and write $\varphi \Leftarrow \psi$ to denote $\psi \Rightarrow \varphi$. Since we do not consider quantifiers, all occurrences of variables in formulas are free (in the usual sense, see [13]).

Each atomic formula is called a *head*. Each formula that is free of symbols for fuzzy implications is called a *tail*. According to [17], a *theory* is a map which assigns to each formula of the language $\mathcal{L}$ a degree from $[0, 1]$. Moreover, a *definite program* is a theory such that

(i) there are only finitely many formulas that are assigned a nonzero degree,

(ii) all the assigned degrees are rational numbers from the unit interval,

(iii) each formula which is assigned a nonzero degree is either a head (a fact) or a formula of the form $\psi \Leftarrow \varphi$ (a rule), where $\psi$ is a head, $\varphi$ is a tail, and $\Leftarrow$ is an arbitrary symbol for implication.

Obviously, definite programs as defined above correspond to finite collections of formulas like (1) with rational degrees from $(0, 1]$ on the top of $\Leftarrow$, optionally with a blank space after $\Leftarrow$ (if the formula stands for a fact).

The declarative meaning of programs is defined using substitutions and models which we introduce here using the following notions. A *substitution* $\theta$ is a set of pairs denoted $\theta = \{\mathbb{X}_1/t_1, \ldots, \mathbb{X}_n/t_n\}$ where each $t_i$ is a term and each $\mathbb{X}_i$ a variable such that $\mathbb{X}_i \neq t_i$ and $\mathbb{X}_i \neq \mathbb{X}_j$ if $i \neq j$. Term/formula $\psi$ results by application of $\theta$ from $\varphi$ if $\psi$ is obtained from $\varphi$ by simultaneously replacing $t_i$ for every free occurrence of $\mathbb{X}_i$ in $\varphi$. We then denote $\psi$ as $\varphi\theta$ and call it an *instance* of $\varphi$. An instance $\varphi\theta$ is called *ground* if $\varphi\theta$ does not have any free occurrences of variables. For substitutions $\theta = \{\mathbb{X}_1/s_1, \ldots, \mathbb{X}_m/s_m\}$ and $\eta = \{\mathbb{Y}_1/t_1, \ldots, \mathbb{Y}_n/t_n\}$, the *composition* $\theta\eta$ is a substitution obtained from $\eta \cup \{\mathbb{X}_1/s_1\eta, \ldots, \mathbb{X}_m/s_m\eta\}$ by removing all $\mathbb{X}_i/s_i\eta$ for which $\mathbb{X}_i = s_i\eta$ and by removing all $\mathbb{Y}_j/t_j$ for which $\mathbb{Y}_j \in \{\mathbb{X}_1, \ldots, \mathbb{X}_m\}$. The composition is a monoidal operation on the set of all substitutions [14].

Let $P$ be a definite program formalized in language $\mathcal{L}$ (we often think of $\mathcal{L}$ as the least language in which all rules $\chi$ such that $P(\chi) > 0$ are correctly written). The set of all ground terms of $\mathcal{L}$ is called a *Herbrand universe* of $P$ and denoted by $\mathcal{U}_P$. The set of all ground atomic formulas of $\mathcal{L}$ is called a *Herbrand base* of $P$ and denoted by $\mathcal{B}_P$. Due to our assumptions on $\mathcal{L}$, $\mathcal{B}_P$ is nonempty. A *structure* for $P$ is any **L**-set in $\mathcal{B}_P$. If $M$ is a structure for $P$, $M(\chi)$ is interpreted as a degree to which the atomic ground formula $\chi$ is true under $M$. The notion of a formula being true in $M$ can be extended to all formulas as follows: We let $M^\sharp$ be an **L**-set of ground formulas defined by

(i) $M^\sharp(\varphi) = M(\varphi)$ if $\varphi$ is a ground atomic formula,

(ii) $M^\sharp(\psi \Leftarrow \varphi) = M^\sharp(\varphi) \to M^\sharp(\psi)$, where both $\varphi$ and $\psi$ are ground and $\to$ is a truth function (a residuum) interpreting $\Rightarrow$; analogously for the other binary connectives and the corresponding truth functions,

(iii) $M^\sharp\big(\text{ag}(\varphi_1, \ldots, \varphi_n)\big) = ag\big(M^\sharp(\varphi_1), \ldots, M^\sharp(\varphi_n)\big)$, where all $\varphi_i$ are ground and $\text{ag}$ is an $n$-ary symbol for aggregation which is interpreted by a monotone function $ag: [0,1]^n \to [0,1]$ which preserves $\{0\}^n$ and $\{1\}^n$.

Moreover, we define $M^\sharp_\lor$ to extend the notion for all formulas as follows:

$$M^\sharp_\lor(\varphi) = \bigwedge\{M^\sharp(\varphi\theta) \mid \theta \text{ is a substitution such that } \varphi\theta \text{ is ground}\}. \qquad (9)$$

Structure $M$ is called a *model* for theory $T$ if $T(\chi) \le M^\sharp_\lor(\chi)$ for each formula $\chi$ of the language $\mathcal{L}$. The collection of all models of $T$ will be denoted by $\mathrm{Mod}(T)$. An important notion of the declarative semantics of definite programs is that of a *correct answer*: A pair $\langle a, \theta \rangle$ consisting of $a \in (0,1]$ and a substitution $\theta$ is a *correct answer* for a definite program $P$ and an atomic formula $\varphi$ (called a query) if $M^\sharp_\lor(\varphi\theta) \ge a$ for each $M \in \mathrm{Mod}(P)$.

## 3    Representing FAIs by Propositional FLPs

Let $\mathbf{L} = \langle L, \wedge, \vee, \otimes, \to, 0, 1\rangle$ be a complete residuated lattice on the real unit interval. We now show that for each finite set $T$ of FAIs $A \Rightarrow B$, where both $A$ and $B$ are finite (i.e., there are finitely many attributes $y \in Y$ such that $A(y) > 0$ and $B(y) > 0$) and all degrees $A(y)$ and $B(y)$ are rational, we can find a corresponding definite program in which the correct answers can be used to describe degrees $||\cdots||_T$ of semantic entailment of FAIs.

First, we consider a language $\mathcal{L}$ with only nullary relation symbols $R = \{y_1, y_2, \ldots, y_k\}$ ($\mathrm{ar}(y_i) = 0$) that correspond to attributes which appear in the antecedents or consequents of FAIs from $T$ to a nonzero degree. Due to our assumptions, $R$ is a finite set. Notice that the Herbrand base of any program written in $\mathcal{L}$ is equal to $R$. Moreover, we consider the following logical connectives and aggregations:

(i) $\Rightarrow$ (interpreted by the residuum $\to$),

(ii) $\wedge$ (interpreted by the infimum $\wedge$),

(iii) a unary aggregation $\text{ts}$ (interpreted by hedge $*$, i.e. $M^\sharp(\text{ts}(\varphi)) = M^\sharp(\varphi)^*$),

(iv) for each rational $a \in (0,1]$ a unary aggregation $\text{sh}_a$ called an $a$-shift aggregation (interpreted by $M^\sharp(\text{sh}_a(\varphi)) = a \to M^\sharp(\varphi)$).

We now make the following observation:

**Theorem 1.** *For each set $T$ of FAIs and $A \Rightarrow B$ there is a definite program $P$ such that $||A \Rightarrow B||_T \ge a > 0$ iff for each attribute $y$ such that $B(y) > 0$, the pair $\langle a \otimes B(y), \emptyset \rangle$ is a correct answer for the program $P$ and $y$.*

*Proof (a sketch).* For any $A \Rightarrow B \in T$ and $y \in Y$ such that $B(y) > 0$ and all attributes $y \in Y$ satisfying $A(y) > 0$ being exactly $z_1, \ldots, z_n$, consider a rule

$$y \Leftarrow \mathsf{ts}\big(\mathsf{sh}_{A(z_1)}(z_1) \wedge \cdots \wedge \mathsf{sh}_{A(z_n)}(z_n)\big). \tag{10}$$

Denote the rule (10) by $y \Leftarrow A$. Note that in a special case if $A = \emptyset$, (10) becomes the fact $y$. Moreover, consider an **L**-set of rules $P_T$ defined by

$$P_T(y \Leftarrow A) = \bigvee \{B(y) \mid B \in L^Y \text{ such that } A \Rightarrow B \in T\}. \tag{11}$$

Clearly, $P_T$ is a definite program in $\mathcal{L}$. The proof then continues by observing that $\|A \Rightarrow B\|_T \geq a > 0$ iff $\|A \Rightarrow a \otimes B\|_T = 1$ iff $\|\emptyset \Rightarrow a \otimes B\|_{T \cup \{\emptyset \Rightarrow A\}} = 1$ iff

$$a \otimes B(y) \leq \|\emptyset \Rightarrow \{^1\!/y\}\|_{T \cup \{\emptyset \Rightarrow A\}}$$

for all $y \in Y$ such that $B(y) > 0$. The latter is true iff for each $y \in Y$ such that $B(y) > 0$, the pair $\langle a \otimes B(y), \emptyset \rangle$ ($\emptyset$ is the empty substitution) is a correct answer for the program $P_{T \cup \{\emptyset \Rightarrow A\}}$ and query $y$. Details of the proof are postponed to a full version of the paper. □

**Theorem 2.** *For each set $T$ of FAIs and $A \Rightarrow B$ there is a definite program $P$ such that $\|A \Rightarrow B\|_T$ is the supremum of all degrees $a \in L$ for which $\langle a \otimes B(y), \emptyset \rangle$ is a correct answer for $P$ and all $y$ satisfying $B(y) > 0$.*

*Proof.* Consequence of Theorem 1. □

We have shown that for $T$ and $A$, we can find a *propositional* fuzzy logic program from which we can express degrees of semantic entailment of FAIs of the form $A \Rightarrow B$. Due to the limitations of FLP, the result is limited to finite theories consisting of finite FAIs, and structures of degrees defined on the real unit interval.

*Remark 2.* Note that regarding [17, Theorem 3], our aggregations $\mathsf{ts}$ and $\mathsf{sh}_a$ are not left-semicontinuous in general. That is, in general one cannot directly apply [17, Theorem 3] and Theorem 2 to obtain a characterization of $\|A \Rightarrow B\|_T$ using computed answers.

## 4    Completeness for FAIs over Infinite Attribute Sets

Before we show the reduction in the opposite direction, we provide a syntactic characterization of $\|\cdots\|_T$ for FAIs over infinite sets of attributes and over arbitrary **L**. An analogous result has been shown in [3], where we have considered finite $Y$. The limitation to finite $Y$ in [3] was mainly for historical reasons because originally FAIs were developed as rules extracted from object-attribute data tables, i.e., the sets of attributes were considered finite. Nevertheless, inspecting the results from [3], one can show that the main results hold if $Y$ is infinite. In addition to that, we present here the completeness results for a simplified set of inference rules.

We consider the following inference rules:

$$(\text{Ax}):\frac{}{A\cup B\Rightarrow A}, \quad (\text{Mul}):\frac{A\Rightarrow B}{c^*\otimes A\Rightarrow c^*\otimes B}, \quad (\text{Cut}_\omega):\frac{A\Rightarrow B, \{B\cup C\Rightarrow D_i \mid i\in I\}}{A\cup C\Rightarrow \bigcup_{i\in I} D_i},$$

where $A,B,C,D_i \in L^Y$ ($i \in I$), and $c \in L$. The first two rules come from [3], $(\text{Cut}_\omega)$ is an *infinitary rule* saying that "from $A\Rightarrow B$ and (in general infinitely many) FAIs $B\cup C\Rightarrow D_i$, infer $A\cup C\Rightarrow\bigcup_{i\in I} D_i$". Proofs are defined as labeled infinitely branching rooted (directed) trees with finite depth [18]. Denoting by $\mathbf{T} = \langle l, Z\rangle$ a tree with root label $l$ (a formula) and subtrees from the set $Z$, we introduce the following notion:

(i) for each $A\Rightarrow B \in T$, tuple $\mathbf{T} = \langle A\Rightarrow B, \emptyset\rangle$ is a proof of $A\Rightarrow B$ from $T$,

(ii) if $\mathbf{T}_i = \langle\varphi_i,\dots\rangle$ ($i \in I$) are proofs from $T$ and if $\varphi$ results from $\varphi_i$ ($i \in I$) by any of the deduction rules (Ax), (Mul), or $(\text{Cut}_\omega)$, then $\mathbf{T} = \langle\varphi, \{\mathbf{T}_i \mid i\in I\}\rangle$ is a proof of $\varphi$ from $T$.

Furthermore, $A\Rightarrow B$ is provable from $T$, written $T \vdash A\Rightarrow B$, if there is a proof $A\Rightarrow B$ from $T$. The degree $|A\Rightarrow B|_T$ to which $A\Rightarrow B$ is provable from $T$ is defined by $|A\Rightarrow B|_T = \bigvee\{c \in L \mid T \vdash A\Rightarrow c\otimes B\}$. We can now prove the following characterization and its consequence:

**Theorem 3 (ordinary-style completeness).** *For any $T$ and $A\Rightarrow B$,*

$$T \vdash A\Rightarrow B \quad \text{iff} \quad \|A\Rightarrow B\|_T = 1.$$

*Proof (a sketch).* Follows from [3] by checking that finite $Y$ can be safely replaced by infinite $Y$ and considering that the ordinary cut and the infinitary rule $(\text{Add}_\omega)$ from [3] are derivable from $(\text{Cut}_\omega)$ and *vice versa*. □

**Corollary 1 (graded-style completeness).** *For any $T$ and $A\Rightarrow B$,*

$$|A\Rightarrow B|_T = \|A\Rightarrow B\|_T.$$

*Proof.* Consequence of Theorem 3. □

## 5   Representing FLPs by FAIs over Herbrand Bases

In this section, we consider $\mathbf{L}$ to be a complete residuated lattice on the real unit interval equipped with $^*$ defined by (7). For a definite program $P$, we consider a theory consisting of FAIs where the set of attributes is represented by the Herbrand base $\mathcal{B}_P$. In general, $\mathcal{B}_P$ is infinite and therefore the FAIs are formulas with infinite antecedents and consequents. Note that in the important case when $F$ consists solely of constants, $\mathcal{B}_P$ is finite and thus we work with FAIs that can be understood as formulas in the usual sense.

The following theorem exploits Theorem 3 and establishes the opposite reduction to that from Section 3.

**Theorem 4.** *For every definite program $P$ there is a set $T$ of FAIs such that for each atomic formula $\varphi$ and substitution $\theta$ there is a crisp $B_\varphi \in L^{\mathcal{B}_P}$ so that $\langle a, \theta \rangle$ is a correct answer for $P$ and $\varphi$ iff $T \vdash \emptyset \Rightarrow a \otimes B_\varphi$.*

*Proof (a sketch).* Considering $P$, for each $A \in L^{\mathcal{B}_P}$ such that $A \neq \emptyset$, we put

$$A^\circ(\chi) = \bigvee \{ P(\psi \Leftarrow \xi) \otimes A^\sharp(\xi\theta) \mid \xi\theta \text{ is ground and } \psi\theta \text{ equals } \chi \}, \qquad (12)$$

for all $\chi \in \mathcal{B}_P$. Note that the multiplication $\otimes$ which appears in (12) is the multiplication which is adjoint to the residuum $\to$ interpreting $\Rightarrow$ (in general, multiple different $\Rightarrow_i$ can be used in $P$ simultaneously). In addition, put

$$\emptyset^\circ(\chi) = \bigvee \{ P(\psi) \mid \psi\theta \text{ equals } \chi \}. \qquad (13)$$

for all $\chi \in \mathcal{B}_P$. Clearly, $A^\circ \in L^{\mathcal{B}_P}$ for all $A \in L^{\mathcal{B}_P}$ and we may let $T$ be the set $T_P = \{ A \Rightarrow A^\circ \mid A \in L^{\mathcal{B}_P} \}$. Moreover, for the atomic formula $\varphi$ we define

$$B_\varphi(\psi) = \begin{cases} 1, & \text{if } \psi \text{ is a ground instance of } \varphi\theta, \\ 0, & \text{otherwise.} \end{cases} \qquad (14)$$

Now, one can show that $\langle a, \theta \rangle$ is a correct answer for $P$ and the query $\varphi$ iff $\|\emptyset \Rightarrow B_\varphi\|_{T_P} \geq a$ which is shown by proving that $\text{Mod}(P) = \text{Mod}(T_P)$, i.e., we get $\|\emptyset \Rightarrow a \otimes B_\varphi\|_{T_P} = 1$ (here $\otimes$ can be arbitrary multiplication since $B_\varphi$ is crisp). The latter is true iff $T_P \vdash \emptyset \Rightarrow a \otimes B_\varphi$ due to Theorem 3. □

Let us conclude this section with a few clarifying remarks.

*Remark 3.* (a) The choice of the adjoint operations in $\mathbf{L}$ is not substantial. The role of $\otimes$ and $\to$ in $\mathbf{L}$ from the point of view of FAIs and degrees $\|\cdots\|_T$ is suppressed by the fact that $^*$ is globalization.

(b) In [17], the author uses various connectives together with the aggregations but, in fact, the aggregations are more universal and the connectives (conjunctions and disjunctions) used therein can be seen as binary aggregations. It is not the case of the residua (which are antitone in the first argument) but their role is different from the other connectives since the (symbols of) implications cannot be used in tails of the rules.

(c) The paper [17] does not introduce semantic entailment from definite programs. Technically, the notion can be introduced and one can prove a syntactico-semantical completeness with respect to the present inference system via the reduction we have shown in this paper. Details are postponed to a full version of the paper.

## 6 Illustrative Example

Let $\mathbf{L}$ be the standard Łukasiewicz structure of truth degrees, i.e., a complete residuated lattice on the unit interval with its genuine ordering $\leq$ and adjoint operations $\otimes$, $\to$ defined by $a \otimes b = \max(0, a+b-1)$ and $a \to b = \min(1, 1-a+b)$. Let $^*$ be the identity.

Furthermore, consider a set of attributes of cars $Y = \{lA, lM, hAT, hFE, hP\}$ which mean: "a car has low age", "has low mileage", "has automatic transmission", "has high fuel economy" and "has high price" respectively. Let $T$ being a set containing the following FAIs over $Y$:

$$\{^{0.7}/lA, {}^{0.9}/lM, {}^{0.4}/hAT\} \Rightarrow \{^{0.6}/hFE, {}^{0.9}/hP\},$$
$$\{^{0.8}/lA\} \Rightarrow \{^{0.7}/lM\}.$$

Using Theorem 1, we can find a FLP $P_T$ that corresponds to FAIs from $T$. The program $P_T$ will contain the following rules:

$$hFE \overset{0.6}{\Leftarrow} \mathsf{ts}\big(\mathsf{sh}_{0.7}(lA) \wedge \mathsf{sh}_{0.9}(lM) \wedge \mathsf{sh}_{0.4}(hAT)\big),$$

$$hP \overset{0.9}{\Leftarrow} \mathsf{ts}\big(\mathsf{sh}_{0.7}(lA) \wedge \mathsf{sh}_{0.9}(lM) \wedge \mathsf{sh}_{0.4}(hAT)\big),$$

$$lM \overset{0.7}{\Leftarrow} \mathsf{ts}\big(\mathsf{sh}_{0.8}(lA)\big).$$

Obviously, the aggregator $\mathsf{ts}$ interpreted by identity can be omitted. Furthermore, all aggregations interpreting $\mathsf{sh}_a(y)$ as well as the function $\wedge$ interpreting conjunctor $\wedge$ are left-semicontinuous in this case. Thus, we can use [17, Theorem 3] and Theorem 2 to characterize $\|A \Rightarrow B\|_T$ using computed answers for program $P_{T \cup \{\emptyset \Rightarrow A\}}$ and queries $y \in Y$ with $B(y) > 0$.

For example, a user asks a question "How much expensive are quite new cars with automatic transmission?", i.e., more precisely "To what degree $a \in L$, is the FAI $\{^{0.6}/lA, {}^1/hAT\} \Rightarrow \{^1/hP\}$ true in $T$?". To get the answer, we first extend $P_T$ to $P_{T \cup \{\emptyset \Rightarrow A\}}$ by adding facts $lA \overset{0.6}{\Leftarrow}$ and $hAT \overset{1}{\Leftarrow}$ to the program. Then, we can easily compute an answer to query $hP$ using the usual admissible rules of FLPs [17] (all substitutions are $\emptyset$):

$hP,$

$0.9 \otimes \big(\mathsf{sh}_{0.7}(lA) \wedge \mathsf{sh}_{0.9}(lM) \wedge \mathsf{sh}_{0.4}(hAT)\big),$

$0.9 \otimes \big(\mathsf{sh}_{0.7}(lA) \wedge \mathsf{sh}_{0.9}(0.7 \otimes \mathsf{sh}_{0.8}(lA)) \wedge \mathsf{sh}_{0.4}(hAT)\big),$

$0.9 \otimes \big(\mathsf{sh}_{0.7}(0.6) \wedge \mathsf{sh}_{0.9}(0.7 \otimes \mathsf{sh}_{0.8}(0.6)) \wedge \mathsf{sh}_{0.4}(1)\big),$

$0.9 \otimes \big(0.7 \to 0.6 \wedge 0.9 \to (0.7 \otimes (0.8 \to 0.6)) \wedge 0.4 \to 1\big),$

$0.5.$

Using [17, Theorem 3], Theorem 2 and the computed answer $\langle 0.5, \emptyset \rangle$, we immediately get $\|\{^{0.6}/lA, {}^1/hAT\} \Rightarrow \{^1/hP\}\|_T = 0.5$.

## 7  Conclusions

We have shown that fuzzy attribute implications (in sense of Belohlavek and Vychodil) and fuzzy logic programs (in sense of Vojtáš) are mutually reducible (with some limitations to structures of degrees) and correct answers for fuzzy logic programs and queries can be described via semantic entailment of fuzzy attribute implications and *vice versa*. Furthermore, we have shown a complete Pavelka-style axiomatization for fuzzy attribute logic (FAL) over arbitrary **L** and infinite sets of attributes using a new deduction system containing an infinitary cut. Together with the reduction we have shown in the paper, this gives us a new syntactic characterization of correct answers in fuzzy logic programming (FLP). The results have shown a new theoretical insight and a link of two branches of rule-based reasoning methods. Future research will focus on various other issues interrelating FLP and FAL.

# References

1. Belohlavek, R.: Fuzzy Relational Systems: Foundations and Principles. Kluwer Academic Publishers, Norwell (2002)
2. Bělohlávek, R., Vychodil, V.: Data Tables with Similarity Relations: Functional Dependencies, Complete Rules and Non-redundant Bases. In: Li Lee, M., Tan, K.-L., Wuwongse, V. (eds.) DASFAA 2006. LNCS, vol. 3882, pp. 644–658. Springer, Heidelberg (2006)
3. Belohlavek, R., Vychodil, V.: Fuzzy attribute logic over complete residuated lattices. Journal of Exp. & Theoretical Artif. Int. 18(4), 471–480 (2006)
4. Belohlavek, R., Vychodil, V.: Query systems in similarity-based databases: logical foundations, expressive power, and completeness. In: ACM Symposium on Applied Computing (SAC), pp. 1648–1655. ACM (2010)
5. Viegas Damásio, C., Moniz Pereira, L.: Monotonic and Residuated Logic Programs. In: Benferhat, S., Besnard, P. (eds.) ECSQARU 2001. LNCS (LNAI), vol. 2143, pp. 748–759. Springer, Heidelberg (2001)
6. Esteva, F., Godo, L.: Monoidal t-norm based logic: towards a logic for left-continuous t-norms. Fuzzy Sets and Systems 124(3), 271–288 (2001)
7. Goguen, J.A.: The logic of inexact concepts. Synthese 19, 325–373 (1979)
8. Hájek, P.: Metamathematics of Fuzzy Logic. Kluwer Academic Publishers, Dordrecht (1998)
9. Hájek, P.: On very true. Fuzzy Sets and Systems 124(3), 329–333 (2001)
10. Klement, E.P., Mesiar, R., Pap, E.: Triangular Norms, 1st edn. Springer (2000)
11. Lloyd, J.W.: Foundations of logic programming. Springer-Verlag New York, Inc., New York (1984)
12. Medina, J., Ojeda-Aciego, M., Vojtáš, P.: A Procedural Semantics for Multi-adjoint Logic Programming. In: Brazdil, P.B., Jorge, A.M. (eds.) EPIA 2001. LNCS (LNAI), vol. 2258, pp. 290–297. Springer, Heidelberg (2001)
13. Mendelson, E.: Introduction to mathematical logic. Chapman and Hall (1987)
14. Nilsson, U., Maluszynski, J.: Logic, Programming, and PROLOG, 2nd edn. John Wiley & Sons, Inc., New York (1995)
15. Pavelka, J.: On fuzzy logic I, II, III. Zeitschrift für Mathematische Logik und Grundlagen der Mathematik 25, 45–52, 119–134, 447–464 (1979)
16. Takeuti, G., Titani, S.: Globalization of intuitionistic set theory. Annals of Pure and Applied Logic 33, 195–211 (1987)
17. Vojtáš, P.: Fuzzy logic programming. Fuzzy Sets and Systems 124(3), 361–370 (2001)
18. Wechler, W.: Universal algebra for computer scientists. EATCS Monographs on Theoretical Computer Science, vol. 25. Springer, Heidelberg (1992)

# Refining Discretizations of Continuous-Valued Attributes

Eva Armengol[1] and Àngel García-Cerdaña[1,2]

[1] Artificial Intelligence Research Institute (IIIA - CSIC),
Campus de la UAB, 08193 Bellaterra, Catalonia, Spain
{eva,angel}@iiia.csic.es

[2] Departament de Lògica, Història i Filosofia de la Ciència,
Universitat de Barcelona, Montalegre 6, 08001 Barcelona, Catalonia, Spain

**Abstract.** The Rand index is a measure commonly used to compare crisp partitions. Campello (2007) and Hüllermeier and Rifqi (2009) respectively, proposed two extensions of this index capable to compare fuzzy partitions. These approaches are useful when continuous values of attributes are discretized using fuzzy sets. In previous works we experimented with these extensions and compared their accuracy with the one of the crisp Rand index. In this paper we propose the $\varepsilon$-procedure, an alternative way to deal with attributes taking continuous values. Accuracy results on some known datasets of the Machine Learning repository using the $\varepsilon$-procedure as crisp discretization method jointly with the crisp Rand index are comparable to the ones given using the crisp Rand index and its fuzzifications with standard crisp and fuzzy discretization methods respectively.

**Keywords:** Machine learning, Classification, Discretization methods.

## 1 Introduction

Knowledge representation of domain objects often involves the use of continuous values. One of the most widely used techniques to deal with continuous values is the discretization, consisting on building intervals of values that should be considered as equivalent. There are two kinds of discretization: crisp and fuzzy. In crisp discretization the range of a continuous value is split into several intervals. Elements of an interval are considered as equivalent and each interval is handled as a discrete value. There are different methods of crisp discretization. For instance, some of them take into account the length of the interval, or the frequency of the values, while others are entropy-based (for more information see [1]). In some domains, the crisp discretization shows some counter-intuitive behavior around the thresholds of the intervals: values around the threshold of two adjacent intervals are considered as different but may be they are not so. For this reason, sometimes it is interesting to build a fuzzy discretization from a crisp one, as it is done for instance in [2].

Given an attribute taking continuous values, let $\alpha_1, \ldots, \alpha_n$ be the thresholds determining the discretization intervals for that attribute. Let $\alpha_0$ and $\alpha_{n+1}$ be

V. Torra et al. (Eds.): MDAI 2012, LNAI 7647, pp. 258–269, 2012.
© Springer-Verlag Berlin Heidelberg 2012

the minimum and maximum of the values that this attribute takes in its range. To fuzzy discretize the attribute, assuming that the fuzzy sets are trapezoidal, the membership vector is calculated in the following way:

$$F_1(x) = \begin{cases} 1, & \text{when } \alpha_0 \leq x \leq \alpha_1 - \delta_1, \\ \frac{\alpha_1 + \delta_1 - x}{2\delta_1}, & \text{when } \alpha_1 - \delta_1 < x < \alpha_1 + \delta_1, \\ 0, & \text{when } \alpha_1 + \delta_1 \leq x. \end{cases}$$

$$F_i(x) = \begin{cases} 0, & \text{when } x \leq \alpha_{i-1} - \delta_{i-1}, \\ \frac{x - (\alpha_{i-1} - \delta_{i-1})}{2\delta_{i-1}}, & \text{when } \alpha_{i-1} - \delta_{i-1} < x < \alpha_{i-1} + \delta_{i-1}, \\ 1, & \text{when } \alpha_{i-1} + \delta_{i-1} \leq x \leq \alpha_i - \delta_i \\ \frac{\alpha_i + \delta_i - x}{2\delta_i}, & \text{when } \alpha_i - \delta_i < x < \alpha_i + \delta_i , \\ 0, & \text{when } \alpha_i + \delta_i \leq x. \end{cases} \qquad (1)$$

$$F_{n+1}(x) = \begin{cases} 0, & \text{when } x \leq \alpha_n - \delta_n, \\ \frac{x - (\alpha_n - \delta_n)}{2\delta_n}, & \text{when } \alpha_n - \delta_n < x < \alpha_n + \delta_n, \\ 1, & \text{when } \alpha_n + \delta_n \leq x \leq \alpha_{n+1}. \end{cases}$$

In these formulas, the parameters $\delta_i$ represents the overlapping degree between contiguous fuzzy sets and they are computed as follows: $\delta_i = p \cdot |\alpha_i - \alpha_{i-1}|$, being the factor $p$ a percentage that we can adjust.

Intuitively, since the representation using fuzzy sets is smooth around the thresholds of discretization intervals, it seems more appropriate than the crisp discretization in some domains. Nevertheless, the use of fuzzy sets implies that for each attribute value it is necessary to deal with the membership of this value to each fuzzy set representing the attribute. As we will see later, for the fuzzy extensions of the Rand index, this situation produces an increment of the run time with respect to the crisp version of the Rand index. For this reason, we have searched for an alternative method of discretization such that:

a) it retains as much as possible the advantages of the discretization methods using fuzzy sets,
b) it can be used with crisp measures such as the Rand index and, therefore,
c) it involves a reduction of the run time associated to the fuzzy measures.

As a concrete alternative, in the current paper we propose the $\varepsilon$-procedure, a method of discretization that induces classical partitions from continuous values of attributes. It consists on a refinement of the crisp discretization obtained by any standard discretization method. From a set of discretization thresholds $\alpha_1, \alpha_2, \ldots, \alpha_n$, the $\varepsilon$-procedure introduces the intervals $(\alpha_i - \delta_i, \alpha_i + \delta_i]$ being $\delta_i$ a parameter that depends on the length of the interval $(\alpha_{i-1}, \alpha_i]$.

In [3] we experimented with the Rand Index and two fuzzy extensions of it: one proposed by Campello [4] and the other proposed by Hüllermeier and Rifqi [5]. These experiments have been performed in the framework of a lazy learning method called LID. For this reason in the present paper we have carried out similar experiments with the $\varepsilon$-procedure in order to compare its behavior with the one obtained using standard discretization methods.

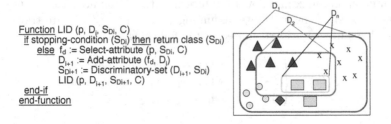

Figure code on the left:

```
Function LID (p, Dᵢ, S_Di, C)
  if stopping-condition (S_Di) then return class (S_Di)
    else  f_d := Select-attribute (p, S_Di, C)
         D_i+1 := Add-attribute (f_d, D_i)
         S_Di+1 := Discriminatory-set (D_i+1, S_Di)
         LID (p, D_i+1, S_Di+1, C)
  end-if
end-function
```

**Fig. 1.** The LID algorithm. On the right there is the intuitive idea of LID.

The paper is organized as follows. Section 2 contains preliminary concepts. In Sec. 3 the $\varepsilon$-procedure is presented. Section 4 contains the explanation of the experiments and a discussion of the results. The last section contains conclusions and future work.

## 2    Preliminary Concepts

In this section we explain the algorithm of the method LID used in the experiments. LID uses a mesure $\Delta$ to compare partitions. In the experiments we have used as $\Delta$ the Rand index and two of its fuzzifications, one proposed by Campello in [4] an the other by Hüllermeier and Rifqi in [5]. In this section we also explain these three measures in some detail.

### 2.1    Lazy Induction of Descriptions

*Lazy Induction of Descriptions* (LID) is a lazy learning method for classification tasks. LID determines which are the most relevant attributes of a problem (i.e., an object to be classified) and searches in a case base for cases sharing these relevant attributes. The problem is classified when LID finds a set of relevant attributes shared by a subset of cases all of them belonging to the same class. We call *similitude term* the description formed by these relevant features and *discriminatory set* the set of cases satisfying the similitude term.

Given a problem for solving $\mathfrak{p}$, LID (Fig. 1) initializes $D_0$ as a description with no attributes, the discriminatory set $S_{D_0}$ as the set of cases satisfying $D_0$, i.e., all the available cases, and $C$ as the set of solution classes into which the known cases are classified. Let $D_i$ be the current similitude term and $S_{D_i}$ be the set of all the cases satisfying $D_i$. When the stopping condition of LID is not satisfied, the next step is to select an attribute for specializing $D_i$. The specialization of $D_i$ is achieved by adding attributes to it. Given the set $F$ of attributes candidate to specialize $D_i$, the next step of the algorithm is the selection of an attribute $f \in F$. Selecting the most discriminatory attribute in $F$ is heuristically done using a measure $\Delta$ to compare each partition $\mathcal{P}_f$ induced by an attribute $f$ with the correct partition $\mathcal{P}_c$. The *correct partition* has as many sets as solution classes. Each attribute $f \in F$ induces in the discriminatory set a partition $\mathcal{P}_f$ with as many sets as the number of different values that $f$ takes in the cases.

Given a measure $\Delta$ and two attributes $f$ and $g$ inducing respectively partitions $\mathcal{P}_f$ and $\mathcal{P}_g$, we say that $f$ is *more discriminatory* than $g$ iff $\Delta(\mathcal{P}_f, \mathcal{P}_c) < \Delta(\mathcal{P}_g, \mathcal{P}_c)$. This means that the partition $\mathcal{P}_f$ is closer to the correct partition than the partition $\mathcal{P}_g$. LID selects the most discriminatory attribute to specialize $D_i$. Let $f_d$ be the most discriminatory attribute in $F$. The specialization of $D_i$ defines a new similitude term $D_{i+1}$ by adding to $D_i$ the attribute $f_d$. The new similitude term $D_{i+1} = D_i \cup \{f_d\}$ is satisfied by a subset of cases in $S_{D_i}$, namely $S_{D_{i+1}}$. Next, LID is recursively called with $S_{D_{i+1}}$ and $D_{i+1}$. The recursive call of LID has $S_{D_{i+1}}$ instead of $S_{D_i}$ because the cases that are not satisfied by $D_{i+1}$ will not satisfy any further specialization. Notice that the specialization reduces the discriminatory set at each step, i.e., we get a sequence $S_{D_n} \subset S_{D_{n-1}} \subset \ldots \subset S_{D_0}$.

LID has two stopping situations: 1) all the cases in the discriminatory set $S_{D_j}$ belong to the same solution class $C_i$, or 2) there is no attribute allowing the specialization of the similitude term. When the stopping condition 1) is satisfied, $\mathfrak{p}$ is classified as belonging to $C_i$. When the stopping condition 2) is satisfied, $S_{D_j}$ contains cases from several classes; in such situation the *majority criteria* is applied, and $\mathfrak{p}$ is classified in the class of the majority of cases in $S_{D_j}$. When there is a tie between two classes, LID gives a multiple solution proposing both classes as the classification for $\mathfrak{p}$.

In our experiments we have taken the Rand index as the measure $\Delta$ that supports the selection of relevant attributes. This index is introduced in the next section.

## 2.2   The Rand Index

The Rand index [6] was conceived to compare clusterings produced by several automatic methods. The basic assumptions for using the Rand index are the following: 1) the clusterings to be compared are crisp in the sense that the set of clusters is a crisp partition of the domain; 2) the clusters are defined by both the objects that they contain and the objects that they do not contain; and 3) all objects are equally important in determining the clustering. From these assumptions it follows that a basic unit of comparison between two clusterings is how pairs of objects are clustered. If a pair of objects are placed together in a class in each one of the two clusterings, or if they are assigned to different classes in both clusterings, this represents a similarity between the clusterings. The opposite case is the one in which a pair of objects are in the same class in one clustering and in different classes in the other one. From this point of view, a measure of the similarity between two clusterings of the same data set can be defined as the number of equal assignments of object pairs normalized by the total number of object pairs.

Let $X$ be a finite set $X = \{x_1, \ldots, x_n\}$; $\mathcal{P} = \{P_1, \ldots, P_k\}$ be a partition of $X$ in $k$ sets; and $\mathcal{Q} = \{Q_1, \ldots, Q_h\}$ a partition of the same set $X$ in $h$ sets. Given two objects $x$ and $x'$ we say that both objects are *paired* in a partition when

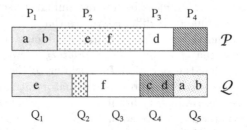

**Fig. 2.** Examples of paired and impaired objects: 1) $a$ and b are paired in both partitions; 2) $e$ and f are paired in $\mathcal{P}$ and impaired in $\mathcal{Q}$; 3) $c$ and d are paired in $\mathcal{Q}$ and impaired in $\mathcal{P}$; 4) $d$ and f are impaired in both partitions.

both objects belong to the same class of the partition (see Fig. 2). Otherwise, we say that both objects are *impaired*.

Now, let us consider the set $C := \{(x_i, x_j) \in X \times X : 1 \leq i < j \leq n\}$ which can be identified with the set of unordered pairs $\{x, y\}$, with $x, y \in X$. The Rand index among the partitions $\mathcal{P}$ and $\mathcal{Q}$ is defined as follows:

$$R(\mathcal{P}, \mathcal{Q}) = \frac{a + d}{a + b + c + d} \tag{2}$$

where,

- $a$ is the number of pairs $(x, x') \in C$ such that $x$ and $x'$ are paired in both partitions.
- $b$ is the number of pairs of objects $(x, x') \in C$ such that $x$ and $x'$ are paired in $\mathcal{P}$ and impaired in $\mathcal{Q}$.
- $c$ is the number of pairs of objects $(x, x') \in C$ such that $x$ and $x'$ are impaired in $\mathcal{P}$ and paired in $\mathcal{Q}$.
- $d$ is the number of pairs of objects $(x, x') \in C$ such that $x$ and $x'$ are impaired in both partitions.

The Rand index is commonly used to compare clusterings formed by automatic systems. It gives a measure of how similar are two clusterings. Inside LID, the Rand index is used to compare the partitions induced by each one of the attributes describing the objects with the correct partition. However, when continuous values are represented by means of fuzzy sets the partitions induced are fuzzy and in this situation the Rand index is not appropriate to make comparisons involving fuzzy partitions. In the next sections we will explain two proposals of fuzzification of the Rand index.

## 2.3   The Campello's Fuzzy Rand Index

In [4] Campello extends the Rand index to make it feasible to compare fuzzy partitions. Given a finite data set $X = \{x_1, \ldots, x_n\}$, a *fuzzy partition* on $X$ (in the sense of Ruspini [7]) is any finite collection $\mathcal{P} = \{P_1, \ldots, P_k\}$ of fuzzy subsets on $X$ such that $\sum_{i=1}^{k} P_i(x_j) = 1$, for each $j$, $1 \leq j \leq n$. To the end to define a fuzzy extension, Campello first rewrites the original formulation of the Rand index in an equivalent form by using basic concepts of set theory. Given the crisp partitions $\mathcal{P}$, with $k$ sets, and $\mathcal{Q}$, with $h$ sets, Campello defines the following sets of pairs:

$V$: pairs $(x, x') \in C$ paired in $\mathcal{P}$,    $W$: pairs $(x, x') \in C$ impaired in $\mathcal{P}$,
$Y$: pairs $(x, x') \in C$ paired in $\mathcal{Q}$,    $Z$: pairs $(x, x') \in C$ impaired in $\mathcal{Q}$,

where $C$ is the set of pairs of elements of $X$ defined in Sec. 2.2.

According to the sets above, the coefficients $a, b, c$ and $d$ of the Rand index in Eq. (2) can be rewritten in the following way: $a = |V \cap Y|$, $b = |V \cap Z|$, $c = |W \cap Y|$, and $d = |W \cap Z|$.

When we consider fuzzy partitions, the sets above are fuzzy sets. Let $P_i(x) \in [0, 1]$ denote the membership degree of the object $x \in X$ to the set $P_i$. Then, Campello defines the fuzzy binary relations $V, W, Y$ and $Z$ on the set $C$ by using the following expressions involving a $t$-norm $\otimes$ and a $t$-conorm $\oplus$:

$$V(x, x') = \bigoplus_{i=1}^{k}(P_i(x) \otimes P_i(x')), \; W(x, x') = \bigoplus_{1 \leq i \neq j \leq k}(P_i(x) \otimes P_j(x')),$$
$$Y(x, x') = \bigoplus_{i=1}^{h}(Q_i(x) \otimes Q_i(x')), \; Z(x, x') = \bigoplus_{1 \leq i \neq j \leq h}(Q_i(x) \otimes Q_j(x')).$$

Now, as it is usually done, Campello calculates the intersection of two fuzzy relations by using the $t$-norm (applied to the membership degrees of each pair to each relation). Then, using the sigma-count principle for defining the fuzzy set cardinality, he obtains the coefficients $a$, $b$, $c$, and $d$ in the following way:

$$a = \left|V \bigcap Y\right| = \sum_{(x,x') \in C}(V(x, x') \otimes Y(x, x'))$$
$$b = \left|V \bigcap Z\right| = \sum_{(x,x') \in C}(V(x, x') \otimes Z(x, x'))$$
$$c = \left|W \bigcap Y\right| = \sum_{(x,x') \in C}(W(x, x') \otimes Y(x, x'))$$
$$d = \left|W \bigcap Z\right| = \sum_{(x,x') \in C}(W(x, x') \otimes Z(x, x'))$$

Then, the fuzzy version of the Rand index is also defined by Eq. (2), giving a measure of the similarity between two partitions. Since LID uses a normalized distance measure, we have to take $1 - R(\mathcal{P}, \mathcal{Q})$. Nevertheless, as Campello himself warns in [4], his fuzzy formulation of the Rand index does not satisfies some basic metric properties and it is properly defined only for the comparison of a fuzzy partition with a non-fuzzy reference partition (see also [5] for a discussion on this subject). However, notice that the correct partition in classification problems is commonly crisp; thus the use of the distance associated to the Rand index of Campello inside LID is justified. From now on, we denote as CI the distance associated with the Campello Rand index.

## 2.4   The Hüllermeier-Rifqi's Fuzzy Rand Index

Hüllermeier and Rifqi proposed in [5] a different fuzzy version for the Rand index which allows the comparison of two fuzzy partitions and satisfies all the desirable metric properties. In the next we recall the definition of this fuzzy version.

Given a fuzzy partition $\mathcal{P} = \{P_1, P_2, \ldots, P_k\}$, each object $x$ is characterized by its membership vector $\mathcal{P}(x) = (P_1(x), P_2(x), \ldots, P_k(x)) \in [0,1]^k$ where $P_i(x)$ is the membership degree of $x$ to the cluster $P_i$. Given two objects $x$ and $x'$ and two fuzzy partitions $\mathcal{P}$ and $\mathcal{Q}$, the *degree of concordance* of both objects in these partitions is defined by means the expression $1 - |E_{\mathcal{P}}(x,x') - E_{\mathcal{Q}}(x,x')|$ where $E_{\mathcal{P}}$ is the fuzzy equivalence relation defined by $E_{\mathcal{P}}(x,x') := 1 - \|\mathcal{P}(x) - \mathcal{P}(x')\|$ being $\|.\|$ a distance on $[0,1]^k$. Thus, two objects are equivalent to a degree 1 when both have the same membership degrees in all the sets of the partition. This fuzzy equivalence is used to define the notion of *concordance* as a fuzzy binary relation, which generalizes the crisp binary relation (induced by a crisp partition) defined on the set $C$ of unordered pairs of objects of $X$ using the notions of *paired* and *unpaired*. Then, a distance measure on fuzzy partitions using the *degree of discordance* is defined as $|E_{\mathcal{P}}(x,x') - E_{\mathcal{Q}}(x,x')|$. Thus, given a data set $X$ of $n$ elements, and two fuzzy partitions $\mathcal{P}$ and $\mathcal{Q}$ on $X$, the distance between both partitions is the normalized sum of degrees of discordance:

$$d(\mathcal{P}, \mathcal{Q}) = \frac{\sum_{(x,x') \in C} |E_{\mathcal{P}}(x,x') - E_{\mathcal{Q}}(x,x')|}{n(n-1)/2}. \tag{3}$$

As it is shown in [5,11], the function (3) is a *pseudometric*, that is, it satisfies reflexivity, simmetry and triangular inequality, but it is not a *metric* because in general it does not satisfies the property of *separation* ($d(\mathcal{P}, \mathcal{Q}) = 0$ implies $\mathcal{P} = \mathcal{Q}$). A fuzzy partition in the sense of Ruspini $\mathcal{P} = \{P_1, \ldots, P_k\}$ is called *normal* if it has a prototypical element, i.e., for every $P_i \in \mathcal{P}$, there exists an $x \in X$ such that $P_i(x) = 1$. Hüllermeier and Rifqi show that for normal partitions, taking the equivalence relation on $X$ defined by

$$E_{\mathcal{P}}(x,x') = 1 - \frac{1}{2} \sum_{i=1}^{k} |P_i(x) - P_i(x')|, \tag{4}$$

the distance defined by Eq. (3) is a metric. From now on, we will call HRI (for Hüllermeier-Rifqui Index) this distance measure.

In addition to the extensions of the Rand Index presented by Campello and Hülermeier and Rifqi, other extensions of the Rand index have been proposed in the literature [8,9,10]. In the article [11] a comparative study of the indices presented in the mentioned papers in relation to the indexes proposed by Campello and Hülermeier and Rifqi is performed.

## 3     The $\varepsilon$-Procedure

Our goal is to design a procedure allowing the discretization of continuous values. The idea is, on the one hand, to keep the advantages provided by the fuzzy set representation and, on the other hand, to generate crisp partitions in order to use standard crisp measures to compare them. There are related works as the one from Ishibuchi and Yamamoto [12], that propose a method to construct fuzzy

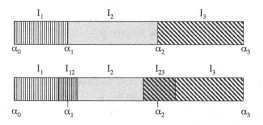

**Fig. 3.** The upper part shows the effect of a crisp discretization using three intervals. The lower part shows the discretization proposed by the $\varepsilon$-procedure.

discretizations from crisp ones. The authors consider that sometimes experts cannot give the discretization thresholds and they propose to dynamically obtain them. In [2] Kuajima et al. analyze the effects of fuzzy discretization on rule-based classifiers performance. The authors define the *fuzzification grade* as the overlap between adjacent fuzzy sets and they conduct experiments taking several of these grades.

The $\varepsilon$-procedure we introduce in the present paper discretizes continuous values by considering intermediate intervals that could be interpreted as the overlapping of two contiguous fuzzy sets. The partition generated in this way is crisp, therefore the standard Rand index can be used. Let $f$ be an attribute taking continuous values, $\alpha_1, \alpha_2, \ldots, \alpha_n$ be the discretization values for $f$, and $\alpha_0$ and $\alpha_{n+1}$ be the minimum and maximum respectively of the values of $f$. The $\varepsilon$-procedure considers the following intervals: $[\alpha_0, \alpha_1 - \delta_1], (\alpha_1 - \delta_1, \alpha_1 + \delta_1], (\alpha_1 + \delta_1, \alpha_2 - \delta_2], (\alpha_2 - \delta_2, \alpha_2 + \delta_2] \ldots (\alpha_n + \delta_n, \alpha_{n+1}]$, where $\delta_i = p \cdot |\alpha_i - \alpha_{i-1}|$, being $p$ an adjustable percentage. In order to avoid some undesired overlappings, the parameter $p$ must respect the following constraint:

$$p \leq \frac{1}{2} \cdot \frac{|\alpha_{i+1} - \alpha_i|}{|\alpha_i - \alpha_{i-1}|}. \tag{5}$$

Notice that whereas with the usual discretization the values $v = \alpha_i - \varepsilon$ and $v' = \alpha_i + \varepsilon$, being $\varepsilon$ sufficiently small, belong to different intervals, using the $\varepsilon$-procedure both $v$ and $v'$ belong to the same partition when $\varepsilon < \delta$. Figure 3 shows an example where a crisp discretization produces three intervals and for the same range of values, the $\varepsilon$-procedure introduces the intervals $I_{12}$ and $I_{23}$ that contain values around the thresholds $\alpha_1$ and $\alpha_2$. These new intervals join values that in the first discretization belong to different intervals.

## 4   Experiments

We have performed experiments with the goal of proving the feasibility of the $\varepsilon$-procedure as an alternative to fuzzy discretization. In previous works [3] we have shown that the fuzzifications of the Rand index have good predictivity but

**Table 1.** The left column shows the discretization method used in each one of the four situations. The central column shows the index used as $\Delta$-measure. The right column corresponds to the label assigned to each situation.

| DISCRETIZATION METHOD | $\Delta$-MEASURE | LABEL |
|---|---|---|
| Crisp intervals given by an standard method | Rand Index | CRI |
| Fuzzy intervals built over the crisp ones | CI | FCI |
| Fuzzy intervals built over the crisp ones | HRI | FHRI |
| Crisp intervals built with the $\varepsilon$-procedure | Rand Index | $\varepsilon$RI |

they also have a high computational cost mainly due to the fact that they need to operate with all the membership degrees of all pairs of objects. Instead, the $\varepsilon$-procedure uses the crisp Rand index.

In the experiments we used four data sets, *iris*, *heart-statlog*, *glass*, and *thyroids* coming from the UCI Repository [13], where objects are described by attributes having continuous values. The discretization thresholds have been obtained with the MDL discretization method proposed by Fayyad and Irani's [14] and we have used the implementation of it given by Weka [15,16]. These thresholds have been taken as basis to define the fuzzy sets used by CI and HRI, and also for the $\varepsilon$-procedure to induce the new discretization intervals. The target of the experiments is to compare the predictivity of LID in the following four situations, which are summarized in Table 1:

1. The attributes with continuous values are discretized by using the thresholds given by a standard discretization method. Then LID runs using the Rand index (RI) as measure $\Delta$ to compare the partitions induced by the attibutes with the correct partition. So, in this case we have a crisp discretization and $\Delta = $ RI. We denote this procedure by CRI (the "C" stands by *Crisp*).

2. The attributes with continuous values are discretized by using the fuzzy sets built from the thresholds given by a standard discretization method. Then LID runs using CI as measure $\Delta$ with the Minimum and the Maximum as $t$-norm and $t$-conorm, respectively. We denote this procedure with the label FCI (the "F" stands by *Fuzzy*).

3. The attributes with continuous values are discretized by using the fuzzy sets built as in the previous procedure. Then LID runs using the Hüllermeier-Rifqi index (HRI) as measure $\Delta$. So, in this case we have a fuzzy discretization and $\Delta = $ HRI. We denote this procedure by FHRI.

4. The attributes with continuous values are discretized by using the thresholds given by a standard discretization method and refined by the $\varepsilon$-procedure. Then LID runs using the Rand index (RI) as measure $\Delta$. So, in this case we have the crisp discretization given by the $\varepsilon$-procedure, and $\Delta = $ CI. We denote this procedure with the label $\varepsilon$RI.

As we have mentioned before, in the formulas (1) each $\delta_i$ is computed by $\delta_i = p \cdot |\alpha_i - \alpha_{i-1}|$, being the factor $p$ a percentage that we can adjust and that must

**Table 2.** Mean accuracy of LID corresponding to the procedures labeled by CRI, FCI, FHRI, and $\varepsilon$RI after seven trials of 10-fold cross-validation. The table shows the best results; the corresponding $p$ is indicated between parenthesis.

| DATASET | CRI | FCI | FHR | $\varepsilon$RI |
|---|---|---|---|---|
| glass | 35.460 | 25.696 (0.10) | 38.644 (0.10) | **49.373** (0.10) |
| heart-statlog | 65.397 | 62.857 (0.05) | 65.026 (0.05) | **65.555** (0.10) |
| iris | 88.780 | 91.917 (0.10) | 94.482 (0.10) | **96.286** (0.10) |
| thyroids | **86.562** | 81.447 (0.05) | 82.331 (0.10) | 84.935 (0.15) |

satisfy the constraint (5). We have experimented with $p = 0.05, 0.10$ and $0.15$. All these values of $p$ satisfy (5) for all the considered data sets. In the fuzzy version of LID, the correct partition is the same than in the crisp case since each object belongs to a unique solution class. However, when the partitions induced by each attribute are fuzzy, an object can belong (to a certain degree) to more than one partition set. The algorithm of the fuzzy LID is the same explained in Sec. 2.1 but using a particular representation for the fuzzy cases.

Table 2 shows the mean accuracy of LID corresponding to the procedures labeled by CRI, FCI, FHRI, and $\varepsilon$RI after seven trials of 10-fold cross-validation. The accuracy depends on the value of $p$ and this may be different for each domain. In the table we show the best results and we indicate the corresponding $p$ between parenthesis. Thus, for instance, the best result on *iris* for FCI is taking $p = 0.10$ whereas the best accuracy for FCI on *heart-statlog* is taking $p = 0.05$. In our experiments we have seen that for each data set there is a value of $p$ that represents an inflection point on the accuracy. For instance, on the *iris* data set using $\varepsilon$RI, the accuracy taking $p = 0.05$ is 93.720 and taking $p = 0.15$ is 95.333. The best value for $p$ is different for each data set and also for each method. So, in practice, we should try with several values of $p$ in order to find the best one.

Experiments show that the $\varepsilon$-procedure gives good predictive results outperforming all the fuzzy versions (FCI and FHRI) in all domains. Moreover, $\varepsilon$RI also outperforms the procedure CRI except for the *thyroids* data set and for the *heart-statlog* where the accuracy of both is not significantly different.

LID can produce two kinds of outputs: the classification in one (correct or incorrect) class or a multiple classification. Multiple classification means that LID has not been capable to classify the input object in only one class, i.e., most of the time it could be considered as a *no answer* of the system. This explains the low accuracy percentage given by all measures on the *glass* domain since most of times LID produces multiple classifications.

The computational complexity of the crisp Rand index is $\mathcal{O}((k + h) \cdot n^2)$, where $k$ and $h$ are the cardinalities of the partitions $\mathcal{P}$ and $\mathcal{Q}$ respectively; the cost of CI is $\mathcal{O}((max(k, h))^2 \cdot n^2)$; and the cost of the HRI is $\mathcal{O}(max(k, h) \cdot n^2)$. Notice that, in the worst case, both the Rand index and HRI have the same cost whereas the cost of CI is higher than them. In practice, the Rand index has lower

**Table 3.** Mean runtime necessary (in seconds) to evaluate a complete trial of 10-fold cross-validation on the datasets *iris* and *thyroids*

| DATASET | CRI | FCI | FHRI | $\varepsilon$RI |
|---------|-----|-----|------|-----|
| iris | 2.624 | 226.773 | 77.463 | 14.358 |
| thyroids | 27.219 | 1092.055 | 383.037 | 165.348 |

cost than HRI mainly due to the lower complexity of the input data since fuzzy representations have to take into account membership degrees. Table 3 shows the mean runtime necessary to evaluate a complete trial of 10-fold cross-validation on the data sets *iris* and *thyroids*. These times have been obtained using a Mac with a processor Intel Core 2 Duo of 2.93 GHz.

## 5    Conclusions and Future Work

In this paper we have introduced a method, called $\varepsilon$-procedure, that constructs classical partitions on the range of an attribute taking continuous values. These partitions can be seen as refinements of the ones given by the expert or the ones given by a standard method of discretization. Moreover, the method can be seen as "similar" to the fuzzy methods of discretization since the $\varepsilon$-procedure takes into account the neighborhood of the thresholds given by the crisp discretization methods. The $\varepsilon$-procedure runs inside the LID method allowing it to deal with cases having attributes that take continuous values. We have carried on experiments with LID comparing its performance when dealing with both cases whose continuous attributes have been discretized and cases whose continuous attributes have been represented as fuzzy sets.

As future work we plan to conduct experiments to analyze in depth the effect of the value of $\delta$ in the accuracy of the $\varepsilon$-procedure. We think that this effect could also depend on the particular characteristics of the domain at hand. A different research line is to experiment with different $\Delta$ measures after the discretization produced by the $\varepsilon$-procedure. In particular we plan to experiment with the López de Mántaras (LM) distance [17] and to compare the results with those produced by a fuzzification of LM proposed in [18].

**Acknowledgments.** This research is partially funded by the Spanish MICINN projects Next-CBR (TIN 2009-13692-C03-01), MTM2011-25747, ARINF (TIN2009-14704-C03-03), TASSAT TIN2010-20967-C04-01, and CONSOLIDER (CSD2007-0022), and the grants 2009-SGR-1433 and 2009-SGR-1434 from the Generalitat de Catalunya. The authors also want to thank Félix Bou and Pilar Dellunde for their helpful comments and suggestions.

# References

1. Yang, Y., Webb, G.I., Wu, X.: Discretization Methods. In: The Data Mining and Knowledge Discovery Handbook, ch. 6, pp. 113–130. Springer (2005)
2. Kuwajima, I., Nojima, Y., Ishibuchi, H.: Effects of constructing fuzzy discretization from crisp discretization for rule-based classifiers. Artificial Life and Robotics 13(1), 294–297 (2008)
3. Armengol, E., García-Cerdaña, À.: Lazy Induction of Descriptions Using Two Fuzzy Versions of the Rand Index. In: Hüllermeier, E., Kruse, R., Hoffmann, F. (eds.) IPMU 2010, Part I. CCIS, vol. 80, pp. 396–405. Springer, Heidelberg (2010)
4. Campello, R.J.G.B.: A fuzzy extension of the Rand index and other related indexes for clustering and classification assessment. Pattern Recognition Letters 28(7), 833–841 (2007)
5. Hüllermeier, E., Rifqi, M.: A fuzzy variant of the Rand index for comparing clustering structures. In: Proceedings of the 2009 IFSA/EUSFLAT Conference, pp. 1294–1298 (2009)
6. Rand, W.M.: Objective criteria for the evaluation of clustering methods. Journal of the American Statistical Association 66(336), 846–850 (1971)
7. Ruspini, E.H.: A new approach to clustering. Information and Control 15(1), 22–32 (1969)
8. Frigui, H., Hwang, C., Rhee, F.C.H.: Clustering and aggregation of relational data with applications to image database categorization. Pattern Recognition 40, 3053–3068 (2007)
9. Brower, R.: Extending the Rand, adjusted Rand and Jaccard indices to fuzzy partitions. Journal of Intelligent Informtion Systems 32, 213–235 (2009)
10. Anderson, D.T., Bezdek, J.C., Popescu, M., Keller, J.M.: Comparing fuzzy, probabilistic, and possibilistic partitions. IEEE Transactions on Fuzzy Systems 18(5), 906–918 (2010)
11. Hüllermeier, E., Rifqi, M., Henzgen, S., Senge, R.: Comparing Fuzzy Partitions: A Generalization of the Rand Index and Related Measures. IEEE Transactions on Fuzzy Systems 20(3), 546–556 (2012)
12. Ishibuchi, H., Yamamoto, T.: Deriving fuzzy discretization from interval discretization. In: Proceedings of FUZZ-IEEE 2003, vol. 1, pp. 749–754 (2003)
13. Asuncion, A., Newman, D.J.: UCI machine learning repository (2007)
14. Fayyad, U.M., Irani, K.B.: Multi-interval discretization of continuous-valued attributes for classification learning. In: Proceedings of IJCAI 1993, pp. 1022–1029 (1993)
15. Witten, I., Frank, E., Trigg, L., Hall, M., Holmes, G., Cunningham, S.: Weka: Practical machine learning tools and techniques with java implementations (1999)
16. Hall, M., Frank, E., Holmes, G., Pfahringer, B., Reutemann, P., Witten, I.H.: The WEKA data mining software: An update. SIGKDD Explorations 11(1), 10–18 (2009)
17. López de Mántaras, R.: A distance-based attribute selection measure for decision tree induction. Machine Learning 6, 81–92 (1991)
18. Armengol, E., Dellunde, P., García-Cerdaña, À.: Towards a Fuzzy Extension of the López de Mántaras Distance. In: Greco, S., Bouchon-Meunier, B., Coletti, G., Fedrizzi, M., Matarazzo, B., Yager, R.R. (eds.) IPMU 2012, Part I. CCIS, vol. 297, pp. 81–90. Springer, Heidelberg (2012)

# Linear Programming with Graded Ill-Known Sets

Shizuya Kawamura and Masahiro Inuiguchi

Graduate School of Engineering Science, Osaka University
1-3 Machikaneyama, Toyonaka, Osaka 560-8531, Japan
{s.kawamura,inuiguti}@inulab.sys.es.osaka-u.ac.jp
http://www-inulab.sys.es.osaka-u.ac.jp/

**Abstract.** In this paper, we investigate linear programming problems with graded ill-known sets (GIS-LP problems). Because a graded ill-known set (GIS) is defined by a possibility distribution on the power set, treatments of GISs are usually complex. To treat them in a simpler way at the expense of precision, the representation by upper and lower approximations have been investigated. Once these approximations are applied, the original GIS is not usually restored. We propose a class of GISs restorable from the approximations. Utilizing the previous results in GISs, we formulate a GIS-LP problem based on the idea of symmetric model in possibilistic programming. We show that the formulated GIS-LP problem is solved by a bisection method together with the simplex method. A simple numerical example is given.

**Keywords:** Graded ill-known set, linear programming, upper approximation, lower approximation.

## 1 Introduction

In the conventional mathematical programming problems, the coefficients of the objective function and constraints are assumed to be real numbers. Moreover, the constraints are assumed to be hard so that any small violation is not allowed. However, in the real world problems, we may come across cases where the knowledge about coefficients is not very clear to specify them as real numbers and cases where constraints are flexible so that small violation is allowed. Fuzzy programming approaches [1] have been proposed to treat ambiguous coefficients as well as soft constraints as fuzzy sets.

Some researchers [2,3] further argue that the exact specifications of membership functions of fuzzy sets in fuzzy programming problems are difficult. They proposed to apply type-2 fuzzy sets [4] and intuitionistic fuzzy sets [5]. In those higher-order fuzzy sets, membership grades can take interval or fuzzy values in [0, 1]. However, the interpretations of interval and fuzzy membership grades are not convincing very much or rather restrictive. In most of these papers, higher-order fuzzy sets are applied only to express soft constraints. Apart from fuzzy programming, the applications of type-2 fuzzy sets and intuitionistic fuzzy

V. Torra et al. (Eds.): MDAI 2012, LNAI 7647, pp. 270–281, 2012.

sets, have become popular. Considering this trend, we consider application of higher-order fuzzy sets to fuzzy programming.

As a model of higher-order fuzzy set, Inuiguchi [6] has proposed to use the graded ill-known set (GIS) [7,8]. The GIS model was originally proposed by Dubois and Prade [7] to express a set whose members are not known exactly. Then originally it was the model to express a set-valued variable. However, Inuiguchi [6] has proposed to use the GIS model to express the variation range of a single-valued variable. In that way, he used the graded ill-known set model for a single-valued variable. He investigated the calculations of GISs.

The GIS has a good interpretation. It can be defined by aggregation of several pieces of information. From these advantages, we apply GISs to mathematical programming under uncertainty. More concretely, we study linear programming with GISs (GIS-LP). Because a graded ill-known set (GIS) is defined by a possibility distribution on the power set, treatments of GISs are usually complex. To treat them in a simpler way at the expense of precision, the representation by upper and lower approximations have been investigated. We propose a class of GISs restorable from the approximations and apply them to linear programming problems. The GIS-LP problems are formulated by using the idea of symmetric models developed in fuzzy programming. It is shown that the formulated GIS-LP problem is solved by a bisection method together with the simplex method. Simple numerical example is given to demonstrate the usage of GIS-LP approach.

In the next section, the results in graded ill-known sets are briefly reviewed. In section 3, we propose a class of GIS which is recoverable and investigate inclusion degrees between two GISs in the class. In section 4, using inclusion degrees, we formulate a linear programming problem with GISs. In section 5, a simple example is given.

## 2   Graded Ill-Known Sets

Let $X$ be a universe. Let $\boldsymbol{A}$ be a crisp set which is ill-known, i.e., there exists at least one element $x \in X$, for which it is not known whether $x$ belongs to $\boldsymbol{A}$ or not. To represent such an ill-known set, collecting possible realizations $A_i \subseteq X$ of $\boldsymbol{A}$, we obtain the following family:

$$\mathcal{A} = \{A_1, A_2, \ldots, A_n\}. \tag{1}$$

Given $\mathcal{A}$, we obtain a set $A^-$ of elements which is a certain member of $\boldsymbol{A}$ and a set $A^+$ of elements which is a possible member of $\boldsymbol{A}$ can be defined as

$$A^- = \bigcap \mathcal{A} = \bigcap_{i=1,\ldots,n} A_i, \quad A^+ = \bigcup \mathcal{A} = \bigcup_{i=1,\ldots,n} A_i. \tag{2}$$

We call $A^-$ and $A^+$ "the lower approximation" of $\boldsymbol{A}$ and "the upper approximation" of $\boldsymbol{A}$, respectively.

In the real world, we may know certain members (lower approximation $A^-$) and certain non-members (complement of upper approximation $A^+$) only. Given $A^-$ and $A^+$ (or equivalently, the complement of $A^+$), we obtain a family $\hat{\mathcal{A}}$ of possible realizations of $\boldsymbol{A}$ by $\hat{\mathcal{A}} = \{A_i \mid A^- \subseteq A_i \subseteq A^+\}$. We note that $A^-$ and

$A^+$ are restored by applying (2) to the family $\hat{\mathcal{A}}$. On the other hand, $\mathcal{A}$ cannot be always restored from $\hat{\mathcal{A}}$ with $A^-$ and $A^+$ defined by (2).

If all $A_i$'s of (1) are not equally possible, we may assign a possibility degree $\pi_{\mathcal{A}}(A_i)$ to each $A_i$ so that $\max_{i=1,\ldots,n} \pi_{\mathcal{A}}(A_i) = 1$. A possibility distribution $\pi_{\mathcal{A}} : 2^X \to [0,1]$ can be seen as a membership function of a fuzzy set $\mathcal{A}$ in $2^X$. Thus, we may identify $\mathbf{A}$ with $\mathcal{A}$. The ill-known set having such a possibility distribution is called "a graded ill-known set (GIS)".

In this case, the lower approximation $A^-$ and the upper approximation $A^+$ are defined by fuzzy sets with the following membership functions:

$$\mu_{A^-}(x) = \inf_{x \notin A_i} n(\pi_{\mathcal{A}}(A_i)), \quad \mu_{A^+}(x) = \sup_{x \in A_i} \varphi(\pi_{\mathcal{A}}(A_i)), \qquad (3)$$

where $n : [0,1] \to [0,1]$ and $\varphi[0,1] \to [0,1]$ are strictly decreasing and strictly increasing functions such that $n(0) = \varphi(1) = 1$ and $n(1) = \varphi(0) = 0$. Function $n$ can be seen as a negation but we do not assume $n(n(a)) = a$. We have

$$\forall x \in X, \ \mu_{A^-}(x) > 0 \text{ implies } \mu_{A^+}(x) = 1. \qquad (4)$$

The specification of possibility distribution $\pi_{\mathcal{A}}$ may need a lot of information. However, as is often the case in real world applications, we know only the lower approximation $A^-$ and the upper approximation $A^+$ as fuzzy sets satisfying (4). There are many possibility distributions $\pi_{\mathcal{A}}$ having given $A^-$ and $A^+$ as their lower and upper approximations. We adopt the following maximal possibility distribution $\pi_{\mathcal{A}}^*(A_i)$ with respect to given $A^-$ and $A^+$:

$$\pi_{\mathcal{A}}^*(A_i) = \min \left( \inf_{x \notin A_i} n^{-1}(\mu_{A^-}(x)), \inf_{x \in A_i} \varphi^{-1}(\mu_{A^+}(x)) \right), \qquad (5)$$

where we define $\inf \emptyset = 1$ and $n^{-1}$ and $\varphi^{-1}$ are inverse functions of $n$ and $\varphi$. We call the imprecise specification of a GIS by upper and lower approximations satisfying (5) "roughly specified GIS".

Possibility degree $\Pi(\mathcal{B}|\mathcal{A})$ and necessity degree $N(\mathcal{B}|\mathcal{A})$ are defined by

$$\Pi(\mathcal{B}|\mathcal{A}) = \sup_{C \subseteq X} \min(\pi_{\mathcal{A}}(C), \pi_{\mathcal{B}}(C)), \quad N(\mathcal{B}|\mathcal{A}) = \inf_{C \subseteq X} I(\pi_{\mathcal{A}}(C), \pi_{\mathcal{B}}(C)), \quad (6)$$

where $I : [0,1] \times [0,1] \to [0,1]$ is an implication function which satisfies (I1) $I(0,0) = I(0,1) = I(1,1) = 1$ and $I(1,0) = 0$, (I2) $I(a,b) \leq I(c,d)$, for any $a,b,c,d$ such that $0 \leq c \leq a \leq 1$, $0 \leq b \leq d \leq 1$, and (I3) $I$ is upper semi-continuous.

For roughly specified GISs we have the following results (see Inuiguchi [8]):

$$\Pi(\mathcal{B}|\mathcal{A}) = \min \left( \inf_{x \in X} \max \left( n^{-1}(\mu_{B^-}(x)), \varphi^{-1}(\mu_{A^+}(x)) \right), \right.$$

$$\left. \inf_{x \in X} \max \left( n^{-1}(\mu_{A^-}(x)), \varphi^{-1}(\mu_{B^+}(x)) \right) \right), \qquad (7)$$

$$N(\mathcal{B}|\mathcal{A}) = \min \left( \inf_{x \in X} I \left( \varphi^{-1}(\mu_{A^+}(x)), \varphi^{-1}(\mu_{B^+}(x)) \right), \right.$$

$$\left. \inf_{x \in X} I \left( n^{-1}(\mu_{A^-}(x)), n^{-1}(\mu_{B^-}(x)) \right) \right). \qquad (8)$$

Inuiguchi [6] has investigated the calculations of GISs in real line $\mathbf{R}$ called "GIS of quantities". The set of GISs of quantities is denoted by $\mathcal{IQ}$. We define a set $\mathcal{IQ}_{\mathrm{ci}} \subseteq \mathcal{IQ}$ of GISs of quantities $\mathcal{A}$ satisfying the following properties:

$$\forall \alpha \in (0,1], \ \hat{A}(\alpha) = \bigcap \{Q \subseteq \mathbf{R} \mid \pi_{\mathcal{A}}(Q) \geq \alpha\} \text{ is nonempty, closed and convex,}$$

and there exist convex sets $Q_j$, $j = 1, 2, \ldots, k$ such that

$$\pi_{\mathcal{A}}(Q_j) \geq \alpha, j = 1, 2, \ldots, k \text{ and } \hat{A}(\alpha) = \bigcap_{j=1,2,\ldots,k} Q_j. \tag{9}$$

We consider the following calculations of GISs (see [6]).

**Definition 1.** *Let $\mathcal{A}_i$, $i = 1, 2, \ldots, m$ be graded ill-known sets of real numbers. Given a function $f : \mathbf{R}^m \to \mathbf{R}$, the image $f(\mathcal{A}_1, \ldots, \mathcal{A}_m)$ is defined by a graded ill-known set of real numbers characterized by*

$$\pi_{f(\mathcal{A}_1,\ldots,\mathcal{A}_m)}(Y)$$
$$= \begin{cases} \sup\limits_{\substack{Q_1,\ldots,Q_m \subseteq \mathbf{R} \\ Y = f(Q_1,\ldots,Q_m)}} \min\left(\pi_{\mathcal{A}_1}(Q_1), \cdots, \pi_{\mathcal{A}_m}(Q_m)\right), \text{ if } f^{-1}(Y) \neq \emptyset, \\ 0, \hspace{4.5cm} \text{ if } f^{-1}(Y) = \emptyset, \end{cases} \tag{10}$$

*where we define $f(Q_1, \ldots, Q_m) = \{f(x_1, \ldots, x_m) \mid x_i \in Q_i, \ i = 1, \ldots, m\}$.*

Definition 1 is obtained from application of the extension principle in fuzzy sets. We have the following theorem about upper approximation (see [6]).

**Theorem 1.** *The upper approximation $f^+(\mathcal{A}_1, \ldots, \mathcal{A}_m)$ of $f(\mathcal{A}_1, \ldots, \mathcal{A}_m)$ can be calculated by upper approximations of $\mathcal{A}_i$, $i = 1, 2, \ldots, m$. More concretely, we obtain*

$$\mu_{f^+(\mathcal{A}_1,\ldots,\mathcal{A}_m)}(y) = \sup_{y \in Y} \varphi(\pi_{f(\mathcal{A}_1,\ldots,\mathcal{A}_m)}(Y))$$
$$= \sup_{\substack{x_1,\ldots,x_m \in \mathbf{R} \\ y = f(x_1,\ldots,x_m)}} t(\mu_{A_1^+}(x_1), \ldots, \mu_{A_m^+}(x_m))) = \mu_{f(A_1^+,\ldots,A_m^+)}(y), \tag{11}$$

*where $\mu_{f^+(\mathcal{A}_1,\ldots,\mathcal{A}_m)}$ is the membership function of $f^+(\mathcal{A}_1, \ldots, \mathcal{A}_m)$ and $\mu_{A_i^+}$ is the membership function of the upper approximation $A_i^+$ of $\mathcal{A}_i$. Similarly, $\mu_{f(A_1^+,\ldots,A_m^+)}$ is the membership function of the image $f(A_1^+, \ldots, A_m^+)$.*

Moreover, we have the following theorem (see Inuiguchi [6]).

**Theorem 2.** *Let $f : \mathbf{R}^m \to \mathbf{R}$ be continuous and monotone Let $\mathcal{A}_i \in \mathcal{IQ}_{\mathrm{ci}}$, $i = 1, 2, \ldots, m$. If $\{\pi_{\mathcal{A}_i}(A) \mid A \subseteq X\}$ is finite for $i = 1, 2, \ldots, m$ then we have*

$$\mu_{f^-(\mathcal{A}_1,\ldots,\mathcal{A}_m)}(y) = \inf_{y \notin Y} n(\pi_{f(\mathcal{A}_1,\ldots,\mathcal{A}_m)}(Y))$$
$$= \sup_{\substack{x_1,\ldots,x_m \in \mathbf{R} \\ y = f(x_1,\ldots,x_m)}} \min(\mu_{A_1^-}(x_1), \ldots, \mu_{A_m^-}(x_m)) = \mu_{f(A_1^-,\ldots,A_m^-)}(y). \tag{12}$$

## 3    A Class of GISs and Inclusions

### 3.1    A Class of GISs

The treatments of GISs are complex because they are possibility distributions on the power set. Therefore, rough treatments using lower and upper approximations are conceivable. However, once those approximations are applied, the original GISs cannot be restored generally. A GIS which can be restored from the application of approximations is called a "restorable GIS". Then, in this subsection, we propose a class of restorable GISs which are obtained from evaluations of multiple experts.

We assume that $m$ experts evaluate the variation ranges of an uncertain parameter $a$ by sets of surely possible values and sets of surely impossible values. Let $Y_k$ and $N_k$ denote the set of surely possible values and the set of surely impossible values evaluated by the $k$-th expert, respectively. Taking the complement $\bar{N}_k$ of set $N_k$, $\bar{N}_k$ can be understood as the set of somehow possible values. We assume that there is no conflict, i.e., $\forall i, j \in \{1, 2, \ldots, m\}$, $Y_i \cap N_j = \emptyset$. Then we define GIS $\mathcal{A}$ showing the variation range of $a$ by the following possibility distribution on the power set:

$$\pi_{\mathcal{A}}(A) = 1 - \sup_{x} \max \left( \frac{|\{i \mid x \notin A, \ x \in Y_i, \ i = 1, 2, \ldots, m\}|}{m}, \right.$$
$$\left. \frac{|\{i \mid x \in A, \ x \in N_i, \ i = 1, 2, \ldots, m\}|}{m} \right). \quad (13)$$

To interpret (13), we consider a statement $ST(A) = $"$A$ is the variation range of $a$". For $x \notin A$, $x \in Y_k$ implies that the $k$-th expert rejects $ST(A)$ because he/she insists the variation range includes $Y_k$. On the other hand, for $x \in A$, $x \in N_k$ implies that the $k$-th expert rejects $ST(A)$ because he/she insists the variation range never intersects $N_k$. Therefore, $\pi_{\mathcal{A}}(A)$ of (13) shows the minimum ratio of experts who do not reject $ST(A)$ by considering all points in the universe.

The lower and upper approximations $A^-$ and $A^+$ of the GIS $\mathcal{A}$ defined by $\pi_{\mathcal{A}}$ of (13) are obtained by the following membership functions:

$$\mu_{A^-}(x) = n \left( 1 - \frac{|\{i \mid x \in Y_i, \ i = 1, 2, \ldots, m\}|}{m} \right), \quad (14)$$

$$\mu_{A^+}(x) = \varphi \left( 1 - \frac{|\{i \mid x \in N_i, \ i = 1, 2, \ldots, m\}|}{m} \right). \quad (15)$$

When $n(h) = 1 - h$ and $\varphi(h) = h$, $\forall h \in [0, 1]$, $\mu_{A^-}(x)$ simply shows the ratio of experts who insist $x$ is surely possible while $\mu_{A^+}(x)$ shows the ratio of experts who do not think $x$ is totally impossible. Applying (5) to (14) and (15), we obtain $\pi_{\mathcal{A}}$ as the maximal possibility distribution, i.e.,

$$\pi_{\mathcal{A}}^*(A) = \min \left( 1 - \sup_{x \notin A} \frac{|\{i \mid x \in Y_i, \ i = 1, 2, \ldots, m\}|}{m}, \right.$$
$$\left. 1 - \sup_{x \in A} \frac{|\{i \mid x \in N_i, \ i = 1, 2, \ldots, m\}|}{m} \right) = \pi_{\mathcal{A}}(A). \quad (16)$$

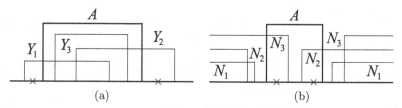

**Fig. 1.** Counter examples

In what follows, because of this good property and the good interpretation, we consider the GISs defined in the same way as GIS $\mathcal{A}$ of (13) with $m$ experts' opinions. We call this class of GISs PwMs-GISs (point-wisely defined GIS with multiple information sources). Note that PwMs-GIS $\mathcal{A}$ satisfies (i) $\mathcal{A} \in \mathcal{IQ}_{ci}$ and (ii) $\{\pi_{\mathcal{A}}(A) \mid A \subseteq X\}$ is finite. Namely, the assumptions on GISs in Theorem 2 are satisfied with PwMs-GISs.

*Remark 1.* Inuiguchi [8] proposed to use the same experts' opinions to define the possibility distribution of GIS $\mathcal{A}$. However, his definition of the possibility distribution is different from (13). He defined the possibility distribution by

$$\pi'_{\mathcal{A}}(A) = \frac{|\{i \mid Y_i \subseteq A,\ A \cap N_i = \emptyset,\ i = 1, 2, \ldots, m\}|}{m}. \tag{17}$$

Here $\pi'_{\mathcal{A}}(A)$ shows the ratio of experts who agree that $A$ is a possible variation range. While the validity of set $A$ is considered in comparisons with $Y_i$ and $N_i$ in the definition of $\pi'_{\mathcal{A}}(A)$ of (17), the validity of $A$ is examined by considerations of all points in the universe in the definition of $\pi_{\mathcal{A}}(A)$ of (13). Namely, $\pi_{\mathcal{A}}(A)$ is the point-wise definition of the ratio of experts supporting $A$ while $\pi'_{\mathcal{A}}(A)$ is the set-based definition of the ratio of experts supporting $A$.

From the point of view that we are considering the variation range of single-valued variable, and thus the realization of the variable is a point, $\pi_{\mathcal{A}}$ can be more suitable than $\pi'_{\mathcal{A}}$. Moreover, taking the lower and upper approximations of $\pi'_{\mathcal{A}}(A)$, we obtain fuzzy sets defined by (14) and (15). Namely, $\pi'_{\mathcal{A}}(A)$ and $\pi_{\mathcal{A}}(A)$ have same lower and upper approximations.

*Remark 2.* We note the following inequalities for any $A \subseteq X$:

$$\frac{|\{i \mid Y_i \subseteq A,\ i = 1, 2, \ldots, m\}|}{m} \leq 1 - \sup_{x \notin A} \frac{|\{i \mid x \in Y_i,\ i = 1, 2, \ldots, m\}|}{m}, \tag{18}$$

$$\frac{|\{i \mid A \cap N_i = \emptyset,\ i = 1, 2, \ldots, m\}|}{m} \leq 1 - \sup_{x \in A} \frac{|\{i \mid x \in N_i,\ i = 1, 2, \ldots, m\}|}{m}. \tag{19}$$

The strict inequalities hold when $Y_i$ and $N_i$, $i = 1, 2, 3$ ($m = 3$) are given as in Figure 1. In Figure 1(a), the left-hand value of (18) becomes $1/3$ while the right-hand value of (18) becomes $2/3$. Similarly, in Figure 1(b) of (19), the left-hand value of (19) becomes $1/3$ while the right-hand value becomes $2/3$.

From (18) and (19), we observe $\pi'_{\mathcal{A}}(A) \leq \pi_{\mathcal{A}}(A)$.

## 3.2   Inclusion Degrees

We investigate the inclusion degree between PwMs-GISs $\mathcal{A}$ and $\mathcal{B}$. To this end, first, we investigate the inclusion degree between a PwMs-GIS and a usual set $C$. The family of sets including $C$ can be represented by an ill-known set $C_{\subseteq}$ defined by $C_{\subseteq} = \{D \mid C \subseteq D \subseteq X\}$. Similarly, the family of sets included in $C$ can be represented by an ill-known set $C_{\supseteq}$ defined by $C_{\supseteq} = \{D \mid \emptyset \subseteq D \subseteq C\}$. Then, for PwMs-GIS $\mathcal{A}$ and $\mathcal{B}$, we can define the following four degrees: (a) a necessity degree that $\mathcal{B}$ includes $C$ by $NES(\mathcal{B} \supseteq C) = N(C_{\subseteq}|\mathcal{B})$, (b) a necessity degree that $\mathcal{A}$ is included in $C$ by $NES(\mathcal{A} \subseteq C) = N(C_{\supseteq}|\mathcal{A})$, (c) a possibility degree that $\mathcal{B}$ includes $C$ by $POS(\mathcal{B} \supseteq C) = \Pi(C_{\subseteq}|\mathcal{B})$ and (d) a possibility degree that $\mathcal{A}$ is included in $C$ by $POS(\mathcal{A} \subseteq C) = \Pi(C_{\supseteq}|\mathcal{A})$.

By definitions, $C_{\subseteq}$ is defined by the lower and upper approximations $C$ and $X$ while $C_{\supseteq}$ is defined by the lower and upper approximations $\emptyset$ and $C$. Because we have $I(a, 1) = 1$ for any $a \in [0, 1]$, $n^{-1}(1) = \varphi^{-1}(0) = 0$ and $n^{-1}(0) = \varphi^{-1}(1) = 1$ from definitions, we obtain $NES(\mathcal{B} \supseteq C) = \inf_{x \in C} I(n^{-1}(\mu_{B^-}(x)), 0)$, $NES$ $(\mathcal{A} \subseteq C) = \inf_{x \notin C} I(\varphi^{-1}(\mu_{A^+}(x)), 0)$, $POS(\mathcal{B} \supseteq C) = \inf_{x \in C} \varphi^{-1}(\mu_{B^+}(x))$ and $POS(\mathcal{A} \subseteq C) = \inf_{x \notin C} n^{-1}(\mu_{A^-}(x))$ owing to (7) and (8).

Now let us extend those degrees to the inclusion degrees between PwMs-GISs $\mathcal{A}$ and $\mathcal{B}$. Using $NES(\mathcal{B} \supseteq C)$, $NES(\mathcal{A} \subseteq C)$, $POS(\mathcal{B} \supseteq C)$ and $POS(\mathcal{A} \subseteq C)$, we define four graded ill-known sets: (a) a family of graded sets necessarily included in PwMs-GIS $\mathcal{B}$ is defined by a GIS $\mathcal{B}_{\supseteq}^N$ with possibility distribution $\pi_{\mathcal{B}_{\supseteq}^N}(C) = NES(\mathcal{B} \supseteq C)$, (b) a family of graded sets necessarily including PwMs-GIS $\mathcal{A}$ is defined by a GIS $\mathcal{A}_{\subseteq}^N$ with possibility distribution $\pi_{\mathcal{A}_{\subseteq}^N}(C) = NES(\mathcal{A} \subseteq C)$, (c) a family of graded sets possibly included in PwMs-GIS $\mathcal{B}$ is defined by a GIS $\mathcal{B}_{\supseteq}^\Pi$ with possibility distribution $\pi_{\mathcal{B}_{\supseteq}^\Pi}(C) = POS(\mathcal{B} \supseteq C)$ and (d) a family of graded sets possibly including PwMs-GIS $\mathcal{A}$ is defined by a GIS $\mathcal{A}_{\subseteq}^\Pi$ with possibility distribution $\pi_{\mathcal{A}_{\subseteq}^\Pi}(C) = POS(\mathcal{A} \subseteq C)$.

Then we obtain eight inclusion degrees: (a) a necessity degree that $\mathcal{A}$ is necessarily included in $\mathcal{B}$ is defined by $NES(\mathcal{A} \subseteq \mathcal{B}) = N(\mathcal{B}_{\supseteq}^N|\mathcal{A})$, (b) a necessity degree that $\mathcal{A}$ is possibly included in $\mathcal{B}$ is defined by $\overline{NEPO}(\mathcal{A} \subseteq \mathcal{B}) = N(\mathcal{B}_{\supseteq}^\Pi|\mathcal{A})$, (c) a possibility degree that $\mathcal{A}$ is necessarily included in $\mathcal{B}$ is defined by $\overline{PONE}(\mathcal{A} \subseteq \mathcal{B}) = \Pi(\mathcal{B}_{\supseteq}^N|\mathcal{A})$, (d) a possibility degree that $\mathcal{A}$ is possibly included in $\mathcal{B}$ is defined by $POS(\mathcal{A} \subseteq \mathcal{B}) = \Pi(\mathcal{B}_{\supseteq}^\Pi|\mathcal{A})$, (e) a necessity degree that $\mathcal{B}$ necessarily includes $\mathcal{A}$ is defined by $NES(\mathcal{B} \supseteq \mathcal{A}) = N(\mathcal{A}_{\subseteq}^N|\mathcal{B})$, (f) a necessity degree that $\mathcal{B}$ possibly includes $\mathcal{A}$ is defined by $NEPO(\mathcal{B} \supseteq \mathcal{A}) = N(\mathcal{A}_{\subseteq}^\Pi|\mathcal{B})$, (g) a possibility degree that $\mathcal{B}$ necessarily includes $\mathcal{A}$ is defined by $\overline{PONE}(\mathcal{B} \supseteq \mathcal{A}) = \Pi(\mathcal{A}_{\subseteq}^N|\mathcal{B})$, (h) a possibility degree that $\mathcal{B}$ possibly includes $\mathcal{A}$ is defined by $POS(\mathcal{B} \supseteq \mathcal{A}) = \Pi(\mathcal{A}_{\subseteq}^\Pi|\mathcal{B})$. We obtain the following theorem.

**Theorem 3.** *When $\mathcal{A}$ and $\mathcal{B}$ are PwMs-GISs, we have*

$$NES(\mathcal{A} \subseteq \mathcal{B}) = \inf_{x \in X} I\left(\varphi^{-1}(\mu_{A^+}(x)), I\left(n^{-1}(\mu_{B^-}(x)), 0\right)\right), \quad (20)$$

$$NEPO(\mathcal{A} \subseteq \mathcal{B}) = \inf_{x \in X} I\left(\varphi^{-1}(\mu_{A^+}(x)), \varphi^{-1}(\mu_{B^+}(x))\right), \quad (21)$$

$$PONE(\mathcal{A} \subseteq \mathcal{B}) = \inf_{x \in X} \max \left(n^{-1}(\mu_{A^-}(x)), I\left(n^{-1}\left(\mu_{B^-}(x)\right), 0\right)\right), \quad (22)$$

$$POS(\mathcal{A} \subseteq \mathcal{B}) = \inf_{x \in X} \max \left(n^{-1}(\mu_{A^-}(x)), \varphi^{-1}\left(\mu_{B^+}(x)\right)\right), \quad (23)$$

$$NES(\mathcal{B} \supseteq \mathcal{A}) = \inf_{x \in X} I\left(n^{-1}\left(\mu_{B^-}(x)\right), I\left(\varphi^{-1}(\mu_{A^+}(x)), 0\right)\right), \quad (24)$$

$$NEPO(\mathcal{B} \supseteq \mathcal{A}) = \inf_{x \in X} I\left(n^{-1}\left(\mu_{B^-}(x)\right), n^{-1}(\mu_{A^-}(x))\right), \quad (25)$$

$$PONE(\mathcal{B} \supseteq \mathcal{A}) = \inf_{x \in X} \max \left(I\left(\varphi^{-1}(\mu_{A^+}(x)), 0\right), \varphi^{-1}(\mu_{B^+}(x))\right), \quad (26)$$

$$POS(\mathcal{B} \supseteq \mathcal{A}) = \inf_{x \in X} \max \left(n^{-1}(\mu_{A^-}(x)), \varphi^{-1}(\mu_{B^+}(x))\right), \quad (27)$$

*where $n$ and $\varphi$ are functions used in (3).*

*Proof.* We prove (20) and (26) because (21), (22) and (23) are proved similarly to (20) and (24), (25) and (27) are proved similarly to (26).

$$NES(\mathcal{A} \subseteq \mathcal{B}) = \inf_{C \subseteq X} I\left(\pi_{\mathcal{A}}(C), \pi_{B^N_{\supseteq}}(C)\right)$$

$$= \inf_{C \subseteq X} I\left(\pi_{\mathcal{A}}(C), \inf_{x \in C} I\left(n^{-1}\left(\mu_{B^-}(x)\right), 0\right)\right)$$

$$= \inf_{C \subseteq X} \inf_{x \in C} I\left(\pi_{\mathcal{A}}(C), I\left(n^{-1}\left(\mu_{B^-}(x)\right), 0\right)\right)$$

$$= \inf_{x \in X} I\left(\sup_{C \ni x} \pi_{\mathcal{A}}(C), I\left(n^{-1}\left(\mu_{B^-}(x)\right), 0\right)\right)$$

$$= \inf_{x \in X} I\left(\varphi^{-1}\left(\varphi\left(\sup_{C \ni x} \pi_{\mathcal{A}}(C)\right)\right), I\left(n^{-1}\left(\mu_{B^-}(x)\right), 0\right)\right)$$

$$= \inf_{x \in X} I\left(\varphi^{-1}\left(\mu_{A^+}(x)\right), I\left(n^{-1}\left(\mu_{B^-}(x)\right), 0\right)\right).$$

Now we prove (26). Using the restorability of $\pi_{\mathcal{A}}$, we obtain

$$PONE(\mathcal{B} \supseteq \mathcal{A}) = \sup_{C \subseteq X} \min \left(\pi_{\mathcal{A}}(C), \pi_{B^N_{\supseteq}}(C)\right)$$

$$= \sup_{C \subseteq X} \min \left(\inf_{x \notin C} n^{-1}(\mu_{A^-}(x)), \inf_{x \in C} \varphi^{-1}(\mu_{A^+}(x)), \inf_{x \in C} I\left(n^{-1}\left(\mu_{B^-}(x)\right), 0\right)\right).$$

We have the following equivalences:

$$\inf_{x \notin C} n^{-1}(\mu_{A^-}(x)) \geq \alpha \Leftrightarrow n^{-1}(\mu_{A^-}(x)) < \alpha \text{ implies } x \in C,$$

$$\inf_{x \in C} \varphi^{-1}(\mu_{A^+}(x)) \geq \alpha \Leftrightarrow x \in C \text{ implies } \varphi^{-1}(\mu_{A^+}(x)) \geq \alpha,$$

$$\inf_{x \in C} I\left(n^{-1}\left(\mu_{B^-}(x)\right), 0\right) \geq \alpha \Leftrightarrow x \in C \text{ implies } I\left(n^{-1}\left(\mu_{B^-}(x)\right), 0\right) \geq \alpha.$$

From (4), we obtain $\mu_{A^-}(x) > n(\alpha)$ implies $\mu_{A^+}(x) \geq \varphi(\alpha)$. Then we have

$$PONE(\mathcal{B} \supseteq \mathcal{A}) \geq \alpha$$

$$\Leftrightarrow n^{-1}(\mu_{A^-}(x)) < \alpha \text{ implies } \varphi^{-1}(\mu_{A^+}(x)) \geq \alpha \text{ and } I\left(n^{-1}\left(\mu_{B^-}(x)\right), 0\right) \geq \alpha$$

$$\Leftrightarrow n^{-1}(\mu_{A^-}(x)) < \alpha \text{ implies } I\left(n^{-1}\left(\mu_{B^-}(x)\right), 0\right) \geq \alpha.$$

Hence, we obtain (26).                                                        □

From Theorem 3, we always have $POS(\mathcal{A} \subseteq \mathcal{B}) = POS(\mathcal{B} \supseteq \mathcal{A})$. When $I$ satisfies $I(a, I(b, 0)) = I(b, I(a, 0))$ for any $a, b \in [0, 1]$, $NES(\mathcal{A} \subseteq \mathcal{B}) = NES(\mathcal{B} \supseteq \mathcal{A})$ holds. Morever, when $I$ satisfies $I(a, b) = \max(I(a, 0), b)$ for any $a, b \in [0, 1]$, $NEPO(\mathcal{A} \subseteq \mathcal{B}) = PONE(\mathcal{B} \subseteq \mathcal{A})$ and $NEPO(\mathcal{B} \supseteq \mathcal{A}) = PONE(\mathcal{A} \subseteq \mathcal{B})$ hold. Furthermore, $I(a, b) = \max(I(a, 0), b)$ implies $I(a, I(b, 0)) = I(b, I(a, 0))$. Then when $I$ satisfies $I(a, b) = \max(I(a, 0), b)$ for any $a, b \in [0, 1]$, inclusion degrees are reduced to four indices.

## 4    Linear Programming with Graded Ill-Known Sets

We consider the following linear programming problem with GISs (we call this problem "GIS-LP problem"):

$$\begin{aligned} \text{maximize} \quad & c_1 x_1 + c_2 x_2, \\ \text{subject to} \quad & a_{11} x_1 + a_{12} x_2 \lesssim_1 b_1, \ a_{21} x_1 + a_{22} x_2 \lesssim_2 b_2, \ x_1, x_2 \geq 0, \end{aligned} \quad (28)$$

where ranges of $a_{ij}$, $c_j$, $i, j = 1, 2$ are estimated by PwMs-GISs $\mathcal{A}_{ij}$ and $\mathcal{C}_j$. Notation '$\lesssim_i b_i$' means "roughly less than $b_i$" and we assume ranges "roughly less than $b_i$", $i = 1, 2$ are represented by PwMs-GISs $\mathcal{B}_i$. Here we treat a problem with two constraints but the approach described in this section can be extended to problems with many constraints.

To treat the objective function, we assume goals represented by '$\gtrsim_0 b_0$' with meaning "roughly greater than $b_0$" which are expressed by PwMs-GIS $\mathcal{B}_0$. Then GIS-LP problem (28) is transformed to the following problem:

$$c_1 x_1 + c_2 x_2 \gtrsim_0 b_0, \ a_{11} x_1 + a_{12} x_2 \lesssim_1 b_1, \ a_{21} x_1 + a_{22} x_2 \lesssim_2 b_2, \ x_1, x_2 \geq 0. \quad (29)$$

This problem is a system of linear inequalities with PwMs-GISs. Similar to the case with fuzzy numbers, we can discuss the satisfaction degree of all inequalities to a solution $\boldsymbol{x} = (x_1, x_2)$. Then we formulate problem (29) as a maximization problem of the satisfaction degree. The variation range of the left-hand side value can be expressed by a PwMs-GIS while the satisfactory range is given also by a PwMs-GIS. Then the satisfaction degree can be represented by some of inclusion degrees (20) to (27) between two PwMs-GISs. For example, when the inclusion degree (20) is used, Problem (29) is formulated as

$$\begin{aligned} \text{maximize} \quad & \alpha, \\ \text{subject to} \quad & Nes(\mathcal{C}_1 x_1 + \mathcal{C}_2 x_2 \subseteq \mathcal{B}_0) \geq \alpha, \ Nes(\mathcal{A}_{11} x_1 + \mathcal{A}_{12} x_2 \subseteq \mathcal{B}_1) \geq \alpha, \quad (30) \\ & Nes(\mathcal{A}_{21} x_1 + \mathcal{A}_{22} x_2 \subseteq \mathcal{B}_2) \geq \alpha, \ x_1, x_2 \geq 0. \end{aligned}$$

Let $I$ satisfy $I(a, b) = \max(I(a, 0), b)$, for any $a, b \in [0, 1]$. Then, for PwMs-GISs $\mathcal{A}$ and $\mathcal{B}$, we have the following property:

$$Nes(\mathcal{A} \subseteq \mathcal{B}) \geq \alpha \Leftrightarrow \mu_{A^+}(x) > \varphi(I^*(\alpha, 0)) \text{ implies } \mu_{B^-}(x) \geq n(I^*(\alpha, 0)), \quad (31)$$

where we define $I^*(a, 0) = \sup\{s \in [0, 1] | I(s, 0) \geq a\}$.

When $f : \mathbf{R}^m \to \mathbf{R}$ is continuous and monotonous, $f(\mathcal{A}_1, \ldots, \mathcal{A}_m)$ is a roughly specified GIS for PwMs-GISs $\mathcal{A}_i$, $i = 1, 2, \ldots, m$. By Theorems 1 and 2, in this case, $f(\mathcal{A}_1, \ldots, \mathcal{A}_m)$ is specified by its lower approximation $f(A_1^-, \ldots, A_m^-)$ and upper approximation $f(A_1^+, \ldots, A_m^+)$. Moreover, we have $(f(A_1^\pm, \ldots, A_m^\pm))_\beta = f((A_1^\pm)_\beta, \ldots, (A_m^\pm)_\beta)$ (double-sign corresponds), where $(D)_\beta = \{r \mid \mu_D(r) > \beta\}$ for a fuzzy set $D$. $\overline{D}^{\mathrm{L}}(\beta)$ and $\overline{D}^{\mathrm{R}}(\beta)$ denote $\inf(D)_\beta$ and $\sup(D)_\beta$, respectively. Similarly, we define $[D]_\beta = \{r \mid \mu_D(r) \geq \beta\}$ for a fuzzy set $D$. $D^{\mathrm{L}}(\beta)$ and $D^{\mathrm{R}}(\beta)$ denote $\inf[D]_\beta$ and $\sup[D]_\beta$, respectively.

Applying (31) to Problem (30), we obtain

maximize $\alpha$,

sub. to $\overline{C}_1^{+\mathrm{L}}(\varphi(I(\alpha,0)))x_1 + \overline{C}_2^{+\mathrm{L}}(\varphi(I(\alpha,0)))x_2 \geq B_0^{-\mathrm{L}}(n(I^*(\alpha,0)))$,

$$\overline{A}_{11}^{+\mathrm{R}}(\varphi(I(\alpha,0)))x_1 + \overline{A}_{12}^{+\mathrm{R}}(\varphi(I(\alpha,0)))x_2 \leq B_1^{-\mathrm{R}}(n(I^*(\alpha,0))), \quad (32)$$

$$\overline{A}_{21}^{+\mathrm{R}}(\varphi(I(\alpha,0)))x_1 + \overline{A}_{22}^{+\mathrm{R}}(\varphi(I(\alpha,0)))x_2 \leq B_2^{-\mathrm{R}}(n(I^*(\alpha,0))),$$

$x_1, x_2 \geq 0$, $\alpha \in [0,1]$.

Problem (32) is solved by a bisection method together with the simplex method.

## 5    A Numerical Example

We consider the following simple problem.

**Problem.** In a factory, the factory manager intends to produce products A and B. Products A and B are made of materials 1 and 2. To produce a unit of product A, $a_{11}$ units of material 1 and $a_{12}$ units of material 2 are needed. To produce a unit of product B, $a_{21}$ units of material 1 and $a_{22}$ units of material 2 are needed. The manager wants consumed amounts of materials 1 and 2 to be less than $b_1$ units and $b_2$ units, respectively. The gross revenue of product A per unit is $c_1$ while the gross revenue of product B per unit is $c_2$. The manager has his preference on the total gross revenue, not to be less than $b_0$. How much amounts of products A and B should be produced ?

The parameters $a_{ij}$, $c_j$, $i = 1, 2$, $j = 1, 2$ are not known exactly. However, two experts has their estimations of variation ranges by sets $Y_{a_{ij}}^k$, $Y_{c_j}^k$ of surely possible values and sets $N_{a_{ij}}^k$, $N_{c_j}^k$ of surely impossible values. They are shown in Table 1. The manager's preferences on the total gross revenue as well as consumed amounts of materials are given by sets $Y_{b_j}^k$ of surely satisfactory values and sets $N_{b_j}^k$ of surely unsatisfactory values. They are given in Table 2.

From Tables 1 and 2, we observe that sets of surely possible values of product A are larger than those of product B. In other words, product A has more ambiguous factors than product B. Applying (13) with $n(a) = 1 - 1$, $\varphi(a) = 1$, $I(a, b) = \max(1-a, b)$, we obtain possibility distributions $\pi_{\mathcal{A}_{ij}}$, $\pi_{\mathcal{C}_j}$ and $\pi_{\mathcal{B}_l}$. The lower and upper approximations of those possibility distributions are obtained by (14) and (15). For example, membership functions of lower approximation $A_{11}^-$ and upper approximation $A_{11}^+$ of $\mathcal{A}_{11}$ are obtained as follows:

**Table 1.** Estimated ranges of coefficients

| $k$ | $Y_{a_{11}}^k$ | $N_{a_{11}}^k$ | $Y_{a_{12}}^k$ | $N_{a_{12}}^k$ |
|---|---|---|---|---|
| 1 | [1.9, 2.1] | $(-\infty, 1), (3, +\infty)$ | [2.4, 2.6] | $(-\infty, 2.3), (2.7, +\infty)$ |
| 2 | [1.95, 2.05] | $(-\infty, 1.1), (2.9, +\infty)$ | [2.45, 2.55] | $(-\infty, 2.35), (2.65, +\infty)$ |

| $k$ | $Y_{a_{21}}^k$ | $N_{a_{21}}^k$ | $Y_{a_{22}}^k$ | $N_{a_{22}}^k$ |
|---|---|---|---|---|
| 1 | [3.8, 4.2] | $(-\infty, 3), (5, +\infty)$ | [1.9, 2.1] | $(-\infty, 1.8), (2.2, +\infty)$ |
| 2 | [3.9, 4.1] | $(-\infty, 3.2), (4.8, +\infty)$ | [1.95, 2.05] | $(-\infty, 1.85), (2.15, +\infty)$ |

| $k$ | $Y_{c_1}^k$ | $N_{c_1}^k$ | $Y_{c_2}^k$ | $N_{c_2}^k$ |
|---|---|---|---|---|
| 1 | [9.7, 10.3] | $(-\infty, 9.5), (10.5, +\infty)$ | [6.8, 7.2] | $(-\infty, 5.5), (8.5, +\infty)$ |
| 2 | [9.8, 10.2] | $(-\infty, 9.6), (10.4, +\infty)$ | [6.9, 7.1] | $(-\infty, 5.7), (8.3, +\infty)$ |

**Table 2.** Satisfactory ranges of goal and constraints

| $k$ | $Y_{b_1}^k$ | $N_{b_1}^k$ | $Y_{b_2}^k$ | $N_{b_2}^k$ | $Y_{b_0}^k$ | $N_{b_0}^k$ |
|---|---|---|---|---|---|---|
| 1 | $(-\infty, 190]$ | $(205, +\infty)$ | $(-\infty, 290]$ | $(305, +\infty)$ | $[510, +\infty)$ | $(-\infty, 495)$ |
| 2 | $(-\infty, 195]$ | $(210, +\infty)$ | $(-\infty, 295]$ | $(310, +\infty)$ | $[505, +\infty)$ | $(-\infty, 490)$ |

$$
\mu_{A_{11}^-}(r)
= \begin{cases}
1, & \text{if } r \in [1.95, 2.05], \\
0.5, & \text{if } r \in [1.9, 1.95) \cup (2.05, 2.1], \\
0, & \text{if } r \in (-\infty, 1.9) \cup (2.1, +\infty),
\end{cases}
$$

$$
\mu_{A_{11}^+}(r)
= \begin{cases}
1, & \text{if } r \in [1.1, 2.9], \\
0.5, & \text{if } r \in [1, 1.1) \cup (2.9, 3], \\
0, & \text{if } r \in (-\infty, 1) \cup (3, +\infty).
\end{cases}
\tag{33}
$$

Using inclusion degree $NES(\mathcal{A} \subseteq \mathcal{B})$ which corresponds to a robust treatment of constraint, Problem (32) are reduced to the following problem:

(NES)  maximize $\alpha$,  subject to
$$
\begin{cases}
9.6x_1 + 5.7x_2 \ge 505, \text{ if } 0 < \alpha \le 0.5, & 9.5x_1 + 5.5x_2 \ge 510, \text{ if } 0.5 < \alpha \le 1, \\
2.9x_1 + 2.65x_2 \le 195, \text{ if } 0 < \alpha \le 0.5, & 3.0x_1 + 2.7x_2 \le 190, \text{ if } 0.5 < \alpha \le 1, \\
4.8x_1 + 2.15x_2 \le 295, \text{ if } 0 < \alpha \le 0.5, & 5.0x_1 + 2.2x_2 \le 290, \text{ if } 0.5 < \alpha \le 1, \\
x_1 \ge 0, \ x_2 \ge 0.
\end{cases}
$$

Let $\alpha = 1$, we obtain an optimal solution $x_1 = 54.53, x_2 = 29.03$. There are many feasible solutions with $\alpha = 1$. Then the manager can enlarge his required total gross revenue to some extent. The solution described above is obtained by maximizing $9.5x_1 + 5.5x_2$ (an estimated total gross revenue).

The differences of feasible regions with $\alpha = 1$ by the selected inclusion indices are depicted in Figure 2. As a reference, we show a feasible region (Center) defined by center values of $\{r \mid \mu_{A_{ij}^-}(r) = 1\}$, $\{r \mid \mu_{C_j^-}(r) = 1\}$ and $\bar{b}_0 = 0.5 \inf\{r \mid \mu_{B_0^-}(r) = 1\} + 0.5 \inf\{r \mid \mu_{B_0^+}(r) = 1\}$, $\bar{b}_i = 0.5 \sup\{r \mid \mu_{B_i^-}(r) = 1\} + 0.5 \sup\{r \mid \mu_{B_i^+}(r) = 1\}$, $i = 1, 2$. As shown in this figure, all feasible regions are included in the region of (Center). From this fact, we confirm the robust treatments by inclusion degress. We observe (NES) is strongest while (POS) is weakest. (NEPO) and (PONE) are between them. In this example, (NEPO) looks stronger than (PONE) but the inclusion relation does not hold.

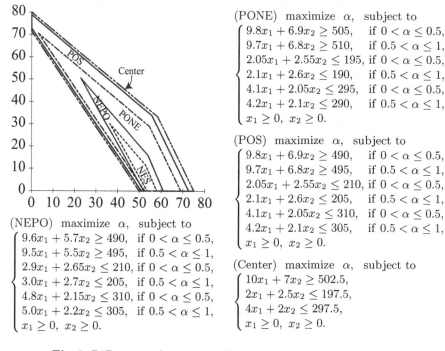

(PONE) maximize $\alpha$, subject to
$$\begin{cases} 9.8x_1 + 6.9x_2 \geq 505, & \text{if } 0 < \alpha \leq 0.5, \\ 9.7x_1 + 6.8x_2 \geq 510, & \text{if } 0.5 < \alpha \leq 1, \\ 2.05x_1 + 2.55x_2 \leq 195, & \text{if } 0 < \alpha \leq 0.5, \\ 2.1x_1 + 2.6x_2 \leq 190, & \text{if } 0.5 < \alpha \leq 1, \\ 4.1x_1 + 2.05x_2 \leq 295, & \text{if } 0 < \alpha \leq 0.5, \\ 4.2x_1 + 2.1x_2 \leq 290, & \text{if } 0.5 < \alpha \leq 1, \\ x_1 \geq 0, \ x_2 \geq 0. \end{cases}$$

(POS) maximize $\alpha$, subject to
$$\begin{cases} 9.8x_1 + 6.9x_2 \geq 490, & \text{if } 0 < \alpha \leq 0.5, \\ 9.7x_1 + 6.8x_2 \geq 495, & \text{if } 0.5 < \alpha \leq 1, \\ 2.05x_1 + 2.55x_2 \leq 210, & \text{if } 0 < \alpha \leq 0.5, \\ 2.1x_1 + 2.6x_2 \leq 205, & \text{if } 0.5 < \alpha \leq 1, \\ 4.1x_1 + 2.05x_2 \leq 310, & \text{if } 0 < \alpha \leq 0.5, \\ 4.2x_1 + 2.1x_2 \leq 305, & \text{if } 0.5 < \alpha \leq 1, \\ x_1 \geq 0, \ x_2 \geq 0. \end{cases}$$

(NEPO) maximize $\alpha$, subject to
$$\begin{cases} 9.6x_1 + 5.7x_2 \geq 490, & \text{if } 0 < \alpha \leq 0.5, \\ 9.5x_1 + 5.5x_2 \geq 495, & \text{if } 0.5 < \alpha \leq 1, \\ 2.9x_1 + 2.65x_2 \leq 210, & \text{if } 0 < \alpha \leq 0.5, \\ 3.0x_1 + 2.7x_2 \leq 205, & \text{if } 0.5 < \alpha \leq 1, \\ 4.8x_1 + 2.15x_2 \leq 310, & \text{if } 0 < \alpha \leq 0.5, \\ 5.0x_1 + 2.2x_2 \leq 305, & \text{if } 0.5 < \alpha \leq 1, \\ x_1 \geq 0, \ x_2 \geq 0. \end{cases}$$

(Center) maximize $\alpha$, subject to
$$\begin{cases} 10x_1 + 7x_2 \geq 502.5, \\ 2x_1 + 2.5x_2 \leq 197.5, \\ 4x_1 + 2x_2 \leq 297.5, \\ x_1 \geq 0, \ x_2 \geq 0. \end{cases}$$

**Fig. 2.** Differences of constraints by the adopted inclusion degrees

# References

1. Inuiguchi, M., Ramík, J.: Possibilistic Linear Programming: A Brief Review of Fuzzy Mathematical Programming and a Comparison with Stochastic Programming in Portfolio Selection Problem. Fuzzy Sets and Systems 111, 29–45 (2000)
2. Garcia, J.C.F.: Solving Fuzzy Linear Programming Problems with Interval Type-2 RHS. In: IEEE International Conference on Systems, Man, and Cybernetics (2009)
3. Dubey, D., Chandra, S., Mehra, A.: Fuzzy linear programming under interval uncertainty based on IFS representation. Fuzzy Sets and Systems 188, 68–87 (2012)
4. Zadeh, L.A.: The Concept of a Linguistic Variable and its Application to Approximate Reasoning-1. Information Sciences 8, 199–249 (1975)
5. Atanassov, K.T.: Intuitionistic fuzzy sets. Fuzzy Sets and Systems 20, 87–96 (1986)
6. Inuiguchi, M.: Ill-Known Set Approach to Disjunctive Variables: Calculations of Graded Ill-Known Intervals. In: Greco, S., Bouchon-Meunier, B., Coletti, G., Fedrizzi, M., Matarazzo, B., Yager, R.R. (eds.) IPMU 2012, Part I. CCIS, vol. 297, pp. 643–652. Springer, Heidelberg (2012)
7. Dubois, D., Prade, H.: Incomplete Conjunctive Information. Comput. Math. Applic. 15(10), 797–810 (1988)
8. Inuiguchi, M.: Rough Representations of Ill-Known Sets and Their Manipulations in Low Dimensional Space. In: Skowron, A., Suraj, Z. (eds.) Rough Sets and Intelligent Systems. ISRL, vol. 42, pp. 309–331. Springer, Heidelberg (2013)

# Sharing Online Cultural Experiences: An Argument-Based Approach

Leila Amgoud[1], Roberto Confalonieri[1], Dave de Jonge[2], Mark d'Inverno[3], Katina Hazelden[3], Nardine Osman[2], Henri Prade[1], Carles Sierra[2], and Matthew Yee-King[3]

[1] Institut de Recherche en Informatique Toulouse (IRIT)
Université Paul Sabatier
118 Route de Narbonne
31062 Toulouse Cedex 9, France
{roberto.confalonieri,amgoud,prade}@irit.fr

[2] Artipcial Intelligence Research Institute, IIIA-CSIC
Spanish National Research Council
Campus de la Universitat Autónoma de Barcelona
08193 Bellaterra, Catalonia, Spain
{davedejonge,sierra,nardine}@iiia.csic.es

[3] Department of Computing
Goldsmiths, University of London,
London SE14 6NW, United Kingdom
{dinverno,exs01kh2,mas01mjy}@gold.ac.uk

**Abstract.** This paper proposes a system that allows a group of human users to share their cultural experiences online, like buying together a gift from a museum or browsing simultaneously the collection of this museum. We show that such application involves two multiple criteria decision problems for choosing between different alternatives (e.g. possible gifts): one at the level of each user, and one at the level of the group for making joint decisions. The former is made manually by the users via the WeShare interface. This interface displays an image with tags reflecting some features (criteria) of the image. Each user expresses then his opinion by rating the image and each tag. A user may change his choices in light of a report provided by his WeShare agent on the opinion of the group. Joint decisions are made in an automatic way. We provide a negotiation protocol which shows how they are reached. Both types of decisions are based on the notion of argument. Indeed, a tag which is liked by a user constitutes an argument pro the corresponding image whereas a tag which is disliked gives birth to a cons argument. These arguments may have different strengths since a user may express to what extent he likes/dislikes a given tag. Finally, the opinion analysis performed by a WeShare agent consists of aggregating the arguments of the users.

## 1 Introduction

Visiting a museum with friends can be considered one of the preferred social shared experiences of people visiting a city. In such a social experience, users can share opinions with other users and together decide which cultural artefacts to see. Unfortunately, museums have been placed under financial pressure by the European economic crisis.

V. Torra et al. (Eds.): MDAI 2012, LNAI 7647, pp. 282–293, 2012.

Consequently, several museums in the UK reduced their opening hours [13] and, as such, physically visiting museums or seeing all the artifacts in a museum is becoming harder [15]. Internet may partly solve this problem since some digital versions of several institutional collections are now available online.

A digital version of a museum is usually represented as a searchable database containing many images of objects held by an institution. For example, the Horniman Museum allows users to express opinions about images by means of *tags*, and to see the opinions generated by others, including a tag cloud view of the complete collection. However, in such a case, the system does not provide a realtime social experience, where the user is aware of the other people online and can interact with them, unlike a visit with a group friends to the physical museum itself. On the other hand, the predominance of social network tools is changing the way in which users can interact with information and with other people. This social network could be exploited to enrich, encourage and enliven engagement with online cultural artefacts and to benefit the museum themselves, by getting new visitors more engaged with a museum so that a physical visit becomes more likely.

These concerns suggest that building a system which enables an online, shared cultural experience for users willing to visit cultural artefacts online is worthy to be investigated. The design and implementation of such a system demand several decisions. For instance, deciding which kind of artefacts to a group of users want to buy or add to a shared collection (which could be later used to produce a guide for when visiting the museum physically), the partners with whom a user wants to share this experience, or which artefact to display next. While some of them can be easily carried out by the users, the choice of a specific artefact from a large collection, taking into account the opinions of a group of users as a whole and not only the individuals can require a more complex decision model. We are in initital stages of an european project and have started user evaluations and a relationship with a local museum in London in order to set out what kinds of user groups would take advantage of the system we are building.

In this paper, we propose the *WeShare* system, a first prototype that allows the consumption of an online shared experience by providing the following functionalities. It allows two or more users to connect to the digital collection of a museum, browse synchronously images, and decide all together which image to add to their joint collection or which one to buy. Such an application involves two multiple criteria decision problems: one at the level of each user for accepting/rejecting a displayed image, and one at the level of the group for making a joint decision about the same image. The former is made manually by the users via the *WeShare* interface. This interface displays an image with *tags* sent by the server of the museum. A tag represents a feature (or a criterion) of an image. Each user expresses then his opinion by rating the image itself and each tag. In addition, he provides various weights expressing to what extent he likes/dislikes the image and the tags. Joint decisions are made in an automatic way. We provide a negotiation protocol which shows how they are reached. Both types of decisions are based on the notion of *argument* which has a particular form in this system. Indeed, unlike existing logical argumentation systems where an argument is a logical proof for a given conclusion (e.g. [2,7,19]), an argument in favor of an image is a pair ((tag, value), image). When a user likes the tag (i.e. the value is positive), then the argument is pro the image. However, when the tag is disliked, the pair is a cons argument. These arguments with varying strengths are thus built by the users through the WeShare interface. Again, unlike argumentation systems for defeasible reasoning where the construction of arguments is monotonic, in our application this is not the case. Actually, a user may revise his opinion about a given tag. Consequently, the initial argument is removed and replaced by the new one. This revision is possible in light of a report provided by the user's WeShare agent on the opinion of the group. Indeed, an analysis of the opinion of the group is performed. It consists of aggregating the arguments of the users. We propose two aggregation operators: one that computes the average value for each tag and the average value of the image. The second operator aggregates in the same way the values of the tags, however applies a multiple criteria procedure for computing the final recommendation of the image.

The paper is organized as follows: Section 2 provides the architecture of our system. Section 3 describes the decision procedure of the server, namely how it selects the next image to browse. Section 4 describes the activities of the human user as well as his assistant agent. Section 5 provides a negotiation protocol that allows the group of users to make a common decision about a given image. The last section concludes.

## 2    System Architecture

The architecture of our system is depicted in Figure 1. It contains two main components: a *Media Server* agent and a *WeShare* agent per human user.

The Media Server agent has access to the database of the museum. It is equipped with a decision model that computes the next image to propose to the users. That image is sent to the WeShare agents which display the image to all the users at the same time. In addition to the image, the Media Server agent provides several tags associated with

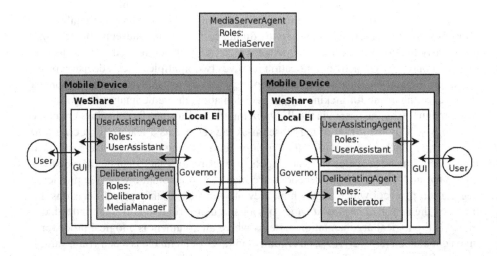

**Fig. 1.** The agents and the roles they play. Arrows indicate exchange of messages.

an image. They represent some particular features of the image (like being *fish* in the image shown in Figure 2).

WeShare agents consist of two different subcomponents: the User Assistant agent and the Deliberating agent. The User Assistant agent forms a layer between the We-Share interface (GUI) and the system. Through the interface (Figure 2), users express whether they like or not the image. Similarly, they can evaluate positively or negatively each tag. For instance, in the case of Figure 2, one may say that he *likes* the image, the fact that it represents a *fish*, but *does not like* that it is a *toy*.

The Deliberating agent is responsible for collecting liked and disliked tags of users in terms of preferences and rejections. It also maintains the preferences and rejections of other agents. It has a decision model for opinion analysis. Moreover, it is equipped with a negotiation protocol which allows the users to decide whether to accept or not a given displayed image.

The communication between the agents in the system is regulated by a lightweight version of a peer-to-peer Electronic Institution (Local EI). Generally speaking, Electronic Institutions allow to model and control agents' interactions. Since they are out of scope of this paper, we omit their description and we refer to [5,10].

The decision models of the Media Server and WeShare agents will be described in the following sections.

## 3   Media Server Agent

The Media Server agent is responsible for answering the queries made by human users. This agent is equipped with an *image archive* which consists of 2500 image files with varying numbers of natural language tags. Throughout the paper, $\mathcal{I} = \{im_1, \ldots, im_n\}$ (with $n = 2500$) is the set of available images where each $im_i \in \mathcal{I}$ is the identifier of an image. Each image is described with a finite set of *tags* or *features*, for instance the color. The set $\mathcal{T} = \{t_1, \ldots, t_k\}$ contains all the available tags. Finally, we assume the availability of a function $\mathcal{F} : \mathcal{I} \rightarrow 2^{\mathcal{T}}$ that returns the tags associated with a given image. Note that the same tag may be associated with more than one image.

The Media Server agent is also equipped with a decision model which defines a preference relation $\succeq$ on a set $\mathcal{I}' \subseteq \mathcal{I}$. The best element with respect to this relation is sent to the WeShare agents for browsing. In case of ties, one of them is chosen randomly. An important question now is how is the relation $\succeq$ defined? The model has three inputs:

1. A set $\mathcal{I}' \subseteq \mathcal{I}$ of images.
2. A set of preferences $\mathcal{P} \subseteq T$ of tags that an image should have.
3. A set of rejections $\mathcal{R} \subseteq T$ of tags that an image should not have.

The two sets $\mathcal{P}$ and $\mathcal{R}$ represent respectively the preferences and the rejections of the group of users. They are provided by the Deliberating agent of a user, the one who is the administrator of a browsing session (as we will see in Section 5). The decision model prefers the image that suits better these preferences and avoids the rejections. This principle is suitable when all the tags are equally important. It is worth pointing out that several principles can be found in [4,9,11] in case of weighted tags.

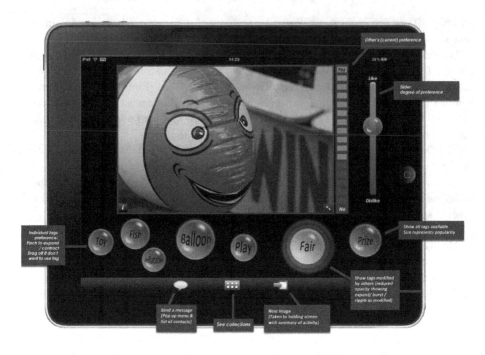

**Fig. 2.** WeShare interface

**Definition 1 (Decision model).** *Let* $\langle \mathcal{P}, \mathcal{R}, \mathcal{I}' \rangle$ *be the input sets. For* $im_i, im_j \in \mathcal{I}'$, $im_i \succeq im_j$ *iff* $|\mathcal{F}(im_i) \cap \mathcal{P}| \geq |\mathcal{F}(im_j) \cap \mathcal{P}|$ *and* $|\mathcal{F}(im_i) \cap \mathcal{R}| \leq |\mathcal{F}(im_j) \cap \mathcal{R}|$.

It is easy to check that the relation $\succeq$ is a partial preorder. Its maximal elements are gathered in the set $\max_{\succeq}$ defined as follows: $\max_{\succeq} = \{im_i \in \mathcal{I}' \text{ s.t. } \nexists im_j \in \mathcal{I}' \text{ with } im_j \succeq im_i\}$. Note also that in case the two sets $\mathcal{P}$ and $\mathcal{R}$ are empty, all the images are equally preferred. Finally, the Media Server sends to the WeShare agents one of the best images wrt. the relation $\succeq$. In case the set $\max_{\succeq}$ is empty, an image is chosen randomly.

**Definition 2 (Best image).** $\text{Best}(\langle \mathcal{P}, \mathcal{R}, \mathcal{I}' \rangle) = \begin{cases} im_i \in \max_{\succeq} & \text{if } \max_{\succeq} \neq \emptyset \\ im_i \in \mathcal{I}' & \text{else} \end{cases}$

To simplify notation we will use $\text{Best}(\mathcal{I}')$ in the rest of the paper.

## 4   Human User and WeShare Agent

Human users interact with the system via the WeShare interface depicted in Figure 2. Each user is responsible for expressing an opinion about each image sent by the Media Server agent. Indeed, he provides an overall rating to the image as well as a (positive or negative) value to each tag associated with the image. Throughout the paper, we assume the availability of a bipolar scale $\mathcal{S} = [-1, 1]$ which is used for evaluating the tags and the image. Assigning a positive value to an image means that the image is recommended. For a given image $im \in \mathcal{I}$, each user $u_i$ provides the following information:

| User/Tags | $t_1$ | ... | $t_j$ | ... | $t_m$ | $im$ |
|:---:|:---:|:---:|:---:|:---:|:---:|:---:|
| $u_i$ | $v_{i,1}$ | ... | $v_{i,j}$ | ... | $v_{i,m}$ | $r_i$ |

where $\mathcal{F}(im) = \{t_1, \ldots, t_m\}$, $v_{i,j} \in \mathcal{S}$ is the value assigned by user $i$ to tag $j$, and $r_i \in \mathcal{S}$ is the overall rating of the image $im$.

A user can revise his opinion in light of a report sent by his WeShare agent about the opinion of the remaining users. Note that the WeShare agent only gives advices to the user and the final decisions are made by the user himself. Finally, a user may engage in a negotiation dialogue with other users in order to persuade them to accept/reject a given image. This part will be described in Section 5.

### 4.1   Arguments

The notion of argument is at the heart of several models developed for reasoning about defeasible information (e.g. [12,17]), decision making (e.g. [4,8]), practical reasoning (e.g. [6]), and modeling different types of dialogues (e.g. [3,18]). An argument is a reason for believing a statement, choosing an option, or doing an action. In most existing works on argumentation, an argument is either considered as an abstract entity whose origin and structure are not defined, or it is a logical proof for a statement where the proof is built from a knowledge base. In our application, arguments are reasons for accepting or rejecting a given image. They are built by the human user when rating the different tags associated with an image. Indeed, a tag which is evaluated positively gives birth to an argument pro the image whereas a tag which is rated negatively induces a con argument against the image. The tuple $\langle \mathcal{I}, \mathcal{T}, \mathcal{S} \rangle$ will be called a *theory*.

**Definition 3 (Argument).** *Let* $\langle \mathcal{I}, \mathcal{T}, \mathcal{S} \rangle$ *be a theory and* $im \in \mathcal{I}$.

- *An* argument pro $im$ *is a pair* $((t, v), im)$ *where* $t \in \mathcal{T}$ *and* $v \in \mathcal{S}$ *and* $v > 0$.
- *An* argument con $im$ *is a pair* $((t, v), im)$ *where* $t \in \mathcal{T}$ *and* $v \in \mathcal{S}$ *and* $v < 0$.

*The pair* $(t, v)$ *is the* support *of the argument and* $im$ *is its* conclusion. *The functions* Tag, Val *and* Conc *return respectively the tag* $t$ *of an argument* $((t, v), im)$, *its value* $v$, *and the conclusion* $im$.

It is well-known that the construction of arguments in systems for defeasible reasoning is monotonic (see [2] for a formal result). Indeed, an argument cannot be removed when the knowledge base from which the arguments are built is extended by new information. This is not the case in our application. When a user revises his opinion about a given tag, the initial argument is removed and replaced by a new one. For instance, if a user assigns value 0.5 to a tag $t$ which is associated with an image $im$, then he decreases the value to 0.3, the argument $((t, 0.5), im)$ is no longer considered as an argument and is completely removed from the set of arguments of the user and is replaced by the argument $((t, 0.3), im)$. To say it differently, the set of arguments of a user contains only one argument per tag for a given image.

## 4.2   Opinion analysis

Opinion analysis is gaining increasing interest in linguistics (see e.g. [1,14]) and more recently in AI (e.g. [16,20]). This is due to the importance of having efficient tools that provide a synthetic view on a given subject. For instance, politicians may find it useful to analyze the popularity of new proposals or the overall public reaction to certain events. Companies are definitely interested in consumer attitudes towards a product and the reasons and motivations of these attitudes. In our application, it may be important for each user to know the opinion of the group about a certain image. This may lead the user to revise his own opinion.

The problem of opinion analysis consists of aggregating the opinions of several agents/users about a particular subject, called *target*. An opinion is a global rating that is assigned to the target, and the evaluation of some features associated with the target. Thus, this amounts to aggregating arguments which have the structure given in Definition 3. Let us illustrate this issue on the following example.

**Example 1.** *Let us consider the following opinion expressed on a digital camera.*

> *"It is a great digital camera for this century. The rotatable lens is great. It has fast response from the shutter. The LCD has increased from 1.5 to 1.8, which gives bigger view. But, it would be better if the model is designed for smaller size. I recommend this camera."*

*The target here is the digital camera, the overall rating is "recommended". The features are: the size, rotatable lens, response from the shutter, size of LCD. For instance, response from the shutter is evaluated positively whereas the size is evaluated negatively.*

In our application, the target is an image sent by the Media Server agent and the features are the associated tags. In what follows, we propose two models that are used by the WeShare agents of the users (in particular by the Deliberating agent component) in order to analyze the opinion of a group of users. Both models take as input the evaluations of the users and provide an aggregated value for the image and an aggregated value for each tag. The first model computes simply the average of existing values. The second model is based on a multiple criteria procedure in which one has to choose between two alternatives: recommending/accepting an image and rejecting it. The model prefers the alternative that satisfies more criteria (tags in our case), i.e. the one with more arguments pros. Note that in the application, all the tags are assumed to be equally important.

**Definition 4 (Opinion aggregation).** *Let $Ag = \{u_1, \ldots u_n\}$ be a set of users, $im \in \mathcal{I}$ where $\mathcal{F}(im) = \{t_1, \ldots, t_m\}$. The next table summarizes the opinions of the n users.*

| Users/Tags | $t_1$ | $\ldots$ | $t_j$ | $\ldots$ | $t_m$ | $im$ |
|---|---|---|---|---|---|---|
| $u_1$ | $v_{1,1}$ | $\ldots$ | $v_{1,j}$ | $\ldots$ | $v_{1,m}$ | $r_1$ |
| $\vdots$ | $\vdots$ | $\vdots$ | $\vdots$ | $\vdots$ | $\vdots$ | $\vdots$ |
| $u_i$ | $v_{i,1}$ | $\ldots$ | $v_{i,j}$ | $\ldots$ | $v_{i,m}$ | $r_i$ |
| $\vdots$ | $\vdots$ | $\vdots$ | $\vdots$ | $\vdots$ | $\vdots$ | $\vdots$ |
| $u_n$ | $v_{n,1}$ | $\ldots$ | $v_{n,j}$ | $\ldots$ | $v_{n,m}$ | $r_n$ |

*The result of the aggregation is:*

| Group | $v_1$ | $\ldots$ | $v_j$ | $\ldots$ | $v_m$ | $r$ |
|---|---|---|---|---|---|---|

*where for all $v_i$,* $v_i = \sum_{j=1,n} v_{j,i}/n$, *and*

**Average operator:** $r = \sum_{j=1,n} r_j/n$

**MCD operator:** $r = \begin{cases} 1 & if\,|\{t_j \mid v_j > 0\}| > |\{t_k \mid v_k < 0\}| \\ 0 & otherwise \end{cases}$

Note that the first aggregation operator assigns a value from the set $S$ to an image while the second one allows only binary values: 1 (for acceptance) and 0 (for rejection).

It is worth mentioning that even if both models aggregate in the same way the values of the tags, they do not necessarily rate in the same way the image. The following example shows a case where one operator accepts an image while the second rejects it.

**Example 2.** *Let us consider the following opinions expressed by four users about an image $im$ where this image is described by four tags.*

| Users/Tags | $t_1$ | $t_2$ | $t_3$ | $t_4$ | $im$ |
|---|---|---|---|---|---|
| $u_1$ | 0.9 | 0.7 | -0.2 | -0.3 | 0.5 |
| $u_2$ | 0.5 | 0.6 | -0.5 | 0.2 | -0.2 |
| $u_3$ | -0.5 | -0.3 | -0.2 | 0.9 | -0.6 |
| $u_4$ | 0.1 | 0.2 | 0.3 | -0.6 | 0 |

*The aggregated values of the tags are respectively: 0.25, 0.3, -0.15, and 0.05. The average operator assigns value -0.075 to the image whereas the MCD operator accepts the image (the overall rating is 1). This discrepancy is due to the fact that the decision model of each user is unknown. Indeed, it is not clear how a user aggregates the values he assigns to tags in order to get an overall rating of an image. For instance, user $u_2$ likes most of the tags, however he rejects the image. This means that either he has in mind other tags which are not considered in the table, or gives a higher importance to tag $t_3$. The second reason of discrepancy is that the second model does not take into account the overall recommendations of the users.*

Finally, it is worth noticing that opinion analysis amounts to aggregating arguments pros/cons a given target into a new argument. In the previous example, the four arguments $((t_1, 0.9), im)$, $((t_1, 0.5), im)$, $((t_1, -0.5), im)$ and $((t_1, 0.1), im)$ are aggregated into a new argument $((t_1, 0.25), im)$. This argument is pro the image $im$ while it is based on argument $((t_1, -0.5), im)$ con the same image.

## 5    Group Decision Making

In the previous sections, we have mainly presented the architecture of the system and described the reasoning part of the users and of the agents in the system. In what follows, we focus on the reasoning of the group. We provide a negotiation protocol that allows agents to make joint decisions. The idea is the following: a session starts when a user invites other users for sharing an experience online. When the invited users accept, a request is sent to the Media Server agent that will compute the best image and send it to the WeShare agents. These agents display the image to all the users. Each user expresses his opinion about the image and the tags via the WeShare interface. WeShare agents provide to their respective users a report on the aggregated opinion of the other agents. Users may consider this information for revising their own opinions. In case all agents agree on the overall rating of the image, then the image is bought (or stored) and the session is over. In case of disagreement, pairs of users may engage in private dialogues where they exchange arguments (as in Definition 3) about the image. The exchanged arguments may be either the ones that are built by the user when introducing his opinion or new ones. Actually, a user may add new tags for an image. When the disagreement persists, the preferences (about tags) are aggregated and the result is sent to the Media Server in order to select a new image that suits better those preferences.

In what follows, $Ag = \{u_1, \ldots, u_n\}$ is a set of users, and $\mathtt{Args}^t(u_i)$ is the set of arguments of user $u_i$ at step $t$. At the beginning of a session, the sets of arguments of all users are assumed to be empty (i.e., $\mathtt{Args}^0(u_i) = \emptyset$). Moreover, the set of images contains all the available images in the database of the museum, that is $\mathcal{I}^0 = \mathcal{I}$. We assume also that a user $u_i$ is interested in having a joint experience with other users. The protocol uses a communication language based on four locutions:

- `Invite`: it is used by a user to invite a set of users either for sharing an experience or for engaging in a dialogue.
- `Send` is used by agents for sending information to other agents.
- `Accept` is used mainly by users for accepting requests made to them by other users.
- `Reject` is used by users for rejecting requests made to them by other users.

**Interaction Protocol:**

1. $\mathtt{Invite}(u_i, G)$ (user $u_i$ sends an invitation to users in $G$ where $G \subseteq Ag$). User $u_i$ is the Administrator of the session.
2. Each user $u_j \in G$ sends either $\mathtt{Accept}(u_j)$ or $\mathtt{Reject}(u_j)$. Let $G' \subseteq G$ be the set of agents who answered positively to the invitation.
3. If $G' = \emptyset$, then either go to Step 4 (in case the user $u_i$ decides to have the experience alone), or the session is over.
4. $\mathtt{Send}(\text{WeShare}_i, \{\text{Media Server}\}, \langle \mathcal{P} = \emptyset, \mathcal{R} = \emptyset, \mathcal{I}^t \rangle)$ (the WeShare agent of user $u_i$ sends a request to the Media Server agent). $\mathtt{Send}(\text{Media Server}, \{WeShare\}_{i=1,\ldots,n}, \mathtt{Best}(\mathcal{I}^t))$ (the Media Server agent computes $\mathtt{Best}(\mathcal{I}^t)$ and sends it to all the WeShare agents).
5. Each WeShare agent displays the image $\mathtt{Best}(\mathcal{I}^t)$ and its tags (i.e., $t_i \in \mathcal{F}(\mathtt{Best}(\mathcal{I}^t))$).

6. Each user $u_j \in G' \cup \{u_i\}$:

   (a) bids the tags and gives an overall rating $\text{Res}_j(\text{Best}(I^t))$ to the image. Let $\text{Args}_j^t = \text{Args}_j^{t-1} \cup \{((t_i, v_i), \text{Best}(I^t)) \mid t_i \in \mathcal{F}(\text{Best}(I^t))\}$ be the set of arguments of user $u_j$ at step $t$.

   (b) The Deliberating agent of $u_j$ computes the opinion of the group $(G' \cup \{u_i\}) \setminus \{u_j\}$ using the average or MCD operator. Let $\langle (t_1, v_1), \ldots, (t_k, v_k), (r, v) \rangle$ be the result of the aggregation.

   (c) The user $u_j$ may change his bids in light of $\langle (t_1, v_1), \ldots, (t_k, v_k), (r, v) \rangle$. Thus, the set $\text{Args}_j^t$ is revised accordingly. All the arguments that are modified will be replaced by the new ones. Let $\mathcal{T}' \subseteq \mathcal{F}(\text{Best}(I^t))$ be the set of tags whose values are modified. Thus, $\text{Args}_j^t = (\text{Args}_j^t \setminus \{((t, v), \text{Best}(I^t)) \in \text{Args}_j^t \mid t \in \mathcal{T}'\}) \cup \{((t, v'), \text{Best}(I^t)) \mid t \in \mathcal{T}'\}$.

   (d) When the user $u_j$ is sure about his bids, he clicks on a 'Send' button (on the WeShare interface).

7. If for all $u_j \in G' \cup \{u_i\}$, $\text{Res}_j(\text{Best}(I^t)) > 0$, then the session is over.

8. If for all $u_j \in G' \cup \{u_i\}$, $\text{Res}_j(\text{Best}(I^t)) < 0$, then go to Step 12.

9. For all $u_j, u_k \in G' \cup \{u_i\}$ such that $\text{Res}_j(\text{Best}(I^t)) > 0$ and $\text{Res}_k(\text{Best}(I^t)) < 0$, then:

   (a) $\text{Invite}(u_j, \{u_k\})$ (user $u_j$ invites user $u_k$ for a private dialogue).

   (b) User $u_k$ utters either $\text{Accept}(u_k)$ or $\text{Reject}(u_k)$.

   (c) If $\text{Accept}(u_k)$, then $\text{Send}(u_j, \{u_k\}, a)$ where $a$ is an argument, $\text{Conc}(a) = \text{Best}(I^t)$ and either $a \in \text{Args}_j^t$ or $\text{Tag}(a) \notin \mathcal{T}$ (i.e., the user introduces a new argument using a new tag).

   (d) User $u_k$ may revise his opinion about $\text{Tag}(a)$. Thus, $\text{Args}_k^t = (\text{Args}_k^t \setminus \{((\text{Tag}(a), v), \text{Best}(I^t))\}) \cup \{((\text{Tag}(a), v'), \text{Best}(I^t)) \mid v' \neq v\}$.

   (e) Go to Step 10(c) with the roles of the agents reversed (the exchange stops either when the users have no more arguments to send or one of the users decides to exit the dialogue).

10. If $\exists u_j, u_k \in G' \cup \{u_i\}$ such that $\text{Res}_j(\text{Best}(I^t)) > 0$ and $\text{Res}_k(\text{Best}(I^t)) < 0$, then Go to Step 12, otherwise Go to Step 8. (In this case, even after a phase of bilateral persuasion, two users still disagree on the final rating of the current image).

11. Go to Step 4 with:

    - $\mathcal{P} = \{t \in \mathcal{F}(\text{Best}(I^t)) \mid \forall u_j \in G' \cup \{u_i\}, \exists a \in \text{Args}_j^t \text{ such that } \text{Tag}(a) = t, \text{Val}(a) > 0, \text{ and } \text{Conc}(a) = \text{Best}(I^t)\}$,
    - $\mathcal{R} = \{t \in \mathcal{F}(\text{Best}(I^t)) \mid \forall u_j \in G' \cup \{u_i\}, \exists a \in \text{Args}_j^t \text{ such that } \text{Tag}(a) = t, \text{Val}(a) < 0, \text{ and } \text{Conc}(a) = \text{Best}(I^t)\}$,
    - $\mathcal{I}^{t+1} = \mathcal{I}^t \setminus \{\text{Best}(\mathcal{I}^t)\}$.

    These sets are computed by the Deliberating agent of the Administrator of the session.

It is worth mentioning that when a user does not express opinion about a given tag, then he is assumed to be indifferent wrt. that tag. Consequently, the value 0 is assigned to the tag.

Note also that the step 10(a) is not mandatory. Indeed, the invitation to dialogue is initiated by users who really want to persuade their friends.

The previous protocol generates dialogues that terminate either when all the images in the database of the museum are displayed, or when users exit, or when they agree on an image; This means also that the outcome of a dialogue may be either an image on which all users agree or a failure.

## 6   Conclusions

This paper proposed a system that allows a group of users to have a shared online cultural experience. In our system several users are provided *synchronously* with images from the digital collection of a museum. Users can then express their own opinions about each image, and finally make joint decisions about whether or not to accept the image (so as, for example, to buy a hardcopy of that image for another friend). The system has two main components: a Media Server agent which connects to the museum and provides images, and WeShare agents through which users interact with the system and with each other. Finally, although not explicitly covered in the paper, a lightweight version of a peer-to-peer Electronic Institution is responsible for the multiple interactions between the two other components.

From a reasoning point of view, the application involves two multiple criteria decision problems: one at the level of each user and one at the level of the group. Both decision problems are about accepting or not an image sent by the museum shop server. Users individually make their decisions in a non-automatic way through the WeShare interface. However, they are assisted by a software agent which provides an aggregated view of the opinion of the group. This may later be taken into account by the user in order to revise his choices. Two aggregation operators are defined: the first one computes the average of the preferences of the different users whereas the second one applies a multiple criteria aggregation. The decision of the group is made after a negotiation phase where each user tries to persuade other users to change their preferences.

We are currently at the beginning of a European Project and plan to improve the system in a number different ways. The first line of our research with respect to the work described in this paper concerns the aggregation operator that may be used in opinion analysis. We are investigating the possibility of using more sophisticated operators such applying a multiple criteria aggregation of the data provided by each user and then to aggregate the result. Another idea consists of considering weighted tags and providing users with the ability to weight their preferences using HCI devices such as the speed or length of time they press a tag during an online session. More future work consists of us extending the negotiation architecture in order to allow the exchanging arguments built from pre-existing domain ontologies.

**Acknowledgment.** This work was supported by the European framework ERA-Net CHIST-ERA, under contract CHRI-001-03, ACE project "Autonomic Software Engineering for online cultural experiences".

# References

1. Albert, C., Amgoud, L., Costedoat, C., Bannay, D., Saint-Dizier, P.: Introdcing argumentation in opinion analysis: Language and reasoning challenges. In: Proc. of Sentiment Analysis where AI meets Psychology (Workshop at IJCNLP 2011) (2011)
2. Amgoud, L., Besnard, P.: Bridging the Gap between Abstract Argumentation Systems and Logic. In: Godo, L., Pugliese, A. (eds.) SUM 2009. LNCS, vol. 5785, pp. 12–27. Springer, Heidelberg (2009)
3. Amgoud, L., Dimopoulos, Y., Moraitis, P.: A unified and general framework for argumentation-based negotiation. In: Proc. of the 6th International Joint Conference on Autonomous Agents and Multiagent Systems (AAMAS 2007), pp. 963–970. ACM Press (2007)
4. Amgoud, L., Prade, H.: Using arguments for making and explaining decisions. Artificial Intelligence Journal 173, 413–436 (2009)
5. Arcos, J.L., Esteva, M., Noriega, P., Rodríguez-Aguilar, J.A., Sierra, C.: Engineering open environments with electronic institutions. Engineering Applications of Artificial Intelligence 18(2), 191–204 (2005)
6. Atkinson, K., Bench-Capon, T., McBurney, P.: Justifying practical reasoning. In: Proc. of the Fourth Workshop on Computational Models of Natural Argument (CMNA 2004), pp. 87–90 (2004)
7. Besnard, P., Hunter, A.: Elements of Argumentation. MIT Press (2008)
8. Bonet, B., Geffner, H.: Arguing for decisions: A qualitative model of decision making. In: Proc. of the 12th Conference on Uncertainty in Artificial Intelligence (UAI 1996), pp. 98–105 (1996)
9. Bonnefon, J.F., Fargier, H.: Comparing sets of positive and negative arguments: Empirical assessment of seven qualitative rules. In: Proc. of the 17th European Conference on Artificial Intelligence (ECAI 2006), pp. 16–20 (2006)
10. d'Inverno, M., Luck, M., Noriega, P., Rodríguez-Aguilar, J.A., Sierra, C.: Communicating open systems. Artificial Intelligence (in print, 2012)
11. Dubois, D., Fargier, H.: Qualitative decision making with bipolar information. In: Proc. of the 10th International Conference on Principles of Knowledge Representation and Reasoning (KR 2006), pp. 175–186 (2006)
12. Dung, P.M.: On the acceptability of arguments and its fundamental role in nonmonotonic reasoning, logic programming and $n$-person games. Artificial Intelligence Journal 77, 321–357 (1995)
13. Higgins, C.: Arts funding cut 30% in spending review (2010), guardian.co.uk
14. Krishna-Bal, B., Saint-Dizier, P.: Towards building annotated resources for analyzing opinions and argumentation in news editorials. In: Proc. of the International Conference on Language Resources and Evaluation (2010)
15. Newman, K., Tourle, P.: The ImpacT of cuTs on uK museums A report for the Museums Association (2011)
16. Osman, N., Sierra, C., Sabater, J.: Propagation of opinions in structural graphs. In: Proc. of the 17th European Conference on Artificial Intelligence, ECAI 2006 (2006)
17. Pollock, J.: How to reason defeasibly. Artificial Intelligence Journal 57, 1–42 (1992)
18. Prakken, H.: Coherence and flexibility in dialogue games for argumentation. Journal of Logic and Computation 15, 1009–1040 (2005)
19. Prakken, H., Sartor, G.: Argument-based extended logic programming with defeasible priorities. Journal of Applied Non-Classical Logics 7, 25–75 (1997)
20. Subrahmanian, V.: Mining online opinions. IEEE Computer, 88–90 (2009)

# Simple Proof of Basic Theorem
# for General Concept Lattices
# by Cartesian Representation*

Radim Belohlavek, Jan Konecny, and Petr Osicka

DAMOL (Data Analysis and Modeling Laboratory)
Department of Computer Science, Palacký University, Olomouc, Czech Republic
{radim.belohlavek,osicka}@acm.org, jan.konecny@upol.cz

**Abstract.** We promote a useful representation of fuzzy sets by ordinary sets, called the Cartesian representation. In particular, we show how the main structures related to a general type of concept lattices may be reduced using this representation to their ordinary counterparts. As a consequence of this representation, we obtain a simple proof of the basic theorem for this type of concept lattices.

## 1 Motivation and Problem Description

Many models involving ordinary sets and relations, including those used in data analysis, decision making, information retrieval, or automated reasoning, have been subject to extensions in which the ordinary sets and relations are replaced by fuzzy sets and fuzzy relations. While the natural impetus for such extensions comes from the need to extend the applicability of the models from binary data to data involving grades (ordinal data), the technical side of the extensions is far from being obvious. In the search for simple yet powerful principles to aid such extensions, various methods have been proposed. Perhaps the best-known concept in this regard is the concept of representation of fuzzy sets by cuts. In this paper, we focus on models based on closure-like structures derived from a binary relation. Such structures include Galois connections, closure and interior operators and systems, concept lattices, as well as various forms of data dependencies, and have been utilized in several areas, most notably for the purpose of data analysis.

The aim of this paper is twofold. First, we provide a simple proof of the so-called basic theorem of a general type of concept lattices that were proposed in the literature and generalize several existing approaches to concept lattices. Second, we promote a useful representation of fuzzy sets we call the Cartesian representation. This representation is the core of the simple proof and, in fact, it enables to reduce several problems regarding structures with degrees such as data

* We acknowledge support by the Grant No. P202/10/0262 of the Czech Science Foundation (Belohlavek, Konecny, Osicka) and by IGA of Palacky University, No. PrF_20124029 (Konecny, Osicka).

V. Torra et al. (Eds.): MDAI 2012, LNAI 7647, pp. 294–305, 2012.

with fuzzy attributes to ordinary structures. Hence, we argue that analogously to the well-known representation by cuts, the Cartesian representation is a useful tool in fuzzy set theory and its applications.

The paper is organized as follows. In Section 2, we provide preliminaries on sup-preserving aggregation structures and the general type of concept lattices used in the rest of our paper. In Section 3, we present the basic theorem, introduce the Cartesian representation of fuzzy sets, present the results on representing the general concept lattices by their ordinary counterparts and, as a consequence of this representation, a simple proof of the basic theorem. In Section 4, we provide concluding remarks.

## 2  Preliminaries

We assume that the reader is familiar with basic notions from fuzzy logic (see e.g. [6, 18, 19]) and concept lattices and other closure structures (see e.g. [13, 14]).

### 2.1  Supremum-Preserving Aggregation Structures

The notion of a sup-preserving aggregation structure has been introduced in [8] and studied further in [9], see also [2, 10, 23–25] for related work, to which we refer for more details.

Let $\mathbf{L}_i = \langle L_i, \leq_i \rangle$ be complete lattices, for $i = 1, 2, 3$. The operations in $\mathbf{L}_i$ are denoted as usual, adding subscript $i$. That is, the infima, suprema, the least, and the greatest element in $\mathbf{L}_2$ are denoted by $\bigwedge_2$, $\bigvee_2$, $0_2$, and $1_2$, respectively; the same for $\mathbf{L}_1$ and $\mathbf{L}_3$. Consider now an operation $\square : L_1 \times L_2 \to L_3$ that commutes with suprema in both arguments. That is, for any $a, a_j \in L_1$ ($j \in J$), $b, b_{j'} \in L_2$ ($j' \in J'$),

$$\left(\bigvee_{1 \, j \in J} a_j\right) \square \, b = \bigvee_{3 \, j \in J}(a_j \square b) \text{ and } a \square \left(\bigvee_{2 \, j' \in J'} b_{j'}\right) = \bigvee_{3 \, j' \in J'}(a \square b_{j'}). \quad (1)$$

A quadruple $\langle \mathbf{L}_1, \mathbf{L}_2, \mathbf{L}_3, \square \rangle$ satisfying (1) is called a (*supremum preserving*) *aggregation structure*. Note that in our approach, $\langle \mathbf{L}_1, \mathbf{L}_2, \mathbf{L}_3, \square \rangle$ plays the role of a structure of truth degrees in fuzzy logic. In fact, it generalizes the notion of a complete residuated lattice [28]. Due to commuting with suprema, the following operations of residuation may be introduced: $\circ_\square : L_1 \times L_3 \to L_2$ and $\square^\circ : L_3 \times L_2 \to L_1$ (adjoints to $\square$) are defined by

$$a_1 \circ_\square a_3 = \bigvee_2 \{a_2 \mid a_1 \square a_2 \leq_3 a_3\}, \quad (2)$$

$$a_3 \square^\circ a_2 = \bigvee_1 \{a_1 \mid a_1 \square a_2 \leq_3 a_3\}. \quad (3)$$

We put indices in $a_1$ and the like for mnemonic reasons. Thus, $a_1$ indicates that $a_1$ is taken from $L_1$ and the like.

*Example 1.* Let $\langle L, \wedge, \vee, \otimes, \to, 0, 1 \rangle$ be a complete residuated lattice with a partial order $\leq$. The following two particular cases, in which $L_i = L$ and $\leq_i$ is either $\leq$ or the dual of $\leq$ (i.e. $\leq_i = \leq$ or $\leq_i = \leq^{-1}$) are important for our purpose.

(a) Let $\mathbf{L}_1 = \langle L, \leq \rangle$, $\mathbf{L}_2 = \langle L, \leq \rangle$, and $\mathbf{L}_3 = \langle L, \leq \rangle$, let $\square$ be $\otimes$. Then, as is well known [17, 28], $\square$ commutes with suprema in both arguments. Namely, due to commutativity of $\otimes$, commuting amounts to $a \otimes \bigvee_{j \in J} b_j = \bigvee_{j \in J} (a \otimes b_j)$. Furthermore,

$$a_1 \circ_\square a_3 = \bigvee \{ a_2 \mid a_1 \otimes a_2 \leq a_3 \} = a_1 \to a_3$$

and, similarly, $a_3 {}_\square\!\circ a_2 = a_2 \to a_3$.

(b) Let $\mathbf{L}_1 = \langle L, \leq \rangle$, $\mathbf{L}_2 = \langle L, \leq^{-1} \rangle$, and $\mathbf{L}_3 = \langle L, \leq^{-1} \rangle$, let $\square$ be $\to$. Then, $\square$ commutes with suprema in both arguments. Namely, the conditions (1) for commuting with suprema in this case become

$$(\bigvee_{j \in J} a_j) \to b = \bigwedge_{j \in J} (a_j \to b) \text{ and } a \to (\bigwedge_{j \in J} b_j) = \bigwedge_{j \in J} (a \to b_j)$$

which are well-known properties of residuated lattices. In this case, we have

$$a_1 \circ_\square a_3 = \bigwedge \{ a_2 \mid a_1 \to a_2 \geq a_3 \} = a_1 \otimes a_3$$

and

$$a_3 {}_\square\!\circ a_2 = \bigvee \{ a_1 \mid a_1 \to a_2 \geq a_3 \} = a_3 \to a_2.$$

*Example 2.* This example is the structure behind crisply generated concept lattices and the one-sided concept lattices, see [10] for more information. Let $L_1 = \{0,1\}$, $L_2 = [0,1]$, $L_3 = [0,1]$, let $\leq_1$, $\leq_2$, $\leq_3$ be the usual total orders on $L_1$, $L_2$, and $L_3$, respectively. Let $\square$ be defined by $a_1 \square a_2 = \min(a_1, a_2)$. Then $\mathbf{L}_1$, $\mathbf{L}_2$, $\mathbf{L}_3$, and $\square$ satisfy (1). In this case,

$$0 \circ_\square a = 1, \ 1 \circ_\square a = a,$$

and

$$a_3 {}_\square\!\circ a_2 = \begin{cases} 0 & \text{for } a_2 > a_3, \\ 1 & \text{for } a_2 \leq a_3. \end{cases}$$

The following are some of the properties of aggregation structures which are generalizations of well-known properties of residuated lattices [9].

$$a_1 \square a_2 \leq_3 a_3 \text{ iff } a_2 \leq_2 a_1 \circ_\square a_3 \text{ iff } a_1 \leq_1 a_3 {}_\square\!\circ a_2, \tag{4}$$

$$a_1 \square (a_1 \circ_\square a_3) \leq_3 a_3, \tag{5}$$

$$(a_3 {}_\square\!\circ a_2) \square a_2 \leq_3 a_3, \tag{6}$$

$$a \circ_\square (\bigwedge_{3 j \in J} c_j) = \bigwedge_{2 j \in J} (a \circ_\square c_j), \tag{7}$$

$$(\bigvee_{1 j \in J} a_j) \circ_\square c = \bigwedge_{2 j \in J} (a_j \circ_\square c), \tag{8}$$

$$c {}_\square\!\circ (\bigvee_{2 j \in J} b_j) = \bigwedge_{1 j \in J} (c {}_\square\!\circ b_j), \tag{9}$$

$$(\bigwedge_{3 j \in J} c_j) {}_\square\!\circ b = \bigwedge_{1 j \in J} (c_j {}_\square\!\circ b). \tag{10}$$

Note that various monotony conditions follow from (7)–(10). For example, (7) and (8) imply that $\circ_\square$ is isotone in the second and antitone in the first argument, respectively.

## 2.2    Concept Lattices over Sup-preserving Aggregation Structures

The following notions have been introduced in [8]. Note that essentially the same notions were studied in [23] and in [10]. In [23], the notion of a generalized concept lattice has been introduced. In [10], it has been shown that the notion of a generalized concept lattice may be introduced within a framework that practically coincides with the notion of a supremum-preserving aggregation structure. Importantly, generalized concept lattices generalize some previously introduced types of concept lattices and we come back to this issue in the next section.

Let $\langle \mathbf{L}_1, \mathbf{L}_2, \mathbf{L}_3, \square \rangle$ be a sup-preserving aggregation structure. Let $\langle X, Y, I \rangle$ be an $\mathbf{L}_3$-context. That is, $X$ and $Y$ are non-empty sets of objects and attributes, respectively, and $I : X \times Y \to L_3$ is a binary $\mathbf{L}_3$-relation between $X$ and $Y$. For $x \in X$ and $y \in Y$, the degree $I(x, y)$ is interpreted as the degree to which the object $x$ has the attribute $y$. Consider the operators $^{\uparrow} : L_1{}^X \to L_2{}^Y$ and $^{\downarrow} : L_2{}^Y \to L_1{}^X$, called the concept-forming operators, defined by

$$A^{\uparrow}(y) = \bigwedge_{x \in X}(A(x) \circ_{\square} I(x, y)), \tag{11}$$

$$B^{\downarrow}(x) = \bigwedge_{y \in Y}(I(x, y) \,_{\square} \circ B(y)), \tag{12}$$

for any $A \in L_1{}^X$ and $B \in L_2{}^Y$. A formal concept of $I$ is then a pair $\langle A, B \rangle$ consisting of an $\mathbf{L}_1$-set $A$ in $X$ and an $\mathbf{L}_2$-set $B$ in $Y$ for which $A^{\uparrow} = B$ and $B^{\downarrow} = A$. Furthermore, $\mathcal{B}(X, Y, I)$ denotes the set of all formal concepts of $I$, i.e.

$$\mathcal{B}(X, Y, I) = \{\langle A, B \rangle \in L_1{}^X \times L_2{}^Y \mid A^{\uparrow} = B,\ B^{\downarrow} = A\}.$$

## 3    Structure of $\mathcal{B}(X, Y, I)$: Simple Proof of Basic Theorem via Cartesian Representation

### 3.1    The Basic Theorem for $\mathcal{B}(X, Y, I)$, and the Relationships to Other Types of Concept Lattices

In this section, we provide the basic theorem for concept lattices over aggregation structures and describe the relationship of these concept lattices to various particular types of concept lattices proposed in the literature. Some of these or close relationships have partially been described in the literature [10, 23, 24] and we provide the relationships and explicit formulas for completeness and reader's convenience.

As usual, let us introduce a partial order $\leq$ on $\mathcal{B}(X, Y, I)$ by

$$\langle A_1, B_1 \rangle \leq \langle A_2, B_2 \rangle \text{ if and only if } A_1 \subseteq_1 A_2 (\text{or, equivalently, } B_2 \subseteq_2 B_1), \tag{13}$$

for every $\langle A_1, B_1 \rangle, \langle A_2, B_2 \rangle \in \mathcal{B}(X, Y, I)$. Here, $\subseteq_i$ denotes the (bivalent) inclusion relation defined for $C, D \in L_i{}^U$ by $C \subseteq_i D$ iff $C(u) \leq_i D(u)$ for each $u \in U$.

The following theorem is the basic theorem for concept lattices over aggregation structures. Note that its form is essentially as the one for $\mathbf{L}$-concept lattices from [4]. Note also that for so-called generalized concept lattices [23], which are very close to the concept lattices over aggregation structures, the basic theorem is present in [22]. The theorem in [22] contains additional conditions for reasons that we omit here.

Note that a subset $K \subseteq V$ of a complete lattice $\mathbf{V}$ is infimally (supremally) dense in $\mathbf{V}$ if every element of $\mathbf{V}$ is the infimum (supremum) of certain subset of $K$.

**Theorem 1.** *Let $\langle \mathbf{L}_1, \mathbf{L}_2, \mathbf{L}_3, \square \rangle$ be a supremum-preserving aggregation structure. Let $\langle X, Y, I \rangle$ be an $\mathbf{L}_3$-context $\langle X, Y, I \rangle$.*

*(1) $\mathcal{B}(X, Y, I)$ equipped with $\leq$ is a complete lattice with infima and suprema described as:*

$$\bigwedge_{j \in J} \langle A_j, B_j \rangle = \left\langle \bigcap_{j \in J} A_j, \left( \bigcup_{j \in J} B_j \right)^{\uparrow \downarrow} \right\rangle, \bigvee_{j \in J} \langle A_j, B_j \rangle = \left\langle \left( \bigcup_{j \in J} A_j \right)^{\downarrow \uparrow}, \bigcap_{j \in J} B_j \right\rangle \quad (14)$$

*(2) Moreover, a complete lattice $\mathbf{V} = \langle V, \leq \rangle$ is isomorphic to $\mathcal{B}(X, Y, I)$ iff there are mappings $\gamma : X \times L_1 \to V$ and $\mu : Y \times L_2 \to V$ such that $\gamma(X \times L_1)$ is supremally dense in $\mathbf{V}$, $\mu(Y \times L_2)$ is infimally dense in $\mathbf{V}$, and $a \square b \leq_3 I(x, y)$ is equivalent to $\gamma(x, a) \leq \mu(y, b)$ for all $x \in X, y \in Y, a \in L_1, b \in L_2$.*

Let us now consider several particular cases.

*Case 1.* Let $\langle \mathbf{L}_1, \mathbf{L}_2, \mathbf{L}_3, \square \rangle$ be defined by $L_i = \{0, 1\}$, $\leq_i$ be the usual orders on $\{0, 1\}$, let $a \square b = \min(a, b)$ (note that this is a particular structure from Example 1 (a)). It is easily seen that upon identifying ordinary sets with their membership functions, $\uparrow$ and $\downarrow$ become

$$A^{\uparrow} = \{ y \in Y \mid \text{ for each } x \in A : \langle x, y \rangle \in I \}, \quad (15)$$
$$B^{\downarrow} = \{ x \in X \mid \text{ for each } y \in B : \langle x, y \rangle \in I \}. \quad (16)$$

These are the ordinary operators used in formal concept analysis (FCA) [14], and hence $\mathcal{B}(X, Y, I)$ is the set of ordinary formal concepts of the ordinary formal context $\langle X, Y, I \rangle$. Clearly, (13) becomes

$$\langle A_1, B_1 \rangle \leq \langle A_2, B_2 \rangle \text{ if and only if } A_1 \subseteq A_2 \text{ (or, equivalently, } B_2 \subseteq B_1),$$

which is the partial order of formal concepts as used in FCA. Note also that the following theorem, due to Wille, is the so-called basic theorem of ordinary concept lattices [14]. (We include a complete statement because we show in Section 3.3 that Theorem 1 may be proved using this theorem.)

**Theorem 2.** *Let $\langle X, Y, I \rangle$ be an (ordinary) context $\langle X, Y, I \rangle$.*

*(1) $\mathcal{B}(X, Y, I)$ equipped with $\leq$ is a complete lattice with infima and suprema described as:*

$$\bigwedge_{j \in J} \langle A_j, B_j \rangle = \left\langle \bigcap_{j \in J} A_j, \left( \bigcup_{j \in J} B_j \right)^{\uparrow \downarrow} \right\rangle, \bigvee_{j \in J} \langle A_j, B_j \rangle = \left\langle \left( \bigcup_{j \in J} A_j \right)^{\downarrow \uparrow}, \bigcap_{j \in J} B_j \right\rangle \quad (17)$$

*(2) Moreover, a complete lattice* $\mathbf{V} = \langle V, \leq \rangle$ *is isomorphic to* $\mathcal{B}(X, Y, I)$ *iff there are mappings* $\gamma : X \to V$ *and* $\mu : Y \to V$ *such that* $\gamma(X)$ *is supremally dense in* $\mathbf{V}$, $\mu(Y)$ *is infimally dense in* $\mathbf{V}$, *and* $\langle x, y \rangle \in I$ *is equivalent to* $\gamma(x) \leq \mu(y)$ *for all* $x \in X, y \in Y$.

Note that Theorem 2 is in fact a consequence of the present instance of Theorem 1 (i.e. the instance given by the aggregation structure of Case 1). Indeed, this is clear for condition (1). Condition (2) of the present instance of Theorem 2 says "there are mappings $\gamma : X \times \{0, 1\} \to V$ and $\mu : Y \times \{0, 1\} \to V$ such that $\gamma(X \times \{0, 1\})$ is supremally dense in $\mathbf{V}$, $\mu(Y \times \{0, 1\})$ is infimally dense in $\mathbf{V}$, and $a \square b \leq I(x, y)$ is equivalent to $\gamma(x, a) \leq \mu(y, b)$ for all $x \in X, y \in Y, a, b \in \{0, 1\}$". Let us see that this condition, denote it by $(2)^{\{0,1\}}$, implies (2) of Theorem 2: Since $0 \otimes b = 0 \leq I(x, y)$ for any $x, y$, the last condition of $(2)^{\{0,1\}}$ implies that for each $x \in X$, $y \in Y$, and $b \in \{0, 1\}$, we have $\gamma(x, 0) \leq \mu(y, b)$, i.e. $\gamma(x, 0) \leq \bigwedge_{y \in Y, b \in \{0,1\}} \mu(y, b)$. Now, since $\mu(Y \times \{0, 1\})$ is infimally dense in $\mathbf{V}$, $\bigwedge_{y \in Y, b \in \{0,1\}} \mu(y, b)$ equals the least element $0_\mathbf{V}$. Hence, $\gamma(x, 0) = 0_\mathbf{V}$ for each $x \in X$. Similarly, using the supremal density of $\gamma(X, \{0, 1\})$, one may check that $\mu(y, 0) = 1_\mathbf{V}$ for each $y \in Y$. Since $K$ is supremally dense in $\mathbf{V}$ iff $K - \{0_\mathbf{V}\}$ is supremally dense in $\mathbf{V}$, and the same for infimal density of $K$ and $K - \{1_\mathbf{V}\}$, it clearly follows that $\gamma(X \times \{1\})$ is supremally dense in $\mathbf{V}$, $\mu(Y \times \{1\})$ is infimally dense in $\mathbf{V}$, and $1 = \min(1, 1) \leq I(x, y)$ (i.e. $\langle x, y \rangle \in I$) is equivalent to $\gamma(x, 1) \leq \mu(y, 1)$ for all $x \in X, y \in Y$. Now, putting $\gamma'(x) = \gamma(x, 1)$ and $\mu'(y) = \mu(y, 1)$, $\gamma'$ and $\mu'$ satisfy condition (2) of Theorem 2. Conversely, from mappings $\gamma'$ and $\mu'$ satisfying (2) of Theorem 2 one obtains mappings $\gamma$ and $\mu$ satisfying $(2)^{\{0,1\}}$ by putting $\gamma(x, 0) = 0_\mathbf{V}$, $\gamma(x, 1) = \gamma'(x)$, $\mu(y, 0) = 1_\mathbf{V}$, and $\mu(y, 1) = \mu'(y)$.

*Case 2.* Let $\langle \mathbf{L}_1, \mathbf{L}_2, \mathbf{L}_3, \square \rangle$ be defined by $L_i = \{0, 1\}$, $\leq_1 = \leq$ and $\leq_2 = \leq_3 = \leq^{-1}$ where $\leq$ is the usual order on $\{0, 1\}$. Let furthermore $\square$ be the truth function of classical implication (again, this is a particular structure from Example 1 (b)). Upon identifying ordinary sets with their membership functions, $^\uparrow$ and $^\downarrow$ of (11) and (12) become the following operators:

$$A^\cap = \{y \in Y \mid \text{ for some } x \in A : \langle x, y \rangle \in I\}, \tag{18}$$

$$B^\cup = \{x \in X \mid \text{ for each } y \in Y : \text{ if } \langle x, y \rangle \in I \text{ then } y \in B\}. \tag{19}$$

These operators were studied in [15] where the basic relationships to the ordinary concept-forming operators (described in Case 1) were established. Note that the relationships are due to the fact that in classical sets one can use the law of double negation due to which these operators and the ordinary operators are mutually definable. Clearly, (13) becomes

$$\langle A_1, B_1 \rangle \leq \langle A_2, B_2 \rangle \text{ if and only if } A_1 \subseteq A_2 \text{ (or, equivalently, } B_1 \subseteq B_2\text{)}.$$

As we show next, Theorem 1 for the present case is equivalent to the following theorem (proved in [15]).

**Theorem 3.** *Let $\langle X, Y, I \rangle$ be an (ordinary) context $\langle X, Y, I \rangle$.*

*(1) $\mathcal{B}(X, Y, I)$ equipped with $\leq$ is a complete lattice with infima and suprema described as:*

$$\bigwedge_{j \in J} \langle A_j, B_j \rangle = \left\langle \bigcap_{j \in J} A_j, \left( \bigcap_{j \in J} B_j \right)^{\cup \cap} \right\rangle, \quad \bigvee_{j \in J} \langle A_j, B_j \rangle = \left\langle \left( \bigcup_{j \in J} A_j \right)^{\cap \cup}, \bigcup_{j \in J} B_j \right\rangle \tag{20}$$

*(2) Moreover, a complete lattice $\mathbf{V} = \langle V, \leq \rangle$ is isomorphic to $\mathcal{B}(X, Y, I)$ iff there are mappings $\gamma : X \to V$ and $\mu : Y \to V$ such that $\gamma(X)$ is supremally dense in $\mathbf{V}$, $\mu(Y)$ is infimally dense in $\mathbf{V}$, and $\langle x, y \rangle \in I$ is equivalent to $\gamma(x) \not\leq \mu(y)$ for all $x \in X, y \in Y$.*

As with Theorem 2, we now show that Theorem 3 is in fact a consequence of the present instance of Theorem 1. This is clear for condition (1). Condition (2) of the present instance of Theorem 1 says "there are mappings $\gamma : X \times \{0, 1\} \to V$ and $\mu : Y \times \{0, 1\} \to V$ such that $\gamma(X \times \{0, 1\})$ is supremally dense in $\mathbf{V}$, $\mu(Y \times \{0, 1\})$ is infimally dense in $\mathbf{V}$, and $a \to b \geq I(x, y)$ is equivalent to $\gamma(x, a) \leq \mu(y, b)$ for all $x \in X, y \in Y, a, b \in \{0, 1\}$". This condition, denote it by $(2)^{\{0,1\}}$, implies (2) of Theorem 3: Since $0 \to b = 1 \geq I(x, y)$ for any $x, y$, the last condition of $(2)^{\{0,1\}}$ implies that for each $x \in X$, $y \in Y$, and $b \in \{0, 1\}$, we have $\gamma(x, 0) \leq \mu(y, b)$, i.e. $\gamma(x, 0) \leq \bigwedge_{y \in Y, b \in \{0,1\}} \mu(y, b)$. The infimal density of $\mu(Y \times \{0, 1\})$ in $\mathbf{V}$ yields $\bigwedge_{y \in Y, b \in \{0,1\}} \mu(y, b) = 0_{\mathbf{V}}$. Hence, $\gamma(x, 0) = 0_{\mathbf{V}}$ for each $x \in X$. Similarly one may check that for each $y \in Y$. As in Case 1, it then follows that $\gamma(X \times \{1\})$ is supremally dense in $\mathbf{V}$, $\mu(Y \times \{0\})$ is infimally dense in $\mathbf{V}$, and $0 = 1_3 = 1_1 \square 1_2 = 1 \to 0 \geq I(x, y)$ (i.e. $\langle x, y \rangle \notin I$) is equivalent to $\gamma(x, 1) = \gamma(x, 1_1) \leq \mu(y, 1_2) = \mu(y, 0)$. That is, $\langle x, y \rangle \in I$ is equivalent to $\gamma(x, 1) \not\leq \mu(y, 0)$ for all $x \in X, y \in Y$. Now, putting $\gamma'(x) = \gamma(x, 1)$ and $\mu'(y) = \mu(y, 0)$, $\gamma'$ and $\mu'$ satisfy condition (2) of Theorem 3. Conversely, from mappings $\gamma'$ and $\mu'$ satisfying (2) of Theorem 3 one obtains mappings $\gamma$ and $\mu$ satisfying $(2)^{\{0,1\}}$ by putting $\gamma(x, 0) = 0_{\mathbf{V}}$, $\gamma(x, 1) = \gamma'(x)$, $\mu(y, 0) = \mu'(y)$, and $\mu(y, 1) = 1_{\mathbf{V}}$.

*Case 3.* Let $\mathbf{L} = \langle L, \wedge, \vee, \otimes, \to, 0, 1 \rangle$ be a complete residuated lattice with a partial order $\leq$ and let $\langle \mathbf{L}_1, \mathbf{L}_2, \mathbf{L}_3, \square \rangle$ be as in Example 1 (a). Then the operators $^{\uparrow}, ^{\downarrow}$ become

$$A^{\uparrow}(y) = \bigwedge_{x \in X} A(x) \to I(x, y), \quad B^{\downarrow}(x) = \bigwedge_{y \in Y} B(y) \to I(x, y). \tag{21}$$

These operators were introduced independently in [3] and [27]. The subconcept-superconcept ordering is defined the same way as in Case 1 only this time we consider the (bivalent) inclusion relation on fuzzy sets instead the ordinary inclusion relation on ordinary sets. Note that the basic theorem proved in [6, 7] and [27] is just a particular instance of Theorem 1.

*Case 4.* Let $\mathbf{L} = \langle L, \wedge, \vee, \otimes, \to, 0, 1 \rangle$ be a complete residuated lattice with a partial order $\leq$ and let $\langle \mathbf{L}_1, \mathbf{L}_2, \mathbf{L}_3, \square \rangle$ be as in Example 1 (b). The concept-forming operators then become precisely the operators

$$A^{\cap}(y) = \bigvee_{x \in X} A(x) \otimes I(x, y), \quad B^{\cup}(x) = \bigwedge_{y \in Y} I(x, y) \to B(y) \tag{22}$$

studied in [16]. The concept ordering is defined similarly as in Case 2, the only difference is that we consider the ordinary subsethood relation between fuzzy sets. Again, the basic theorem proved in [16] is just a particular case of Theorem 1. Note also that the relational products related to these operators were extensively studied [1, 20, 21].

## 3.2 Cartesian Representation of Fuzzy Sets

The Cartesian representation of fuzzy sets is investigated in [6]. It has been utilized for the first time in the context of formal concept analysis of data with attributes in [5, 27], see also [11]. The basic ideas are as follows. For a complete lattice $\mathbf{L} = \langle L, \leq \rangle$ and a fuzzy set $A$ in $X$ with truth degrees in $L$, we put

$$\lfloor A \rfloor = \{\langle x, a \rangle \in X \times L \mid a \leq A(x)\}$$

For an ordinary set $A' \subseteq X \times L$ define an **L**-set $\lceil A' \rceil$ in $X$ by

$$\lceil A' \rceil (x) = \bigvee \{a \mid \langle x, a \rangle \in A'\}.$$

That is, $\lfloor A \rfloor$ is the "area below the membership function", see Fig. 1, while $\lceil A' \rceil$ may be thought of an "envelope" of $A'$. We say that $A' \subseteq X \times L$ is **L**-*set-representative* if

- if $\langle x, a \rangle \in A'$ and $b \leq a$ then $\langle x, b \rangle \in A'$.
- $\{a \mid \langle x, a \rangle \in A'\}$ has a greatest element.

The following lemma is easy to verify.

**Lemma 1.** *(a)* $\lfloor A \rfloor$ *is* **L**-*set-representative for any* $A \in L^X$.
   *(b)* $A' = \lfloor \lceil A' \rceil \rfloor$ *for each* **L**-*set-representative* $A' \subseteq X \times L$.
   *(c)* $A = \lceil \lfloor A \rfloor \rceil$ *for each* $A \in L^X$.

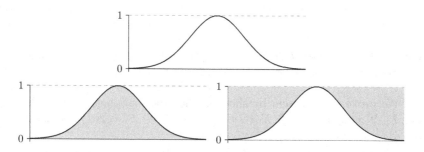

**Fig. 1.** A $\langle [0,1], \leq \rangle$-set $A$ (top), a corresponding $\lfloor A \rfloor$ for $\langle [0,1], \leq \rangle$ (bottom left) and $\lfloor A \rfloor$ for $\langle [0,1], \leq^{-1} \rangle$ (bottom right)

### 3.3  Representation Theorems and Simple Proof of Basic Theorem

In this section we present a useful representation of formal contexts and concept lattices over aggregation structures and an application of these these representations, namely, a simple proof of Theorem 1. We need the following lemma [8, Claim].

**Lemma 2.** *For every* $\mathbf{L}_3$*-context* $\langle X, Y, I \rangle$ *the following conditions hold.*

(a) *If* $\langle A, B \rangle \in \mathcal{B}(X, Y, I)$ *then* $A(x) \square B(y) \leq_3 I(x, y)$ *for every* $x \in X$, $y \in Y$. *Moreover, there is no* $\langle A', B' \rangle$ *with* $A' \supseteq_1 A$, $B' \supseteq_2 B$ *other than* $\langle A, B \rangle$ *such that* $A'(x) \square B'(y) \leq_3 I(x, y)$ *for every* $x \in X$, $y \in Y$.

(b) *If* $A(x) \square B(y) \leq_3 I(x, y)$ *for every* $x \in X$, $y \in Y$ *then there exists* $\langle A', B' \rangle \in \mathcal{B}(X, Y, I)$ *such that* $A \subseteq_1 A'$ *and* $B \subseteq_2 B'$.

For a sup-preserving aggregation structure $\langle \mathbf{L}_1, \mathbf{L}_2, \mathbf{L}_3, \square \rangle$ and an $\mathbf{L}_3$-context $\langle X, Y, I \rangle$, consider the ordinary context $\langle X \times L_1, Y \times L_2, I^{\times} \rangle$, where $I^{\times} \subseteq (X \times L_1) \times (Y \times L_2)$ is defined by

$$\langle \langle x, a \rangle, \langle y, b \rangle \rangle \in I^{\times} \text{ iff } a \square b \leq_3 I(x, y).$$

Furthermore, denote by $\langle ^{\Uparrow}, ^{\Downarrow} \rangle$ the ordinary concept-forming operators induced by $\langle X \times L_1, Y \times L_2, I^{\times} \rangle$ (as in (15) and (16)).

**Lemma 3.** *For every* $\langle A', B' \rangle \in \mathcal{B}(X \times L_1, Y \times L_2, I^{\times})$, $A'$ *and* $B'$ *are* $\mathbf{L}_1$- *and* $\mathbf{L}_2$-*set representative, respectively.*

*Proof.* Since $A' = B'^{\Downarrow}$, we have $\langle x, a \rangle \in A'$ iff for each $\langle y, b \rangle \in B'$: $\langle \langle x, a \rangle, \langle y, b \rangle \rangle \in I^{\times}$, i.e. iff for each $\langle y, b \rangle \in B'$: $a \square b \leq_3 I(x, y)$, i.e. iff for each $\langle y, b \rangle \in B'$: $a \leq_1 I(x, y) \square^{\circ} b$. It is now clear that, first, if $\langle x, a \rangle \in A'$ and $c \leq_1 a$, then $\langle x, c \rangle \in A'$ and, second, the largest $a$ for which $\langle x, a \rangle \in A'$ is $a = \bigwedge_{1 \langle y, b \rangle \in B'} (I(x, y) \square^{\circ} b)$. Therefore, $A'$ is $\mathbf{L}_1$-set representative. The proof for $B'$ is analogous. $\square$

The following theorem provides a reduction result which, as we show below, enables us to obtain a simple proof of Theorem 1.

**Theorem 4.** *The concept lattice* $\mathcal{B}(X, Y, I)$ *over* $\langle \mathbf{L}_1, \mathbf{L}_2, \mathbf{L}_3, \square \rangle$ *is isomorphic to the ordinary concept lattice* $\mathcal{B}(X \times L_1, Y \times L_2, I^{\times})$.

*Proof.* We prove the theorem by showing that the mappings $\varphi \colon \mathcal{B}(X, Y, I) \to \mathcal{B}(X \times L_1, Y \times L_2, I^{\times})$, $\psi \colon \mathcal{B}(X \times L_1, Y \times L_2, I^{\times}) \to \mathcal{B}(X, Y, I)$ defined by

$$\varphi(\langle A, B \rangle) = \langle \lfloor A \rfloor, \lfloor B \rfloor \rangle, \tag{23}$$

$$\psi(\langle A', B' \rangle) = \langle \lceil A' \rceil, \lceil B' \rceil \rangle \tag{24}$$

for $\langle A, B \rangle \in \mathcal{B}(X, Y, I)$, $\langle A', B' \rangle \in \mathcal{B}(X \times L_1, Y \times L_2, I^{\times})$ are well-defined, mutually inverse, order-preserving bijections.

Lemma 2 implies that $\varphi(\langle A, B \rangle) \in \mathcal{B}(X \times L_1, Y \times L_2, I^\times)$ and $\psi(\langle A', B' \rangle) \in \mathcal{B}(X, Y, I)$. Indeed, for $\langle A, B \rangle \in \mathcal{B}(X, Y, I)$ we have

$\langle y, b \rangle \in \lfloor A \rfloor^{\Uparrow}$ iff

for each $\langle x, a \rangle \in \lfloor A \rfloor$ we have $\langle \langle x, a \rangle, \langle y, b \rangle \rangle \in I^\times$ iff

for each $x \in X$ and $a \leq_1 A(x)$ we have $a \square b \leq_3 I(x, y)$ iff

for all $x \in X$ we have $A(x) \square b \leq_3 I(x, y)$ iff

$b \leq_2 B(y)$, proving $\lfloor A \rfloor^{\Uparrow} = \lfloor B \rfloor$.

The proof of $\lfloor B \rfloor^{\Downarrow} = \lfloor A \rfloor$ is analogous. This proves that $\varphi$ is well defined. For $\langle A', B' \rangle \in \mathcal{B}(X \times L_1, Y \times L_2, I^\times)$ we have

$$\lceil A' \rceil^{\uparrow}(y) = \bigwedge\nolimits_{2x \in X}(\lceil A' \rceil(x) \circ_\square I(x, y))$$
$$= \bigvee\nolimits_2 \{ b \mid b \leq_2 (\lceil A' \rceil(x) \circ_\square I(x, y)) \text{ for each } x \in X \}$$
$$= \bigvee\nolimits_2 \{ b \mid \lceil A' \rceil(x) \square b \leq_3 I(x, y) \text{ for each } x \in X \}$$
$$= \bigvee\nolimits_2 \{ b \mid a \square b \leq_3 I(x, y) \text{ for each } \langle x, a \rangle \in A' \}$$
$$= \bigvee\nolimits_2 \{ b \mid \langle \langle x, a \rangle, \langle y, b \rangle \rangle \in I^\times \text{ for each } \langle x, a \rangle \in A' \}$$
$$= \bigvee\nolimits_2 \{ b \mid \langle y, b \rangle \in A'^{\Uparrow} \} = \bigvee\nolimits_2 \{ b \mid \langle y, b \rangle \in B' \} = \lceil B' \rceil(y).$$

The proof of $\lceil B' \rceil^{\downarrow}(x) = \lceil A' \rceil(x)$ is analogous. This proves that $\psi$ is well defined.

Lemma 3 and Lemma 1 now imply that $\varphi$ and $\psi$ are mutually inverse bijections. In addition, it is immediate to observe that both $\varphi$ and $\psi$ are order preserving.                                                                    □

Now, Theorem 4 can be utilized to prove Theorem 1 by reduction from Theorem 2—the basic theorem of ordinary concept lattices [14].

*Proof of Theorem 1* Condition (1) follows directly from the fact that the pair of mappings $\langle \uparrow, \downarrow \rangle$ forms a Galois connection between the complete lattices $\langle L_1{}^X, \subseteq_1 \rangle$ and $\langle L_2{}^Y, \subseteq_2 \rangle$ by the well-known description of infima and suprema in the lattice of fixpoints of Galois connections [12, 13, 26].

(2) "⇒": Assume that **V** is isomorphic to $\mathcal{B}(X, Y, I)$. By Theorem 4 **V** is isomorphic to $\mathcal{B}(X \times L_1, Y \times L_2, I^\times)$ which is an ordinary concept lattice. Applying to this ordinary concept lattice the basic theorem of ordinary concept lattice, i.e. Theorem 2, we get mappings $\gamma$ and $\mu$ such that $\gamma(X \times L_1)$ and $\mu(Y \times L_2)$ are supremally and infimally dense in **V**, respectively, and $\gamma(x, a) \leq \mu(y, b)$ iff $\langle \langle x, a \rangle, \langle y, b \rangle \rangle \in I^\times$. Taking into account that by definition of $I^\times$, $\langle \langle x, a \rangle, \langle y, b \rangle \rangle \in I^\times$ is equivalent to $a \square b \leq_3 I(x, y)$, we see that $\gamma$ and $\mu$ are the required mappings.

"⇐": Assume that $\gamma$ and $\mu$ have the required properties. Theorem 2 (2) yields that **V** is isomorphic to $\mathcal{B}(X \times L_1, Y \times L_2, I^\times)$. Hence, **V** is isomorphic to $\mathcal{B}(X, Y, I)$ due to Theorem 4.                                              □

# 4   Conclusions

We provided a simple proof of the basic theorem for a general type of concept lattices. We utilized a useful, not well-known representation of fuzzy sets using ordinary sets, called the Cartesian representation. As a side-effect, we explained in detail the relationship of the general type of concept lattices to the main existing types of concept lattices. An interesting topic for future research is to explore further applications of the Cartesian representation in systems based on fuzzy sets and relations, in particular in the context of mathematical fuzzy logic (i.e. fuzzy logic in the narrow sense).

# References

1. Bandler, W., Kohout, L.J.: Semantics of implication operators and fuzzy relational products. Int. J. Man-Machine Studies 12, 89–116 (1980)
2. Bartl, E., Belohlavek, R.: Sup-t-norm and inf-residuum are a single type of relational equations. Int. Journal of General Systems 40(6), 599–609 (2011)
3. Belohlavek, R.: Fuzzy Galois connections. Math. Logic Quarterly 45, 497–504 (1999)
4. Belohlavek, R.: Lattices of fixed points of fuzzy Galois connections. Math. Logic Quarterly 47(1), 111–116 (2001)
5. Belohlavek, R.: Reduction and a simple proof of characterization of fuzzy concept lattices. Fundamenta Informaticae 46(4), 277–285 (2001)
6. Belohlavek, R.: Fuzzy Relational Systems: Foundations and Principles. Kluwer, Academic/Plenum Publishers, New York (2002)
7. Belohlavek, R.: Concept lattices and order in fuzzy logic. Annals of Pure and Applied Logic 128(1-3), 277–298 (2004)
8. Belohlavek, R.: Optimal decompositions of matrices with entries from residuated lattices. J. Logic and Computation (September 7, 2011), doi: 10.1093/logcom/exr023
9. Belohlavek, R.: Sup-t-norm and inf-residuum are one type of relational product: unifying framework and consequences. Fuzzy Sets and Systems 197(16), 45–58 (2012)
10. Belohlavek, R., Vychodil, V.: What is a fuzzy concept lattice? In: Proc. CLA 2005, 3rd Int. Conference on Concept Lattices and Their Applications, pp. 34–45 (2005)
11. Belohlavek, R., Vychodil, V.: Formal concept analysis and linguistic hedges. Int. J. General Systems 41(5), 503–532 (2012)
12. Birkhoff, G.: Lattice Theory, vol. 25. AMS Colloq. Publ. (1940)
13. Davey, B.A., Priestley, H.A.: Introduction to Lattices and Order. Cambridge University Press (2002)
14. Ganter, B., Wille, R.: Formal Concept Analysis. Mathematical Foundations. Springer, Berlin (1999)
15. Gediga, G., Düntsch, I.: Modal-style operators in qualitative data analysis. In: Proc. IEEE ICDM 2002, pp. 155–162 (2002)
16. Georgescu, G., Popescu, A.: Non-dual fuzzy connections. Archive for Mathematical Logic 43, 1009–1039 (2004)
17. Goguen, J.A.: L-fuzzy sets. J. Math. Anal. Appl. 18, 145–174 (1967)
18. Gottwald, S.: A Treatise on Many-Valued Logics. Research Studies Press, Baldock (2001)

19. Hájek, P.: Metamathematics of Fuzzy Logic. Kluwer, Dordrecht (1998)
20. Kohout, L.J., Bandler, W.: Relational-product architectures for information processing. Information Sciences 37, 25–37 (1985)
21. Kohout, L.J., Kim, E.: The role of BK-products of relations in soft computing. Soft Computing 6, 92–115 (2002)
22. Krajči, S.: The basic theorem on generalized concept lattices. In: Proc. CLA 2004, pp. 25–33 (2004)
23. Krajči, S.: A generalized concept lattice. Logic J. of IGPL 13, 543–550 (2005)
24. Medina, J., Ojeda-Aciego, M., Ruiz-Claviño, J.: Formal concept analysis via multi-adjoint concept lattices. Fuzzy Sets and Systems 160, 130–144 (2009)
25. Morsi, N.N., Lotfallah, W., El-Zekey, M.S.: The logic of tied implications, part 1: Properties, applications and representation; part 2: Syntax. Fuzzy Sets and Systems 157, 647–669, 2030–2057 (2006)
26. Ore, O.: Galois connexions. Trans. Amer. Math. Soc. 55, 493–513 (1944)
27. Pollandt, S.: Fuzzy Begriffe. Springer, Berlin (1997)
28. Ward, M., Dilworth, R.P.: Residuated lattices. Trans. Amer. Math. Soc. 45, 335–354 (1939)
29. Zadeh, L.A.: Fuzzy sets. Information and Control 8, 338–353 (1965)

# On Some Properties of the Negative Transitivity Obtained from Transitivity

Susana Díaz[1], Susana Montes[1], and Bernard De Baets[2]

[1] Dept. of Statistics and O.R., University of Oviedo
Calvo Sotelo s/n, 33007 Oviedo, Spain
http://unimode.uniovi.es/
{diazsusana,montes}@uniovi.es
[2] Dept. of Appl. Math., Biometrics and Process Control, Ghent University
Coupure Links 653, B-9000 Gent, Belgium
bernard.debaets@ugent.be

**Abstract.** For crisp relations the transitivity of a relation and the negative transitivity of its dual are equivalent conditions. Particularly, a crisp complete large preference relation is transitive if and only if its associated strict preference relation is negatively transitive. In this contribution we focus on one of those implications for fuzzy relations. Recall that in the context of fuzzy relations there are multiple ways of obtaining the strict preference relation from the large preference relation, and also multiple ways for defining transitivity. We analyze the type of negative transitivity we can assure for the strict preference relation, departing from a large preference relation that satisfies almost any kind of transitivity. We recall the general expression we obtained and study some interesting properties. Finally, we pay special attention to the particular case of the minimum t-norm both in the role of generator and in the role of conjunctor defining the transitivity of the original reflexive relation.

**Keywords:** fuzzy relation, transitivity, negative transitivity, conjunctor, disjunctor.

## 1 Introduction

Coherence is a permanent goal in decision making. Many different properties have been introduced with the aim of formalizing this condition. In this contribution we study in depth the connection between two of the most important ones.

In the context of preference modeling the comparison of alternatives is usually carried out pairwisely. The decision maker is asked to compare every two alternatives. In the crisp context, this is, when no degrees of preference of indifference are allowed, the results of the comparison by pairs can be summarized by a crisp reflexive relation, denoted $R$, called large or weak preference relation. It connects an alternative $a$ to another $b$ if the first one is considered by the decision maker at least as good as the second one. In parallel with the large preference relation,

V. Torra et al. (Eds.): MDAI 2012, LNAI 7647, pp. 306–317, 2012.
© Springer-Verlag Berlin Heidelberg 2012

the answers of the decision maker can be formalized by a preference structure. It is a triplet of binary relations. The first one of those relations is usually called strict preference relation and connects two alternatives when the first one is considered strictly better than the second one; the second relation is the indifference relation, that connects two alternatives when they are equally good for the decision maker; and the third relation is the incomparability relation, that connects two alternatives when the decision maker cannot order them. When the decision maker is able to order all the pairs of alternatives, the incomparability relation does not connect any pair and we talk about completeness.

Obviously, since the large preference relation and the preference structure associated to a decision maker summarize the same information, they are related. In fact, they are equivalent: if we know the large preference relation associated to the answers of a decision maker, we can build the three relations of the preference structure and the other way around.

Going back to coherence, when binary relations are involved, transitivity is probably the most important and most employed property for implementing coherence. It is well known that under completeness, the transitivity of a large preference relation is characterized by the negative transitivity of the corresponding strict preference relation. In fact, the equivalence holds in a wider context: it is known that for any relation, it is transitive if and only if its dual relation is negatively transitive. Under completeness, the large preference and the strict preference relations associated to the answers of a decision maker become dual relations and the result applies.

This contribution is devoted to study one of those implications when degrees of preferences are allowed. The main drawback of crisp relations is that they lack of flexibility. They do not express accurately real life situations. Preferences expressed by humans usually involve imprecise answers. Expressions as "slightly preferred", "more or less the same" are frequent in natural speaking. Fuzzy relations were introduced to tackle these situations. They allow the decision maker to express an intermediate degree of connection. Since fuzzy relations are a generalization of crisp relations, a first effort was developed to extend notions as large preference relation and preference structure to the new context. Many different proposals appeared for the definition of fuzzy preference structure. See [3] for a historical account of its development. We will deal with additive fuzzy preference structures. This definition seems to be the best for generalizing the spirit of (crisp) preference structures.

As in the crisp case, in the fuzzy sets context transitivity plays an important role when modeling rationality. Traditionally, transitivity of a fuzzy relation is defined with respect to a triangular norm and negative transitivity with respect to a triangular conorm. In this setting, it is also known (see e.g. [13]) that a valued binary relation is $T$-transitive if and only if its dual relation is negatively $S$-transitive, where $S$ is the dual triangular conorm of $T$. So, also in this context, if the strict preference relation is obtained from the weak preference relation as its dual, the connection is guaranteed. However, for fuzzy relations, there are multiple ways to obtain the preference structure from the large preference relation. Also, there are different ways to define completeness of a fuzzy relation. We are interested in the connection between the transitivity of the large preference relation and the negative transitivity of any associated strict preference relation. The study is carried out under weak completeness. Also, following the general study we have carried out in previous works [6–8, 10], we have not restricted to t-norms to define transitivity, but we have considered a wider family of operators: conjunctors. In this general setting, we study the strongest type of negative transitivity we can assure for the strict preference relation.

The paper is divided in six sections. In Section 2 we recall the definition of large preference relation and preference structure when only crisp relations are considered and later we recall the same notions in the general framework of fuzzy relations. In Section 3 we include the definition of conjunctor, the operator that generalizes t-norms. We also introduce other operators involved in our study. Section 4 contains the general result connecting transitivity and negative transitivity and some properties. In Section 5 we focus on the main t-norm, the minimum and try to simplify the general expression for this operator. Section 6 contains some conclusions and open problems.

## 2   Preference Structures

As commented in the introduction, preference structures summarize the answers of a decision maker when he/she is confronted with a set of alternatives and must compare them two by two.

We will denote the set of alternatives $A$ and $A^2$ will be the set of ordered pairs of alternatives in $A$. Given two alternatives, the decision maker can act in one of the following three ways: (i) he/she clearly prefers one to the other; (ii) the two alternatives are indifferent to him/her; (iii) he/she is unable to compare the two alternatives. According to these cases, we can define three (binary) relations on $A$: the strict preference relation $P$, the indifference relation $I$ and the incomparability relation $J$. Thus, for any $(a, b) \in A^2$, we classify:

$$(a, b) \in P \quad \Leftrightarrow \quad \text{he/she prefers } a \text{ to } b;$$
$$(a, b) \in I \quad \Leftrightarrow \quad a \text{ and } b \text{ are indifferent to him/her;}$$
$$(a, b) \in J \quad \Leftrightarrow \quad \text{he/she is unable to compare } a \text{ and } b.$$

Observe that we are identifying relations with subsets of $A^2$. The notation $(a, b) \in Q$ stands for $a$ connected to $b$ by $Q$, this is, $aQb$. This can also be expressed by means of its characteristic mapping, $Q(a, b) = 1$. We recall that for a relation $Q$ on $A$, its converse is defined as $Q^t = \{(b, a) \mid (a, b) \in Q\}$, its complement as $Q^c = \{(a, b) \mid (a, b) \notin Q\}$ and its dual as $Q^d = (Q^t)^c$. One easily verifies that $P$, $I$, $J$ and $P^t$ establish a particular partition of $A^2$ [15], this is, they cover all the possible answers of the decision maker and for any ordered pair of alternatives $(a, b)$ only one of those answers can be given. The underlying idea of a preference structure is to write mathematically the preferences of a decision maker. The formal definition is the following one.

**Definition 1.** *A preference structure on $A$ is a triplet $(P, I, J)$ of relations on $A$ that satisfies:*

*(i) $P$ is irreflexive, $I$ is reflexive and $J$ is irreflexive;*
*(ii) $P$ is asymmetric, $I$ and $J$ are symmetric;*
*(iii) $P \cap I = \emptyset$, $P \cap J = \emptyset$ and $I \cap J = \emptyset$;*
*(iv) $P \cup P^t \cup I \cup J = A^2$.*

A preference structure $(P, I, J)$ on $A$ is characterized by the reflexive relation $R = P \cup I$, called large preference relation, in the following way:

$$(P, I, J) = (R \cap R^d, R \cap R^t, R^c \cap R^d) .$$

Conversely, for any reflexive relation $R$ on $A$, the triplet $(P, I, J)$ constructed in this way from $R$ is a preference structure on $A$ such that $R = P \cup I$. As $R$ is the union of the strict preference relation and the indifference relation, $(a, b) \in R$ means that $a$ is at least as good as $b$.

A relation $Q$ on $A$ is called complete if $(a, b) \in Q \vee (b, a) \in Q$, for all $(a, b) \in A^2$. In the crisp sets context, the completeness of the large preference relation is characterized by the absence of incomparability in the associated preference structure. The large preference relation is complete if and only if its associated incomparability relation $J$ is empty.

A relation $Q$ on $A$ is called transitive if $((a, b) \in Q \wedge (b, c) \in Q) \Rightarrow (a, c) \in Q$, for any $(a, b, c) \in A^3$. A relation $Q$ on $A$ is called negatively transitive if $(a, c) \in Q \Rightarrow ((a, b) \in Q \vee (b, c) \in Q)$, for any $(a, b, c) \in A^3$. This property can be also expressed as $((a, b) \notin Q \wedge (b, c) \notin Q) \Rightarrow (a, c) \notin Q)$.

The transitivity of the large preference relation $R$ can be characterized as follows [1].

**Theorem 1.** *For any reflexive complete relation $R$ with associated preference structure $(P, I, J)$ it holds that*

$$R \text{ is transitive } \Leftrightarrow P \text{ is negatively transitive.}$$

Finally, we recall an important characterization of preference structures. Let us identify relations with their characteristic mappings, then Definition 1 can be written in the following minimal way [5]: $I$ is reflexive and symmetric, and for any $(a, b) \in A^2$ it holds that

$$P(a, b) + P^t(a, b) + I(a, b) + J(a, b) = 1 .$$

Classical, also called crisp, preference structures can therefore also be considered as Boolean preference structures, employing 1 and 0 for describing presence or absence of strict preference, indifference and incomparability.

## 2.1   Additive Fuzzy Preference Structures

A serious drawback of classical preference structures is their inability to express intensities. In contrast, in fuzzy preference modelling, relations are a matter of degree. Fuzzy relations can take any value in the unit interval $[0, 1]$ and those values are used for capturing the intermediate intensities of the relations.

The intersection of fuzzy relations is defined pointwisely based on some triangular norm (t-norm for short), *i.e.* an increasing, commutative and associative binary operation on $[0, 1]$ with neutral element 1. The three most important t-norms are the minimum operator $T_{\mathbf{M}}(x, y) = \min(x, y)$, the algebraic product $T_{\mathbf{P}}(x, y) = x \cdot y$ and the Łukasiewicz t-norm $T_{\mathbf{L}}(x, y) = \max(x + y - 1, 0)$. According to the usual ordering of functions, the above t-norms can be ordered as follows: $T_{\mathbf{L}} \leq T_{\mathbf{P}} \leq T_{\mathbf{M}}$. In fact, the minimum operator is greater than any other t-norm, it is the greatest t-norm.

Similarly, the union of fuzzy relations is based on a t-conorm, *i.e.* an increasing, commutative and associative binary operation on $[0, 1]$ with neutral element 0. T-norms and t-conorms come in dual pairs: to any t-norm $T$ there corresponds a t-conorm $S$ through the relationship $S(x, y) = 1 - T(1 - x, 1 - y)$. For the above three t-norms, we thus obtain the maximum operator $S_{\mathbf{M}}(x, y) = \max(x, y)$, the probabilistic sum $S_{\mathbf{P}}(x, y) = x + y - xy$ and the Łukasiewicz t-conorm (bounded sum) $S_{\mathbf{L}}(x, y) = \min(x + y, 1)$. For more background on t-norms and t-conorms and the notations used in this paper, we refer to [14].

T-conorms are used to define completeness. A fuzzy relation $Q$ on $A$ is $S$-complete if $S(Q(a, b), Q(b, a)) = 1$ for all $(a, b) \in A^2$ (see for example [2]). The two most important types of completeness are defined by the Łukasiewicz and maximum t-conorm:

– A fuzzy relation $Q$ on $A$ is called weakly complete if it is $S_{\mathbf{L}}$-complete: $Q(a, b) + Q(b, a) = 1$ for all $(a, b) \in A^2$.
– A fuzzy relation $Q$ on $A$ is called strongly complete if it is $S_{\mathbf{M}}$-complete: $max(Q(a, b), Q(b, a)) = 1$.

In this work, we will focus on weak completeness as it shows an important property we recall in Proposition 1.

The definition of a fuzzy preference structure has been a topic of debate during several years (see e.g. [13, 16, 17]). Accepting the *assignment principle* — for any pair of alternatives $(a, b)$ the decision maker is allowed to assign at least one of the degrees $P(a, b)$, $P(b, a)$, $I(a, b)$ and $J(a, b)$ freely in the unit interval — has finally led to a fuzzy version of Definition 1 with intersection based on the Łukasiewicz t-norm and union based on the Łukasiewicz t-conorm.

Another topic of controversy has been how to construct such a fuzzy preference structure from a reflexive fuzzy relation. The most recent and most successful approach is that of De Baets and Fodor based on (indifference) generators [4].

**Definition 2.** *A generator $i$ is a commutative binary operation on the unit interval $[0, 1]$ that is bounded by the Łukasiewicz t-norm $T_{\mathbf{L}}$ and the minimum operator $T_{\mathbf{M}}$, i.e. $T_{\mathbf{L}} \leq i \leq T_{\mathbf{M}}$.*

Note that despite they have neutral element 1 and are bounded between two t-norms, generators are not necessarily t-norms. For any reflexive fuzzy relation $R$ on $A$ it holds that the triplet $(P, I, J)$ of fuzzy relations on $A$ defined by:

$$P(a, b) = R(a, b) - i(R(a, b), R(b, a)),$$

$$I(a, b) = i(R(a, b), R(b, a)),$$
$$J(a, b) = i(R(a, b), R(b, a)) - (R(a, b) + R(b, a) - 1).$$

is an additive fuzzy preference structure on $A$ such that $R = P \cup_{S_L} I$, i.e.
$R(a, b) = P(a, b) + I(a, b)$.

Recall that a binary operation $f : [0, 1]^2 \to [0, 1]$ is 1-Lipschitz continuous if

$$|f(x_1, y_1) - f(x_2, y_2)| \leq |x_1 - x_2| + |y_1 - y_2|,$$

for any $(x_1, x_2, y_1, y_2) \in [0, 1]^4$. We proved in [8] that the 1-Lipschitz property
plays an important role in the study of the propagation of the transitivity from
a weak preference relation to its associated strict preference and indifference
relation. In this contribution, it plays again an important role. Let us recall
that the two most employed generators, the Łukasiewicz and the minimum t-
norms, are 1-Lipschitz. The Łukasiewicz operator plays a very special role as the
following result shows.

**Proposition 1.** *Consider an additive fuzzy preference structure $(P, I, J)$ obtained from a reflexive fuzzy relation $R$ by means of a generator $i$. Then*

$$R \text{ is weakly complete and } i = T_L \Leftrightarrow J = \emptyset.$$

Observe also that for this particular generator, the additive fuzzy preference
structure obtained from a reflexive relation $R$ is

$$(P, I) = (R^d, R + R^t - 1) \tag{1}$$

The usual way of defining transitivity for fuzzy relations is by means of a t-norm
$T$. A fuzzy relation $Q$ is said $T$-transitive if

$$Q(a, c) \geq T(Q(a, b), Q(b, c)), \qquad \forall a, b, c \in A.$$

Similarly, the usual definition of negative transitivity depends on a t-conorm $S$.
A fuzzy relation $Q$ is said negatively $S$-transitive if

$$Q(a, c) \leq S(Q(a, b), Q(b, c)), \qquad \forall a, b, c \in A.$$

The equivalence between the transitivity of a fuzzy relation and the negative
transitivity of its dual relation is a well known result (see for example [13]),

$$Q \text{ is } T\text{-transitive} \Leftrightarrow Q^d \text{ is negatively } S\text{-transitive}, \tag{2}$$

where $S$ is the dual t-conorm of $T$.

If we consider in particular a weakly complete large preference relation and
the Łukasiewicz t-norm as generator, the associated strict preference relation is
the dual of $R$ (see Eq. 1). Then, in this particular case the previous equivalence
holds. Our aim is to generalize the implication from left to right and see what
happens when other generators are involved.

# 3    Conjunctors and Other Operators

In this section we deal with some operators involved in the study of the connection between transitivity and negative transitivity. The first important operators are conjunctors and disjunctors that allow to generalize the classical definitions of transitivity and negative transitivity in the fuzzy sets context.

## 3.1    Generalizing $T$-transitivity

The usual way of defining the transitivity of a fuzzy relation is by means of a t-norm $T$. However, the restriction to t-norms is questionable. On the one hand, even when the large preference relation $R$ is $T$-transitive with respect to a t-norm $T$, the transitivity of the generated $P$ and $I$ cannot always be expressed with respect to a t-norm [7, 9, 10]. On the other hand, the results presented in the following sections also hold when $R$ is transitive with respect to a more general operation. From the point of view of fuzzy preference modelling, it is not that surprising that the class of t-norms is too restrictive, as a similar conclusion was drawn when identifying suitable generators, as was briefly explained in the previous section. There, continuity, *in casu* the 1-Lipschitz property, was more important than associativity. As discussed in [9, 10], suitable operations for defining the transitivity of fuzzy relations are conjunctors.

**Definition 3.** *A conjunctor $f$ is an increasing binary operation on $[0,1]$ that coincides on $\{0,1\}^2$ with the Boolean conjunction.*

The smallest conjunctor $c_S$ and greatest conjunctor $c_G$ are given by

$$c_S(x,y) = \begin{cases} 0 & \text{, if } \min(x,y) < 1, \\ 1 & \text{, otherwise,} \end{cases}$$

and

$$c_G(x,y) = \begin{cases} 0 & \text{, if } \min(x,y) = 0, \\ 1 & \text{, otherwise.} \end{cases}$$

Given a conjunctor $f$, we say that a fuzzy relation $Q$ on $A$ is $f$-transitive if $f(Q(a,b), Q(b,c)) \leq Q(a,c)$ for any $(a,b,c) \in A^3$. Clearly, for two conjunctors $f$ and $g$ such that $f \leq g$, it holds that $g$-transitivity implies $f$-transitivity. Restricting our attention to reflexive fuzzy relations only, such as large preference relations, not all conjunctors are suitable for defining transitivity. Indeed, for a reflexive fuzzy relation $R$, we should consider conjunctors upper bounded by $T_M$ only (see [8]).

In the same way as we have generalized classical t-norms, we can generalize t-conorms.

**Definition 4.** *A disjunctor is an increasing binary operation on $[0,1]$ that coincides on $\{0,1\}^2$ with the Boolean disjunction.*

As t-norms and t-conorms, disjunctors and conjunctors are dual operators. For any conjunctor $f$, the operator $g(x,y) = 1 - f(1-x, 1-y)$ is a disjunctor and the converse also holds.

## 3.2   Fuzzy Implications and Related Operations

Given a t-norm $T$, the associated fuzzy implication (also called $R$-implication or $T$-residuum) is a binary operation on $[0, 1]$ defined by (see e.g. [13, 14]):

$$\mathcal{I}_T(x, y) = \sup\{z \in [0, 1] \mid T(x, z) \leq y\}.$$

When $T$ is left-continuous it holds that $T(x, z) \leq y \Leftrightarrow z \leq \mathcal{I}_T(x, y)$, and $\mathcal{I}_T$ is called the residual implicator of $T$. The definition can be easily generalized to conjunctors.

**Definition 5.** *With a given commutative conjunctor $f$ we associate a binary operation $\mathcal{I}_f$ on the unit interval defined by*

$$\mathcal{I}_f(x, y) = \sup\{z \in [0, 1] \mid f(x, z) \leq y\}.$$

The above definition could also be extended to non-commutative conjunctors, but in that case we should distinguish between left and right operators. In this work we will only consider the case of commutative operators (commutative conjunctors or generators). Clearly, $\mathcal{I}_f$ is decreasing in its first argument and increasing in its second argument.

**Definition 6.** *An implicator $f$ is a binary operation on $[0, 1]$ that is decreasing in its first argument, increasing in its second argument and that coincides on $\{0, 1\}^2$ with the Boolean implication.*

**Proposition 2.** *Consider a commutative conjunctor $f$, then $\mathcal{I}_f$ is an implicator if and only if $f(1, y) > 0$, for any $y > 0$.*

The condition in the preceding proposition is obviously fulfilled when $f$ has 1 as neutral element.

In this paper, we associate another binary operators with any generator. This operator will play a key role in the characterization of the negative transitivity of the strict preference relation.

**Definition 7.** *With a given commutative conjunctor $f$ we associate a binary operation $\mathcal{K}_f$ on the unit interval defined by*

$$\mathcal{K}_f(x, y) = \sup\{z \in [1 - x, 1] \mid z - f(x, z) = y\}.$$

Despite the previous definition is given for any commutative conjunctor, we will only use it for a particular type of generators. Observe that the set $\{z \in [1 - y, 1] \mid z - f(x, z) = y\}$ is not guaranteed to be non-empty in general. However, under suitable conditions, the set is not empty.

**Lemma 1.** *Let $i$ be a 1-Lipschitz increasing generator and let $(x, y)$ satisfy $y \geq 1 - x - i(x, 1 - x)$, then $\{z \in [1 - x, 1] \mid z - f(x, z) = y\}$ is a non-empty set that admits maximum. Moreover, $\mathcal{K}_i(x, y)$ is increasing on its first argument and decreasing on its second argument.*

This lemma is employed to assure that the operator obtained in the following section is well defined.

## 4    General Implication

In this section we will recall the general result obtained concerning the negative transitivity that can be assured for the strict preference relation $P$ associated to a weak preference relation $R$ by a generic generator $i$.

**Theorem 2.** [11] *Consider a 1-Lipschitz increasing generator $i$ and a commutative conjunctor $h$ upper bounded by the minimum. For any reflexive fuzzy relation $R$ with corresponding strict preference relation $P$ generated by means of $i$, it holds that*

$$R \text{ is } h\text{-transitive} \quad \Rightarrow \quad P \text{ is negatively } j_h^i\text{-transitive}$$

*where*

$$j_h^i(x, y) = \sup_{\substack{u \leq 1-x \\ u+i(u, 1-u) \geq 1-x \\ v \leq 1-y \\ v+i(v, 1-v) \geq 1-y}} f(x, y, u, v) - i(f(x, y, u, v), h(u, v))$$

*for*

$$f(x, y, u, v) = \min(\mathcal{I}_h(u, \mathcal{K}_i(v, y)), \mathcal{I}_h(v, \mathcal{K}_i(u, x))).$$

*Moreover, this is the strongest result possible.*

At the sight of Equivalence (2), the first idea is that if $R$ is $f$-transitive for some conjunctor $f$, the obtained strict preference relation $P$ will be negatively $g$-transitive for some disjunctor $g$. However, this is not always the case. The operator obtained $j_h^i$ is not always a disjunctor. It is easy to check that three of the boundary conditions are satisfied for any $h$ and $i$.

**Lemma 2.** *For any conjunctor $h$ and for any generator $i$ the operator $j_h^i$ satisfies that*

$$j_h^i(1, 0) = j_h^i(0, 1) = j_h^i(1, 1) = 1$$

But the fourth boundary condition $j_h^i(0, 0) = 0$ is not always guaranteed as we show in the following example.

*Example 1.* Let us consider the weakly complete large preference relation $R$ defined on a set of three alternatives $\{a, b, c\}$ as:

| $R$ | $a$ | $b$ | $c$ |
|---|---|---|---|
| $a$ | 1 | 0.5 | 1 |
| $b$ | 0.5 | 1 | 0.5 |
| $c$ | 0 | 0.5 | 1 |

This relation is $T_M$-transitive, but the associated strict preference relation $P$ obtained as $R - T_M(R, R^t)$, this is, by means of the generator $i = T_M$,

| $P$ | $a$ | $b$ | $c$ |
|---|---|---|---|
| $a$ | 0 | 0 | 1 |
| $b$ | 0 | 0 | 0 |
| $c$ | 0 | 0 | 0 |

is not negatively $g$-transitive for any disjunctor $g$ since

$$P(a, c) = 1 \not\leq 0 = g(0, 0) = g(P(a, b), P(b, c)).$$

Therefore, when the generator is the minimum t-norm, the operator $j_h^{T_M}$ may not satisfy the boundary condition $j_h^{T_M}(0, 0) = 0$. However this problem does not appear if the generator is strictly smaller than the minimum t-norm in $\{(x, 1 - x) | x \in (0, 1)\}$.

**Lemma 3.** *For any commutative conjunctor $h$ upper bounded by the minimum and any generator $i$ such that $i(x, 1 - x) < T_M(x, 1 - x)$ for $x \in (0, 1)$, it holds that $j_h^i$ satisfies the boundary conditions:*

$$j_h^i(0, 0) = 0 \qquad j_h^i(1, 0) = 1 = j_h^i(0, 1) = j_h^i(1, 1).$$

In particular, we can recall that for $i = T_L$, the operator $j_h^i$ becomes the dual disjunctor of the conjunctor $h$:

**Corollary 1.** [11] *Let us consider a commutative conjunctor $h$ and the generator $i = T_L$. For any reflexive fuzzy relation $R$ with corresponding strict preference relation $P$ generated by means of $i = T_L$, it holds that*

$$R \text{ is } h\text{-transitive} \;\Rightarrow\; P \text{ is negatively } h^d\text{-transitive}$$

*where*

$$h^d(x, y) = 1 - h(1 - x, 1 - y).$$

We are presently studying some other general properties of the operator $j_h^i$. The results obtained and the proof of Theorem 2 will be submitted to a special number of the journal Fuzzy Sets and Systems.

## 5    The Minimum t-norm

This section is devoted to the study of the role of the minimum t-norm. We will consider both situations: when it is used as generator and when it is employed to define the transitivity of the large preference relation $R$.

We first focus on the minimum t-norm as generator. Let us recall that this means that the strict preference relation is obtained from the weak preference relation as $P = \max(R - R^t, 0)$. We have already shown an example where the negative transitivity guaranteed for this type of $P$ may be defined by an operator that is not a disjunctor. The general expression when we fix the minimum as generator is the following one.

**Proposition 3.** *Let $h$ be a commutative conjunctor upper bounded by the minimum t-norm. Let $R$ be a large preference relation and $P = \max(R - R^t, 0)$ the strict preference relation obtained from $R$ by means of the generator $i = T_M$. Then,*

$$R \text{ is } h\text{-transitive} \;\Rightarrow\; P \text{ is negatively } j_h^{T_M}\text{-transitive}$$

*where*

$$j_h^{T_M}(x, y) = 1 - h\left(\frac{1 - x}{2}, \frac{1 - y}{2}\right)$$

We can observe again that $j_{T_L}^{T_M}$ is not a disjunctor, since $j_{T_L}^{T_M}(0,0) = 1$. In general, the operator $j_h^{T_M}$ is not a disjunctor for any $h \leq T_M$, since in this case

$$j_h^{T_M}(0,0) = 1 - h\left(\frac{1-0}{2}, \frac{1-0}{2}\right) \geq 1 - T_M\left(\frac{1}{2}, \frac{1}{2}\right) = \frac{1}{2} > 0.$$

The minimum is the most common operator employed to define the transitivity of a fuzzy relation. If we depart from the transitivity defined by the minimum and we consider any generator $i$, the general expression does not look much simpler. Since the operator $\mathcal{I}_{T_M}(x,y)$ takes the value 1 if $x \leq y$ and it is $y$ otherwise, the operator $f(x,y,u,v)$ admits different expressions for different values of $u$ and $v$ once fixed $x, y$, so it is not easy at all to provide an explicit expression of $j_{T_M}^i$ for a generic $i$ much simpler than the one presented in Theorem 2. Our next step will be to simplify the general expression for some important specific generators as the Frank operators.

# 6    Conclusion

In the classical case, the transitivity of a complete large preference relation is equivalent to the negative transitivity of the associated strict preference relation. For a particular case of fuzzy relations, this is, when the strict preference relation is obtained from the large preference relation by means of the Łukasiewicz t-norm, it is known that the equivalence still holds under weak completeness. In this contribution we focus on the implication from transitivity to negative transitivity. We have provided a general expression that defines the negative transitivity assured for the strict preference relation, when the original large preference relation is weakly complete and satisfies a generic type of transitivity. The study is valid for any strict preference relation obtained by a 1-Lipschitz generator from the large preference relation. We have shown with a counterexample that the obtained operator is not necessarily a disjunctor, as it was expected to be. We have also discussed the aspect of the general expression when the minimum t-norm is involved either as generator or as a conjunctor. We have discussed that it is not trivial at all to provide a simpler formula for the general expression when the large preference relation is min-transitive. However, in future works we would like to study more properties of the general expression and get a simpler aspect of the operator for the most common generators and types of transitivity.

**Acknowledgement.** This work has been partially supported by Project MTM2010-17844.

# References

1. Arrow, K.J.: Social Choice and Individual Values. Wiley (1951)
2. Bodenhofer, U., Klawonn, F.: A formal study of linearity axioms for fuzzy orderings. Fuzzy Set Syst. 145, 323–354 (2004)

3. De Baets, B., Fodor, J.: Twenty years of fuzzy preference structures (1978-1997). Belg. J. Oper. Res. Statist. Comput. Sci. 37, 61–82 (1997)
4. De Baets, B., Fodor, J.: Additive fuzzy preference structures: the next generation. In: De Baets, B., Fodor, J. (eds.) Principles of Fuzzy Preference Modelling and Decision Making, pp. 15–25. Academia Press (2003)
5. De Baets, B., Van de Walle, B.: Minimal definitions of classical and fuzzy preference structures. In: Proceedings of the Annual Meeting of the North American Fuzzy Information Processing Society, pp. 299–304. Syracuse, New York (1997)
6. Díaz, S., De Baets, B., Montes, S.: Additive decomposition of fuzzy pre-orders. Fuzzy Set Syst. 158, 830–842 (2007)
7. Díaz, S., De Baets, B., Montes, S.: On the compositional characterization of complete fuzzy pre-orders. Fuzzy Set Syst. 159, 2221–2239 (2008)
8. Díaz, S., De Baets, B., Montes, S.: General results on the decomposition of transitive fuzzy relations. Fuzzy Optim. Decis. Making 9, 1–29 (2010)
9. Díaz, S., Montes, S., De Baets, B.: Transitive decomposition of fuzzy preference relations: the case of nilpotent minimum. Kybernetika 40, 71–88 (2004)
10. Díaz, S., Montes, S., De Baets, B.: Transitivity bounds in additive fuzzy preference structures. IEEE Trans. Fuzzy Syst. 15, 275–286 (2007)
11. Díaz, S., De Baets, B., Montes, S.: Transitivity and negative transitivity in the fuzzy setting. In: Proceedings of the Workshop on Fuzzy Methods for Based-Knowledge Systems, pp. 91–100. Regua, Portugal (2011)
12. Fishburn, P.C.: Utility Theory for Decision Making. Wiley, New York (1970)
13. Fodor, J., Roubens, M.: Fuzzy Preference Modelling and Multicriteria Decision Support. Kluwer Academic Publishers, Dordrecht (1994)
14. Klement, E.P., Mesiar, R., Pap, E.: Triangular Norms. Kluwer Academic Publishers (2000)
15. Roubens, M., Vincke, P.: Preference Modelling. Lecture Notes in Economics and Mathematical Systems, vol. 76. Springer (1998)
16. Van de Walle, B., De Baets, B., Kerre, E.: A plea for the use of Łukasiewicz triplets in the definition of fuzzy preference structures. Part 1: General argumentation. Fuzzy Set Syst. 97, 349–359 (1998)
17. Van de Walle, B., De Baets, B., Kerre, E.: Characterizable fuzzy preference structures. Ann. Oper. Res. 80, 105–136 (1998)

# Multi Criteria Operators
# for Multi-attribute Auctions

Albert Pla[1], Beatriz Lopez[1], and Javier Murillo[2]

[1] Unviersity of Girona, Girona, Spain
[2] Newronia, Girona, Spain
{albert.pla,beatriz.lopez}@udg.edu, javier.murillo@newronia.com

**Abstract.** Multi-attribute auctions allow agents to sell and purchase goods and services taking into account more attributes besides the price (e.g. service time, tolerances, qualities, etc.). The coexistence of different attributes in the auction mechanism increases the difficulty of determining the winner and its payment. multi-criteria functions can be used to deal with the problem of determining the auction winner. However, in order to make the payment possible, multi criteria functions must fulfill certain conditions. In this paper we discuss which properties must satisfy a multi-criteria function so it can be used to determine the winner of a multi-attribute auction and we experimentally show how the valuation function choice conditions the behavior of the auction mechanism.

**Keywords:** Multi-attribute auctions, Resource allocation.

## 1 Introduction

Resource allocation in dynamic production environments is becoming a more complex task as the number of actors and types of resources involved in the process increases. In certain domains the production cannot be known in advance, moreover the status of the production resources (e.g. technicians, transports, services, etc.) is unknown as they can be managed by different departments inside the organization or even by outsourcing companies which are in charge of dealing with certain parts of the productin process. Thus, the resource allocation process needs to be adapted to be performed under demand and taking into account the possible confrontation between managers, which try to obtain the lowest resource price at the higher quality, and the internal or external resources which try to maximize their occupation and benefits while keeping their information in private.

Auction mechanisms, offer the possibility to allocate the resources in a market, competitive framework, while optimizing the outcome from all of the participants (production process owners and resource providers, both, either internal or external to the organization) [1]. Thus, given a production task, resource providers bid for it, and the winner bid is the one that best fits the required resource specifications. However, the proposed bid may not correspond to the real ones of the

V. Torra et al. (Eds.): MDAI 2012, LNAI 7647, pp. 318–328, 2012.

bidder, forcing the poduction process owners to use mechanism which encourage the resource providers to bid honestly, revealing their true values.

There are several auction models, most of them focus on the resource price as the attribute which determines the winning bid. However, when allocating resources to production process, the cost of the resources is not the only relevant aspect to be taken into account. Attributes such as service time, distance among providers, ecological footprint, etc. can play an important part in the process of the determining which suppliers best suits the production needs. Therefore, it is important to find a compromise between all the elements that condition the resource in order to obtain a satisfying allocation. Multi-attribute auctions offer the chance to consider different aspects besides the price, becoming an ideal option for the problem we are dealing with.

In the multi-attribute auction mechanism we discuss in this paper, a multi-criteria function evaluates the different attributes provided by the resources and determines the winner of the auction. The multi-criteria function used to determine the winner of the auction, known as evaluation function, becomes a critical point in the mechanism as not only conditions the winner of the auction but also the payment it will receive, the strategies the bidders will follow and the kind of attributes that will became more relevant during the allocation process.

The contribution of this paper is the analysis and the definition of the characteristics that a multi-criteria function must fulfill in order to be used in a multi-attribute auction mechanism,

This paper is structured as follows: first we introduce some basics regarding auctions and multi-attribute auctions in order to facilitate the understanding of the paper; in Section 3 we comment some previous work related to our research; afterwards, in Section 4, we present a multi-attribute mechanism for business process resource allocation, we define the characteristics a multi-criteria function must satisfy in order to be used as evaluation function and we present some examples. In section 5 we describe an experiment performed to illustrate differences between the evaluation functions. We end the paper in Section 6 with our conclusions and the future work.

## 2   Background

This section provides the basic concepts related to this paper: First we introduce some basics about auctions and, second, the particularities about the multiattribute ones.

### 2.1   Auctions

An auction is a method for buying and selling goods using a bid system in which the winner bids obtain the auctioned goods [2]. Some of the basic concepts related to auctions are the following:

- The *utility* is the measurement of the satisfaction received by the participants of an auction, either the bidders or the auctioneers [3]. It can be defined as $U(B_i)$ given a bid $B_i$.

- The value of an item is the score or the price which the participants of an auction assign to a certain item. Can be defined using an evaluation function $V(B_i)$ given a bid $B_i$.
- Given a set of bids, the *winner determination problem* (WDP) is the problem to compute the winner bid that maximizes the utility of the auctioneer [4].
- The *payment mechanism* is the process of deciding which is the price $p$ and payout for the auctioneers and the bidders.
- When the bid price is $p$, the utility or *revenue* of the auctioneer becomes $U(B_i) = V(B_i) - p$.
- The utility or *profit* of the winner bidder is $U(B_i) = p - V(B_i)$. [1]
- Each bidder follows a *bidder's policy* in order to maximize their profit.

A desirable property that an auction mechanism should provide is to ensure that bidders provide truthful bids (incentive compatible mechanism). That means, that bidders obtain a better profit by revealing their real attributes than by cheating. The most popular mechanism that guarantees true bidding is the Vickrey-Clarke-Groves (VCG) one [6]: bidders bid in private, in a sealed bid, so only the auctioneer knows what are the other bids; when only one item is auctioned, the winner pays the price offered by the second-best bid. For example, given three bids, $b_1, b_2, b_3$ with prices $b_1 = 1, b_2 = 2, b_3 = 3$ the winner is $b_3$ (highest price offered) and pays 2 (second best offer).

Another common assumption in auctions is the absence of externalities, which means agents do not take care of which are the other agents winning the auction [7]. Regarding the winner determination problem, it consists on, given a set of bids $b_i$ to $b_n$, selecting the bid $b_i$ with the best valuation function. In the simplest auction mechanism, bids consist on the price and the valuation functions are the value of the price $V(B_i) = b_i$. This kind of auction protocol is also known as the contract net [8]. When the bid contains other information than prices, the auctions are known to be multidimensional [9], and the winner determination problem becomes much more complex.

## 2.2   Multiattribute Auctions

When the bid contains other information than prices, the auctions are known to be multidimensional [9]. There are two main kind of multidimensional auctions: multiattribute and combinatorial. In the former case, the dimensions correspond to the qualifications of the items to be sold. In the latter case, there are several items for selling and the dimensions corresponds to the bundles of goods each bidder is interested in.

We are interested in multiattribute auctions, in which each bid is characterized by a set of attributes in addition to price: $B = (b, AT)$ where $AT = (at^1, ..., at^n)$. The winner determination problem (WDP) consist on finding the optimal bid regarding price but also the other attributes. As an optimization problem, results

---

[1] $V(B_i)$ is private to each participant [5]. Thus, the auctioneer and bidders have a different one.

depend on the goal of the auctioneer or its objective optimization function, also known as scoring rule [5] or evaluation function $V(B_i)$ [10]. To simplify the notation, we can use $V(b_i, AT_i)$ to denote that function. Thus, the WDP consists in maximizing $V$.

$$argmax_i(V(b_i, AT_i)) \qquad (1)$$

In consequence, the auctioneer needs to make the scoring rule public so the bidders can maximize their chances to win. This will also ensure transparency during the winner determination process. Note that in single-attribute auctions, the rule is implicit: the cheaper, the best. However, multiattribute auctions require a more complex solution such as a multicriteria function.

Multiattribute auctions make the payment method difficult. That is, in a second-price auction, the winner gets the good and pays the price of the second best. However, when several attributes are involved, there is a discussion about how the attributes, different from price, should be provided. [5] demonstrates that an incentive compatible schema follows a second-score mechanism, in which the winner is allowed to provide the other attributes according to the second-best bid, but not exactly the same. This means that the attributes do not need to be the same as in the second highest bid but the valuation must be, at least, as good as in the second best bid. A solution is to provide a set of attributes in such a way that its valuation is not lower than the second best bid: e.g. keeping the bid attributes but modifying the economical value in order to equal the second best bid.

## 3    Related Work

The key work in multi-attribute auctions is [5], where the author describe different scenarios regarding the payment rule and demonstrate that the attributes should match the second best bid, but not exactly, to be incentive compatible. In a posterior work, [10] proposes an adaptation of the Vickrey-Clarke-Groves [6] (VCG) for multi-attribute auctions under an iterative schema. That means that bidders are allowed to modify their bids in response to the bids from other agents. In the approach this paper deals with, iteration is not allowed due to the dynamics of the problem domain. [11] presents an mechanism for auctions with temporal constraints based on VCG with a new payment method. Time constraints are used to filter the participating bids, but time is not considered when evaluating the bids, leaving aside whether time improves a bid or not. Multi-attribute auctions have been also used in the electronic advertisement markets [12,13], e.g. [14] proposes the adaptation of the GSP auctions in order to include an extra *quality* attribute, however this attribute is provided by the auctioneer itself, not by the bidder. This approach could be similar to trust-based approaches as [15]. Almost all these approaches compare bids with more than one attribute but none of them defines the characteristics their evaluation function must fulfill.

## 4   Multicriteria Methods on Multi-attribute Auctions

As many attributes are involved in determining which bid is the one which best fits to the auctioneer requirements, multicriteria functions can be considered appropriate evaluation functions. However, they must fulfill a set of conditions within the domain of the attributes in order to be used as evaluation function an unambiguous auction mechanis. The evaluation function contributes in two parts of the mechanism: the winner determination problem and the payment mechanism.

On the one hand, the auctioneer uses an evalutiona function $V(B_i) = V((b_i, AT_i))$ which evaluates the bid price $b_i$ and the bundle of attributes $AT_i$ of each bid $B_i$. Then, the auctioneer ranks them from the highest to the lowest value, being $j$ the bid ranking index $(B_j = (b_j, AT_j))$. Thus, the bid with the highest value is the winner of the auction.As $AT_i$ can contain more than one attribute, $V(b_i, AT_i)$ should be a multi-criteria function which express the auctioneer preferences.

On the other hand, a second price auction means that the winner bid receives just the necessary amount to beat the second highest bid (Equation 2).

$$V(p, AT_1)) = V(b_2, AT_2)) \tag{2}$$

$$p = V'(V(b_2, AT_2), AT_1) \tag{3}$$

Where $p$ is the payment of the single winner in our mechanism, $AT_1$ the attributes of the winner bid, $b_2, AT_2$ the components of the second best bid and $V'(x, AT_i) = b_i$ the anti-function of $V(b_i, AT_i) = x$.

However, this strategy does not prevent the bidders to lie regarding their attributes since including a false attribute could increase the chances to win the auction while not being penalized in the payment. For example, a bidder could submit a bid saying that it will finish its task in 10 minutes when actually it will finish the task in 15 minutes. This lie would have increased the chances of the bidder to win the auction despite breaking the contract agreement. Therefore, when a bidder breaks an agreement and does not commit with the bid attributes, the payment the bidder receives corresponds to the amount it should have bid to win the auction with final delivered set of attributes $AT^v$

$$V(p, AT_1^v) = V(b_1, AT_1) \tag{4}$$

$$p = V'(V(b_1, AT_1), AT_1^v) \tag{5}$$

Summarizing, Equation 6 shows the payment mechanism when bidders respect the contract agreement and when they break it.

$$p = \begin{cases} V'(V(b_2, AT_2), AT_1) & \text{if } AT_1 = AT_1^v \\ V'(V(b_1, AT_1), AT_1^v) & \text{if } AT_1 \neq AT_1^v \end{cases} \tag{6}$$

## 4.1   Multicriteria Function as Evaluation Function: Requirements

In order to use a multicriteria function as evaluation function, it must fulfill a set of conditions within the range of the attributes. First, the functions used for the evaluation must return a real number so the different bids can be analytically compared and ranked from the best to the worst; then, the functions must be monotonic, giving a better score for a better bid; finally, so the payment can be calculated, the evaluation function must be bijective for the price attribute.

### Real-Valued Function

Given a set of bids, the evaluation function must return a real number evaluation for each bid so the bids can be ranked and compared. As the auction payment involves the score obtained by the second best bidder and does not directly correspond to the price $b_1$ bid by the winner, multi-criteria methods which result in ranked lists without a score for each item cannot be used since the payment cannot be calculated.

### Monotonicity

The evaluation function must be monotonic. If one of the attributes of a bid is improved, the result of the evaluation function will change consequently, granting that a better bid will not obtain a worse evaluation. This property also implies that, for every possible value inside the domain attribute, the evaluation function will return a value. It is important to notice that the monotonicity requirement is applied only to the range of values that an attribute can take, allowing functions which are only monotonic in the attribute range to be used as evaluation functions. E.g. in a situation where all the attributes take values inside the positive numbers domain, the euclidean norm could be used as evaluation function.

### Bijection

In order to allow the mechanism to calculate the payment, the evaluation function must have a bijective behavior regarding the price attribute. This means that, given the bid attributes values and the result of the evaluation function, the cost attribute of the bid can take only one possible value. If this condition is not fulfilled, then, the auctioneer would be unable to calculate the payment which the winner should receive as Equation 6 would have more than one solution. In other words, given the function $V(b, AT) = x$ the antifunction will be $V'(x, AT) = b$ where $b$ can take just one value.

## 4.2   Multicriteria Function as Evaluation Function: Examples

There is a wide range of functions which, given a well defined domain, can act as evaluation functions. n this section we present some examples assuming that all the attribute values belong to the domain of real numbers and are normalized.

## Product and Weighed Sum

The product(Equation 7) and the weighed sum (Equation 9) can be used as evaluation functions. Due to the simplicity of these functions, the payment functions which are derived from them are also very simple and practical to use (Equations 8 and 10).

$$V(b_i, AT_i) = b_i * \prod_{j=1}^{j=n} at_i^j \tag{7}$$

$$p = \begin{cases} \dfrac{b2*\prod_{j=1}^{j=n} at_2^j}{\prod_{j=1}^{j=n} at1^j} & \text{if } AT_1 = AT_1^v \\[2ex] \dfrac{b1*\prod_{j=1}^{j=n} at_1^j}{\prod_{j=1}^{j=n} at_1^{vj}}) & \text{if } AT_1 \neq AT_1^v \end{cases} \tag{8}$$

$$V(b_i, AT_i) = \mu_0 b_i + \sum_{j=1}^{n} \mu_j at_i^j \tag{9}$$

$$p = \begin{cases} \dfrac{\mu_0 b2 + \sum_{j=1}^{n} \mu_j (at_2^j - at_1^j)}{\mu_0} & \text{if } AT_1 = AT_1^v \\[2ex] \dfrac{\mu_0 b1 + \sum_{j=1}^{n} \mu_j (at_1^j - at_1^{vj})}{\mu_0} & \text{if } AT_1 \neq AT_1^v \end{cases} \tag{10}$$

where $\mu^j \in [0,1]$ is the weight of each summation term and $\mu_0 + \sum_{j=1}^{n} \mu^j = 1$.

## Mathematical Norms

Another example of possible evaluation function are certain mathematical norms. E.g, assuming that the attribute domain belongs to the positive numbers plus 0, the euclidean norm (Equation 11) can be used. In contrast to the weighted sum, this evaluation function would favour bids with more balanced attributes (see Section 5.3. The correspondent payment function is given in Equation 12. However, it is important to remark that not all the norms can be used as evaluation function. E.g, Chebyshev norm cannot be used as evaluation function since it is not a bijective function and the payment could not be calculated.

$$V(b_i, AT_i) = \sqrt[2]{b_i^2 + \sum_{j=1}^{n} at_i^{j^2}} \tag{11}$$

$$p = \begin{cases} \sqrt[2]{b_2^2 + \sum_{j=1}^{n} (at_2^{j^2} - at_1^{j^2})} & \text{if } AT_1 = AT_1^v \\[2ex] \sqrt[2]{b_1^2 + \sum_{j=1}^{n} (at_1^{j^2} - at_1^{vj^2})} & \text{if } AT_1 \neq AT_1^v \end{cases} \tag{12}$$

## Weighted Sum of Functions

All the functions commented above treat the different attributes in the same way, however, in some domains, the attributes require an individual treatment and modeling. For these cases, a multicriteria function which fits the evaluation function requirements is the weighted sum of functions (WSF). Equation 13 shows, each attribute is individually evaluated using a function $f_j(x)$ and the different results are then aggregated using a weighted function. This multicriteria method presents the advantage of being highly adaptable to the domain, however, in order to be used as evaluation function all the involved functions $f_j(x)$ must fulfill the requirements presented in section 4.1. Moreover, the payment function will depend on the the attribute functions $f_j(x)$ and their antifunctions.

$$V(b_i, AT_i) = \mu_0 f_0(b_i) + \sum_{j=1}^{n} \mu_j f_j(at_i^j) \qquad (13)$$

$$p = \begin{cases} \frac{f_0'\left(\mu_0 f_0(b_2) + \sum_{j=1}^{n} \left(\mu_j f_j(at_2^j) - \mu_j f_j(at_i^j)\right)\right)}{\mu_0} & \text{if } AT_1 = AT_1^v \\ \frac{f_0'\left(\mu_0 f_0(b_i) + \sum_{j=1}^{n} \left(\mu_j f_j(at_i^{vj}) - \mu_j f_j(at_i^j)\right)\right)}{\mu_0} & \text{if } AT_1 \neq AT_1^v \end{cases} \qquad (14)$$

where $f_j(x)$ are the functions which evaluate the attributes, $\mu_j$ are the weights of each attribute function and $f_0'(x)$ is the antifunction of $f_0(x)$.

## 5    Experimentation

In this section we pretend to illustrate the behavior of the auction mechanism when different evaluation functions are used and how, if assuming that the different auctions are independent, the best strategy for the bidders is to bid truthfully despite using different evaluation functions during the auction process. In order to do so, we present a simple experiment: we estudied the execution of different business process composed by different tasksusing a multi-agent system simulation framework [16].

In this simulation different business process are executed concurrently and all of them share a finite set of resources which are suplied by different resource providers. Both the business process and the resources are repreented by agents: Bussiness process agents (BP) and resource providers agents (RP). When a BP requires a resource, the BP summons a second price reverse auction [17] and the interested RP bid for providing the required resource.

### 5.1    Experimental Set Up

To test the performance of our auction mechanism we defined three business process to be simulated. Each of these business process is composed by six different tasks which have a duration compressed between 10 an 15 time units. Each task of a business process needs one or none resource of a randomly assigned category (between A and D). In consequence, each business process has a full duration (from the first to the last task) between 60 and 90 time units and requires between 1 and 4 different resources. There is a random probability of $\rho = 0.07$ in each unit time that a business process is enacted.

There are three BP agents, one per kind of business process. We consider two attributes, time to deliver the task $t$ (minutes) plus an error margin $e$ (%)so

**Table 1.** Example of resources skills randomly generated per each agent: (Resource type, deliver time, tolerance to errors)

| Resources | RP0 | RP1 | RP2 | RP3 | RP4 | RP5 | RP6 | RP7 |
|---|---|---|---|---|---|---|---|---|
| Skills | A,15,2 | B,15,1 | C,10,8 | D,14,2 | A,12,6 | B,13,4 | C,12,5 | D,13,4 |
| | C,14,3 | D,14,2 | A,14,2 | B,10,1 | C,13,4 | D,15,4 | A,13,5 | B,13,5 |
| | B,10,9 | A,10,9 | D,15,1 | A,14,2 | D,12,5 | C,12,5 | D,12,6 | A,12,6 |

($AT = \{t, e\}$, $15 > t > 10$, $10 > e > 0$), plus price. There are 8 RP bidders. Each resource provider agent has the skills of performing three different tasks, with different qualifications (see Table 1). Two of the resource provider agents cheat and follow the strategy of not revealing their trutful values ($RP3$ and $RP7$), while the others bid truthfully and adapt their bids as auctions progresses [18]. This situation will be useful to point the incentive compatibility of the system Moreover, providers from $RP0$ to $RP3$ have unbalanced attributes (a good time, a good error margin or a good price) while resources $RP4$ to $RP7$ have balanced attributes (an average price, error margin and service time). These differences will be useful to analyze the different behavior of the mechanism when different evaluation functions are used.

## 5.2 Scenarios

The presented situation will be simulated in three different scenarios. In each one, a different evaluation function is used in the auction: The weighted sum ($\mu_0 = 0.5, \mu_1 = 0.3, \mu_2 = 0.2$), the product and the euclidean norm. Each scenario is repeated 100 times so significant data can be extracted from the simulations.

## 5.3 Results

Figure 1 shows the experiments of the different simulations. In it we can see how the different evaluation functions have conditioned the behavior of the auctions.

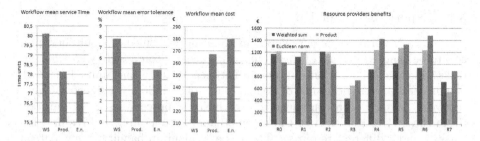

**Fig. 1.** 1) Business process mean service time with different evaluation functions (WS: Weighted Sum, Prod: Product, E.n: Euclidean norm) 2) Business process mean error tolerance 3) Busines process mean cost 4) Resource providers benefits when using different evaluation functions

E.g, we can see how when the weighted sum have been used the mean cost for each business process is lower than with the other functions, however if we compare the service time and the error margin we can see how their results are worse than with the Euclidean norm and with the product. This can be explained for the weights used in the evaluation function which gives more importance to the price than to the other attributes.

If we analyze the results of the product and the Euclidean norm we can observe how both evaluation functions have provoqued a higher business process mean price but with a lower service time and error margin as none of the evaluation functions give priority to one or to other attributes. If we observe the providers benefits, we can see how using the product all the providers (except the cheaters) have a similar amount of benefits, however, when the Euclidean norm have been used, the providers which offered balanced resources (RP4 to RP7) obtain a higher revenue as the Euclidean norm gives lower values when the components of a vector are equilibrated.

Finally, we can see how the incentive compatibility of the mechanism has been preserved despite the evaluation function used as in all the scenarios the cheater agents (RP3 and RP7) obtain lower benefits than the honest ones.

# 6   Conclusions

This paper deals with the problem of resource allocating using multi-attribute auctions in a decentralized environment where business process instances are executed concurrently and without a previously known production scheduling. In particular, it focus on the problem of determining which requirements multi-criteria functions must fulfill in order to be used as evaluation functions.

The evaluation function of a multi-attribute auction is strongly dependent of the auction domain, as attributes may drastically vary from one auction to another. However, all of them must satisfy certain conditions in order to grant a proper behavior of the mechanism. In this paper we have shown that, when using a multicriteria function as evaluation function, it must be a monotonic function within the range of the attributes and it must be a bijective function regarding the price attribute. As an illustrative example, these function requirements have been applied to a second price reverse multi-attribute auction mechanism, however this properties are extensible to other mechanisms which require the use of a numeric evaluation function (e.g. the mechanism presented in [10]).

As a future work we plan to extend the auction mechanism presented in Section 4.1 with fairness properties in order to deal with sequential auctions and to include. Another path of research, is the study of the relation between the decision of the evaluation function and the incentive compatibility of the auction mechanism.

**Acknowledgments.** This research project has been partially funded through the projects labeled TIN2008-04547 and DPI2011-24929, and BR10/18 Scholarship of the University of Girona granted to Albert Pla.

# References

1. Chevaleyre, Y., Dunne, P., Endriss, U., Lang, J., Lemaïtre, M., Maudet, N., Padget, J., Phelps, S., Rodrïguez-aguilar, J., Sousa, P.: Issues in multiagent resource allocation. Informatica 30 (2006)
2. Krishna, V.: Auction Theory. Academic Press (March 2002)
3. Neumann, J.V., Morgenstern, O.: Theory of Games and Economic Behavior. Princeton University Press (1944)
4. Lehman, D., Muller, R., Sandholm, T.: The Winner Determination Problem, ch. 12. MIT Press (2006)
5. Che, Y.K.: Design competition through multidimensional auctions. The RAND Journal of Economics 24(4), 668–680 (1993)
6. MacKie-Mason, J.K., Varian, H.R.: Generalized vickrey auctions (1994)
7. Conitzer, V.: Algorithms and theory of computation handbook, p. 16. Chapman & Hall/CRC (2010)
8. Smith, R.: The contract net protocol: High-level communication and control in a distributed problem solver. IEEE Transactions on Computers C-29(12), 1104–1113 (1980)
9. Parsons, S., Rodriguez-Aguilar, J.A., Klein, M.: Auctions and bidding: A guide for computer scientists. ACM Comput. Surv. 43(2), 10:1–10:59 (2011)
10. Parkes, D.C., Kalagnanam, J.: Iterative multiattribute vickrey auctions. Management Science 51, 435–451 (2005)
11. Zhao, D., Zhang, D., Perrussel, L.: Mechanism design for double auctions with temporal constraints. In: IJCAI, pp. 472–477 (2011)
12. Athey, S., Ellison, G.: Position auctions with consumer search. Forthcoming Quarterly Journal of. Economics 126(3), 1213–1270 (2011)
13. Krishna, V.: Auction Theory. Academic Press/Elsevier (2009)
14. Varian, H.R.: Position auctions. International Journal of Industrial Organization 25(6), 1163–1178 (2007)
15. Ramchurn, S.D., Mezzetti, C., Giovannucci, A., Rodriguez-Aguilar, J.A., Dash, R.K., Jennings, N.R.: Trust-based mechanisms for robust and efficient task allocation in the presence of execution uncertainty. J. Artif. Int. Res. 35, 119–159 (2009)
16. Pla, A., Lopez, B., Melendez, J., Gay, P.: Petri net based agents for coordinating resources in a workflow management system. In: ICAART, Rome, Italy, pp. 514–523 (February 2011)
17. Pla, A., López, B.: Truthful bidding prove for multiatribute auctions. Research report IIiA 12-01-rr, Institute of Informatics and Applications, University of Girona (in press, 2012)
18. Lee, J., Szymanski, B.: A novel auction mechanism for selling time-sensitive e-services. In: IEEE Conference on ECommerce Technology (CEC 2005), pp. 75–82 (2005)

# Finding Patterns in Large Star Schemas at the Right Aggregation Level

Andreia Silva and Cláudia Antunes

Department of Computer Science and Engineering
Instituto Superior Técnico – Technical University of Lisbon
Lisbon, Portugal
{andreia.silva,claudia.antunes}@ist.utl.pt

**Abstract.** There are many stand-alone algorithms to mine different types of patterns in traditional databases. However, to effectively and efficiently mine databases with more complex and large data tables is still a growing challenge in data mining. The nature of data streams makes streaming techniques a promising way to handle large amounts of data, since their main ideas are to avoid multiple scans and optimize memory usage. In this paper we propose in detail an algorithm for finding frequent patterns in large databases following a star schema, based on streaming techniques. It is able to mine traditional star schemas, as well as stars with degenerate dimensions. It is able to aggregate the rows in the fact table that relate to the same business fact, and therefore find patterns at the right business level. Experimental results show that the algorithm is accurate and performs better than the traditional approach.

**Keywords:** Pattern Mining, Multi-Relational Data Mining, Data Streams, Star Schema, Degenerate Dimensions.

## 1 Introduction

A growing challenge in data mining is the ability to deal with complex, large and dynamic data. In many real world applications, complex data is organized in multiple, related and large database tables. There are many stand-alone algorithms for finding frequent patterns, but they are only able to deal with traditional databases, composed of a singular table. To join all tables into one is usually a very time consuming process, possibly impracticable, that can easily lead to the loss of information. Multi-relational data mining (MRDM) is a fairly recent area that aims for learning from multiple tables, in their original structure.

The nature of data streams makes streaming techniques a promising way to handle large amounts of data, since their main ideas are to avoid multiple scans of the entire datasets, optimize memory usage and use a small constant time per record. Existing techniques keep only the needed information in some summary data structure and maintain it updated. Most of the existing algorithms for mining data streams are designed for a single data table ([4,8]).

A commonly used structure for databases is a *star schema*, which is composed of a central fact table linking a set of dimension tables. In a star schema, data

V. Torra et al. (Eds.): MDAI 2012, LNAI 7647, pp. 329–340, 2012.
© Springer-Verlag Berlin Heidelberg 2012

is modeled as a set of facts, each describing an occurrence, characterized by a particular combination of dimensions. In turn, each dimension aggregates a set of attributes for a same domain property [7]. In traditional transactional star schemas, each row in the fact table corresponds to a business transaction (or business fact). However, it is very common to have a control number in the fact table, such as an order or transaction number. They are usually stored in the fact table as a *degenerate dimension*, i.e. they act like dimension keys, but do not have a physical dimension table associated. Nevertheless, they can be a separate dimension (normally with only one attribute). These control numbers provide a way to group the rows in the fact table that were generated as a part of the same order or transaction. Instead of simply considering each row as one different fact, a more interesting challenge is to aggregate the rows of each business fact, in order to analyze the star at the right business level. For example, in a sales domain, if we group the rows belonging to each sale, we can find patterns of items bought together. Otherwise, we can only find frequent singular products.

In this paper we propose in detail an algorithm for finding frequent patterns in large databases following a star schema, based on streaming techniques, that is able to find patterns at the right business level.

## 2    Problem Statement

Frequent pattern mining aims for enumerating all frequent patterns that conceptually represent relations among items. Depending on the complexity of these relations, different types of patterns arise, being transactional patterns the most common. A *transactional pattern* is a set of items that occur together frequently.

Let $S$ be a tuple $(D_1, D_2, ... D_n, FT)$ representing a data warehouse modeled as a star schema, with $D_i$ corresponding to each dimension table and $FT$ to the fact table. Also, let $I = \{i_1, i_2, \ldots, i_m\}$ be a set of distinct literals, called *items*. In the context of a database, an *item* corresponds to a proposition of the form (*attribute, value*), and a subset of items is denoted as an *itemset*. $T = (tid, X)$ is a tuple where *tid* is a tuple-id (corresponding to a primary key) and $X$ is an itemset in $I$. Each dimension table in $S$, is a set of these tuples. If there is a degenerate dimension (DD), it consists in just a key in the fact table that identifies the rows corresponding to the same business fact. It can be seen as an aggregating dimension or key. Rows on the fact table are sets of $n$ *tids*: tuples of the form $(tid_{DD}, tid_{D_1}, tid_{D_2}, ... tid_{D_n})$. In traditional transactional star schemas, a business fact corresponds to one row in the fact table. With degenerate dimensions, a *business fact* is a set of rows in the fact table that share the same degenerate key.

The support of a foreign key $tid_{D_i}$ is the number of business facts where they occur. The support of an Itemset $X$ of the dimension $D_i$ consists on the number of business facts that contain the $tid_{D_i}$s that have $X$ in $D_i$. The problem of multi-relational frequent pattern mining over star schemas is to mine all itemsets whose support is greater or equal than $\sigma \times BF$ where $\sigma \in ]0, 1]$ is the user defined *minimum support threshold*, and $BF$ is the number of business facts.

Let us now consider that the tables are data streams, where new business facts arrive sequentially in the form of continuous streams. Let the *fact stream* $FS = B_1 \cup B_2 \cup ...B_k$ be a sequence of batches, where $B_k$ is the current batch, $B_1$ the oldest one, and each batch is a set of business facts. Additionally, let $N$ be the current length of the stream, i.e. the number of business facts seen so far. As it is unrealistic to hold all streaming data in the limited main memory, data streaming algorithms have to sacrifice the correctness of their results by allowing some counting errors. These errors are bounded by a user defined *maximum error threshold*, $\epsilon \in [0, 1]$, such that $\epsilon \ll \sigma$. Thus, the support calculated for each item is an approximate value, which at most has an error of $\epsilon N$. The problem of multi-relational frequent pattern mining over star streams consists in finding all itemsets whose estimated support is greater or equal to $(\sigma - \epsilon) \times N$.

# 3    Mining Star Streams

*Star FP-Stream* is a MRDM algorithm that is able to find approximate frequent relational patterns in large databases following a star schema. It is able to deal with degenerate dimensions, and to aggregate the rows of the fact table into business facts, making possible the mining of the star at the right business level. It is also able to mine multiple relational data streams, assuming that patterns are measured from the start of the stream up to the current moment (landmark model). The algorithm is a complement of *Star FP-Stream* [11] that combines the strategies of two algorithms: *Star FP-Growth* [10] (MRDM algorithm over star schemas) and FP-Streaming [4] (data streaming algorithm), both based on the traditional algorithm FP-Growth [5]. It does not materialize the join of the tables, making use of the star properties, and it processes one batch of data at a time, maintaining and updating frequent itemsets (patterns) in a pattern-tree. A *pattern-tree* is a compact data structure based on the FP-tree [5] that maintains crucial, quantitative information only about patterns, instead of any itemset.

## 3.1    How Star FP-Stream Deals with Star Schemas?

As referred in [7], dimensions are, by definition, smaller than the fact table. Therefore, we assume that all dimension tables are kept in memory, and only the fact table (the *fact stream*) is arriving in batches. However, dimensions can be data streams as well: if new transactions arrive for some dimension (transactions can only be added, not deleted nor changed), just add them to the respective dimension table in memory, before their *tids* first appearance in the fact stream, since these transactions are only read if their *tids* are seen in the current batch.

To mine a star stream, the algorithm uses *Star FP-Growth* [10] techniques. The idea is to build, in each batch, the DimFP-Trees for each dimension as new business facts arrive, with the respective occurring transactions (local mining step). When a batch is completed, the DimFP-Trees are combined to form a Super FP-Tree (global mining step), which will contain the itemsets of all dimension that co-occur in the current batch.

When counting the support and combining the trees, we have to guarantee that each *tid* and also each itemset do not count more than once per business fact. For example, if the fact table has $(\{t1, client1, product1\}, \{t1, client1, product2\})$, it means that client 1 bought product 1 and product 2 in the same transaction, and therefore client 1 should only count as appearing once in this business fact.

## 3.2   How Star FP-Stream Deals with Fact Streams?

To deal with fact streams, the algorithm follows the strategies of *FP-Streaming* [4], where arriving transactions are stored in a new FP-tree structure, and at each batch boundary, the frequent patterns are extracted from the tree by means of FP-Growth, and stored in a pattern-tree structure, which is then pruned to remove infrequent patterns. Each node in this tree represents a pattern (from the root to the node), stored along with its current estimated frequency and its maximum error. However, while FP-Streaming is an algorithm for mining time sensitive data streams (keeping frequencies for several time intervals), Star FP-Stream aims to find patterns over the entire data stream (landmark model).

In Star FP-Stream the fact stream is conceptually divided into $k$ batches of $\lceil 1/\epsilon \rceil$ business facts each, so that the batch id $(1..k)$ exactly refers to the threshold $\epsilon N$ (the maximum error allowed is $k$, one per batch). Note that the number of *facts* of each batch is fixed, but the number of rows is not, because $n$ business facts may need more than $n$ rows in the fact table.

All items that appear in more than one business fact in a batch are frequent with respect to that batch, and potentially frequent with respect to the entire stream. As for items that appear just in one business fact in a batch and are not in the pattern-tree, they are *infrequent* and can be discarded, because even if they reappear later and become frequent, the loss of support will not affect significantly the calculated support (the error is less than $k$). Considering that an itemset $I$ first occurs in batch $B_j$, let us denote $f$ its real frequency and $\hat{f}$ its estimated frequency after the current batch $B_i$ (with $j \leq i \leq k$), and $\Delta = j - 1$ its maximum error (i.e. the number of times it could have appeared before $j$). Frequent itemsets since the first batch have $\Delta = 0$ and $f = \hat{f}$. Otherwise they can have been discarded in the first $\Delta$ batches. Therefore, $f \leq \hat{f} + \Delta$. And since $\Delta \leq i - 1 \leq \epsilon N$, we can state that $f \leq \hat{f} + \Delta \leq f + \epsilon N$. If we want itemsets whose $f \geq \sigma N$, getting all itemsets whose $\hat{f} + \Delta \geq \sigma N$ guarantees that all patterns are returned. Similarly, for all patterns, $f + \epsilon N \geq \sigma N \Leftrightarrow f \geq (\sigma - \epsilon)N$.

The update of the pattern-tree structure is done only when enough incoming business facts have arrived to form a new batch $B_i$. We have three pruning strategies, like FP-Streaming [4]. For the current batch $B_i$ and item $I$:

**Type I Pruning:** If $I$ only occurs in one business fact in $B_i$ and it is not in the pattern-tree, we do not insert it in the pattern-tree and we can stop mining the supersets of $I$, because it is infrequent. (anti-monotone property)

**Type II Pruning:** While mining $I$, if its $\hat{f} + \Delta \leq i$, it will be deleted later because it is infrequent, therefore we can stop mining the supersets of $I$.

**Tail Pruning:** After mining $B_i$ and updating the pattern-tree, we can prune all items in the tree whose $\hat{f} + \Delta \leq i$.

### 3.3    Algorithm Star FP-Stream

The detailed algorithm is presented in Algorithms 1 and 2.

---

**Algorithm 1.** Star FP-Stream Pseudocode

---

**Input:** Star Stream S, error rate $\epsilon$
**Output:** Approximate frequent items with threshold $\sigma$, whenever the user asks
  $i = 1, |B| = 1/\epsilon$, *flist* and *ptree* are empty
  $B_1 \leftarrow$ the first $|B|$ *facts*
  $L \leftarrow$ **StarFP-Growth**($B_1$, support $= \epsilon|B| + 1$)
  *flist* $\leftarrow$ frequent items in $B_1$, sorted by minimum support
  **for all** patterns $P \in L$ **do**
    insert $P$ in the *ptree* with max error $i - 1$
  $N = |B|$, discard $B_1$ and $L$
  // prepare next batch
  $i = i + 1$, initialize $n$ *DimFP-trees* to empty
  **for all** arriving business fact (set of at least one $(tid_{D_1}, ..., tid_{D_n})$) **do**
    $N = N + 1$
    **for all** Dimension $D_j$ **do**
      **for all Different** foreign key of the business fact, $tid_{D_j}$ **do**
        $T \leftarrow$ transaction of $D_j$ with $tid_{D_j}$
        insert $T$ in the *DimFP-tree$_j$*
      *flist* $\leftarrow$ append new items introduced by all $T$s
    **if** all facts of $B_i$ arrived **then**
      *super-tree* $\leftarrow$ **combineDimFP-trees**(*DimFP-trees*, $B_i$)
      **FP-Growth-for-streams**(*super-tree*, $\emptyset$, *ptree*, $i$)
      discard the *super-tree*
      **tail-pruning**(*ptree.Root*, $i$)
      // prepare next batch
      $i = i + 1$, initialize $n$ *DimFP-trees* to empty

**combineDimFP-trees**(DimFP-trees *dim-trees*, Batch of facts $B_i$)
  *fptree* $\leftarrow$ new FP-tree
  **for all** business fact $f \in B_i$ **do**
    **for all** Dimension $D_j$ **do**
      **for all Different** foreign key of the business fact, $tid_{D_j}$ **do**
        $T \leftarrow$ append branch of *DimTree$_j$* with $tid_{D_j}$
    sort $T$ accordingly to *flist* and **remove duplicates**
    insert $T$ in *fptree*
  **return** *fptree*

**tail-pruning**(Pattern-tree node $R$, Batch id $i$)
  **for all** children $C$ of $R$ **do**
    **if** $C.support + C.error \leq i$ **then**
      remove $C$ from the tree
    **else**
      **tail-pruning**($C$, $i$)

---

The first batch is processed separately with *Star FP-Growth* [10]. In this batch, itemsets that occur only once can be discarded, because they never occurred before. All resulting patterns are stored in the pattern tree, and frequent items are also stored to fix the items' order through all batches.

Next, for each arriving business fact, the respective occurring transactions are inserted in the corresponding compact DimFP-tree (*Local Mining*). At each batch boundary, i.e. when $|B|$ business facts have arrived, the DimFP-trees are combined into the *Super FP-tree* (*Global Mining*), containing all possible relational patterns in that batch (with items ordered accordingly to $flist$).

---

**Algorithm 2.** FP-Growth-for-streams Pseudocode

---

**Input:** FP-tree *fptree*, Itemset $\alpha$, Pattern-tree *ptree*, Current batch id $i$

  if $fptree = \emptyset$ then
    return
  else if *fptree* contains a single path $P$ then
    for all $\beta \in \mathbb{P}(P)$ do
      **processPattern**(*ptree*, $\alpha \cup \beta$: min[support(nodes$\in \beta$)], $i$)
  else
    for all $a \in$ Header(*fptree*) do
      $\beta \leftarrow \alpha \cup a$ : $a$.support
      if **processPattern**(*ptree*, $\beta$, $i$) = **false** then
        proceed to the next $a$
      else
        $tree_\beta \leftarrow$ conditional fptree on $a$
        **FP-Growth-for-streams**($tree_\beta$, $\beta$, *ptree*, $i$)

  **processPattern**(Pattern-tree *ptree*, Itemset $I$, Batch id $i$)
  if $I \in ptree$ then
    $P \leftarrow$ last node of $I$ in *ptree*
    $P$.support $\leftarrow$ increment by $I$.support
    if $P$.support $+ P$.error $\leq i$ then
      return **false**// Type II Pruning
  else if $I$.support $> \epsilon|B|$ then
    insert $I$ in *ptree* with support = $I$.support and maximum error = $i - 1$
  else
    return **false**// Type I pruning
  return **true**

---

The Super FP-Tree is then mined using the modified FP-Growth algorithm, presented in Algorithm 2, which differs from the original in how they deal with each found itemset $I$ (see function **processPattern**): If it is in the pattern-tree, update its frequency, by adding the number of occurrences in $B_i$, and test *Type II Pruning*. If $I$ is not in the pattern-tree, test *Type I Pruning*.

After mining the batch, we can discard the Super FP-tree and prune the pattern tree by *Tail Pruning*. The pattern tree is now updated, and contains all approximate frequent itemsets until that batch.

If there are no more batches, or every time a user asks for the current patterns, scan the pattern-tree and return all itemsets with $\hat{f} + \Delta \geq \sigma N$.

## 4   Performance Evaluation

The main goal of these experiments is to analyze our algorithm in the presence of a degenerate dimension. We evaluate the accuracy, time and memory usage, and show that: (1) Star FP-Stream has a good accuracy and does not miss any real pattern; and (2) mining directly the star is better than joining before mining.

We assume that we are facing a landmark model, where all patterns are equally relevant, regardless of when they appear in the data. Therefore, we test Star FP-Stream over an adaptation of FP-Streaming for landmark models, which we will call Simple FP-Stream, that stores only one counter in each node of the pattern tree (instead of one per time window). Since Simple FP-Stream does not deal with stars directly, it denormalizes each business fact when it arrives (i.e. it goes to every dimension and join all the transactions corresponding to the *tids* of the business fact in question, with no duplicates), before mining it.

Experiments were conducted varying both minimum support and maximum error thresholds: $\sigma \in \{50\%, 40\%, 30\%, 20\%\}$ and $\epsilon \in \{10\%, 5\%, 4\%, 3\%, 2\%\}$.[1] Note that the course of the mining process of streaming algorithms does not depend on the minimum support defined, only on the maximum error allowed. The support only influences the pattern extraction from the pattern-tree, which is ready for the extraction of patterns that surpass any asked support ($\sigma \gg \epsilon$).

We tested the algorithms with a sample of the *AdventureWorks 2008 Data Warehouse*[2], from a ficticious multinational manufacturing company. In this work we will analyze a sample of the star *Internet sales*, which contains information about individual customer Internet sales orders, from July 2001 to July 2004. Dimension tables were kept in memory and the fact table is read as new facts are needed. We consider four dimension tables: *Product, Date, Customer* and *SalesTerritory*, so that we are able to relate who bought what, when and where. The fact table has the keys of those dimensions (other attributes were removed), and each dimension has only one primary key and other attributes (no foreign keys). Numerical attributes were excluded (except *year* and *semester* in dimension *Date*) as well as translations and other personal textual attributes, like addresses, names and descriptions. We also used the degenerate dimension corresponding to the sales order number, that indicates which products were bought together (by the same customer, in the same date and place).

The computer used to run the experiments was an Intel Xeon E5310 1.60GHz (Quad Core), with 2GB of RAM. The operating system used was GNU/Linux amd64 and the algorithms were implemented using the Java Programming language (Java Virtual Machine version 1.6.0_24).

---

[1] A common way to define the error is $\epsilon = 0.1\sigma$ [8]. Additionally, we use a larger error to see how worse the results are.

[2] AdventureWorks Sample Data Warehouse:
http://sqlserversamples.codeplex.com/

### 4.1 Experimental Results

**Accuracy.** The accuracy of the results is influenced by both error and support thresholds. Note that the resulting patterns of Star FP-Stream and Simple FP-Stream are the same (the algorithms only differ in how they manipulate the data). The exact patterns were given by FP-Growth (with all rows of the fact table as input) and were compared with the approximate ones.

Fig. 1 shows the number of patterns returned, and Fig. 2 presents the precision as the support varies. Precision measures the rate of real patterns over the patterns returned by the streaming algorithm.

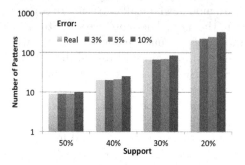
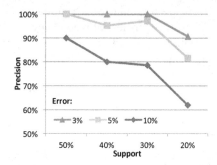

**Fig. 1.** Number of patterns returned    **Fig. 2.** Precision variation per support

As the minimum support decreases, the number of patterns increases, since we require fewer occurrences of an item for it to be frequent. And as the maximum error increases, the number of patterns returned also tends to increase, but the precision decreases, because although we can discard more items, we have to return more possible patterns to make sure we do not miss any real one. Nevertheless, the overall results show that precision was always above 60%.

The *recall* is proved theoretically to be 100% (there are no false positives, i.e. there are no real patterns that the algorithm considers infrequent).

**Pattern Tree.** The pattern tree is the key element of these algorithms, since it is the summary structure that holds all the possible patterns. The maximum error and the characteristics of data influence its size, which in turn influence the time and memory needed. The minimum support only counts to extract the patterns of the pattern tree, and it does not influence its size. Since both algorithms use the same rules to construct the pattern tree, it will be equivalent on both cases.

Fig. 3 reveals, for each error, the average size of the pattern tree after processing a batch. There we confirm that, as the error decreases, the size of the pattern tree increases. This is explained by the fact that for higher errors, the number of business facts in each batch is smaller and the algorithms can discard much more possible patterns than for lower errors. Although being a summary structure, it still is a very large structure, with thousands of nodes.

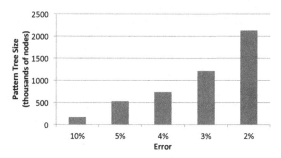

**Fig. 3.** Average pattern tree size

**Time and Memory.** Processing time was analyzed in terms of the time needed to process one batch (update time). It consists on the elapsed time from the reading of a transaction to the update of the pattern tree, and it depends both on the size of the batches, of the pattern-tree and of the characteristics of data. To analyze the maximum memory per batch, we measured the memory used by the algorithms for each batch, right before discarding the Super FP-Tree and doing the pruning step.

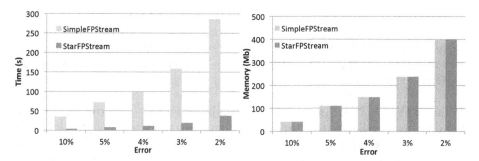

**Fig. 4.** Average update time             **Fig. 5.** Average maximum memory

Fig. 4 shows the average update time and Fig. 5 the average maximum memory per batch, of both algorithms for all errors. For consistency, we do not take into account the time or memory needed to process the first batch, since it is processed separately. We can state in the first figure that Simple FP-Stream demands, on average, much more time than Star FP-Stream. This demonstrates that, for star streams, denormalize before mining takes more time than mining directly the star schema, corroborating our goal and one of the goals of MRDM. In terms of memory, in the second figure we can note that the algorithms perform very similar and need the same amount of memory.

Both the update time and memory needed are required to be constant and not depend on the number of transactions. This can be verified in Fig. 6 and Fig. 7, that show in detail the time and memory needed per batch, respectively. We can verify that both tend to be constant and do not increase as more batches are processed. Star FP-Stream always outperformed the other in terms of time, and matched in terms of memory.

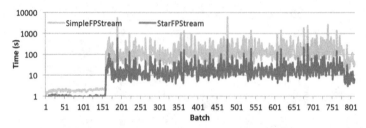

**Fig. 6.** Update time per batch, for 3% of error

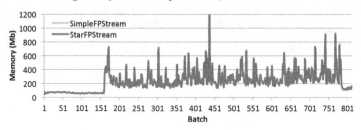

**Fig. 7.** Maximum memory per batch for 3% of error

In the beginning, both algorithms need less time and memory because, until the business transaction number 5400, customers only bought one product, and therefore there are fewer business facts to process. The ups and downs in time and memory correspond respectively to the entrance of more patterns in the pattern tree, and the process of smaller trees after pruning.

**What If We Do Not Aggregate the Business Facts?** To better understand the difference, while there are around 60 thousands rows in the fact table, there are less than a half business facts, around 27 thousand. For example, for 30% of support, mining each row as a singular fact asks for patterns that appear in more than 18 thousand rows. By aggregating per degenerate key, e.g. we can find the products and sets of products that are bought together in more than 8100 real sales. This leads to the increase of the number of patterns returned (Fig. 8), since items appearing more than 8100 times but less than 18 thousand, are infrequent in the first case, but frequent when aggregating. The Super FP-Tree, as well as the pattern-tree are also substantially different (Fig. 9): when aggregating, their size increases a lot, and they have less paths, but longer ones, because of the co-occurrences of items of the same table in the same transaction.

With a similar idea, 3% of error divides the fact table into batches of 34 business facts, which results in around 1760 batches of rows, or around 800 batches of business facts. While the first processes each batch faster because it has less rows, the second takes more time to process one batch of business facts. Similarly, the second needs much more memory, since it has to store bigger trees. However, the second is able to return more useful patterns in a business perspective, since they are mined at the right aggregation level.

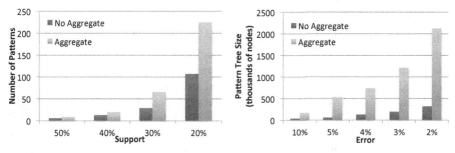

**Fig. 8.** Average update time (3% error)     **Fig. 9.** Average maximum memory

# 5  Related Work

There are many stand-alone algorithms to mine different types of patterns in traditional databases, with *FP-growth* [5] one of the most efficient. This algorithm follows a pattern-growth philosophy and represents the data into a compact tree structure, called *FP-tree*, to facilitate counting the support of each set of items and to avoid expensive, repeated database scans. It then uses a *depth-first search* approach to traverse the tree and find the patterns.

Some of the traditional algorithms have been extended to the multi-relational case. In this work we will focus on frequent pattern mining over star schemas.

The first multi-relational methods have been developed by the *Inductive Logic Programming* (ILP) community about ten years ago (WARMR [2]), but they are usually not scalable with respect to the number of relations and attributes in the database and they need all data in the form of prolog tables. An apriori-based algorithm was introduced in [1], wich first generates frequent tuples in each single table, and then looks for frequent tuples whose items belong to different tables via a multi-dimensional count array; [9] proposed an algorithm that mines first each table separately, and then two tables at a time; [12] presented *MultiClose*, that first converts all dimension tables to a vertical data format, and then mines each of them locally, with a closed algorithm. The patterns are stored in two-level hash trees, which are then traversed in pairs to find multi-table patterns; StarFP-Growth, proposed in [10], is a pattern-growth method based on FP-Growth [5]. Its main idea is to construct an FP-Tree for each dimension (*DimFP-Tree*), accordingly to the global support of its items, and then, build a Super FP-Tree, combining the FP-Trees of each dimension, accordingly to the facts. In the end it runs FP-Growth on this tree to find multi-relational patterns.

To the best of our knowledge, there are only two works on multi-relational frequent pattern mining over data streams, *SWARM* [3] and *RFPS* [6], both based on the multi-relational ILP algorithm WARMR [2]. These approaches are apriori-based and are able to find relational patterns over a sliding time window.

# 6    Conclusions

In this paper, we propose a new algorithm for mining very large data warehouses, modeled in a star schema, at the right aggregating level. Particularly, it is able to deal with degenerate dimensions by aggregating the rows in the fact table corresponding to the same business fact, and still mining directly the star.

Experiments on Adventure Works showed that Star FP-Stream is accurate, achieving a good precision and 100% of recall. The pattern-tree tends to be very large, but its size tends to be stable, and it is able to return the patterns for every minimum support $\sigma \gg \epsilon$, anytime. The time and memory needed by the algorithm tend to be constant and do not depend on the total number of transactions processed so far, but only on the size of the batches and on the size of the current pattern tree, which in turn depends on the characteristics of the data. Star FP-Stream greatly outperforms Simple FP-Stream in terms of time.

**Acknowledgment.** This work is partially supported by FCT – Fundação para a Ciência e a Tecnologia, under research project educare (PTDC/EIA-EIA/110058/2009) and PhD grant SFRH/BD/64108/2009.

# References

1. Crestana-Jensen, V., Soparkar, N.: Frequent itemset counting across multiple tables. In: PADKK 2000: Proc. of the 4th Pacific-Asia Conf. on Knowl. Discovery and Data Mining, Current Issues and New Applications, London, pp. 49–61 (2000)
2. Dehaspe, L., Raedt, L.: Mining Association Rules in Multiple Relations. In: Džeroski, S., Lavrač, N. (eds.) ILP 1997. LNCS, vol. 1297, pp. 125–132. Springer, Heidelberg (1997)
3. Fumarola, F., Ciampi, A., Appice, A., Malerba, D.: A Sliding Window Algorithm for Relational Frequent Patterns Mining from Data Streams. In: Gama, J., Costa, V.S., Jorge, A.M., Brazdil, P.B. (eds.) DS 2009. LNCS, vol. 5808, pp. 385–392. Springer, Heidelberg (2009)
4. Giannella, C., Han, J., Pei, J., Yan, X., Yu, P.S.: Mining frequent patterns in data streams at multiple time granularities: Next generation data mining (2003)
5. Han, J., Pei, J., Yin, Y.: Mining frequent patterns without candidate generation. In: Proc. of the 2000 ACM SIGMOD, pp. 1–12. ACM, New York (2000)
6. Hou, W., Yang, B., Xie, Y., Wu, C.: Mining multi-relational frequent patterns in data streams. In: BIFE 2009: Proc. of the Second Intern. Conf. on Business Intelligence and Financial Engineering, pp. 205–209 (2009)
7. Kimball, R., Ross, M.: The Data warehouse Toolkit, 2nd edn. John Wiley & Sons, Inc., New York (2002)
8. Liu, H., Lin, Y., Han, J.: Methods for mining frequent items in data streams: an overview. Knowl. Inf. Syst. 26, 1–30 (2011)
9. Ng, E., Fu, A., Wang, K.: Mining association rules from stars. In: ICDM 2002: Proc. of the 2002 IEEE Intern. Conf. on DM, Japan, pp. 322–329. IEEE (2002)
10. Silva, A., Antunes, C.: Pattern Mining on Stars with FP-Growth. In: Torra, V., Narukawa, Y., Daumas, M. (eds.) MDAI 2010. LNCS, vol. 6408, pp. 175–186. Springer, Heidelberg (2010)
11. Silva, A., Antunes, C.: Mining Patterns from Large Star Schemas Based on Streaming Algorithms. In: Lee, R. (ed.) Computer and Information Science 2012. SCI, vol. 429, pp. 139–150. Springer, Heidelberg (2012)
12. Xu, L.-J., Xie, K.-L.: A novel algorithm for frequent itemset mining in data warehouses. Journal of Zhejiang University - Science A 7(2), 216–224 (2006)

# Application of Quantitative MCDA Methods for Parameter Setting Support of an Image Processing System

Lionel Valet and Vincent Clivillé

LISTIC, Université de Savoie, Domaine Universitaire, BP 80439
74944 Annecy le Vieux Cedex France
{lionel.valet,vincent.cliville}@univ-savoie.fr

**Abstract.** This paper proposes to use quantitative methods to identify a preference model reflecting the overall satisfaction of the user according to the numerous parameters of a complex fusion system. The studied fusion system is devoted to 3D image interpretation and it works in interaction with experts who have knowledge and experience of the concerned applications. Such a system involves many sub-parts and each of them has many parameters that must be adjusted to obtain interesting detections. The link between the parameters and the overall satisfaction expressed by the experts is *a priori* unknown and it is a key issue to better interact with the system. After the presentation of the preference model relevance with the problematic, three model identifications (multivariate, UTA+ and MACBETH) are attempted in this paper to find an interesting set of parameters according to the available overall satisfaction. Obtained results show the complexity of this kind of identification, mainly because of the non monotonicity of the parameter utilities.

**Keywords:** Fusion system, 3D image, Overall satisfaction, MCDA, Multiple variable regression, UTA+, MACBETH.

## 1 Parameter Adjustment of a Complex Fusion System

Fusion systems for 3D image interpretation are a complete processing chains that start from the local information measurement and attempts to deliver useful synthetic information for the end-users. Generally, the end-users are experts in scientific domains not directly in connection with those involved in the fusion system. In order to give us more confidence in the obtained results, they are implicated in the process by giving some samples of what they are looking for in the images. Such a system thus becomes a so-called supervised system. Supervised fusion systems for 3D image interpretation are complex systems and they need a strong collaboration with the experts. The system complexity is due to (1) the important number of operations (stages) needed to compute the results, (2) the non-linearity of many of them, (3) the non-analytic expression of the global transfer function, and (4) the important computation time.

Image interpretation consists of identifying typical regions within the 3D images to better understand a complex phenomenon. The concerned fusion system was designed and applied on two main applications: (1) an analysis of 3D tomographic images for

V. Torra et al. (Eds.): MDAI 2012, LNAI 7647, pp. 341–354, 2012.

part quality control in collaboration with Schneider Electric [4], (2) an analysis of 3D seismic images for oil prospection in collaboration with the Federal University of Rio de Janeiro [5]. The use of the system on these two real applications has highlighted the difficulty of adjusting the numerous parameters of the fusion system. There is a need to help the designers but also the end-users of the system to find interesting sets of parameters with a fewer number of experimentations.

The proposed fusion system is composed of four main levels presented in figure 1. Firstly, the extraction level computes some attributes thanks to image processing techniques. Each attribute has the same size as the input image and it contains a numerical quantity for each voxel within the image. According to the chosen measurement, these attributes can inform on the local texture that exists in the image, on the local organisation of the grey level intensities or on the form of local objects (like porosities for example). The main parameters are concentrated on this level. Secondly, the attributes are represented in a common space to be comparable. Similarity cards are computed thanks to possibility theory. Thirdly, a Choquet integral is evaluated for each region of interest to give a global similarity for each voxel of the image to the regions. The fuzzy measures used to compute the Choquet integral are learnt thanks to the samples (called regions of reference) that are given by the experts. The last level is in charge of taking the final decision for each voxel on which region it belongs according to its global similarity degrees.

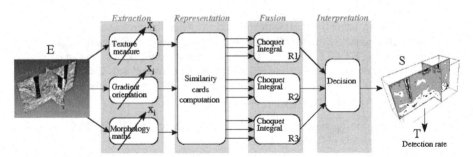

**Fig. 1.** Synoptic of the fusion system

There is a need for the end-users of such systems (but also for the designers) to have information on how the parameters must be adjusted in order to obtain a useful interpretation. The parameters are mainly concentrated on the extraction level and their number depends on the chosen attributes. The next section will present in detail a real case where three attributes (for a total of 18 parameters) are used to interpret a synthetic 3D image. The result of the fusion is called a "detection" which refers to a particular configuration of the parameters. Previous works have tried to better understand the system behaviour: a local evaluation based on the separability index [1] has been proposed to quantify each subpart of the fusion system. At the same time, a global detection rate $T$ is computed on the result by comparing the regions of reference given by the experts to the obtained detection. With these quantities, an optimisation approach based on genetic algorithms has been applied to find some optimal parameter settings. The obtained results presented in [5] and [6] have shown that the detection rate $T$ can effectively be increased. Nevertheless, $T$ is only a partial evaluation of the result and it does not reflect the overall satisfaction of the experts. The optimisation process is highly sensitive to the objective function and it can let some part of the image (out of the references) remain undetected to focus only on the regions used for the global rate computation. Moreover, the computation time needed for by such an optimisation is also very important.

This communication analyses the MCDA methods opportunity to support this parameter setting problem. The parameter adjustment problem can be seen as a decision problem. The parameter vector composes a set of alternatives and the decision maker has to select an alternative that leads to an interesting preference on the obtained result (i.e. an output detection that is globally good and not only the detection rate $T$). The objective is also to avoid the computation of many fusions because of the entire time consuming process MCDA has shown its interest in many decision problem areas by the building of a preference model which reflects the decision maker (DM) preferences [7]. The objective, in this paper, is to test different preference models to link the decision data (alternatives, attributes) with the preference information in order to identify an interesting set of parameters.

## 2    System Description and Expert Available Knowledge

### 2.1    Fusion System for Image Interpretation

The fusion system is applied in this paper to a synthetic image (presented on figure 2) where three regions of interest are investigated. To detect the sought-after regions, three attributes are extracted from the original image: (1) $A_1$ measures the local organisation in the third direction obtained with the PCA (Principal Component Analysis) of the gradient vectors computed on each voxel, (2) $A_2$ measures the local organisation using the second direction and (3) $A_3$ is a texture measurement based on the co-occurrence matrix. More information on these well-known image processing techniques can be found in [4]. The computation of the attribute $A_1$ requires the following parameters:

- $x_1$ / *alpha* : the Derich filter coefficient for noise reduction [0...1].

- $x_2$ / *gradient window size* : the size of the cubic windows used for the gradient evaluation.

- $x_3, x_4, x_5$ / *PCA window sizes* : the three dimensions of the windows used for the PCA.

- $x_6$ / *dynamic adjustment* .

The parameters for attribute $A_2$ are the same as those for $A_1$. It leads to 6 other attributes noted $x_7$ to $x_{12}$ $(x_7$ is the *alpha* coefficient, etc.). Only the computation of the attribute is different.

The parameters for attribute $A_3$ are:

- $x_{13}, x_{14}, x_{15}$ / the windows sizes for the texture analysis
- $x_{16}, x_{17}, x_{18}$ / the *a priori* texture direction.

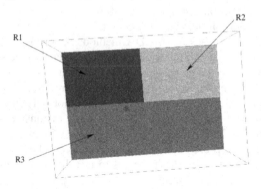

**Fig. 2.** The 3D synthetic images

So a given detection is characterised by a vector of 18 parameters $(x_1...x_i...x_{18})$ noted $P$ coming from the extraction stage. The parameters of the other levels have less impact on the output and they are set to default values. It also prevents the parameters from being in series in the processing chain and to interact with each other during the processing. The objective is thus to adjust the vector $P$ to obtain an interesting detection of the three sought-after regions. The number of possible combinations is very high (18 parameters, about 5 to 10 values) and the duration of a full treatment by the fusion system is about 30 minutes. Thus it can be very interesting to identify an approximate model which links the parameter vector and the overall satisfaction.

The output of the system, noted $S$, is evaluated by a detection rate $T$. Figure 3 presents 14 detections $P^j \in \{P^1...P^{14}\}$ obtained for different vectors $(x_1^j...x_i^j...x_n^j)$ (table 2 in the appendix) chosen by the expert as potential interesting detections. The global rate $T^j$ is given in the same figure. Note that only 2D views corresponding to the 100th section of the 3D image are presented for a better visualisation comfort.

These images clearly show that the $T^j$ does not take into account the global appearance (fragmentation, false area recognition ...) of the detection, whereas the expert will do it. So the expert wishes to complete this information by an overall satisfaction score noted $U^j$ on a scale [0...20] which evaluates the areas detection quality. $U^j$ is also given in figure 3.

Knowing that the goal of the expert is to determine a parameter vector corresponding to a satisfactory overall satisfaction $U$, the problem is the identification of such a vector. Thus the idea is to learn the complex relationship between $P$ and $U$ from a reference set of $P$. Then an approximate model of the expert preferences could be deduced from this learning to identify satisfactory detection vectors, $(U \geq 18)$, and perhaps a totally satisfactory $(U \geq 20)$. In this sense MCDA methods are going to be applied for this approximate model identification.

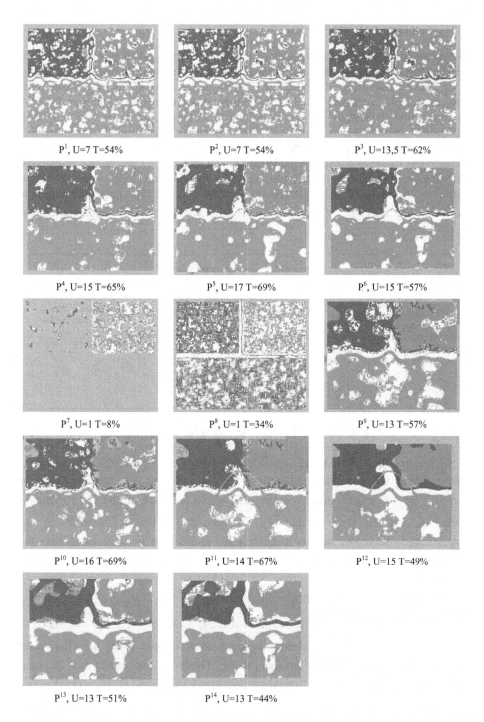

**Fig. 3.** Detection and satisfaction for 14 vectors

## 2.2    Interest of MCDA Methods

The considered problem can be seen as the characterisation of expert satisfaction. It can be dealt using the definition of an overall utility resulting from the aggregation of a set of marginal utilities. In this context using MCDA methods seems to be convenient. As recalled in figure 4, adapted for the considered decision problem, MCDA methods allow the DM to express his preferences between alternatives under the form of cardinal or ordinal information knowing a given decision problem. Therefore it is necessary to identify the alternatives, then to describe them according to a set of attributes and finally to identify a preference model able to give this ordinal or cardinal information. The model identification requires specific knowledge which can take different forms according to the retained method (Fig. 4). The parameter vector is noted, $(x_1...x_i...x_{18})$ and corresponds to a detection noted $P^j$. The overall satisfaction of the detection $P^j$ is noted $U^j$.

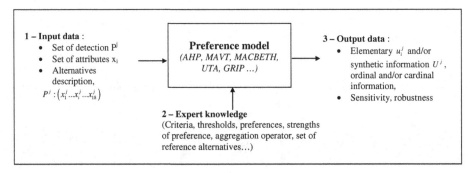

**Fig. 4.** Synoptic of a preference model for decision support

Concerning the knowledge requirements, the expert user of the 3D fusion system is able to quantify the overall satisfaction $U^j$ resulting from a given parameter vector $(x_1^j...x_i^j...x_n^j)$ on a scale defined on $[0...20]$ with a precision of about $\pm 1$. This information will be used to identify the general relationship between $U$ and $(x_1...x_i...x_{18})$. Moreover he has to express his preferences in order to identify the weights of each parameter and marginal utilities defined according to these parameters.

Thus, knowing the availability of numerous couples $\left[\left(x_1^j...x_i^j...x_n^j\right);\left(U^j\right)\right]$ the idea of MCDA approach is to define marginal utilities, noted $\left(u_1^j...u_i^j...u_n^j\right)$ associated to the parameter vector, then to identify an aggregation function which links the vector $\left(u_1^j...u_i^j...u_n^j\right)$ to the overall satisfaction $U^j$. However the conditions of non-redundancy, exhaustivity and coherence between criteria have to be ensured [3].

# 3    Model Identification Attempt

Three model identification attempts are tested in this communication. The first one is the multi variable analysis which needs strict hypotheses concerning the variables [12]. The two other ones are MCDA approaches [3], respectively the UTA+ method and the MACBETH method, which allow the expert to identify a synthesising utility $U$ depending on cardinal information about description of alternatives [2].

## 3.1    Multi Variable Analysis

Firstly a simple relation is tested thanks to the Multivariate Statistics and more precisely the multiple variable linear regression. The reference set of detections and associated vectors is given in table 2 and 3. It consists of the 14 previous detections completed by 38 other detections. The main assumptions of the multiple regression are verified (mainly the matrix calculus ones) according to the available data. The model can be identified as follows: $U = \sum_{i=1}^{18} a_i x_i + a_0$ . The obtained coefficients $a_i$ are given in table 1.

**Table 1.** Multi variable coefficient values

| $a_0$ | $a_1$ | $a_2$ | $a_3$ | $a_4$ | $a_5$ | $a_6$ | $a_7$ | $a_8$ | $a_9$ | $a_{10}$ | $a_{11}$ | $a_{12}$ | $a_{13}$ | $a_{14}$ | $a_{15}$ | $a_{16}$ | $a_{17}$ | $a_{18}$ |
|---|---|---|---|---|---|---|---|---|---|---|---|---|---|---|---|---|---|---|
| 16.7 | 2.6 | -0.16 | 0.05 | 0.05 | 0.12 | -0.07 | 5.04 | -0.18 | -0.01 | 0.05 | 0.01 | -0.03 | 0.09 | -0.03 | -0.13 | -0.27 | -0.35 | -0.27 |

The obtained model is not completely satisfactory because the differences (residues) between the multiple linear model and the expert values are too high to be exploited. Indeed, the determinant coefficient $(R^2) = 0.22$ which signifies a weak explicative model.

One possible reason for the weak model adaptation could be the non-linearity of the relation between the parameter value and the satisfaction value. The researched model could also be non-additive. After a residue examination these explanations can be envisaged and, in a first time, the non linearity should be taken into account. However, it is decided to use the obtained linear model in order to identify the potentially best parameter vector. Knowing the parameters definition on a discrete set of values, it is easy to choose the lowest value when the related coefficient is negative and the highest value when it is positive. It corresponds to the vector $(0.8,3,21,21,21,20,0.8,3,21,3,21,20,3,3,3,1,1,6)$ which leads to a predictive value $U = 22$. After the extraction treatment according to these values, the expert global satisfaction is $U = 18$ which seems to be an interesting value for the expert even if the difference with the predictive value is important. Note that the expert only gives satisfaction between [0...20]. Figure 5 presents the obtained detection that has effectively a better quality than the previous one (less non detected points, less fragmentation).

**Fig. 5.** Detection obtained by the multivariate model, $U = 18$

## 3.2     UTA Method

UTA is an indirect method to determine the additive value function linking for a considered vector of variables, the overall utility and the marginal ones. *"The UTA method proposed by Jacquet-Lagreze and Siskos [8] aims at inferring one or more additive value functions from a given ranking on the reference set $A^R$. The method uses special linear programming techniques to assess these functions so that the ranking(s) obtained through these functions on $A^R$ is (are) as consistent as possible with the given one"* [9].

In the UTA method the decision problem is the identification of a vector which gives a satisfactory detection corresponding to an overall utility as near as possible to 1 (the initial values defined on [0...20] are linearly converted into the interval [0,1]). According to the expert, the method must give at least ordinal information on the new considered alternative regarding the previous ones. Among the numerous methods, the disaggregation method has a high interest because knowledge about a reference actions set is available. In this sense it is proposed to use the UTA+ method which allows to define marginal utilities. These utilities are then summed into an overall one. So the relationship between marginal and overall utilities can be written as follows:

$U^j = \sum_{i=1}^{n} u_i^j$ , where $u_i^j \geq 0$ and $\sum_{i=1}^{n} u_i^{max} = 1$ where $u_i^{max}$ can be viewed as the

relative importance of $u_i / U$

The first step is the definition of the set of criteria. In this problem as said in §2.1, the criteria are the 18 parameters $\left(x_1^j ... x_i^j ... x_n^j\right)$ specified in §2.2. Then the reference set of detections is specified through about 20 detections ($P^1$ to $P^{20}$ in table 2 and 3). Finally the ranking of the reference actions is defined: the expert gives a total pre-order of the set of detections. Unfortunately, no solution can be identified by the UTA+ software edited by the LAMSADE laboratory[1]. With the first set of reference detections being arbitrary, UTA+ method is not able to find a solution, *i.e.* the marginal utility functions cannot be identified from the given data. It can be due to the non monotonic form of the marginal utilities function or the non-additivity of these

---

[1] http://www.lamsade.dauphine.fr/spip.php?rubrique69

marginal utilities to give the overall one. Nevertheless a second set with 31 additional detections, ($P^1$ to $P^{51}$ in table 2 and 3), achieved by changing only one parameter value from the detection $P^{18}$ has been also tested. Results are the same and no function can be found which leaves the previous hypothesis open. Indeed, according to the expert, many marginal utilities do not respect the monotony condition that makes the use of UTA method impossible. For instance the window sizes have a utility that depends on the research object. It leads to non decreasing or non increasing function. So UTA+ is not relevant to deal with this problem. Indeed taking non-monotonic utility functions into account is rarely considered in MCDA domain [10].

### 3.3    MACBETH Method

The MACBETH (Measuring Attractiveness by a Categorical Based Evaluation TecHnique), is another MCDA method which authorises the non-monotony of the marginal utility [11]. It is a MAUT method which is based on the comparisons between different actions (which identify the context) made by the decision-makers. MACBETH describes these actions with, on the one hand, elementary performance expressions, and on the other hand, aggregated ones. In the MACBETH view, the definition of the aggregated performance is based on the weighted mean and is made progressively and interactively. The principle is to translate the qualitative information generally available, thanks to the human expertise of the DMs, into quantitative information [13].

However MACBETH is not a disaggregation method. It needs another type of knowledge which separates the identification of, on the one hand, the marginal utilities, and on the second hand, the weights of the Weighted Arithmetic Mean (WAM) operator. This identification must also be made thanks to the expert knowledge. Thus, it is possible to compute the marginal utilities of a given detection knowing the value of the parameters and then the overall utility.

In this case, the expert knowledge is already available thanks to the previous couples $\left[\left(x_1^j...x_i^j...x_n^j\right);\left(U^j\right)\right]$. But MACBETH authorises the expert to separately compare the values corresponding to a given criterion. For instance the identification of a linear piecewise of the marginal utility $u(x_1)$ can be made thanks to the comparison between the $P^{15}$ and $P^{16}$ vectors of table 3 in the appendix as shown in figure 6.

$$U^1 = \sum_{i=1}^{n} w_i \times u_i^1 , \ U^2 = \sum_{i=1}^{n} w_i \times u_i^2 \text{ and}$$

$u_i^1 = u_i^2$ for all the values of $i$ except one : i=1, then

$$U^1 - U^2 = w_1 \times (u_1^1 - u_1^2)$$

This kind of comparison should be made at least one time for each marginal utility. To identify a more complex model it is necessary to have more comparisons. In our case four piecewises can be identified for each marginal utility. Moreover, it the problem of the identification of the *neutral* and *good* levels remains. Ideally it is only necessary to identify the vectors $\left(... u(x_i^{good}) ...\right)$ and $\left(... u(x_i^{neutral}) ...\right)$ which allow the DM to respectively write the following relations:

$$U^{good} = 1 \Leftrightarrow u(x_i^{good}) = 1 \text{ supposing that } u(x_i^j) \leq 1$$
$$U^{neutral} = 0 \Leftrightarrow u(x_i^{neutral}) = 0 \text{ supposing that } u(x_i^j) \geq 0$$

The identification of these vectors is not obvious because in the previous data, the best and worst detections correspond to: $U^{worst} = 1/20 = 0.05$, and $U^{best} = 17/20 = 0.85$. Such values can be retained knowing that it could be possible to have a value exceeding 1 or smaller than 0. Moreover, the overall utility $U^j$ must be computed again to take into account the new limits of the interval. To avoid these difficulties, the traditional way of MACBETH is used with the independent definition for each attribute of the neutral and good values. Concerning the neutral value, the expert hesitates and perhaps this value should be determined through a few iterations according to the MACBETH[2] interactive way. Finally for each attribute a marginal utility can be computed as shown in figure 6. For instance the marginal utility corresponding to the value $x_3 = 3\%$ is $u_3 = 0.29$ when the marginal utility corresponding to the value $x_3 = 21$ is $u_3 = 0.71$.

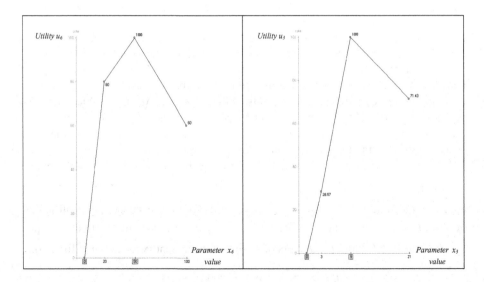

**Fig. 5.** Marginal utilities function for two parameters

The weight determination uses the same procedure based on the comparisons between fictive detections where all the utilities take the neutral level except one which takes the good level. This task is very awkward and in firstly, it is decided to balance the weights. The overall utility can be now computed (table 4 in the appendix). This way seems unsatisfactory because the overall utilities do not correspond to the weighted mean computed by MACBETH.

---

[2] Computation is supported by M-MACBETH software available at
http://www.m-macbeth.com/en/m-home.html

The computation of the overall utilities can now be done and obviously the computation of the best one which corresponds to the highest value according to all the criteria. It corresponds to the following vector $\left(0.3,9,9,9,9,50,0.5,9,21,21,21,100,9,15,9,2,2,2\right)$ which gives, after the fusion, an overall satisfaction $U = 8$ that is obviously not the best. Figure 7 presents the obtained detection. Even if regions R2 and R3 are well detected, region R1 remains completely undetected.

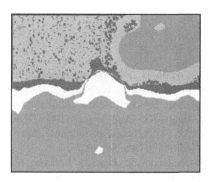

**Fig. 6.** Detection obtained by the MACBETH model, U=8

So, for the MACBETH preference model, the best predictive detection is not right. In this case the asking procedure for both the marginal utilities and the weights identification must be examined. Indeed it is really difficult for the expert to express preferences between pairwise parameter vectors. Comparisons can be really meaningful between two detections [11].

# 4    Conclusions and Perspectives

The parameter adjustment of the extraction stage of a complex fusion system is a difficult task which needs time and expertise. The computation of a global rate is a possible solution but it does not ensure that the highest rate corresponds to the best detection for the experts. In this paper, the parameter adjustment of a fusion system devoted to 3D image interpretation, has been expressed in the form of a decision making problem. Indeed it was shown that the identification problem of a satisfactory detection can be viewed as a characterisation problem. In this sense, three different methods have been applied to identify an approximate model allowing the expert to deduce the overall satisfaction from a parameter vector. Then, the obtained model is used to determine some interesting parameter vectors leading to high estimated overall satisfactions. The tested methods were the multivariate analysis (a statistics approach), and two MCDA methods, UTA+ and MACBETH. Globally the results obtained for the multivariate analysis are interesting because a better detection than the previous best one has been identified. Concerning MACBETH, results are not satisfactory enough and the identification of the preference model remains a difficult task for the expert. Indeed, building independent marginal utilities and determining their weights is not a common way for him. Concerning UTA, the monotonic condition is too restrictive in our case but the use of a relevant variant could overcome this limitation.

This work leads to other perspectives: (1) a deeper exploitation of the multivariate analysis could be attempted. For instance a finer residues analysis can show the possible non linearity; (2) a reasoning on an alternative way for the expert questioning the avoidance of the fictive detections use; (3) the introduction of an non-additive aggregation operator as the Choquet integral could give a more relevant preference model [14]; (4) and more globally, taking into account the expert imperfect knowledge with a relevant method like GRIP, or the robust ordinal regression, could also be interesting for this kind of application [15].

# References

1. Lamallem, A., Valet, L., Coquin, D.: Performance Evaluation of a Fusion System Devoted to Image Interpretation. In: Hüllermeier, E., Kruse, R., Hoffmann, F. (eds.) IPMU 2010. LNCS, vol. 6178, pp. 464–473. Springer, Heidelberg (2010)
2. Clivillé, V., Berrah, L., Mauris, G.: Deploying the ELECTRE III and MACBETH multi-criteria ranking methods for SMEs tactical performance improvements. Journal of Modelling in Management, 24 (edition in October 2012)
3. Roy, B.: Paradigms and Challenges. In: Figueira, J., Greco, S., Ehrgott, M. (eds.) MCDA. Multiple Criteria Decision Analysis. State of the Art Surveys, pp. 3–24. Springer (2005)
4. Jullien, S., Valet, L., Mauris, G., Bolon, P., Teyssier, S.: An attribute fusion system based on the choquet integral to evaluate the quality of composite parts. IEEE Trans. on Instrumentation and Measurement 57(4), 755–762 (2008)
5. Lamallem, A., Valet, L., Coquin, D., de Lima, B.S.L.P., Galichet, S.: Symbolic evaluation of a fusion system devoted to 3D image interpretation. In: 32th Iberian Latin American Congress on Computational Methods in Engineering (CILAMCE), Ouro Preto, Brazil, 15 pages. CDROM (November 2011)
6. Beckmann, M., Valet, L., De Lima, B.S.L.P.: Choquet Integral Parameter Optimization for a Fusion System Devoted to Image Interpretation. In: Greco, S., Bouchon-Meunier, B., Coletti, G., Fedrizzi, M., Matarazzo, B., Yager, R.R. (eds.) IPMU 2012, Part I. CCIS, vol. 297, pp. 531–540. Springer, Heidelberg (2012)
7. Figueira, J., Greco, S., Ehrgott, M.: Multiple Criteria Decision Analysis. State of the Art Surveys, 1045 p. Springer (2005)
8. Jacquet-Lagreze, E., Siskos, Y.: Assessing a set of additive utility functions for multicriteria decision-making, the UTA method. European Journal of Operational Research 10(2), 151–164 (1982)
9. Jacquet-Lagreze, E., Siskos, Y.: Preference disaggregation: 20 years of MCDA experience. European Journal of Operational Research 130, 233–245 (2001)
10. Doumpos, M.: Learning non-monotonic additive value functions for multicriteria decision making. OR Spectrum 34(1), 89–106 (2012)
11. Bana e Costa, C., Vansnick, J.-C.: Applications of the MACBETH approach in the framework of an additive aggregation model. Journal of Multi-Criteria Decision Analysis 6(2), 107–114 (1997)
12. Bourbonnais, R.: Econométrie, 2nd edn., Dunod. Manuel et exercices corrigés (1998) (in French)
13. Clivillé, V., Berrah, L., Mauris, G.: Quantitative expression and aggregation of performance measurements based on the MACBETH multi-criteria method. International Journal of Production Economics 105(1), 171–189 (2007)
14. Labreuche, C., Grabisch, M.: The Choquet integral for the aggregation of interval scales in multicriteria decision making. Fuzzy Sets and Systems 137, 11–26 (2003)

15. Figueira, J., Greco, S., Slowinski, R.: Building a set of additive value functions representing a reference preorder and intensities of preference: GRIP method. European Journal of Operational Research 195(2), 460–486 (2009)

# Appendix

**Table 2.** Arbitrary parameter vector (…$x_i$…) and overall utility function $U$

| | $x_1$ | $x_2$ | $x_3$ | $x_4$ | $x_5$ | $x_6$ | $x_7$ | $x_8$ | $x_9$ | $x_{10}$ | $x_{11}$ | $x_{12}$ | $x_{13}$ | $x_{14}$ | $x_{15}$ | $x_{16}$ | $x_{17}$ | $x_{18}$ | U |
|---|---|---|---|---|---|---|---|---|---|---|---|---|---|---|---|---|---|---|---|
| $P^1$ | 0,5 | 5 | 5 | 5 | 5 | 100 | 0,6 | 3 | 6 | 9 | 9 | 70 | 15 | 15 | 15 | 1 | 1 | 4 | 7 |
| $P^2$ | 0,5 | 5 | 5 | 5 | 5 | 100 | 0,6 | 3 | 9 | 9 | 9 | 70 | 15 | 15 | 15 | 1 | 1 | 4 | 7 |
| $P^3$ | 0,5 | 5 | 5 | 5 | 5 | 100 | 0,5 | 5 | 9 | 9 | 9 | 50 | 15 | 15 | 15 | 1 | 1 | 4 | 13,5 |
| $P^4$ | 0,2 | 7 | 9 | 9 | 9 | 50 | 0,5 | 5 | 9 | 9 | 9 | 50 | 15 | 15 | 15 | 1 | 1 | 4 | 15 |
| $P^5$ | 0,2 | 7 | 9 | 9 | 9 | 50 | 0,5 | 5 | 9 | 9 | 9 | 50 | 9 | 9 | 9 | 1 | 1 | 1 | 17 |
| $P^6$ | 0,8 | 3 | 3 | 3 | 3 | 80 | 0,5 | 5 | 9 | 9 | 9 | 50 | 21 | 21 | 21 | 2 | 2 | 2 | 15 |
| $P^7$ | 0,1 | 3 | 3 | 3 | 3 | 100 | 0,5 | 5 | 9 | 9 | 9 | 50 | 21 | 21 | 21 | 2 | 2 | 2 | 1 |
| $P^8$ | 0,3 | 9 | 15 | 15 | 15 | 60 | 0,3 | 9 | 21 | 21 | 21 | 50 | 27 | 27 | 27 | 2 | 2 | 6 | 1 |
| $P^9$ | 0,3 | 9 | 15 | 15 | 15 | 60 | 0,6 | 3 | 9 | 9 | 9 | 70 | 9 | 9 | 9 | 1 | 1 | 1 | 13 |
| $P^{10}$ | 0,3 | 9 | 15 | 15 | 15 | 60 | 0,5 | 5 | 9 | 9 | 9 | 50 | 9 | 9 | 9 | 1 | 1 | 1 | 16 |
| $P^{11}$ | 0,3 | 9 | 15 | 15 | 15 | 60 | 0,3 | 9 | 21 | 21 | 21 | 50 | 9 | 9 | 9 | 1 | 1 | 1 | 14 |
| $P^{12}$ | 0,3 | 9 | 15 | 15 | 15 | 60 | 0,3 | 9 | 21 | 21 | 21 | 50 | 27 | 27 | 27 | 2 | 2 | 6 | 15 |
| $P^{13}$ | 0,2 | 7 | 9 | 9 | 9 | 50 | 0,6 | 3 | 9 | 9 | 9 | 70 | 21 | 21 | 21 | 2 | 2 | 2 | 13 |
| $P^{14}$ | 0,2 | 7 | 9 | 9 | 9 | 50 | 0,6 | 3 | 9 | 9 | 9 | 70 | 27 | 27 | 27 | 2 | 2 | 6 | 13 |

**Table 3.** Parameter vector with minimal changes and overall utility function $U$

| | $x_1$ | $x_2$ | $x_3$ | $x_4$ | $x_5$ | $x_6$ | $x_7$ | $x_8$ | $x_9$ | $x_{10}$ | $x_{11}$ | $x_{12}$ | $x_{13}$ | $x_{14}$ | $x_{15}$ | $x_{16}$ | $x_{17}$ | $x_{18}$ | U |
|---|---|---|---|---|---|---|---|---|---|---|---|---|---|---|---|---|---|---|---|
| $P^{15}$ | 0,3 | 9 | 9 | 9 | 9 | 50 | 0,3 | 9 | 9 | 9 | 9 | 50 | 9 | 9 | 9 | 2 | 2 | 2 | 11 |
| $P^{16}$ | 0,1 | 9 | 9 | 9 | 9 | 50 | 0,3 | 9 | 9 | 9 | 9 | 50 | 9 | 9 | 9 | 2 | 2 | 2 | 15 |
| $P^{17}$ | 0,8 | 9 | 9 | 9 | 9 | 50 | 0,3 | 9 | 9 | 9 | 9 | 50 | 9 | 9 | 9 | 2 | 2 | 2 | 9 |
| $P^{18}$ | 0,3 | 3 | 9 | 9 | 9 | 50 | 0,3 | 9 | 9 | 9 | 9 | 50 | 9 | 9 | 9 | 2 | 2 | 2 | 9 |
| $P^{19}$ | 0,3 | 21 | 9 | 9 | 9 | 50 | 0,3 | 9 | 9 | 9 | 9 | 50 | 9 | 9 | 9 | 2 | 2 | 2 | 8 |
| $P^{20}$ | 0,3 | 9 | 3 | 9 | 9 | 50 | 0,3 | 9 | 9 | 9 | 9 | 50 | 9 | 9 | 9 | 2 | 2 | 2 | 6 |
| $P^{21}$ | 0,3 | 9 | 21 | 9 | 9 | 50 | 0,3 | 9 | 9 | 9 | 9 | 50 | 9 | 9 | 9 | 2 | 2 | 2 | 9 |
| $P^{22}$ | 0,3 | 9 | 9 | 3 | 9 | 50 | 0,3 | 9 | 9 | 9 | 9 | 50 | 9 | 9 | 9 | 2 | 2 | 2 | 8 |
| $P^{23}$ | 0,3 | 9 | 9 | 21 | 9 | 50 | 0,3 | 9 | 9 | 9 | 9 | 50 | 9 | 9 | 9 | 2 | 2 | 2 | 10 |
| $P^{24}$ | 0,3 | 9 | 9 | 9 | 3 | 50 | 0,3 | 9 | 9 | 9 | 9 | 50 | 9 | 9 | 9 | 2 | 2 | 2 | 8 |
| $P^{25}$ | 0,3 | 9 | 9 | 9 | 21 | 50 | 0,3 | 9 | 9 | 9 | 9 | 50 | 9 | 9 | 9 | 2 | 2 | 2 | 11 |
| $P^{26}$ | 0,3 | 9 | 9 | 9 | 9 | 20 | 0,3 | 9 | 9 | 9 | 9 | 50 | 9 | 9 | 9 | 2 | 2 | 2 | 10 |
| $P^{27}$ | 0,3 | 9 | 9 | 9 | 9 | 100 | 0,3 | 9 | 9 | 9 | 9 | 50 | 9 | 9 | 9 | 2 | 2 | 2 | 10 |
| $P^{28}$ | 0,3 | 9 | 9 | 9 | 9 | 50 | 0,1 | 9 | 9 | 9 | 9 | 50 | 9 | 9 | 9 | 2 | 2 | 2 | 11 |
| $P^{29}$ | 0,3 | 9 | 9 | 9 | 9 | 50 | 0,8 | 9 | 9 | 9 | 9 | 50 | 9 | 9 | 9 | 2 | 2 | 2 | 11 |
| $P^{30}$ | 0,3 | 9 | 9 | 9 | 9 | 50 | 0,3 | 3 | 9 | 9 | 9 | 50 | 9 | 9 | 9 | 2 | 2 | 2 | 10 |

**Table 3.** (*continued*)

| | | | | | | | | | | | | | | | | | | | |
|---|---|---|---|---|---|---|---|---|---|---|---|---|---|---|---|---|---|---|---|
| P$^{31}$ | 0,3 | 9 | 9 | 9 | 9 | 50 | 0,3 | 21 | 9 | 9 | 9 | 50 | 9 | 9 | 9 | 2 | 2 | 2 | 10 |
| P$^{32}$ | 0,3 | 9 | 9 | 9 | 9 | 50 | 0,3 | 9 | 3 | 9 | 9 | 50 | 9 | 9 | 9 | 2 | 2 | 2 | 12 |
| P$^{33}$ | 0,3 | 9 | 9 | 9 | 9 | 50 | 0,3 | 9 | 21 | 9 | 9 | 50 | 9 | 9 | 9 | 2 | 2 | 2 | 13 |
| P$^{34}$ | 0,3 | 9 | 9 | 9 | 9 | 50 | 0,3 | 9 | 9 | 3 | 9 | 50 | 9 | 9 | 9 | 2 | 2 | 2 | 11 |
| P$^{35}$ | 0,3 | 9 | 9 | 9 | 9 | 50 | 0,3 | 9 | 9 | 21 | 9 | 50 | 9 | 9 | 9 | 2 | 2 | 2 | 14 |
| P$^{36}$ | 0,3 | 9 | 9 | 9 | 9 | 50 | 0,3 | 9 | 9 | 9 | 3 | 50 | 9 | 9 | 9 | 2 | 2 | 2 | 12 |
| P$^{37}$ | 0,3 | 9 | 9 | 9 | 9 | 50 | 0,3 | 9 | 9 | 9 | 21 | 50 | 9 | 9 | 9 | 2 | 2 | 2 | 14 |
| P$^{38}$ | 0,3 | 9 | 9 | 9 | 9 | 50 | 0,3 | 9 | 9 | 9 | 9 | 20 | 9 | 9 | 9 | 2 | 2 | 2 | 15 |
| P$^{39}$ | 0,3 | 9 | 9 | 9 | 9 | 50 | 0,3 | 9 | 9 | 9 | 9 | 100 | 9 | 9 | 9 | 2 | 2 | 2 | 12 |
| P$^{40}$ | 0,3 | 9 | 9 | 9 | 9 | 50 | 0,3 | 9 | 9 | 9 | 9 | 50 | 3 | 9 | 9 | 2 | 2 | 2 | 10 |
| P$^{41}$ | 0,3 | 9 | 9 | 9 | 9 | 50 | 0,3 | 9 | 9 | 9 | 9 | 50 | 27 | 9 | 9 | 2 | 2 | 2 | 13 |
| P$^{42}$ | 0,3 | 9 | 9 | 9 | 9 | 50 | 0,3 | 9 | 9 | 9 | 9 | 50 | 9 | 3 | 9 | 2 | 2 | 2 | 11 |
| P$^{43}$ | 0,3 | 9 | 9 | 9 | 9 | 50 | 0,3 | 9 | 9 | 9 | 9 | 50 | 9 | 27 | 9 | 2 | 2 | 2 | 11 |
| P$^{44}$ | 0,3 | 9 | 9 | 9 | 9 | 50 | 0,3 | 9 | 9 | 9 | 9 | 50 | 9 | 9 | 3 | 2 | 2 | 2 | 11 |
| P$^{45}$ | 0,3 | 9 | 9 | 9 | 9 | 50 | 0,3 | 9 | 9 | 9 | 9 | 50 | 9 | 9 | 27 | 2 | 2 | 2 | 9 |
| P$^{46}$ | 0,3 | 9 | 9 | 9 | 9 | 50 | 0,3 | 9 | 9 | 9 | 9 | 50 | 9 | 9 | 9 | 1 | 2 | 2 | 8 |
| P$^{47}$ | 0,3 | 9 | 9 | 9 | 9 | 50 | 0,3 | 9 | 9 | 9 | 9 | 50 | 9 | 9 | 9 | 6 | 2 | 2 | 10 |
| P$^{48}$ | 0,3 | 9 | 9 | 9 | 9 | 50 | 0,3 | 9 | 9 | 9 | 9 | 50 | 9 | 9 | 9 | 1 | 1 | 2 | 9 |
| P$^{49}$ | 0,3 | 9 | 9 | 9 | 9 | 50 | 0,3 | 9 | 9 | 9 | 9 | 50 | 9 | 9 | 9 | 1 | 6 | 2 | 11 |
| P$^{50}$ | 0,3 | 9 | 9 | 9 | 9 | 50 | 0,3 | 9 | 9 | 9 | 9 | 50 | 9 | 9 | 9 | 1 | 2 | 1 | 9 |
| P$^{51}$ | 0,3 | 9 | 9 | 9 | 9 | 50 | 0,3 | 9 | 9 | 9 | 9 | 50 | 9 | 9 | 9 | 2 | 2 | 6 | 10 |

**Table 4.** Vector of marginal utilities and overall utility function $U$ given by MACBETH

| Options | Global | x1 | x2 | x3 | x4 | x5 | x6 | x7 | x8 | x9 | x10 | x11 | x12 | x13 | x14 | x15 | x16 | x17 | x18 |
|---|---|---|---|---|---|---|---|---|---|---|---|---|---|---|---|---|---|---|---|
| P1 | 67.46 | 46.67 | 71.43 | 52.38 | 60.00 | 60.00 | 75.00 | 100.00 | 75.00 | 71.43 | 75.00 | 57.14 | 42.50 | 61.11 | 100.00 | 86.67 | 40.00 | 60.00 | 80.00 |
| P2 | 68.89 | 46.67 | 71.43 | 52.38 | 60.00 | 60.00 | 75.00 | 100.00 | 75.00 | 75.00 | 57.14 | 75.00 | 57.14 | 42.50 | 61.11 | 100.00 | 86.67 | 40.00 | 100.00 |
| P3 | 66.85 | 46.67 | 71.43 | 52.38 | 60.00 | 60.00 | 75.00 | 100.00 | 83.33 | 57.14 | 75.00 | 57.14 | 37.50 | 61.11 | 100.00 | 86.67 | 40.00 | 60.00 | 80.00 |
| P4 | 77.85 | 77.78 | 85.71 | 100.00 | 100.00 | 100.00 | 100.00 | 100.00 | 83.33 | 57.14 | 75.00 | 57.14 | 37.50 | 61.11 | 100.00 | 86.67 | 40.00 | 60.00 | 80.00 |
| P5 | 78.98 | 77.78 | 85.71 | 100.00 | 100.00 | 100.00 | 100.00 | 100.00 | 83.33 | 57.14 | 75.00 | 57.14 | 37.50 | 66.67 | 100.00 | 100.00 | 40.00 | 60.00 | 80.00 |
| P6 | 83.47 | 77.78 | 85.71 | 100.00 | 100.00 | 100.00 | 100.00 | 100.00 | 83.33 | 57.14 | 75.00 | 57.14 | 37.50 | 55.56 | 100.00 | 73.33 | 100.00 | 100.00 | 100.00 |
| P7 | 63.13 | 33.33 | 57.14 | 28.57 | 40.00 | 40.00 | 85.00 | 100.00 | 75.00 | 57.14 | 75.00 | 57.14 | 42.50 | 66.67 | 100.00 | 100.00 | 40.00 | 60.00 | 80.00 |
| P8 | 66.98 | 100.00 | 57.14 | 28.57 | 40.00 | 40.00 | 85.00 | 100.00 | 75.00 | 57.14 | 75.00 | 57.14 | 42.50 | 66.67 | 100.00 | 100.00 | 40.00 | 60.00 | 80.00 |
| P9 | 76.65 | 55.56 | 100.00 | 85.72 | 90.00 | 100.00 | 95.00 | 100.00 | 75.00 | 57.14 | 75.00 | 57.14 | 42.50 | 66.67 | 100.00 | 100.00 | 40.00 | 60.00 | 80.00 |
| P10 | 77.11 | 55.56 | 100.00 | 85.72 | 90.00 | 100.00 | 95.00 | 100.00 | 83.33 | 57.14 | 75.00 | 57.14 | 42.50 | 66.67 | 100.00 | 100.00 | 40.00 | 60.00 | 80.00 |
| P11 | 89.91 | 55.56 | 100.00 | 85.72 | 90.00 | 100.00 | 95.00 | 100.00 | 100.00 | 100.00 | 100.00 | 100.00 | 37.50 | 66.67 | 100.00 | 100.00 | 40.00 | 60.00 | 80.00 |
| P12 | 85.21 | 55.56 | 100.00 | 85.72 | 90.00 | 100.00 | 95.00 | 100.00 | 100.00 | 100.00 | 100.00 | 100.00 | 37.50 | 80.00 | 100.00 | 60.00 | 100.00 | 100.00 | 60.00 |
| P13 | 83.29 | 77.78 | 85.71 | 100.00 | 100.00 | 100.00 | 100.00 | 100.00 | 75.00 | 57.14 | 75.00 | 57.14 | 42.50 | 55.56 | 100.00 | 73.33 | 100.00 | 100.00 | 100.00 |
| P14 | 80.02 | 77.78 | 85.71 | 100.00 | 100.00 | 100.00 | 100.00 | 100.00 | 75.00 | 57.14 | 75.00 | 57.14 | 42.50 | 80.00 | 100.00 | 60.00 | 100.00 | 100.00 | 60.00 |
| P15 | 82.09 | 55.56 | 100.00 | 28.57 | 100.00 | 100.00 | 100.00 | 100.00 | 100.00 | 57.14 | 75.00 | 57.14 | 37.50 | 66.67 | 100.00 | 100.00 | 40.00 | 100.00 | 100.00 |
| P40 | 90.50 | 55.56 | 100.00 | 100.00 | 100.00 | 100.00 | 100.00 | 100.00 | 100.00 | 57.14 | 75.00 | 57.14 | 37.50 | 66.67 | 100.00 | 100.00 | 40.00 | 60.00 | 100.00 |
| P16 | 86.53 | 100.00 | 100.00 | 100.00 | 100.00 | 100.00 | 100.00 | 100.00 | 100.00 | 57.14 | 75.00 | 57.14 | 75.00 | 66.67 | 100.00 | 100.00 | 100.00 | 100.00 | 100.00 |
| P46 | 82.72 | 55.56 | 100.00 | 100.00 | 100.00 | 100.00 | 100.00 | 100.00 | 100.00 | 57.14 | 75.00 | 57.14 | 37.50 | 66.67 | 100.00 | 100.00 | 100.00 | 100.00 | 100.00 |
| P20 | 82.09 | 55.56 | 100.00 | 28.57 | 100.00 | 100.00 | 100.00 | 100.00 | 100.00 | 57.14 | 75.00 | 57.14 | 37.50 | 66.67 | 100.00 | 100.00 | 100.00 | 100.00 | 100.00 |
| P25 | 86.06 | 55.56 | 100.00 | 100.00 | 100.00 | 100.00 | 100.00 | 100.00 | 100.00 | 57.14 | 75.00 | 57.14 | 37.50 | 66.67 | 100.00 | 100.00 | 100.00 | 100.00 | 100.00 |
| P30 | 84.67 | 55.56 | 100.00 | 100.00 | 100.00 | 100.00 | 100.00 | 100.00 | 75.00 | 57.14 | 75.00 | 57.14 | 37.50 | 66.67 | 100.00 | 100.00 | 100.00 | 100.00 | 100.00 |
| P35 | 87.45 | 55.56 | 100.00 | 100.00 | 100.00 | 100.00 | 100.00 | 100.00 | 100.00 | 57.14 | 75.00 | 57.14 | 37.50 | 66.67 | 100.00 | 100.00 | 100.00 | 100.00 | 100.00 |
| P40 | 87.92 | 55.56 | 100.00 | 100.00 | 100.00 | 100.00 | 100.00 | 100.00 | 100.00 | 57.14 | 75.00 | 57.14 | 37.50 | 100.00 | 100.00 | 100.00 | 100.00 | 100.00 | 100.00 |
| P45 | 83.83 | 55.56 | 100.00 | 100.00 | 100.00 | 100.00 | 100.00 | 100.00 | 100.00 | 57.14 | 75.00 | 57.14 | 37.50 | 66.67 | 100.00 | 60.00 | 100.00 | 100.00 | 100.00 |
| P50 | 81.51 | 55.56 | 100.00 | 100.00 | 100.00 | 100.00 | 100.00 | 100.00 | 100.00 | 57.14 | 75.00 | 57.14 | 37.50 | 66.67 | 100.00 | 100.00 | 40.00 | 100.00 | 80.00 |
| [ toutes sup ] | 100.00 | 100.00 | 100.00 | 100.00 | 100.00 | 100.00 | 100.00 | 100.00 | 100.00 | 100.00 | 100.00 | 100.00 | 100.00 | 100.00 | 100.00 | 100.00 | 100.00 | 100.00 | 100.00 |
| [ toutes inf ] | 0.00 | 0.00 | 0.00 | 0.00 | 0.00 | 0.00 | 0.00 | 0.00 | 0.00 | 0.00 | 0.00 | 0.00 | 0.00 | 0.00 | 0.00 | 0.00 | 0.00 | 0.00 | 0.00 |
| Poids : | | 0.0556 | 0.0556 | 0.0556 | 0.0556 | 0.0556 | 0.0556 | 0.0556 | 0.0556 | 0.0556 | 0.0556 | 0.0556 | 0.0556 | 0.0556 | 0.0556 | 0.0556 | 0.0556 | 0.0556 | 0.0556 |

# Inductive Clustering
# and Twofold Approximations
# in Nearest Neighbor Clustering

Sadaaki Miyamoto[1] and Satoshi Takumi[2]

[1] Department of Risk Engineering, Faculty of Systems and Information Engineering
University of Tsukuba, 1-1-1 Tennodai, Tsukuba, Ibaraki 305-8573, Japan
miyamoto@risk.tsukuba.ac.jp
[2] Graduate School of Systems and Information Engineering
University of Tsukuba, 1-1-1 Tennodai, Tsukuba, Ibaraki 305-8573, Japan
s1120627@u.tsukuba.ac.jp

**Abstract.** The aim of this paper is to study the concept of inductive clustering and two approximations in nearest neighbor clustering induced thereby. The concept of inductive clustering means that natural classification rules are derived as the results of clustering, a typical example of which is the Voronoi regions in $K$-means clustering. When the rule of nearest prototype allocation in $K$-means is replaced by nearest neighbor classification, we have inductive clustering related to the single linkage in agglomerative hierarchical clustering. The latter method naturally derives two approximations that can be compared to lower and upper approximations for rough sets. We thus have a method of inductive clustering with twofold approximations related to nearest neighbor classification. Illustrative examples show implications and significances of this concept.

**Keywords:** hierarchical clustering, nearest neighbor classification, $K$-means, inductive clustering, upper and lower approximations.

## 1 Introduction

Many studies have been done on semi-supervised learning and as a result we have two concepts of the inductive learning and the transductive learning [2,15]. Given a set of objects $x_k$ with the respective class labels $y_k$, $k = 1, \ldots, N$, and another set of *unlabeled* objects $x'_\ell$, $\ell = 1 \ldots, L$, the transductive learning derives a rule to provide labels over $x'_\ell$, $\ell = 1 \ldots, L$ using the knowledge of $\{(x_k, y_k),\ k = 1, \ldots, N\}$. In contrast, the inductive learning requires classification rules over the entire space of objects beyond unlabeled data [15].

When we consider various methods of clustering, it is interesting to ask whether or not a parallel concept to the inductive learning and the transductive learning is worth to be studied. We answer positively to this question and introduce the concept of inductive clustering and consider what we have from this

V. Torra et al. (Eds.): MDAI 2012, LNAI 7647, pp. 355–366, 2012.

concept. We already proposed two concepts of inductive clustering and transductive clustering [11], where emphasis is on transductive clustering. In contrast, we study a new method of inductive clustering in this paper, or more precisely, inductive features that exist in well-known methods of $K$-means and of the single linkage in agglomerative hierarchical clustering.

For the latter method we have two classification rules that can be compared to upper and lower approximations for rough set [12,13], and hence we call the derived rules as upper and lower classification rules derived from *inductive* single linkage alias nearest neighbor clustering. Or shortly, we can call it twofold classification rules derived from nearest neighbor clustering.

We have two motivations for this proposal. First is theoretical: we can derive theoretical properties for inductive methods of clustering, while it is more difficult to study a theory of non-inductive clustering. Second is more practical: functions derived from inductive clustering can be used for applications, e.g., Ichihashi [10] proposes a function for classification derived from fuzzy $c$-means.

The rest of this paper is organized as follows. After preliminary consideration in Section 2, We discuss $K$-means and fuzzy $c$-means as methods of inductive clustering in Section 3. Section 4 is devoted to the discussion of inductive feature of the nearest neighbor clustering and how the two approximations are derived. Section 5 then shows illustrative examples that show implications of the inductive properties of the above methods. Finally, Section 6 concludes the paper.

Throughout the paper, the proofs of the propositions are easy and omitted.

## 2    Preliminary Consideration

Let $X = \{x_1, \ldots, x_N\}$ be the set of objects for clustering, and each object $x \in X$ be a point of $p$-dimensional Euclidean space ($X \subset \mathbf{R}^p$). The squared Euclidean distance denoted by

$$\|x - y\|^2 = \sum_{j=1}^{p} (x^j - y^j)^2 \tag{1}$$

for $x = (x^1, \ldots, x^p), y = (y^1, \ldots, y^p) \in \mathbf{R}^p$ is used for the standard dissimilarity measure. Sometimes clusters are denoted by $G_i$, $i = 1, \ldots, K$. Generally they are a crisp partition of $X$, but we moreover consider fuzzy clusters. Distance between clusters are denoted by $D(G_i, G_j)$, which will be used for an agglomerative algorithm described below.

Membership matrix $U = (u_{ki})$ is used in which $u_{ki}$ is the membership of $x_k$ to cluster $i$ (or $G_i$); $u_{ki}$ may either be crisp or fuzzy. Moreover cluster centers denoted by $v_i$ ($i = 1, \ldots, K$) are used for $K$-means and fuzzy $c$-means [1,10].

A number of standard clustering techniques are shown below.

### 2.1    Agglomerative Hierarchical Clustering

The agglomerative hierarchical clustering [4,9] starts from each object and merge a pair of clusters at a time, and finally it ends with the one cluster of the whole sets. We, however, extract clusters formed at a specific level of dissimilarity. We hence specify the input and output to the next **AHC** algorithm:

**AHC: Agglomerative Hierarchical Clustering**
**Input:** Set $X$ and a positive threshold parameter $\beta$;
**Output:** Clusters $\mathcal{G} = \{G_1, \ldots, G_K\}$ formed at the level $\beta$.

1. Let each object be the initial clusters: $G_i = \{x_i\}$, and the number of clusters $K = N$.
2. Find the pair of clusters with minimum distance:

$$(G_p, G_q) = \arg\min_{i,j} D(G_i, G_j) \tag{2}$$

If $\quad \min_{i,j} D(G_i, G_j) > \beta \quad$ then output clusters $\{G_1, \ldots, G_K\}$ and stop;
else merge $G_r = G_p \cup G_q$.
Reduce the number of clusters: $K = K - 1$.
3. If $K = 1$, output $\{X\}$ and stop, else update distance $D(G_r, G_j)$, for all other clusters $G_j$. Go to step 2).

Five linkage methods for the definition of the distance between clusters are known: they are the single linkage, complete linkage, average linkage, centroid method, and the Ward method [4]. We mainly consider the single linkage method in this paper.

**Single Linkage.** The initial dissimilarity measure is assumed to be given in some way for the single linkage. Generally, there is no limitation to the definition of the dissimilarity $D(x, y)$ but we assume the dissimilarity to be the Euclidean distance herein.

The distance between two clusters in the single linkage is then defined by the following:

$$D(G_i, G_j) = \min_{x \in G_i, y \in G_j} D(x, y). \tag{3}$$

The formula for updating in step 3) of the above algorithm then is as follows:

$$D(G_r, G_j) = \min\{D(G_p, G_j), D(G_q, G_j)\}. \tag{4}$$

For convenience, output of **AHC** with the single linkage is sometimes written as

$$\mathcal{G}(\beta) = \{G_1(\beta), \ldots, G_K(\beta)\},$$

or simply, $G_1, \ldots, G_K$ when $\beta$ can be omitted. Moreover we write the same output as **AHC_SL**$(X, \beta)$, i.e.,

$$\mathcal{G}(\beta) = \textbf{AHC\_SL}(X, \beta),$$

showing the input $X$ and $\beta$ explicitly.

*Note 1.* In comparison with the single linkage, we show the definition of the complete linkage:

$$D(G_i, G_j) = \max_{x \in G_i, y \in G_j} D(x, y). \tag{5}$$

although this method is not discussed in detail.

## 2.2   Voronoi Region

The Voronoi regions are important in discussing inductive properties herein. We define a different definition of the Voronoi regions from those in literature, e.g., [5].

Given a set of points $Y = \{y_1, \ldots, y_l\}$, the collection of Voronoi regions with centers $y_1, \ldots, y_l$ denoted by

$$\mathcal{V} = \{V(y_1), \ldots, V(y_l)\},$$

where the members are *fuzzy sets*, are defined as follows:

For an arbitrarily given $x \in \mathbf{R}^p$, if there exists only one $i$ such that

$$i = \arg \min_{1 \le j \le l} \|x - y_j\|^2, \tag{6}$$

then $\mu_{V(y_i)}(x) = 1$.

Otherwise, if there are more than one symbols $i_1, \ldots, i_h$ such that

$$i_k = \arg \min_{1 \le j \le l} \|x - y_j\|^2, \quad k = 1, \ldots, h \tag{7}$$

then we put

$$\mu_{V(y_{i_1})}(x) = \frac{1}{h}. \tag{8}$$

We denote the crisp subset of $V(y_i)$ by

$$\hat{V}(y_i) = \{y \in \mathbf{R}^p : \mu_{V(y_i)}(x) = 1\}.$$

and moreover let the closure of $\hat{V}(y_i)$ be $\bar{V}(y_i)$.

If (6) holds, then $x \in \hat{V}(y_j)$. Otherwise if (7) holds, then $x$ is on the boundary of $\hat{V}(y_{i_1}), \ldots, \hat{V}(y_{i_h})$, or $x \in \bar{V}(y_{i_1}) \cap \cdots \cap \bar{V}(y_{i_h})$, and To summarize, the interior $\hat{V}(y_i)$ of the Voronoi region is crisp, but the boundary of it is *fuzzified* here.

Voronoi regions with fuzzified boundaries are more appropriate than ordinary Voronoi regions for discussing inductive clustering of $K$-means and fuzzy $c$-means, as well as the single linkage.

## 2.3   $K$-means or Crisp $c$-means

Let the number of clusters be $c$ that is fixed in the $K$-means. The method of $K$-means alias crisp $c$-means is the following iterative procedure:

- (I) Assume that the initial $c$ clusters are randomly generated. Let the centroid of cluster $i$ be $v_i$ $(i = 1, \ldots, c)$.
- (II) For $k = 1, \ldots, N$, allocate $x_k$ to the cluster of the nearest center using the allocation rule (6), (7), and (8) by putting $x = x_k$ and $y_1 = v_1, \ldots, y_c = v_c$ $(l = c)$.

- (III) Update the centroid $v_i$ for cluster $i$ $(i = 1, \ldots, c)$. If clusters are convergent, stop. Else go to step (II).

*Note 2.* The allocation rule (II) is different from the ordinary $K$-means on the boundary, but they are the same inside the Voronoi regions. Since a point on the boundary is exceptional, they are essentially the same algorithm and most points are crisply allocated.

### 2.4  Fuzzy $c$-means

There have been many studies on $c$-means and its variations, One of the best known variations is the fuzzy $c$-means [1,10].

In the method of fuzzy $c$-means by Dunn and Bezdek [1], an alternate optimization algorithm is proposed. Given a random partition or initial cluster centers, the iterative updating formula of the membership matrix and cluster centers is as follows:

$$u_{ki} = \frac{\frac{1}{D(x_k,v_i)^{\frac{1}{m-1}}}}{\sum_{j=1}^{K} \frac{1}{D(x_k,v_j)^{\frac{1}{m-1}}}}, \tag{9}$$

$$v_i = \frac{\sum_{k=1}^{N}(u_{ki})^m x_k}{\sum_{k=1}^{N}(u_{ki})^m}, \tag{10}$$

where $D(x_k, v_i)$ is given by the squared Euclidean distance (1). The calculation of (9) and (10) is repeated until convergence.

### 2.5  Mixture of Gaussian Distributions

The mixture of Gaussian distributions has commonly been used for classification and clustering. The Expectation and Maximization (EM) algorithm [8] is used for clustering. The EM algorithm is omitted here, but it is an iterative calculation of the prior distribution $P(C_i)$ of class $i$ $(i = 1, \ldots, K)$ and parameters for multi-normal distributions. As a result, the posterior probability $P(C_i|x_k)$ for allocating $x_k$ to class $i$ is calculated. It has moreover been shown that the solutions by the EM algorithm is closely related to the solution of fuzzy clustering based on an entropy criterion [10].

## 3  Inductive Properties of Non-hierarchical Clustering

Before introducing the concept of inductive clustering, we note again that clustering in general implies that we should have a classification rule $F\colon X \to \{1, \ldots, c\}$: the classification function is defined on the set of objects and the value is onto the set of labels $\{1, \ldots, c\}$.

*Inductive clustering* means that a classification rule on the entire space of objects are directly derived from the result of clustering. More specifically, suppose that a method of clustering has a rule for classification for $X$. If we can use *the same rule* for classifying an arbitrary point $x \in \mathbf{R}^p$, we call that the method is inductive clustering. In other words, a classification rule $F$ is defined on $\mathbf{R}^p$; $F \colon \mathbf{R}^p \to \{1, \ldots, c\}$, where $\{1, \ldots, c\}$ is the set of labels. The inductive clustering moreover should imply that the extension of the classification rule should bring additional and useful information.

In contrast, *non-inductive clustering* can classify only the given objects $X$, and it is not obvious how to extend the classification rule to the entire space $\mathbf{R}^p$. Thus the classification rule is defined on $X$ alone ($F \colon X \to \{1, \ldots, K\}$) and the extension to $\mathbf{R}^p$ is not derived. This implies that even if we extend the classification rule to the entire space in some way, we do not have useful information.

The method of $K$-means is thus an inductive clustering, since the allocation rule (II) defined on $X$ onto $\{1, \ldots, c\}$ is directly extended to the whole space by using (6), (7), and (8). Actually, the only difference is the substitution $x = x_k$. In other words, the allocation rule is derived by replacing object $x_k$ by variable $x$.

Such replacement of object $x_k$ by variable $x$ is applied to the mixture of distributions and fuzzy $c$-means. The former application is simple, since the clustering rule uses $P(C_i|x_k)$ which is a specialization of $P(C_i|x)$; the latter is a probabilistic allocation rule for the whole space.

For fuzzy $c$-means, we have the rules for generic $x \in \mathbf{R}^p$:

$$U_i(x) = \frac{\dfrac{1}{D(x,v_i)^{\frac{1}{m-1}}}}{\sum_{j=1}^{K} \dfrac{1}{D(x,v_j)^{\frac{1}{m-1}}}}.$$

Note that function $U_i \colon \mathbf{R}^p \to \{1, \ldots, c\}$ is derived from $u_{ki}$ by replacing $x_k$ by $x$. The above equation has a singular point at $x = v_i$ and hence the precise formula should be

$$U_i(x) = \frac{1}{1 + \sum_{j \neq i} \left(\dfrac{D(x,v_i)}{D(x,v_j)}\right)^{\frac{1}{m-1}}}. \tag{11}$$

The inductive property in these methods is thus observed by having the classification rules on the whole space. These classification rules show theoretical properties of the above methods more clearly than the original rules defined on $X$. Discussions are found in [10]. For example, we have the followings.

**Proposition 1.** *The allocation rule (II) of the $K$-means divides the space $\mathbf{R}^p$ into the Voronoi regions [5] with centers $v_i$, which have piecewise linear boundaries.*

This proposition implies that clusters with nonlinear boundaries are not separated by the $K$-means. Although this property is well-known, a simple variation leads to a new observation on constrained clustering.

## 3.1 Constrained Clustering

A standard method of constrained clustering is the COP $K$-means [14], which is a variation of $K$-means. The method of COP $K$-means is based on the $K$-means procedure with the constraints of must-link set $ML = \{(x_k, x_l)\}$ and cannot-link set $CL = \{(x_h, x_i)\}$. It returns *success* if all constraints are satisfied, whereas it returns *failure* if some constraint is broken. Note that the COP $K$-means uses the same allocation rule as the $K$-means except the constrained pairs of objects. Hence we have the next proposition.

**Proposition 2.** *The classification rule of the COP $K$-means divides the space $\mathbf{R}^p$ into the Voronoi regions with centers $v_i$ except on the objects with constraints.*

Thus the COP $K$-means cannot generate a nonlinear cluster boundary even when it has many constraints, although the constraints affect the positions of centers $v_1, \ldots, v_c$. Even when the set of points has natural clusters with linear boundaries, the COP $K$-means may produce unsatisfactory results, e.g., see first example in Section 5.1.

# 4 Inductive Features in Nearest Neighbor Clustering

Let us consider the agglomerative hierarchical clustering. We concentrate the single linkage alias *nearest neighbor clustering*.

*Nearest neighbor allocation* means that, given an arbitrary $x \in \mathbf{R}^p$ and clusters $G_1, \ldots, G_K$ derived from $X$, the following rule is used:

1. The rule (6), (7), and (8) is used with $Y = X$.
2. Let the set of nearest neighbors be $Z = \{x_{i_1}, \ldots, x_{i_h}\}$,

$$\mu_{G_j}(x) = \frac{|Z \cap G_j|}{|Z|}, \tag{12}$$

where $|Z|$ is the number of elements in $Z$.

Actually, this allocation rule is essentially the same as the allocation rule of the Voronoi regions except that we allocate a point to a cluster or clusters instead of a Voronoi region.

The above definition is not directly concerned with the single linkage, but the single linkage obviously uses the nearest neighbor distance (3). The allocation rule $F \colon \mathbf{R}^p \to \{1, \ldots, c\}$ is therefore derived by the last procedure of nearest neighbor allocation 1) and 2). In other words, we have the next proposition.

**Proposition 3.** *Assume that $V(x_1), \ldots, V(x_N)$ are the Voronoi regions with $Y = X$. Let $G_1, \ldots, G_K$ be clusters from* **AHC** *using the single linkage. Then the set of points in $\mathbf{R}^p$ allocated to $G_j$ by the nearest neighbor rule is:*

$$W(G_j) = \bigoplus_{x_k \in G_j} V(x_k), \tag{13}$$

*where $\oplus$ means the bounded sum.*

We note that the collection $W(G_1), \ldots, W(G_K)$ forms a partition of the whole space in the sense that

$$\bigoplus_{i=1}^{K} W(G_i) = \boldsymbol{R}^p,$$

$$\hat{W}(G_i) \cap \hat{W}(G_j) = \emptyset \quad (i \neq j), \tag{14}$$

where $\hat{W}(G_i)$ is the set of $x$ such that $\mu_{W(G_i)}(x) = 1$.

*Note 3.* The bounded sum is: $a \oplus b = \min\{1, a+b\}$.

## 4.1   Twofold Classification Rules

We introduce another classification rule in addition to $W(G_j)$ that may be compared to lower approximation in rough sets [12,13].

Let us consider a closed sphere with center $x_k$ and radius $\beta$:

$$B(x_k; \beta) = \{x \in \boldsymbol{R}^p : \|x - x_k\| \leq \beta\}.$$

and let

$$WL(G_j) = \hat{W}(G_j) \cap \left\{ \bigcup_{x_k \in X} B(x_k; \beta) \right\}. \tag{15}$$

Let us remind that $\hat{W}(G_j)$ is the region to which a point is allocated to cluster $G_i$, but the level for the allocation is not specified. On the other hand, $WL(G_j)$ is the region to which a point is allocated to $G_j$ and the level of allocation is below $\beta$. Precisely, we have the next proposition to observe the difference between $\hat{W}(G_j)$ and $WL(G_j)$.

**Proposition 4.** *Assume that $\beta$ is fixed, If $x \in WL(G_1)$, then*

$$\mathbf{AHC\_SL}(X \cup \{x\}, \beta) = \{G_1 \cup \{x\}, G_2, \ldots, G_K\}$$

*while if $x \in \hat{W}(G_1) - WL(G_1)$, then*

$$\mathbf{AHC\_SL}(X \cup \{x\}, \beta) = \mathbf{AHC\_SL}(X, \beta) \cup \{\{x\}\}$$
$$= \{G_1, G_2, \ldots, G_K, \{x\}\}.$$

We thus have a region of *a lower classification* $WL(G_1), \ldots, WL(G_K)$ and *an upper classification* region $\hat{W}(G_1), \ldots, \hat{W}(G_K)$.

This proposition moreover states that the addition of a point $x$ after clusters have been derived is essentially the same as clustering of $X$ with $x$ added beforehand.

Precisely, we denote the nearest allocation of an additional point $x$ after deriving clusters by

$$\mathbf{AHC\_SL}(X, \beta) \leftarrow x, \tag{16}$$

where the allocation rule is given by Proposition 4.

We then have

$$(\mathbf{AHC\_SL}(X, \beta) \leftarrow x) = \mathbf{AHC\_SL}(X \cup \{x\}, \beta). \tag{17}$$

*Note 4.* The proof of the property (17) is not difficult, but it is based on an equivalence theorem [9] between the single linkage and the connected components of a fuzzy graph and hence requires several more definitions. Hence it is omitted, but readers can see that this property is valid by checking simple examples.

## 4.2 Relation to DBSCAN

DBSCAN [3] is a well-known algorithm for density seeking clustering. It is closely related to the nearest neighbor clustering. Actually, we can use the above property (17) and the idea of 'core points' to have an algorithm that is similar to DBSCAN. The algorithm consists of the next three steps.

1. Extract 'core points' from the given set of data. The core points may be defined as those having a sufficient number of points in its neighborhood of a fixed radius, as in DBSCAN.
2. Carry out the single linkage to the set of core points. As is well-known, fast algorithms of the minimum spanning tree that is equivalent to the single linkage [9] can be used for the clustering.
3. Non-core points are allocated after clusters of core points are obtained as above.

*Note 5.* The original DBSCAN algorithm form clusters of core and non-core points simultaneously, but the above procedure allocates non-core points after clusters have been formed. Thus the property (17) is used.

## 5 Illustrative Examples

We show two illustrative examples; the first shows a negative result related to COP $K$-means, and the second illustrates how the twofold approximation in nearest neighbor clustering works.

### 5.1 A Typical Example in Semi-supervised Classification

Figure 1 shows a set of points in two-dimensions which consists of two 'crescentic' shapes. Similar examples of point configurations are often found in literature on semi-supervised learning [2]. Note that the two crescents are linearly separable at a horizontal line of level 0.4, which is not drawn in this figure.

Let us suppose that all points are with the constraints $ML$ and $CL$. Then it is obvious that we have perfect clusters separating the two crescents. Nevertheless, this prefect separation is unsatisfactory from the viewpoint of inductive clustering.

Figure 2 shows the two clusters separating crescents without an error. As a result we have two centroids on this figure, and the line separating the two Voronoi regions. This line and the Voronoi regions show a part of points are in misclassified regions, even if they appear to be correctly classified by the constraints. In other words, correctly classified points in the misclassified regions are nonsense. The regions cannot satisfactorily be separated by the $K$-means rule. This example thus shows a problem in a well-known method in constrained clustering.

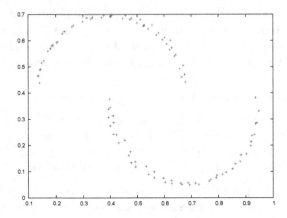

**Fig. 1.** Two linearly separable crescents

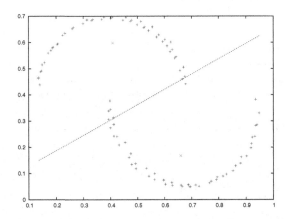

**Fig. 2.** Two linearly separable crescents cannot be separated by Voronoi regions

## 5.2   Twofold Classification in the Single Linkage

The second example concerns the nearest neighbor clustering. The eleven points in Fig. 3 were merged one by one using the Euclidean distance, and three clusters were generated by applying $\beta = 2.4$. Then the boundaries of the regions $W(G_i)$ ($i = 1, 2, 3$) are shown by the thin dotted line. The 'lower approximation' $WL(G_i)$ ($i = 1, 2, 3$) are shown by the red curves.

The dendrogram corresponding to this example is shown in Fig. 4, where the value of $\beta$ is shown by the red line. It is clear that an additional point is given in $WL(G_i)$, it will be given as an additional branch below the value of $\beta$, whereas if another point in $W(G_i) - WL(G_i)$ is given, it will be added to the dendrogram above $\beta$.

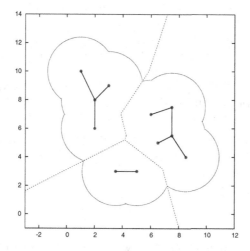

**Fig. 3.** Voronoi diagram and the lower approximations

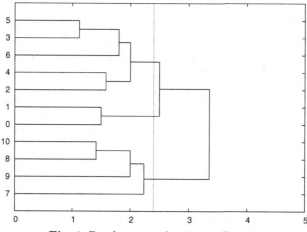

**Fig. 4.** Dendrogram of points in Fig. 3

## 6    Conclusions

The idea of inductive clustering in both $K$-means and the single linkage in agglomerative clustering has been proposed. The discussion here is essentially methodological; the purpose is to clarify implications of clusters derived from these methods. Theoretical properties of fuzzy $c$-means have been omitted, as they are described in [10] in detail.

The examples given here are for illustrating purpose, but they are useful for considering properties of clustering in real and huge examples.

To use classification rules derived from inductive clustering is easy: just replace object symbol $x_k$ in fuzzy $c$-means by variable $x$. Then one obtains a set of classification rules. In contrast, the classification in the single linkage does not have a closed functional form. We can use **AHC_SL**$(X \cup \{x\}, \beta)$, the output from **AHC** algorithm.

As to the twofold clustering, the idea is related to the idea of rough $K$-means [6,7] but the method is totally different. The present method is superior to the rough $K$-means in the sense that theoretical properties are made clear.

Future studies include application to large-scale real data. Fast algorithms suggested in relation to DBSCAN should further be studied.

**Acknowledgement.** This work has partly been supported by the Grant-in-Aid for Scientific Research, Japan Society for the Promotion of Science, No. 23500269.

# References

1. Bezdek, J.C.: Pattern Recognition with Fuzzy Objective Function Algorithms. Plenum Press, New York (1981)
2. Chapelle, O., Schölkopf, B., Zien, A. (eds.): Semi-Supervised Learning. MIT Press, Cambridge (2006)
3. Ester, M., Kriegel, H.-P., Sander, J., Xu, X.W.: A Density-Based Algorithm for Discovering Clusters in Large Spatial Databases with Noise. In: Proc. of 2nd International Conference on Knowledge Discovery and Data Mining (KDD 1996), pp. 226–231. AAAI Press (1996)
4. Everitt, B.S.: Cluster Analysis, 3rd edn. Arnold, London (1993)
5. Kohonen, T.: Self-Organization and Associative Memory. Springer, Heiderberg (1989)
6. Lingras, P., West, C.: Interval set clustering of web users with rough $K$-means. J. of Intel. Informat. Sci. 23(1), 5–16 (2004)
7. Lingras, P., Peters, G.: Rough clustering. In: WIREs Data Mining Knowl. Discov. 2011, pp. 64–72. Wiley (2011)
8. McLachlan, G.J., Krishnan, T.: The EM algorithms and Extensions. Wiley, New York (1997)
9. Miyamoto, S.: Fuzzy Sets in Information Retrieval and Cluster Analysis. Kluwer, Dordrecht (1990)
10. Miyamoto, S., Ichihashi, H., Honda, K.: Algorithms for Fuzzy Clustering. Springer (2008)
11. Miyamoto, S., Terami, A.: Inductive vs. Transductive Clustering Using Kernel Functions and Pairwise Constraints. In: Proc. of 11th Intern. Conf. on Intelligent Systems Design and Applications (ISDA 2011), Cordoba, Spain, November 22-24, pp. 1258–1264 (2011)
12. Pawlak, Z.: Rough Sets. International Journal of Parallel Programming 11(5), 341–356 (1982)
13. Pawlak, Z.: Rough Sets. Kluwer, Dordrecht (1991)
14. Wagstaff, K., Cardie, C., Rogers, S., Schroedl, S.: Constrained K-means Clustering with Background Knowledge. In: Proc. of the 9th ICML, pp. 577–584 (2001)
15. Zhu, X., Goldberg, A.B.: Introduction to Semi-Supervised Learning. Morgan and Claypool (2009)

# Marginality: A Numerical Mapping for Enhanced Exploitation of Taxonomic Attributes

Josep Domingo-Ferrer

Universitat Rovira i Virgili
Dept. of Computer Engineering and Mathematics
UNESCO Chair in Data Privacy
Av. Països Catalans 26
E-43007 Tarragona, Catalonia
josep.domingo@urv.cat

**Abstract.** Hierarchical attributes appear in taxonomic or ontology-based data (*e.g.* NACE economic activities, ICD-classified diseases, animal/plant species, etc.). Such taxonomic data are often exploited as if they were flat nominal data without hierarchy, which implies losing substantial information and analytical power. We introduce marginality, a numerical mapping for taxonomic data that allows using on those data many of the algorithms and analytical techniques designed for numerical data. We show how to compute descriptive statistics like the mean, the variance and the covariance on marginality-mapped data. Also, we define a mathematical distance between records including hierarchical attributes that is based on marginality-based variances. Such a distance paves the way to re-using on taxonomic data clustering and anonymization techniques designed for numerical data.

**Keywords:** Hierarchical attributes, Classification, Taxonomic data, Ontologies, Descriptive statistics, Numerical mapping, Anonymization.

## 1 Introduction

Taxonomic attributes are common in economic, medical or biological data sets and, more generally, in ontology-based data sets. For example, data about companies often include an attribute "Economic activity" which takes values in a standard classification, like NACE [12] or ISIC [9]; data about employees include their position within the company's hierarchy; data about patients include an attribute "Diagnosis" which takes values in some classification of diseases, like ICD9 [8]; data about plants or animals include the name of the plant or animal in the Linnaean taxonomy [11,14], etc.

Statistical analyses tend to treat taxonomic data as if they came from flat nominal attributes without hierarchy, thereby disregarding their hierarchical semantics and losing useful information. Such a wasteful approach can be explained

V. Torra et al. (Eds.): MDAI 2012, LNAI 7647, pp. 367–381, 2012.

by the lack of analytical techniques and algorithms specifically designed for taxonomic data. Indeed, numerical data are the type of data for which a greatest choice of techniques exists; categorical ordinal data are often mapped to integers and treated like numerical data; nominal data, whether drawn from a flat or hierarchical taxonomy, are most of the time treated as flat.

The situation described in the previous paragraph repeats itself for statistical disclosure control (SDC, [6,7,17,5,10]), a.k.a. data anonymization and sometimes as privacy-preserving data mining. SDC aims at making possible the publication of statistical data in such a way that the individual responses of specific users cannot be inferred from the published data and background knowledge available to intruders. If the data set being published consists of records corresponding to individuals, usual SDC methods operate by masking original data (via perturbation or detail reduction), by generating synthetic (simulated) data preserving some statistical features of the original data or by producing hybrid data obtained as a combination of original and synthetic data. The choice of SDC methods is greatest for numerical data.

The attributes in a data set can be classified depending on their range and the operations that can be performed on them:

1. *Numerical.* An attribute is considered numerical if arithmetical operations can be performed on it. Examples are income and age.
2. *Categorical.* An attribute is considered categorical when it takes values over a finite set and standard arithmetical operations on it do not make sense. Two main types of categorical attributes can be distinguished:
   (a) *Ordinal.* An ordinal attribute takes values in an ordered range of categories. Thus, the $\leq$, max and min operators can still be used on this kind of data. The instruction level and the political preferences (left-right) are examples of ordinal attributes.
   (b) *Nominal.* A nominal attribute takes values in an unordered range of categories. The only possible operator is comparison for equality. Nominal attributes can further be divided into two types:
      i. *Hierarchical.* A hierarchical nominal attribute takes values from a hierarchical classification. For example, plants are classified using Linnaeus's taxonomy, the type of a disease is also selected from a hierarchical taxonomy, and the type of an attribute can be selected from the hierarchical classification we propose in this section.
      ii. *Non-hierarchical.* A non-hierarchical nominal attribute takes values from a flat taxonomy. Examples of such attributes could be the preferred soccer team, the address of an individual, the civil status (married, single, divorced, widow/er), the eye color, etc.

This paper focuses on finding a numerical mapping for taxonomic data. Such a mapping can be used to obtain richer descriptive statistics, inspired on those for numerical data. It also makes it possible to use on taxonomic data techniques designed for numerical data (*e.g.* clustering, SDC).

Assuming a hierarchy is less restrictive than it would appear, because very often a non-hierarchical attribute can be turned into a hierarchical one if its flat hierarchy can be developed into a multilevel hierarchy. For instance, the preferred soccer and the address of an individual have been mentioned as non-hierarchical attributes; however, a hierarchy of soccer teams by continent and country could be conceived, and addresses can be hierarchically clustered by neighborhood, city, state, country, etc. Furthermore, well-known approaches to anonymization, like $k$-anonymity [15], assume that any attribute can be generalized, *i.e.* that an attribute hierarchy can be defined and values at lower levels of the hierarchy can be replaced by values at higher levels.

## 1.1   Contribution and Plan of This Paper

We propose to associate a number to each categorical value of a hierarchical nominal attribute, namely a form of centrality of that category within the attribute's taxonomy. We show how this allows computation of centroids, variances and covariances of hierarchical nominal data.

Section 2 gives background on the variance of hierarchical nominal attributes. Section 3 defines a tree centrality measure called marginality and presents the numerical mapping. Section 4 exploits the numerical mapping to compute means, variances and covariances of hierarchical nominal data. Section 5 contains a discussion and conclusions.

## 2   Background

We next recall the variance measure for hierarchical nominal attributes introduced in [4]. To the best of our knowledge, this is the first measure which captures the variability of a sample of values of a hierarchical nominal attribute by taking into account the semantics of the hierarchy. The intuitive idea is that a set of nominal values belonging to categories which are all children of the same parent category in the hierarchy has smaller variance that a set with children from different parent categories.

**Algorithm 1 (Nominal variance in [4])**

1. *Let the hierarchy of categories of a nominal attribute $X$ be such that $b$ is the maximum number of children that a parent category can have in the hierarchy.*
2. *Given a sample $T_X$ of nominal categories drawn from $X$, place them in the tree representing the hierarchy of $X$. Prune the subtrees whose nodes have no associated sample values. If there are repeated sample values, there will be several nominal values associated to one or more nodes (categories) in the pruned tree.*
3. *Label as follows the edges remaining in the tree from the root node to each of its children:*
    - *If $b$ is odd, consider the following succession of labels $l_0 = (b-1)/2$, $l_1 = (b-1)/2-1$, $l_2 = (b-1)/2+1$, $l_3 = (b-1)/2-2$, $l_4 = (b-1)/2+2$, $\cdots$, $l_{b-2} = 0$, $l_{b-1} = b-1$.*

- If $b$ is even, consider the following succession of labels $l_0 = (b-2)/2$, $l_1 = (b-2)/2+1$, $l_2 = (b-2)/2-1$, $l_3 = (b-2)/2+2$, $l_4 = (b-2)/2-2$, $\cdots$, $l_{b-2} = 0$, $l_{b-1} = b-1$.
  - Label the edge leading to the child with most categories associated to its descendant subtree as $l_0$, the edge leading to the child with the second highest number of categories associated to its descendant subtree as $l_1$, the one leading to the child with the third highest number of categories associated to its descendant subtree as $l_2$ and, in general, the edge leading to the child with the $i$-th highest number of categories associated to its descendant subtree as $l_{i-1}$. Since there are at most $b$ children, the set of labels $\{l_0, \cdots, l_{b-1}\}$ should suffice. Thus an edge label can be viewed as a $b$-ary digit (to the base $b$).
4. Recursively repeat Step 3 taking instead of the root node each of the root's child nodes.
5. Assign to values associated to each node in the hierarchy a node label consisting of a $b$-ary number constructed from the edge labels, more specifically as the concatenation of the $b$-ary digits labeling the edges along the path from the root to the node: the label of the edge starting from the root is the most significant one and the edge label closest to the specific node is the least significant one.
6. Let $L$ be the maximal length of the leaf $b$-ary labels. Append as many $l_0$ digits as needed in the least significant positions to the shorter labels so that all of them eventually consist of $L$ digits.
7. Let $T_X(0)$ be the set of $b$-ary digits in the least significant positions of the node labels (the "units" positions); let $T_X(1)$ be the set of $b$-ary digits in the second least significant positions of the node labels (the "tens" positions), and so on, until $T_X(L-1)$ which is the set of digits in the most significant positions of the node labels.
8. Compute the variance of the sample as

$$Var_H(T_X) = Var(T_X(0)) + b^2 \cdot Var(T_X(1)) + \cdots$$

$$+ b^{2(L-1)} \cdot Var(T_X(L-1)) \tag{1}$$

where $Var(\cdot)$ is the usual numerical variance.

In Section 4.2 below we will show that an equivalent measure can be obtained in a simpler and more manageable way.

## 3   A Numerical Mapping for Nominal Hierarchical Data

Consider a nominal attribute $X$ taking values from a hierarchical classification. Let $T_X$ be a sample of values of $X$. Each value $x \in T_X$ can be associated two numerical values:

- The sample frequency of $x$;
- Some centrality measure of $x$ within the hierarchy of $X$.

While the frequency depends on the particular sample, centrality measures depend both on the attribute hierarchy and the sample. Known tree centralities attempt to determine the "middle" of a tree [13]. We are rather interested in finding how far from the middle is each node of the tree, that is, how marginal it is. We next propose an algorithm to compute a new measure of the marginality of the values in the sample $T_X$.

## Algorithm 2 (Marginality of hierarchical values)

1. *Given a sample $T_X$ of hierarchical nominal values drawn from $X$, place them in the tree representing the hierarchy of $X$. There is a one-to-one mapping between the set of tree nodes and the set of categories where $X$ takes values. Prune the subtrees whose nodes have no associated sample values. If there are repeated sample values, there will be several nominal values associated to one or more nodes (categories) in the pruned tree.*
2. *Let $L$ be the depth of the pruned tree. Associate weight $2^{L-1}$ to edges linking the root of the hierarchy to its immediate descendants (depth 1), weight $2^{L-2}$ to edges linking the depth 1 descendants to their own descendants (depth 2), and so on, up to weight $2^0 = 1$ to the edges linking descendants at depth $L - 1$ with those at depth $L$. In general, weight $2^{L-i}$ is assigned to edges linking nodes at depth $i - 1$ with those at depth $i$, for $i = 1$ to $L$.*
3. *For each nominal value $x_j$ in the sample, its marginality $m(x_j)$ is defined and computed as*

$$m(x_j) = \sum_{x_l \in T_X - \{x_j\}} d(x_j, x_l) \qquad (2)$$

*where $d(x_j, x_l)$ is the sum of the edge weights along the path from the tree node corresponding to $x_j$ and the tree node corresponding to $x_l$.*

*Note 1 (On distances and marginality).* The above construction of marginality can be generalized by allowing other distance functions to be used in Expression (2), not necessarily based on edge weights. For example, in [3] it is suggested to use the semantic distance proposed in [16], in which the distance between two categories in a taxonomy is a function of the number of non-common ancestors divided by the total number of ancestors of the category pair.

Clearly, the greater $m(x_j)$, the more marginal (*i.e.* the less central) is $x_j$. We give next a toy running example to illustrate the computation of marginality.

*Example 1.* Assume a hierarchical attribute *"Diagnosis"*, for which a sample is available whose nominal values can be hierarchically classified as shown in Figure 1. The hierarchy is a pruned one, so that only leaves with some value in the sample are depicted. The sample has one element for each diagnostic category, except for "Epilepsy" and "Nose cold", for each of which there are two elements. Figure 1 also shows the weights assigned by Algorithm 2 to each edge in the hierarchy tree.

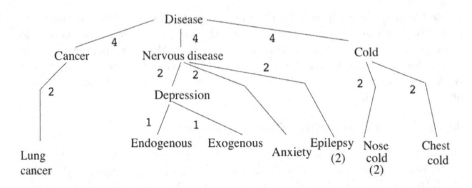

**Fig. 1.** Example pruned hierarchy of a sample of a *"Diagnosis"* attribute

Label the elements in the sample as follows: $x_1$ (lung cancer), $x_2$ (endogenous depression), $x_3$ (exogenous depression), $x_4$ (anxiety), $x_5$ (first epilepsy element), $x_6$ (second epilepsy element), $x_7$ (first nose cold element), $x_8$ (second nose cold element) and $x_9$ (chest cold). The distance matrix between elements is given below, where component $(j, l)$ represents the sum $d(x_j, x_l)$ of edge weights along the path between $x_j$ and $x_l$ (only the upper diagonal matrix is represented):

$$\begin{pmatrix} 0 & 13 & 13 & 12 & 12 & 12 & 12 & 12 & 12 \\ & 0 & 2 & 5 & 5 & 5 & 13 & 13 & 13 \\ & & 0 & 5 & 5 & 5 & 13 & 13 & 13 \\ & & & 0 & 4 & 4 & 12 & 12 & 12 \\ & & & & 0 & 0 & 12 & 12 & 12 \\ & & & & & 0 & 12 & 12 & 12 \\ & & & & & & 0 & 0 & 4 \\ & & & & & & & 0 & 4 \\ & & & & & & & & 0 \end{pmatrix}$$

The marginality $m(x_j)$ of element $x_j$ can be obtained by adding all distances in the $j$-th row of the above matrix. Marginalities for all elements are shown in Table 1. It turns out that $x_1$ (lung cancer) is the most marginal element, which is consistent with the layout of the hierarchy in Figure 1. On the other hand, $x_5$ and $x_6$ are the least marginal elements, due to both the central position of epilepsy in the hierarchy and the fact that there are two epilepsy elements. In fact, the higher frequency of epilepsy is what makes the marginality of $x_5$ and $x_6$ lower than the marginality of $x_4$ (anxiety); otherwise, epilepsy and anxiety have equally central positions in the hierarchy. This illustrates that marginality is a function of both the hierarchy of categories and their frequency in the sample.

Some properties are next stated which illustrate the rationale of the distance and the weights used to compute marginalities.

**Table 1.** Marginalities of elements in the "Diagnosis" sample of Figure 1

| $x_j$ | $m(x_j)$ |
|---|---|
| $x_1$ | $0 + 13 + 13 + 12 + 12 + 12 + 12 + 12 + 12 = 98$ |
| $x_2$ | $13 + 0 + 2 + 5 + 5 + 5 + 13 + 13 + 13 = 69$ |
| $x_3$ | $13 + 2 + 0 + 5 + 5 + 5 + 13 + 13 + 13 = 69$ |
| $x_4$ | $12 + 5 + 5 + 0 + 4 + 4 + 12 + 12 + 12 = 66$ |
| $x_5$ | $12 + 5 + 5 + 4 + 0 + 0 + 12 + 12 + 12 = 62$ |
| $x_6$ | $12 + 5 + 5 + 4 + 0 + 0 + 12 + 12 + 12 = 62$ |
| $x_7$ | $12 + 13 + 13 + 12 + 12 + 12 + 0 + 0 + 4 = 78$ |
| $x_8$ | $12 + 13 + 13 + 12 + 12 + 12 + 0 + 0 + 4 = 78$ |
| $x_9$ | $12 + 13 + 13 + 12 + 12 + 12 + 4 + 4 + 0 = 82$ |

**Lemma 1.** $d(\cdot, \cdot)$ *is a distance in the mathematical sense.*

Being the length of a path, it is immediate to check that $d(\cdot, \cdot)$ satisfies reflexivity, symmetry and subadditivity. The rationale of the above exponential weight scheme is to give more weight to differences at higher levels of the hierarchy; specifically, the following property is satisfied.

**Lemma 2.** *The distance between any non-root node $n_j$ and its immediate ancestor is greater than the distance between $n_j$ and any of its descendants.*

**Proof:** Let $L$ be the depth of the overall tree and $L_j$ be the depth of $n_j$. The distance between $n_j$ and its immediate ancestor is $2^{L-L_j}$. The distance between $n_j$ and its most distant descendant is

$$1 + 2 + \cdots + 2^{L-L_j-1} = 2^{L-L_j} - 1$$

$\square$

**Lemma 3.** *The distance between any two different nodes at the same depth is greater than the longest distance within the subtree rooted at each node.*

**Proof:** Let $L$ be the depth of the overall tree and $L_j$ be the depth of the two nodes. The distance between two different nodes is shortest when they have the same parent and it is

$$2 \cdot 2^{L-L_j} = 2^{L-L_j+1}.$$

The longest distance within any of the two subtrees rooted at the two nodes at depth $L_j$ is the length of the path between two leaves at depth $L$, which is

$$2 \cdot (1 + 2 + \cdots + 2^{L-L_j-1}) = 2(2^{L-L_j} - 1) = 2^{L-L_j+1} - 2$$

$\square$

## 4 Statistical Analysis of Numerically Mapped Nominal Data

In the previous section we have shown how a nominal value $x_j$ can be associated a marginality measure $m(x_j)$. In this section, we show how this numerical magnitude can be used in statistical analysis.

## 4.1  Mean

The mean of a sample of nominal values cannot be computed in the standard sense. However, it can be reasonably approximated by the least marginal value, that is, by the sample centroid.

**Definition 1 (Marginality-based approximated mean).** *Given a sample $T_X$ of a hierarchical nominal attribute $X$, the marginality-based approximated mean is defined as*

$$Mean_M(T_X) = \arg \min_{x_j \in T_X} m(x_j)$$

*if one wants the mean to be a nominal value, or*

$$Num\_mean_M(T_X) = \min_{x_j \in T_X} m(x_j)$$

*if one wants a numerical mean value.*

*Example 2.* It can be seen from Table 1 that, for the sample of Example 1, the marginality-based mean is "Epilepsy" (which is the least marginal value) and the numerical marginality-based mean is 62.

## 4.2  Variance

In Section 2 above, we recalled a measure of variance of a hierarchical nominal attribute proposed in [4] which takes the semantics of the hierarchy into account. Interestingly, it turns out that the average marginality of a sample is an equivalent way to capture the same notion of variance.

**Definition 2 (Marginality-based variance).** *Given a sample $T_X$ of $n$ values drawn from a hierarchical nominal attribute $X$, the marginality-based sample variance is defined as*

$$Var_M(T_X) = \frac{\sum_{x_j \in T_X} m(x_j)}{n}$$

*Example 3.* It can be seen from Table 1 that, for the sample of Example 1, the marginality-based variance is

$$\frac{98 + 69 + 69 + 66 + 62 + 62 + 78 + 78 + 82}{9} = 73.78$$

The following lemma is proven in the Appendix.

**Lemma 4.** *The $Var_M(\cdot)$ measure and the $Var_H(\cdot)$ specified by Algorithm 1 in Section 2 are equivalent.*

## 4.3   Covariance Matrix

It is not difficult to generalize the sample variance introduced in Definition 2 to define the sample covariance of two nominal attributes.

**Definition 3 (Marginality-based covariance).** *Given a bivariate sample* $T_{(X,Y)}$ *consisting of n ordered pairs of values* $\{(x_1, y_1), \cdots, (x_n, y_n)\}$ *drawn from the ordered pair of nominal attributes* $(X, Y)$, *the marginality-based sample covariance is defined as*

$$Covar_M(T_{(X,Y)}) = \frac{\sum_{j=1}^{n} \sqrt{m(x_j) m(y_j)}}{n}$$

The above definition yields a non-negative covariance whose value is higher when the marginalities of the values taken by $X$ and $Y$ are positively correlated: as the values taken by $X$ become more marginal, so become the values taken by $Y$.

Given a multivariate data set $T$ containing a sample of $d$ nominal attributes $X^1, \cdots, X^d$, using Definitions 2 and 3 yields a covariance matrix $\mathbf{S} = \{s_{jl}\}$, for $1 \le j \le d$ and $1 \le l \le d$, where $s_{jj} = Var_M(T_j)$, $s_{jl} = Covar_M(T_{jl})$ for $j \ne l$, $T_j$ is the column of values taken by $X^j$ in $T$ and $T_{jl} = (T_j, T_l)$.

## 4.4   Variance-Based Distance

Based on variances (whether plain numerical or marginality-based), we can define the following distance for records with numerical, hierarchical or flat nominal attributes.

**Definition 4 (S-distance).** *The S-distance between two records* $\mathbf{x}_1$ *and* $\mathbf{x}_2$ *in a data set with d attributes is*

$$\delta(\mathbf{x}_1, \mathbf{x}_2) = \sqrt{\frac{(S^2)_{12}^1}{(S^2)^1} + \cdots + \frac{(S^2)_{12}^d}{(S^2)^d}} \tag{3}$$

*where* $(S^2)_{12}^l$ *is the variance of the l-th attribute over the group formed by* $\mathbf{x}_1$ *and* $\mathbf{x}_2$, *and* $(S^2)^l$ *is the variance of the l-th attribute over the entire data set.*

We prove in the Appendix the following two theorems stating that the distance above satisfies the properties of a mathematical distance.

**Theorem 1.** *The S-distance on multivariate records consisting of hierarchical attributes based on the hierarchical variance computed as per Definition 2 is a distance in the mathematical sense.*

**Theorem 2.** *The S-distance on multivariate records consisting of ordinal or numerical attributes based on the usual numerical variance is a distance in the mathematical sense.*

By combining the proofs of Theorems 1 and 2, the next corollary follows.

**Corollary 1.** *The S-distance on multivariate records consisting of attributes of any type, where the hierarchical variance is used for hierarchical and flat nominal attributes and the usual numerical variance is used for ordinal and numerical attributes, is a distance in the mathematical sense.*

The above distance can be used for a variety of purposes, including clustering. Specifically, it allows microaggregating hierarchical data [1,2] in view of anonymization.

## 5    Discussion and Conclusions

We have presented a centrality-based mapping of hierarchical nominal data to numbers. We have shown how such a numerical mapping allows computing means, variances and covariances of nominal attributes, and distances between records containing any kind of attributes.

Such enhanced flexibility of manipulation can be used to adapt methods intended for numerical data to the treament of hierarchical attributes. If reverse mapping to nominal categories is required at the end of the treatment, two situations arise:

- *Each numerical output of the method exactly equals one of the input marginalities.* E.g. this happens for SDC methods that involve swapping input values that are within a certain distance of each other. In this case, each numerical output $m$ is mapped back to the nominal category having marginality $m$.
- *Numerical outputs do not correspond to marginalities.* E.g. such is the case if numerical outputs are the result of applying a regression model on the input marginalities. In this case, a reasonable option is to map each numerical output $m$ back to the category having marginality closest to $m$.

Reverse mapping may be problematic if there are categories which are semantically very different and have similar marginalities or the same marginality. For example, if the nose colds are suppressed from the sample depicted in Figure 1, then chest cold and lung cancer would have exactly the same marginality. A way to prevent semantic confusion in reverse mapping is to use blocking, that is, to split the hierarchy tree into several subtrees based on semantic criteria and treat each subtree separately: *e.g* divide the sample of Example 1 into a subsample of cancers, a subsample of nervous diseases and a subsample of colds, and treat subsamples separately to avoid big confusions during reverse mapping (we are assuming that confusing two categories within the same subtree is tolerable).

Future research will involve developing real-life applications of marginality, for example data anonymization of hierarchical attributes using SDC methods intended for numerical data (like multiple imputation or microaggregation).

## Appendix

**Proof (Lemma 4):** We will show that, given two samples $T_X = \{x_1, \cdots, x_n\}$ and $T'_X = \{x'_1, \cdots, x'_n\}$ of a nominal attribute $X$, both with the same cardinality $n$, it holds that $Var_M(T_X) < Var_M(T'_X)$ if and only if $Var_H(T_X) < Var_H(T'_X)$.

Assume that $Var_M(T_X) < Var_M(T_X')$. Since both samples have the same cardinality, this is equivalent to

$$\sum_{j=1}^{n} m(x_j) < \sum_{j=1}^{n} m(x_j')$$

By developing the marginalities, we obtain

$$\sum_{j=1}^{n} \sum_{x_l \in T_X - \{x_j\}} d(x_j, x_l) < \sum_{j=1}^{n} \sum_{x_l' \in T_X' - \{x_j'\}} d(x_j', x_l')$$

Since distances are sums of powers of 2, from 1 to $2^{L-1}$, we can write the above inequality as

$$d_0 + 2d_1 + \cdots + 2^{L-1}d_{L-1} < d_0' + 2d_1' + \cdots + 2^{L-1}d_{L-1}' \tag{4}$$

By viewing $d_{L-1} \cdots d_1 d_0$ and $d_{L-1}' \cdots d_1' d_0'$ as binary numbers, it is easy to see that Inequality (4) implies that some $i$ must exist such that $d_i < d_i'$ and $d_{\hat{i}} \leq d_{\hat{i}}'$ for $i < \hat{i} \leq L - 1$. This implies that there are less high-level edge differences associated to the values of $T_X$ than to the values of $T_X'$. Hence, in terms of $Var_H(\cdot)$, we have that $Var(T_X(i)) < Var(T_X'(i))$ and $Var(T_X(\hat{i})) \leq Var(T_X'(\hat{i}))$ for $i < \hat{i} \leq L - 1$. This yields $Var_H(T_X) < Var_H(T_X')$.

If we now assume $Var_H(T_X) < Var_H(T_X')$, we can prove $Var_M(T_X) < Var_M(T_X')$ by reversing the above argument. □.

**Lemma 5.** *Given non-negative $A, A', A'', B, B', B''$ such that $\sqrt{A} \leq \sqrt{A'} + \sqrt{A''}$ and $\sqrt{B} \leq \sqrt{B'} + \sqrt{B''}$ it holds that*

$$\sqrt{A + B} \leq \sqrt{A' + B'} + \sqrt{A'' + B''} \tag{5}$$

**Proof (Lemma 5):** Squaring the two inequalities in the lemma assumption, we obtain

$$A \leq (\sqrt{A'} + \sqrt{A''})^2$$
$$B \leq (\sqrt{B'} + \sqrt{B''})^2$$

Adding both expressions above, we get the square of the left-hand side of Expression (5)

$$A + B \leq (\sqrt{A'} + \sqrt{A''})^2 + (\sqrt{B'} + \sqrt{B''})^2$$

$$= A' + A'' + B' + B'' + 2(\sqrt{A'A''} + \sqrt{B'B''}) \tag{6}$$

Squaring the right-hand side of Expression (5), we get

$$(\sqrt{A' + B'} + \sqrt{A'' + B''})^2$$

$$= A' + B' + A'' + B'' + 2\sqrt{(A' + B')(A'' + B'')} \tag{7}$$

Since Expressions (6) and (7) both contain the terms $A' + B' + A'' + B''$, we can neglect them. Proving Inequality (5) is equivalent to proving

$$\sqrt{A'A''} + \sqrt{B'B''} \leq \sqrt{(A' + B')(A'' + B'')}$$

Suppose the opposite, that is,

$$\sqrt{A'A''} + \sqrt{B'B''} > \sqrt{(A' + B')(A'' + B'')} \tag{8}$$

Square both sides:

$$A'A'' + B'B'' + 2\sqrt{A'A''B'B''} >$$
$$(A' + B')(A'' + B'') = A'A'' + B'B'' + A'B'' + B'A''$$

Subtract $A'A'' + B'B''$ from both sides to obtain

$$2\sqrt{A'A''B'B''} > A'B'' + B'A''$$

which can be rewritten as

$$(\sqrt{A'B''} - \sqrt{B'A''})^2 < 0$$

Since a real square cannot be negative, the assumption in Expression (8) is false and the lemma follows.    □

**Proof (Theorem 1):** We must prove that the S-distance is non-negative, reflexive, symmetrical and subadditive (*i.e.* it satisfies the triangle inequality).

*Non-negativity.* The S-distance is defined as a non-negative square root, hence it cannot be negative.

*Reflexivity.* If $\mathbf{x}_1 = \mathbf{x}_2$, then $\delta(\mathbf{x}_1, \mathbf{x}_2) = 0$. Conversely, if $\delta(\mathbf{x}_2, \mathbf{x}_2) = 0$, the variances are all zero, hence $\mathbf{x}_1 = \mathbf{x}_2$.

*Symmetry.* It follows from the definition of the S-distance.

*Subadditivity.* Given three records $\mathbf{x}_1$, $\mathbf{x}_2$ and $\mathbf{x}_3$, we must check whether

$$\delta(\mathbf{x}_1, \mathbf{x}_3) \overset{?}{\leq} \delta(\mathbf{x}_1, \mathbf{x}_2) + \delta(\mathbf{x}_2, \mathbf{x}_3)$$

By expanding the above expression using Expression (3), we obtain

$$\sqrt{\frac{(S^2)^1_{13}}{(S^2)^1} + \cdots + \frac{(S^2)^d_{13}}{(S^2)^d}} \overset{?}{\leq}$$

$$\sqrt{\frac{(S^2)^1_{12}}{(S^2)^1} + \cdots + \frac{(S^2)^d_{12}}{(S^2)^d}} + \sqrt{\frac{(S^2)^1_{23}}{(S^2)^1} + \cdots + \frac{(S^2)^d_{23}}{(S^2)^d}} \tag{9}$$

Let us start with the case $d = 1$, that is, with a single attribute, *i.e.* $\mathbf{x}_i = x_i$ for $i = 1, 2, 3$. To check Inequality (9) with $d = 1$, we can ignore the variance in the denominators (it is the same on both sides) and we just need to check

$$\sqrt{S^2_{13}} \overset{?}{\leq} \sqrt{S^2_{12}} + \sqrt{S^2_{23}} \tag{10}$$

We have

$$S_{13}^2 = Var(\{x_1, x_3\}) = \frac{m(x_1) + m(x_3)}{2}$$

$$= \frac{d(x_1, x_3)}{2} + \frac{d(x_3, x_1)}{2} = d(x_1, x_3) \tag{11}$$

Similarly $S_{12}^2 = d(x_1, x_2)$ and $S_{23}^2 = d(x_2, x_3)$. Therefore, Expression (10) is equivalent to subadditivity for $d(\cdot, \cdot)$ and the latter holds by Lemma 1. Let us now make the induction hypothesis for $d - 1$ and prove subadditivity for any $d$. Call now

$$A := \frac{(S^2)_{13}^1}{(S^2)^1} + \cdots + \frac{(S^2)_{13}^{d-1}}{(S^2)^{d-1}}$$

$$A' := \frac{(S^2)_{12}^1}{(S^2)^1} + \cdots + \frac{(S^2)_{12}^{d-1}}{(S^2)^{d-1}}$$

$$A'' := \frac{(S^2)_{23}^1}{(S^2)^1} + \cdots + \frac{(S^2)_{23}^{d-1}}{(S^2)^{d-1}}$$

$$B := \frac{(S^2)_{13}^d}{(S^2)^d}; \quad B' := \frac{(S^2)_{12}^d}{(S^2)^d}; \quad B'' := \frac{(S^2)_{23}^d}{(S^2)^d}$$

Subadditivity for $d$ amounts to checking whether

$$\sqrt{A + B} \overset{?}{\le} \sqrt{A' + B'} + \sqrt{A'' + B''} \tag{12}$$

which holds by Lemma 5 because, by the induction hypothesis for $d - 1$, we have $\sqrt{A} \le \sqrt{A'} + \sqrt{A''}$ and, by the proof for $d = 1$, we have $\sqrt{B} \le \sqrt{B'} + \sqrt{B''}$. $\square$

**Proof (Theorem 2):** Non-negativity, reflexivity and symmetry are proven in a way analogous as in Theorem 1. As to subaddivity, we just need to prove the case $d = 1$, that is, the inequality analogous to Expression (10) for numerical variances. The proof for general $d$ is the same as in Theorem 1. For $d = 1$, we have

$$S_{13}^2 = \frac{(x_1 - x_3)^2}{2}; \quad S_{12}^2 = \frac{(x_1 - x_2)^2}{2}; \quad S_{23}^2 = \frac{(x_2 - x_3)^2}{2}$$

Therefore, Expression (10) obviously holds with equality in the case of numerical variances because

$$\sqrt{S_{13}^2} = \frac{x_1 - x_3}{\sqrt{2}} = \frac{(x_1 - x_2) + (x_2 - x_3)}{\sqrt{2}} = \sqrt{S_{12}^2} + \sqrt{S_{23}^2}$$

$\square$

**Acknowledgments and Disclaimer.** Thanks go to Klara Stokes for help with Lemma 5. This work was partly supported by the Government of Catalonia under grant 2009 SGR 1135, by the Spanish Government through projects TSI2007-65406-C03-01 "E-AEGIS", TIN2011-27076-C03-01 "CO-PRIVACY" and CONSOLIDER INGENIO 2010 CSD2007-00004 "ARES", and by the European Comission under FP7 projects "DwB" and "Inter-Trust". The author is partially supported as an ICREA Acadèmia researcher by the Government of Catalonia. The author is with the UNESCO Chair in Data Privacy, but he is solely responsible for the views expressed in this paper, which do not necessarily reflect the position of UNESCO nor commit that organization.

# References

1. Domingo-Ferrer, J., Mateo-Sanz, J.M.: Practical data-oriented microaggregation for statistical disclosure control. IEEE Transactions on Knowledge and Data Engineering 14(1), 189–201 (2002)
2. Domingo-Ferrer, J., Torra, V.: Ordinal, continuous and heterogeneous $k$-anonymity through microaggregation. Data Mining and Knowledge Discovery 11(2), 195–212 (2005)
3. Domingo-Ferrer, J., Sánchez, D., Rufian-Torrell, G.: Anonymization of clinical data based on semantic marginality (manuscript, 2012)
4. Domingo-Ferrer, J., Solanas, A.: A measure of nominal variance for hierarchical nominal attributes. Information Sciences 178(24), 4644–4655 (2008); Erratum in Information Sciences 179(20), 3732 (2009)
5. Duncan, G.T., Elliot, M., Salazar-González, J.-J.: Statistical Confidentiality: Principles and Practice. Springer, New York (2011)
6. Hundepool, A., Domingo-Ferrer, J., Franconi, L., Giessing, S., Lenz, R., Longhurst, J., Schulte-Nordholt, E., Seri, G., DeWolf, P.-P.: Handbook on Statistical Disclosure Control (version 1.2). ESSNET SDC Project (2010), http://neon.vb.cbs.nl/casc
7. Hundepool, A., Domingo-Ferrer, J., Franconi, L., Giessing, S., Schulte Nordholt, E., Spicer, K., De Wolf, P.P.: Statistical Disclosure Control. Wiley, New York (2012)
8. ICD9 - International Classification of Diseases, 9th Revision, Clinical Modification, 6th edn., October 1 (2008), http://icd9cm.chrisendres.com/
9. ISIC Rev. 4 - International Standard Industrial Classification of All Economic Activities, United Nations Statistics Division, http://unstats.un.org/unsd/cr/registry/regcst.asp?Cl=27&prn=yes
10. Lenz, R.: Methoden der Geheimhaltung wirtschaftsstatistischer Einzeldaten und ihre Schutzwirkung. Statistik und Wissenschaft, vol. 18. Statistisches Bundesamt, Wiesbaden (2010)
11. McNeill, J., et al. (eds.): International Code of Botanical Nomenclature (Vienna Code). International Association for Plant Taxonomy (2006), http://ibot.sav.sk/icbn/main.htm
12. NACE Rev. 2 - Statistical Classification of Economic Activities in the European Community, Rev. 2. Eurostat, European Commission (2008), http://epp.eurostat.ec.europa.eu/cache/ITY_OFFPUB/KS-RA-07-015/EN/KS-RA-07-015-EN.PDF

13. Reid, K.B.: Centrality measures in trees. In: Kaul, H., Mulder, H.M. (eds.) Advances in Interdisciplinary Applied Discrete Mathematics, pp. 167–197. World Scientific eBook (2010)
14. Ride, W.D.L., et al. (eds.): International Code of Zoological Nomenclature, 4th edn., January 1. International Union of Biological Sciences (2000), http://www.nhm.ac.uk/hosted-sites/iczn/code/
15. Samarati, P.: Protecting respondents' identities in microdata release. IEEE Transactions on Knowledge and Data Engineering 13(6), 1010–1027 (2001)
16. Sánchez, D., Batet, M., Isern, D., Valls, A.: Ontology-based semantic similarity: a new feature-based approach. Expert Systems with Applications 39(9), 7718–7728 (2012)
17. Willenborg, L., DeWaal, T.: Elements of Statistical Disclosure Control. Springer, New York (2001)

# Introducing Incomparability
# in Modeling Qualitative Belief Functions

Amel Ennaceur[1], Zied Elouedi[1], and Eric Lefevre[2]

[1] LARODEC, University of Tunis, Institut Supérieur de Gestion, Tunisia
amel_naceur@yahoo.fr, zied.elouedi@gmx.fr
[2] Univ. Lille Nord of France, UArtois EA 3926 LGI2A, France
eric.lefevre@univ-artois.fr

**Abstract.** This paper investigates a new model for generating belief functions from qualitative preferences. Our approach consists in constructing appropriate quantitative information from incomplete preferences relations. It is able to combine preferences despite the presence of incompleteness and incomparability in their preference orderings. The originality of our model is to provide additional interpretation values to the existing methods based on strict preferences and indifferences only.

## 1 Introduction

When solving problems dealing with belief function theory, expert is usually required to provide precise numerical values, for determining the portion of belief committed exactly to an event in a particular domain. However, when handling with such situation, the main difficulty is how to quantify these numeric values, therefore linguistic assessments could be used instead. So, the expert is then asked to express his opinions qualitatively, based on knowledge and experience that he provides in response to a given question rather than direct quantitative information.

However, in some cases, the decision maker may be unable to express his opinions due to his lack of knowledge. He is then forced to provide incomplete or even erroneous information. Obviously, rejecting this difficulty in eliciting the expert preference is not a good practice.

Besides, in preference modeling, expert may express preferences among a pair of alternatives in distinct ways: either an expert has a strict preference of one alternative compared to the other, or is indifferent between both alternatives. However, these two interpretations are possible because we made the assumption of complete and sure information. They do not cover all possible situations that a decision maker can be faced with. Consider now the situations in which the expert has symmetrically lack of information and also excess of information in the sense that he has contradictory inputs. He may then introduce two new situations: incompleteness and incomparability. The intuition is that the expert cannot compare apple and cheese because they are too different. For instance, he may consider that alternatives may be incomparable because the expert does not wish very dissimilar alternatives to be compared. Incompleteness, on the other hand, represents simply an absence of knowledge about the preference of certain pairs

V. Torra et al. (Eds.): MDAI 2012, LNAI 7647, pp. 382–393, 2012.

of alternatives. It arises when we have not fully elicited an expert's preferences or when expert do not has the full information.

To deal with such situations, a more realistic solution should be proposed, that is able to efficiently imitate the expert reasoning using belief function theory. In this paper, our main aim is then to elaborate on how may be incomparability and incompleteness represented in the belief function framework as understood in the Transferable Belief Model (TBM) [11]. Some researchers have already dealt with this problem and generate associated quantitative belief functions like [1] [4] [7] [12]. However, these approaches are only based on strict preference and indifference relations.

In this work, we focused on Ben Yaghlane et al. [1] [2] approach in order to construct quantitative belief functions from qualitative preferences by transforming these opinions into basic belief assessment. This approach is chosen since it does not require that the expert supplies the preference relations between all pairs of propositions. In fact, it allows the generation of belief functions using incomplete qualitative preference relations. Besides, it has been noted that this method handles the issue of inconsistency in the pair-wise comparisons. In this research, we propose a general model for constructing belief functions, which takes into account all information contained within the expert. With this approach focused on incompleteness and incomparability, the originality of our model is to allow the expert to easily express his preferences and to provide a convenience framework for constructing quantitative belief functions from qualitative assessments. Thus, we present a model that is able to combine the expert assessments despite the presence of incompleteness and incomparability in their preference orderings.

In order to do this, the paper is set out as follows: Section 2 provides a brief description of basics of belief function theory. In section 3, we describe the preference modeling approach. Then, Section 4 reviews some existing approaches. Next in Section 5, our suggested solution will be approached. Section 6 presents an example to illustrate our method. Finally, Section 7 ends this work.

## 2 Belief Function Theory

In this section, we briefly review the main concepts underlying the belief function theory as interpreted by the Transferable Belief Model (TBM). Details can be found in [9] [11].

### 2.1 Basic Concepts

The TBM is a model to represent quantified belief functions [11]. Let $\Theta$ be the frame of discernment representing a finite set of elementary hypotheses related to a problem domain. We denote by $2^\Theta$ the set of all the subsets of $\Theta$ [9].

The impact of a piece of evidence on the different subsets of the frame of discernment $\Theta$ is represented by the so-called basic belief assignment (bba), called initially by Shafer, basic probability assignment [9].

$$\sum_{A \subseteq \Theta} m(A) = 1 \tag{1}$$

The value $m(A)$, named a basic belief mass (bbm), represents the portion of belief committed exactly to the event $A$. The events having positive bbm's are called focal elements. Let $\mathcal{F}(m) \subseteq 2^\Theta$ be the set of focal elements of the bba $m$.

Associated with $m$ is the belief function is defined for $A \subseteq \Theta$ and $A \neq \emptyset$ as:

$$bel(A) = \sum_{\emptyset \neq B \subseteq A} m(B) \text{ and } bel(\emptyset) = 0 \qquad (2)$$

The degree of belief $bel(A)$ given to a subset $A$ of the frame $\Theta$ is defined as the sum of all the basic belief masses given to subsets that support $A$ without supporting its negation.

The plausibility function $pl$ expresses the maximum amount of specific support that could be given to a proposition $A$ in $\Theta$. It measures the degree of belief committed to the propositions compatible with $A$. $pl(A)$ is then obtained by summing the bbm's given to the subsets $B$ such that $B \cap A \neq \emptyset$ [9]:

$$pl(A) = \sum_{B \cap A \neq \emptyset} m(B), \quad \forall A \subseteq \Theta \qquad (3)$$

### 2.2 Decision Making

The TBM considers that holding beliefs and making decision are distinct processes. Hence, it proposes a two level model:

- The credal level where beliefs are entertained and represented by belief functions.
- The pignistic level where beliefs are used to make decisions and represented by probability functions called the pignistic probabilities, denoted $BetP$ [10]:

$$BetP(A) = \sum_{B \subseteq \Theta} \frac{|A \cap B|}{|B|} \frac{m(B)}{(1 - m(\emptyset))}, \forall A \in \Theta \qquad (4)$$

### 2.3 Uncertainty Measures

In this section, we introduce some uncertainty measures defined in the belief function theory, which quantify the information content or the degree of uncertainty of a piece of information [5] [6].

**Nonspecificity Measures.** The nonspecificity measure is introduced by Dubois and Prade in order to measure the nonspecificity of a normal bba by a function $N$ defined as [5] [6]:

$$N(m) = \sum_{A \in \mathcal{F}(m)} m(A) \log_2 |A| \qquad (5)$$

The bba $m$ is all the most imprecise (least informative) that $N(m)$ is large. The minimum $(N(m) = 0)$ is obtained when $m$ is a Bayesian bba (focal elements are singletons) and the maximum $(N(m) = \log_2 |A|)$ is reached when $m$ is a vacuous bba $(m(\Theta) = 1)$.

**Conflict Measures.** Conflict measures are a generalization of the Shannon's entropy and they were expressed as follows [5] [6]:

$$conflict(m) = - \sum_{A \in \mathcal{F}(m)} m(A) \log_2 f(A) \qquad (6)$$

where $f$ is, respectively, $pl$, $bel$ or $BetP$ and the conflict measures are called, respectively, Dissonance (E), Confusion (C) and Discord (D).

**Composite Measures.** Different measures have been defined by Pal, Bezdek and Hemasinha [5] [6] such as:

$$H(m) = \sum_{A \in \mathcal{F}(m)} m(A) \log_2 \left( \frac{|A|}{m(A)} \right) \qquad (7)$$

$$EP(m) = - \sum_{\omega \in \Omega} BetP(\omega) \log_2 BetP(\omega) \qquad (8)$$

The interesting feature of $H(m)$ is that it has a unique maximum.

## 3 Preference Relations

The preference structure is a basic step of preference modeling. Given two alternatives, decision maker defines three binary relations: preference ($P$ :$\succ$), indifference ($I$ :$\sim$) and incomparability ($J$ :?) [8].

A preference structure is a basic concept of preference modeling. In a classical preference structure, a decision maker makes three decisions for any pair $(a, b)$ from the set $A$ of all alternatives. His decision defines a triplet $P, I, J$ of crisp binary relations on $A$:

1. $a$ is prefered to $b$ ($(a, b) \in P$) iff $(a \succ b) \land \neg(b \succ a)$
2. $a$ is indifferent to $b$ ($(a, b) \in I$) iff $(a \succ b) \land (b \succ a)$
3. $a$ is incomparable to $b$ ($(a, b) \in J$) iff $\neg(a \succ b) \land \neg(b \succ a)$

However, $P, I$ and $J$ must satisfy some rather basic additional conditions. For instance, any couple of alternatives belongs to exactly one of the relations $P, P^t$ (the transpose of $P$), $I$ or $J$. More formally, a preference structure is defined as follows.

1. $I$ is reflexive and $J$ is irreflexive;
2. $P$ is asymmetrical;
3. $I$ and $J$ are symmetrical;
4. $P \cap I = \emptyset, P \cap J = \emptyset$ and $J \cap I = \emptyset$
5. $P \cup P^t \cup I \cup J = A^2$

Property (1) means that the user is always indifferent between $a$ and $a$, and that $a$ can always be compared to itself; (2) is the property that a user cannot prefer $a$ to $b$ and $b$ to $a$ at the same time; (3) means that when a user is indifferent between $a$ and $b$, he is equally so to $b$ and $a$, and that when $a$ and $b$ are incomparable, so are $b$ and $a$. Property (4) states that a pair $(a, b)$ cannot belong to two of the relations $P, I$ and $J$ at the same time. Finally, (5) is the property that a pair $(a, b)$ always belongs to one of the relations $P$, $P^t$, $I$ or $J$, and to no other. Note that the asymmetry of $P$ implies the irreflexivity of $P$.

# 4   Qualitative Belief Functions Methods

The problem of eliciting qualitatively expert opinions and generating basic belief assignments have been addressed by many researchers [1] [4] [7] [12]. In this section, we provide an overview of some existing approaches.

## 4.1   Wong and Lingras' Method

Wong and Lingras [12] proposed a method for generating quantitative belief functions from qualitative preference assessments. So, given a pair of propositions, experts may express which of the propositions is more likely to be true. Thus, they defined two binary relations preference $\succ$ and indifference $\sim$ defined on $2^\Theta$ such as:

$$a \succ b \text{ is equivalent to } bel(a) > bel(b) \tag{9}$$

$$a \sim b \text{ is equivalent to } bel(a) = bel(b) \tag{10}$$

where $a, b \in 2^\Theta$.

This approach is based on two steps. The first one consists in considering that all the propositions that appear in the preference relations are potential focal elements. However, some propositions are eliminated according to the following condition: if $a \sim b$ for some $a \subset b$, then $a$ is not a focal element.

After that, the basic belief assignment is generated using the two presented Equations 9 and 10. This formulation has multiple belief functions that are consistent with the input qualitative information, and so their procedure only generates one of them.

It should be noted that Wong and Lingras' approach do not address the issue of inconsistency in the pair-wise comparisons. For example, the expert could specify the apparently inconsistent preference relationships: $bel(a) > bel(b)$, $bel(a) > bel(c)$, and $bel(c) > bel(a)$.

## 4.2   Bryson et al.' Method

Qualitative discrimination process (QDP), a model for generating belief functions from qualitative preferences, was presented by Bryson, et al. [4].

This method is based on the following steps. First, using this QDA approach, each proposition is assigned into a Broad category bucket, then to a corresponding Intermediate bucket, and finally to a corresponding Narrow category bucket. The qualitative scoring is done using a table where each Broad category is a linguistic quantifier in the sense of Parsons [7]. He considers that linguistic quantifiers could provide a useful approach to representing beliefs vaguely, and that mass and hence bba should be represented using numeric intervals. Then, in step 2, the previous table is used to identify and remove non focal propositions. For each superset proposition, determine if the expert is indifferent in his strength of belief, in the truthfulness of the given proposition and any of its subset propositions in the same or lower Narrow category bucket.

Step 3 is called "imprecise pair-wise comparisons" because the expert is required to provide numeric intervals to express his beliefs on the relative truthfulness of the propositions. In step 4, the consistency of the belief information provided by the expert is checked. Then, the belief function is generated in step 5 by providing a bba interval for each focal element. Finally, in step 6, the expert examines the generated belief functions and stops the QDP if it is acceptable, otherwise the process is repeated.

### 4.3 Ben Yaghlane et al.'s Method

Ben Yaghlane et al. proposed a method for generating optimized belief functions from qualitative preferences [1].

So giving two alternatives, an expert can usually express which of the propositions is more likely to be true, thus they used two binary preference relations: the preference and the indifference relations. The objective of this method is then to convert these preferences into constraints of an optimization problem whose resolution, according to some uncertainty measures (UM) (nonspecificity measures, conflict measures, composite measures), allows the generation of the least informative or the most uncertain belief functions defined as follows:

$$a \succ b \Rightarrow bel(a) - bel(b) \geq \varepsilon \tag{11}$$

$$a \sim b \Rightarrow |bel(a) - bel(b)| \leq \varepsilon \tag{12}$$

where $\varepsilon$ is considered to be the smallest gap that the expert may discern between the degrees of belief in two propositions $A$ and $B$. Note that $\varepsilon$ is a constant specified by the expert before beginning the optimization process.

Ben Yaghlane et al. proposed a method that requires that propositions be represented in terms of focal elements, and they assume that $\Theta$ (where $\Theta$ is the frame of discernment) should always be considered as a potential focal element. Then, a mono-objective technique was used to solve such constrained optimization problem:

$$
\begin{aligned}
& Max_m UM(m) \\
& s.t. \\
& bel(a) - bel(b) \geq \varepsilon \\
& bel(a) - bel(b) \leq \varepsilon \\
& bel(a) - bel(b) \geq -\varepsilon \\
& \sum_{a \in \mathcal{F}(m)} m(a) = 1, m(a) \geq 0, \forall a \subseteq \Theta; m(\emptyset) = 0
\end{aligned} \tag{13}
$$

where the first, second and third constraints are derived from Eqs 11 and 12, representing the quantitative constraints corresponding to the qualitative preference relations.

Furthermore, the proposed method addresses the problem of inconsistency. In fact, if the preference relations are consistent, then the optimization problem is feasible. Otherwise no solutions will be found. Thus, the expert may be guided to reformulate his preferences.

An extension of the proposed solution is also presented. In fact, the authors suggested to use the goal programming, a multiobjective method, in order to take into account simultaneously several objectives in the formulation of the problem. So, the idea behind the use of this method is to be able to integrate additional information about the belief functions to be generated.

It should be noted that this method does not address the issue of incomparability in the pair-wise comparisons. In fact, this proposed method treats incomparability as incompleteness. However, we believe that this interpretation is not appropriate. If an expert is unable to compare two alternatives then this situation should be reflected in the preference relation not as an incomplete situation, but with an entry for that particular pair of alternatives. So, in the following section, we present our method that deal with this problem.

## 5    Modelling Belief Functions Using Qualitative Preferences

As presented above, representing efficiently the expert preferences is a crucial task in elaborating the necessary data for a considered problem. Therefore, we propose a realistic solution that is able to efficiently imitate the expert reasoning. In fact, our main aim is then to elaborate on how may be incomparability and incompleteness represented in qualitative belief functions. The solution we suggest is then a qualitative model for constructing belief functions from elicited expert opinions when dealing with qualitative preference relations. In this section, we start by identifying sufficient conditions of introducing these imperfect preferences. Then, we consider the computational procedure.

### 5.1    Incompleteness in the Belief Function Theory

Incomparability and incompleteness represent very different concepts. In this subsection, we try also to differentiate incomplete preferences from incomparable ones. This situation is illustrated by complete ignorance, missing information, lack of knowledge or an ongoing preference elicitation process. Incompleteness represents then simply an absence of knowledge about the relationship between these pairs of alternatives.

Given such considerations, it may perhaps be useful at times to take incomplete order as the primitive of analysis. Besides, expert is freely allowed to assign this belief to any pairs of alternatives. In other words, a partial order allows some relations between pairs of alternatives to be unknown.

**Example.** Given three alternatives $\Theta = \{a, b, c\}$, an incomplete order can be for example: $(a \succ c, b?c)$ or $(c \succ a, a \sim b)$, where some relations between pairs of alternatives are unknown.

### 5.2    Incomparability in the Belief Function Theory

A missing value in a linguistic preference relation is not always equivalent to a lack of preference of one alternative over another. A missing value can also be the result of the incapacity of an expert to compare one alternative over another because they are too different. In such cases, the expert may not put his opinion forward about certain aspects of the problem, he would not be able to efficiently express his preference between two or more of the available alternative. As a result, he may find some of them to be incomparable and thus has an incomplete preference ordering, i.e., he neither prefers one alternative over the other nor finds them equally as good. Therefore, it would be of great importance to provide the expert with tools that allow them to efficiently model his preferences.

In order to model this situation, we first consider how to represent the incomparability relation. In fact, our problem here is that incomparability is expressed entirely in terms of negations:

$$a?b \text{ iff } \neg(a \succ b) \wedge \neg(b \succ a) \qquad (14)$$

By definition, a couple of alternatives $(a, b)$ belongs to the incomparability relation $J$ if and only if the expert is unable to compare $a$ and $b$. Furthermore, it is hard to see what kind of behavior could correspond to Equation 14. If neither $a$ nor $b$ is chosen, then the expert may not be able to tell which alternative is better, since not $aPb$, not $bPa$ and not $aIb$. In other terms, we apply incomparability when the preference profiles of two alternatives are severely conflicting.

The question now is how to formalize this situation in the belief function framework. In order to build this new preference relation, we may accept that there exist positive reasons which support the relation $\neg(a \succ b)$ and also there exist sufficient negative information to establish the relation $(a \succ b)$. These two assumptions can properly model the contradictory information. Besides, we can surely establish that "$a$ is preferred to $b$" as there are not sufficient reasons supporting the opposite and there are sufficient information against it, while we can also surely establish that "$b$ is preferred to $a$" for the same reasons. Therefore, $a$ and $b$ are in conflicting position. On the other hand, and based on the belief function framework and as defined by Boujelben et al. [3], the incomparability situation appears between two alternatives when their evaluations given by basic belief assessments differ significantly.

In the following, our objective is to represent the case of incomparability with the belief function theory. Consider two alternatives $a$ and $b$, as proved in Wong et al. [12] the belief function exists since the preference relation $\succ$ satisfies the following axioms:

1. Asymmetry: $a \succ b \Rightarrow \neg(b \succ a)$
2. Negative Transitivity: $\neg(a \succ b)$ and $\neg(b \succ c) \Rightarrow \neg(a \succ c)$
3. Dominance: For all $a, b \in 2^\Theta, a \supseteq b \Rightarrow a \succ b$ or $a \sim b$
4. Partial monotonicity: For all $a, b, c \in 2^\Theta$, if $a \supset b$ and $a \cap c = \emptyset \Rightarrow (a \cup c) \succ (b \cup c)$

So, Wong et al. have justified the existing of the following relation:

$$a \succ b \Leftrightarrow bel(a) > bel(b) \qquad (15)$$

In other words, Wong et al. have proved that it may exist functions other than the belief functions, which are also compatible with a preference relation such that for every $a, b \in 2^\Theta$.

$$a \succ b \Leftrightarrow f(a) > f(b) \qquad (16)$$

if and only if the relation $\succ$ satisfies the previous axioms.

Similarly to this idea, we can prove that the plausibility function also exists since the preference relation $\succ$ satisfies the previous axioms. Besides, we can conclude that it exists a plausibility function $pl: 2^\Theta \rightarrow [0, 1]$ such as:

$$a \succ b \Leftrightarrow pl(a) > pl(b) \qquad (17)$$

To summarize, we can get the following relation:

$$a \succ b \Leftrightarrow bel(a) > bel(b) \text{ and } pl(a) > pl(b) \tag{18}$$

As we have defined previously, the incomparability situation appears between two alternatives when their preference profiles are severely conflicting. That is when their evaluations given by basic belief assessments differ significantly. We can then intuitively conclude from Equation 18 that, if $a$ is incomparable with $b$, then:

$$a?b \Leftrightarrow bel(a) \geq bel(b) \text{ and } pl(a) \leq pl(b) \tag{19}$$

The first part of Equation 19 supports the assumption "$a$ is preferred to $b$" however the second one supporting the opposite affirmation. Also, the second part of the Equation supports the assumption "$b$ is preferred to $a$" and the first part affirms the opposite assumption.

Consequently, our purpose is then to prove the existing of the previous Equation 19 in order to correctly represent the bba relative to the incomparability relation.

*Proof. According to the definition of the plausibility function, we have:*

$$pl(a) = bel(\Theta) - bel(\bar{a}).$$

*We start from the second part of the Equation 19, our assumption is:* $pl(a) \leq pl(b)$
$\Leftrightarrow bel(\Theta) - bel(\bar{a}) \leq bel(\Theta) - bel(\bar{b})$
$\Leftrightarrow - bel(\bar{a}) \leq -bel(\bar{b})$
$\Leftrightarrow bel(\bar{a}) \geq bel(\bar{b})$
*Using the Equation 9, we can therefore conclude that:* $\bar{a} \succ \bar{b}$, *which means that* $\neg(a \succ b)$. *This contradicts with the assumption "if* $pl(a) \leq pl(b)$ *then* $a \succ b$". *Hence, if we have* $pl(a) \leq pl(b)$, *then* $b \succ a$. *However, from the first part of the assumption, we have:* $a \succ b$.

As a conclusion, such representation of incomparability (Equation 18), enables us to correctly express the conflicting information produced by the alternative $a$ and the alternative $b$. In fact, the first part of the Equation 19 "$bel(a) \geq bel(b)$" implies that $a$ is preferred to $b$. Then, the plausibility function is used since it expresses the maximum amount of specific support that could be given to a proposition $a$. However, when we define the second part of the Equation 19, we propose to assume that $pl(a) \leq pl(b)$ which means that $b$ is preferred to $a$. This contradicts the first assumption, and can properly express the conflicting information's produced by $a$ and $b$.

## 5.3    Computational Procedure

Now and after modeling the incompleteness and the incomparability preferences, we propose to extend Ben Yaghlane et al. method [1]. We transform these preferences relations into constraints as presented in section 4.1. We get:

$$Max_m UM(m)$$
$$s.t.$$
$$bel(a) - bel(b) \geq \varepsilon$$
$$bel(a) - bel(b) \leq \varepsilon$$
$$bel(a) - bel(b) \geq -\varepsilon \qquad (20)$$
$$bel(a) \geq bel(b)$$
$$pl(a) \leq pl(b)$$
$$\sum_{a \in \mathcal{F}(m)} m(a) = 1; m(a) \geq 0; \forall a \subseteq \Theta; m(\emptyset) = 0$$

where the first, second and third constraints of the model are derived from the preference and indifference relations. The fourth and fifth constraints correspond to the
incomparability relation. $\varepsilon$ is a constant specified by the expert before beginning the
optimization process.

A crucial step is needed before beginning the task of generating belief functions, is
the identification of the candidate focal elements. Thus, as applied in the existing approaches, we may initially assume that prepositions which may appear in the preference
relationships are considered as focal elements. Then, other focal elements could be appear or also eliminated. The next phase of our procedure consists of establishing the
local preference relations between each pair of two alternatives. Finally, these obtained
relations are transformed into constraints to obtain the quantitative belief function.

## 6   Example

Let us consider a problem of eliciting the weight of the candidate criteria. The problem
involves six criteria: $\Omega = \{a, b, c, d, e, f\}$. The focal elements are: $F1 = \{a\}$, $F2 = \{a, b, c\}$, $F3 = \{b, e\}$, $F4 = \{e, f\}$ and $F5 = \{a, e, d\}$.

Next, the expert opinions should be elicitated. For this purpose, an interview with
the expert is realized in order to model his preferences. Consequently, he has validated
the following relations:

$$F2 \succ F1 \, , \, F1?F3 \, , \, F4 \succ F1$$
$$F5 \succ F1 \, , \, F3 \sim F2 \, , \, F5 \succ F4,$$

After eliciting the expert preferences, the following step is to identify the candidate
focal elements. So, we get:

$$\mathcal{F}(m) = \{F1, F2, F3, F4, F5, \Theta\}$$

Next, these obtained relations are transformed into optimization problem according to
our proposed method. We assume that $\varepsilon = 0.01$ and the uncertainty measures is $H$
since it has a unique maximum as defined in Equation 7.

Table 1 gives the results of all ordered couples on the basis of their preference relation. Besides, we are interested in obtaining their corresponding quantitative bba.

In fact, there are different ways to obtain a result by aggregating the obtaining binary
relation. But, the existing approaches present the inconvenience of eliminating useful

**Table 1.** The obtained bba using our proposed model

| Criteria | $\{a\}$ | $\{a,b,c\}$ | $\{b,e\}$ | $\{e,f\}$ | $\{a,e,d\}$ | $\{b,c,d,e,f\}$ | $\{a,c,d,f\}$ | $\Theta$ |
|---|---|---|---|---|---|---|---|---|
| bba | 0.069 | 0.079 | 0.069 | 0.079 | 0.119 | 0.198 | 0.194 | 0.193 |
| bel | 0.069 | 0.148 | 0.069 | 0.079 | 0.119 | 0.346 | 0.263 | 1 |
| pl | 0.535 | 0.069 | 0.46 | 0.047 | 0.312 | 0.391 | 0.391 | 1 |

information as the incompleteness and incomparability. However, by applying our presented solution, it is easy to see that our method aggregates all the elicited data. Here, in the present example, all the incomparabilities are detected. We obtain for example $F1?F3$.

Once the preferences relations are defined, the corresponding bba (Table 1) should be constructed. We suggest to transform our problem into a constrained optimization model in order to choose the optimal solution and to get the previous result.

Now we propose to apply Ben Yaghlane et al. method. By using this model, we assume that the incomparability and the incompleteness are modeled in the same way. In other words, the relation $F1?F3$ will be eliminated and we get the following Table 2.

**Table 2.** The obtained bba using Ben Yaghlane et al. method

| Criteria | $\{a\}$ | $\{a,b,c\}$ | $\{b,e\}$ | $\{e,f\}$ | $\{a,e,d\}$ | $\Theta$ |
|---|---|---|---|---|---|---|
| bba | 0.063 | 0.159 | 0.149 | 0.126 | 0.189 | 0.315 |
| bel | 0.063 | 0.222 | 0.149 | 0.126 | 0.252 | 1 |

In absence of incomparabilities, we note the couple of alternatives $\{b,c,d,e,f\}$ and $\{a,c,d,f\}$ do not appears because the incomparability relation has been assigned to other relation: the incompleteness. Observing the two obtaining results, it is possible to see that in spite of the use of two different models, we get almost the same partial order.

## 7    Conclusion

In this study, the incomplete linguistic preference relations are used to derive quantitative belief function. By presenting our method, a new model for constructing belief functions from elicited expert opinions has been defined, that takes into account the incomplete and even the incomparable alternatives. The originality of our model is then to provide additional interpretation values to the existing methods based on strict preferences and indifferences only.

Under this perspective the paper introduces a new method based on Ben Yaghlane et al. approach [1]. Our work makes it possible to separate incomparability from incompleteness. Then, we suggest to extend Ben Yaghlane et al. method to take into account these distinct levels of preferences. Finally, our method transforms the preference relations provided by the expert into constraints of an optimization problem.

An interesting future work is to make our method able to explore uncertainty preferences. For instance, an expert may not be able to say if a couple of alternatives $(a, b)$ belongs to the preference or indifference relation. Besides, we propose to apply our proposed method in multi-criteria decision making field, which can be interesting in eliciting expert judgments.

# References

1. Ben Yaghlane, A., Denoeux, T., Mellouli, K.: Constructing belief functions from expert opinions. In: Proceedings of the 2nd International Conference on Information and Communication Technologies: from Theory to Applications (ICTTA 2006), Damascus, Syria, pp. 75–89 (2006)
2. Ben Yaghlane, A., Denoeux, T., Mellouli, K.: Elicitation of expert opinions for constructing belief functions. In: Proceedings of IPMU, Paris, France, pp. 403–411 (2006)
3. Boujelben, M.A., Smet, Y.D., Frikha, A., Chabchoub, H.: A ranking model in uncertain, imprecise and multi-experts contexts: The application of evidence theory. International Journal of Approximate Reasoning 52, 1171–1194 (2011)
4. Bryson, N., Mobolurin, A.: A process for generating quantitative belief functions. European Journal of Operational Research 115, 624–633 (1999)
5. Pal, N., Bezdek, J., Hemasinha, R.: Uncertainty measures for evidential reasoning I: A review. International Journal of Approximate Reasoning 7, 165–183 (1992)
6. Pal, N., Bezdek, J., Hemasinha, R.: Uncertainty measures for evidential reasoning II: A review. International Journal of Approximate Reasoning 8, 1–16 (1993)
7. Parsons, S.: Some qualitative approaches to applying the DS theory. Information and Decision Technologies 19, 321–337 (1994)
8. Roubens, M., Vincke, P.: Preference modelling. Springer, Berlin (1985)
9. Shafer, G.: A Mathematical Theory of Evidence. Princeton University Press (1976)
10. Smets, P.: The application of the Transferable Belief Model to diagnostic problems. International Journal of Intelligent Systems 13, 127–158 (1998)
11. Smets, P., Kennes, R.: The Transferable Belief Model. Artificial Intelligence 66, 191–234 (1994)
12. Wong, S., Lingras, P.: Representation of qualitative user preference by quantitative belief functions. IEEE Transactions on Knowledge and Data Engineering 6, 72–78 (1994)

# On Rough Set Based Non Metric Model

Yasunori Endo[1,4], Ayako Heki[2], and Yukihiro Hamasuna[3]

[1] Faculty of Eng., Info. and Sys., University of Tsukuba
Tennodai 1-1-1, Tsukuba, 305-8573 Ibaraki, Japan
endo@risk.tsukuba.ac.jp
[2] Graduate School of Sys. and Info. Eng., University of Tsukuba
Tennodai 1-1-1, Tsukuba, 305-8573 Ibaraki, Japan
s1120638@u.tsukuba.ac.jp
[3] Department of Informatics, Kinki University
3-4-1, Kowakae, Higashiosaka, 577-8502 Osaka, Japan
yhama@info.kindai.ac.jp
[4] International Institute for Applied Systems Analysis (IIASA)
Schlossplatz 1, A-2361 Laxenburg, Austria

**Abstract.** Non metric model is a kind of clustering method in which belongingness or the membership grade of each object to each cluster is calculated directly from dissimilarities between objects and cluster centers are not used.

By the way, the concept of rough set is recently focused. Conventional clustering algorithms classify a set of objects into some clusters with clear boundaries, that is, one object must belong to one cluster. However, many objects belong to more than one cluster in real world, since the boundaries of clusters overlap with each other. Fuzzy set representation of clusters makes it possible for each object to belong to more than one cluster. On the other hand, the fuzzy degree sometimes may be too descriptive for interpreting clustering results. Rough set representation could handle such cases. Clustering based on rough set representation could provide a solution that is less restrictive than conventional clustering and less descriptive than fuzzy clustering.

This paper shows two type of Rough set based Non Metric model (RNM). One algorithm is Rough set based Hard Non Metric model (RHNM) and the other is Rough set based Fuzzy Non Metric model (RFNM). In the both algorithms, clusters are represented by rough sets and each cluster consists of lower and upper approximation. Second, the proposed methods are kernelized by introducing kernel functions which are a powerful tool to analize clusters with nonlinear boundaries.

## 1 Introduction

Computer system data has become large-scale and complicated in recent years due to progress in hardware technology, and the importance of data analysis techniques has been increasing accordingly. Clustering, which means a data classification method without any external criterion, has attracted many researchers as a significant data analysis technique.

V. Torra et al. (Eds.): MDAI 2012, LNAI 7647, pp. 394–407, 2012.
© Springer-Verlag Berlin Heidelberg 2012

Bezdek et al. proposed Fuzzy c-Means (FCM) [1, 2], which can be regarded as the fuzzification of Hard c-Means (HCM). In FCM, cluster centers are introduced to obtain belongingness or the membership grade of each object to each cluster. Roubens proposed a clustering algorithm without cluster centers [3] in which membership grades are calculated with only dissimilarities between objects. The algorithm is then called a Fuzzy Non Metric model (FNM) which means fuzzified Hard Non Metric model (HNM). The FNM has a property enabling membership grades to be calculated directly from dissimilarities, and thereby object space need not necessarily be Euclidean space. After that, Bezdek et al. discussed relations between FCM and FNM [4, 5]. One of us proposed Entropy based FNM (EFNM) by the different fuzzification of HNM from Roubens, and discussed some applications [6]. In these algorithms, fuzzy set concept plays very important role.

On the other hand, it is pointed out that the fuzzy set representation sometimes may be too descriptive for interpreting clustering results [7]. In such cases, rough set representation becomes a useful and powerful tool [8, 9]. The basic concept of the rough representation is based on two definitions of lower and upper approximations of a set. The lower approximation means that "an object surely belongs to the set" and the upper one means that "an object possibly belongs to the set". Clustering based on rough set representation could provide a solution that is less restrictive than conventional clustering and less descriptive than fuzzy clustering [10, 7], and therefore the rough set based clustering has attracted increasing interest of researchers [11–16, 7].

In this paper, we will mainly develop theoretical discussion for a new non metric model based on rough set representation. We will first construct two type of Rough set based Non Metric model (RNM). One algorithm is Rough set based Hard Non Metric model (RHNM) and the other is Rough set based Fuzzy Non Metric model (RFNM). In RNM, clusters are represented by rough sets and each cluster consists of lower and upper approximation.

RFNM is constructed by fuzzifying RHNM with the entropy regularized term. The purpose of fuzzification of RHNM is to overcome a problem of RHNM that if an object does not belong to any lower approximation it belongs to only two upper approximation, not three or more. To fuzzify RHNM, we introduce an entropy regularized term which was considered by Miyamoto et al. as another way to fuzzify HCM [17, 18] from by Bezdek. Bezdek introduced a fuzzification parameter on membership grades in the objective function of HCM, while Miyamoto et al. introduced an entropy regularization term into the function, instead of the fuzzification parameter. In comparison with FCM, EFCM has, among other properties, one in which the classification function converges to 0 or 1 as the dissimilarity between a cluster center and an object goes to infinity.

RHNM is more simple than RFNM so that it is expected that the calculation cost of RHNM is less than RFNM. On the other hand, RFNM is more flexible than RFNM in the meaning that each object can belong to two or more upper approximation.

Second, we will kernelize the proposed methods by introducing kernel functions. Vapnik proposed the Support Vector Machine (SVM) [19, 20]. The SVM is one of supervised methods, and it can recognize nonlinear boundaries between clusters. Kernel functions play a very important role with the SVM. Objects are mapped from the object space to high dimensional feature space and the inner product of data in high dimensional feature space can be easily calculated by kernel functions. Some clustering algorithms have been proposed with kernel functions [21–23] to classify a given dataset into clusters with nonlinear boundaries and the usefulness is verified.

## 2   Preliminaries

### 2.1   Rough Sets

Let $U$ be the universe and $R \subseteq U \times U$ be an equivalence relation on $U$. $R$ is also called indiscernibility relation. The pair $X = (U, R)$ is called an approximation space. If $x$, $y \in U$ and $(x, y) \in R$, we say that $x$ and $y$ are indistinguishable in $X$.

Equivalence classes of the relation $R$ is called elementary sets in $X$. A family of all elementary sets is denoted by $U/R$. The empty set is also elementary in every $X$.

Since it is impossible to distinguish the elements in the same equivalence class, we may not be able to get a precise representation for an arbitrary subset $A \subset U$. Instead, any $A$ can be represented by its lower and upper bounds. The upper bound $\overline{A}$ is the least composed set in $X$ containing $A$, called the best upper approximation or, in short, the upper approximation. The lower bound $\underline{A}$ is the greatest composed set in $X$ containing $A$, called the best lower approximation or, briefly, the lower approximation. The set $\mathrm{Bnd}(A) = \overline{A} - \underline{A}$ is called the boundary of $A$ in $X$.

The pair $(\underline{A}, \overline{A})$ is the representation of an ordinary set $A$ in the approximation space $X$, or simply the rough set of $A$. The elements in the lower approximation of $A$ definitely belong to $A$, while elements in the upper bound of $A$ may or may not belong to $A$.

From the above description of rough sets, we can define the following conditions for clustering:

(C1) An object $x$ can be part of at most one lower approximation.
(C2) If $x \in \underline{A}$, $x \in \overline{A}$.
(C3) An object $x$ is not part of any lower approximation if and only if $x$ belongs to two or more boundaries.

### 2.2   Notations and HNM

For any objects $x_k = (x_{k1}, \ldots, x_{kp})^T \in \Re^p$ $(k = 1, \ldots, n)$, $\nu_{ki}$ and $u_{ki}$ $(i = 1, \ldots, c)$ mean belongingness of an object $x_k$ to a lower approximation of $A_i$ and a boundary of the $i$-th cluster $A_i$, respectively. Partition matrices of $\nu_{ki}$ and $u_{ki}$ are denoted by $N = (\nu_{ki})_{k=1\ldots n, i=1,\ldots,c}$ and $U = (u_{ki})_{k=1\ldots n, i=1,\ldots,c}$, respectively.

The original HNM algorithm minimizes the following objective function:

$$J(N,U) = \sum_{i=1}^{c}\sum_{k=1}^{n}\sum_{t=1}^{n} \nu_{ki}\nu_{ti}r_{kt}. \tag{1}$$

## 3  RHNM

### 3.1  Objective Function and Optimal Solutions

To construct a new relational clustering algorithm based on rough set representation, we define the following objective function based on (1):

$$J(N,U) = \underline{w}\sum_{i=1}^{c}\sum_{k=1}^{n}\sum_{t=1}^{n} \nu_{ki}\nu_{ti}r_{kt} + \overline{w}\sum_{i=1}^{c}\sum_{k=1}^{n}\sum_{t=1}^{n} u_{ki}u_{ti}r_{kt}, \tag{2}$$

here $\underline{w}+\overline{w} = 1$. $r_{kt}$ means a dissimilarity between $x_k$ and $x_t$. One of the examples is Euclidean norm:

$$r_{kt} = \|x_k - x_t\|^2.$$

We consider the following conditions for $\nu_{ki}$ and $u_{ki}$:

$$\nu_{ki} \in \{0,1\}, \qquad u_{ki} \in \{0,1\}.$$

From (C1), (C2), and (C3) in Subsection 2.1, we can derive the following constraints:

$$\sum_{i=1}^{c}\nu_{ki} \in \{0,1\}, \qquad \sum_{i=1}^{c}u_{ki} \neq 1,$$

$$\sum_{i=1}^{c}\nu_{ki} = 1 \iff \sum_{i=1}^{c}u_{ki} = 0.$$

From the above constraints, we can derive the following relation for any $k$.

$$\sum_{i=1}^{c}\nu_{ki} = 0 \iff \sum_{i=1}^{c}u_{ki} \geq 2$$

It is obvious that these relations are equivalent to (C1), (C2), and (C3) in Subsection 2.1.

Optimal solutions to $\nu_{ki}$ and $u_{ki}$ can be obtained by comparing the following two cases for each $x_k$:

**Case 1:** $x_k$ belongs to the lower approximation $\underline{A}_{p_k}$.
**Case 2:** $x_k$ belongs to the boundaries of two clusters $\overline{A}_{q_k^1}$ and $\overline{A}_{q_k^2}$.

We describe the detail of each case as follows:

**Case 1:** Let assume that $x_k$ belongs to the lower approximation $\underline{A}_{p_k}$. $p_k$ is derived as follows:

The objective function $J$ can be rewritten as follows:

$$J(N,U) = \underline{w} \sum_{i=1}^{c} \underline{J}_i + \overline{w} \sum_{i=1}^{c} \sum_{l=1}^{n} \sum_{t=1}^{n} u_{li} u_{ti} r_{lt},$$

here

$$\underline{J}_i = \left( 2\nu_{ki} \sum_{t=1}^{n} \nu_{ti} r_{kt} + \sum_{l=1,l\neq k}^{n} \sum_{t=1,t\neq k}^{n} \nu_{li}\nu_{ti}r_{lt} \right).$$

Note that $r_{kk} = 0$ and $r_{kt} = r_{tk}$. Therefore,

$$p_k = \arg\min_i \sum_{t=1}^{n} \nu_{ti} r_{kt}. \tag{3}$$

It means the following relations:

$$\nu_{ki} = \begin{cases} 1, & (i = p_k) \\ 0, & (\text{otherwise}) \end{cases}$$

$$u_{ki} = 0. \quad (\forall i)$$

In this case, the value of the objective function can be calculated as follows:

$$J(N,U) = \underline{w} \left( 2 \sum_{t=1}^{n} \nu_{tp_k} r_{kt} + \sum_{i=1}^{c} \sum_{l=1,l\neq k}^{n} \sum_{t=1,t\neq k}^{n} \nu_{li}\nu_{ti}r_{lt} \right)$$

$$+ \overline{w} \sum_{i=1}^{c} \sum_{l=1}^{n} \sum_{t=1}^{n} u_{li} u_{ti} r_{lt}$$

$$= 2J_k^\nu + J_c,$$

here

$$J_k^\nu = \sum_{t=1}^{n} \left( \underline{w}\nu_{tp_k} + \sum_{i=1}^{c} \overline{w} u_{ki} u_{ti} \right) r_{kt} = \underline{w} \sum_{t=1}^{n} \nu_{tp_k} r_{kt}, \tag{4}$$

$$J_c = \underline{w} \sum_{i=1}^{c} \sum_{l=1,l\neq k}^{n} \sum_{t=1,t\neq k}^{n} \nu_{li}\nu_{ti}r_{lt} + \overline{w} \sum_{i=1}^{c} \sum_{l=1,l\neq k}^{n} \sum_{t=1,t\neq k}^{n} u_{li} u_{ti} r_{lt}.$$

**Case 2:** Let assume that $x_k$ belongs to the boundaries of two clusters $\overline{A}_{q_k^1}$ and $\overline{A}_{q_k^2}$. $q_k^1$ and $q_k^2$ are derived as follows:

The objective function $J$ can be rewritten as follows:

$$J(N,U) = \overline{w} \sum_{i=1}^{c} \overline{J}_i + \underline{w} \sum_{i=1}^{c} \sum_{l=1}^{n} \sum_{t=1}^{n} \nu_{ki}\nu_{ti}r_{lt},$$

here

$$\overline{J}_i = \left(2u_{ki}\sum_{t=1}^{n}u_{ti}r_{kt} + \sum_{l=1,l\neq k}^{n}\sum_{t=1,t\neq k}^{n}u_{li}u_{ti}r_{lt}\right).$$

Therefore,

$$q_k^1 = \arg\min_{i}\sum_{t=1}^{n}u_{ti}r_{kt}, \tag{5}$$

$$q_k^2 = \arg\min_{i,i\neq q_k^1}\sum_{t=1}^{n}u_{ti}r_{kt}. \tag{6}$$

It means the following relations:

$$\nu_{ki} = 0, \quad (\forall i)$$

$$u_{ki} = \begin{cases} 1, & (i = q_k^1 \vee i = q_k^2) \\ 0. & (\text{otherwise}) \end{cases}$$

In this case, the value of the objective function can be calculated as follows:

$$J(N,U) = \underline{w}\sum_{i=1}^{c}\sum_{l=1}^{n}\sum_{t=1}^{n}\nu_{li}\nu_{ti}r_{lt}$$

$$+ \overline{w}\left(2\sum_{t=1}^{n}(u_{tq_k^1} + u_{tq_k^2})r_{kt} + \sum_{i=1}^{c}\sum_{l=1,l\neq k}^{n}\sum_{t=1,t\neq k}^{n}u_{li}u_{ti}r_{lt}\right)$$

$$= 2J_k^u + J_c,$$

here

$$J_k^u = \sum_{t=1}^{n}\left(\sum_{i=1}^{c}\underline{w}\nu_{ki}\nu_{ti} + \overline{w}(u_{tq_k^1} + u_{tq_k^2})\right)r_{kt}$$

$$= \overline{w}\sum_{t=1}^{n}(u_{tq_k^1} + u_{tq_k^2})r_{kt}. \tag{7}$$

In comparison with $J_k^\nu$ and $J_k^u$, we determine $\nu_{ki}$ and $u_{ki}$ as follows:

$$\nu_{ki} = \begin{cases} 1, & (J_k^\nu < J_k^u \wedge i = p_k) \\ 0, & (\text{otherwise}) \end{cases}$$

$$u_{ki} = \begin{cases} 1, & \left(J_k^\nu \geq J_k^u \wedge (i = q_k^1 \vee i = q_k^2)\right) \\ 0. & (\text{otherwise}) \end{cases}$$

## 3.2   RHNM Algorithm

From the above discussion, we show the RHNM algorithm as Algorithm 1. The proposed algorithm is constructed based on iterative optimization.

---

**Algorithm 1. RHNM**

**RHNM1** The iteration number $L = 0$. Give $r_{kt}$ and set initial values of $\nu_{ki}^{(0)}$ and $u_{ki}^{(0)}$.

**RHNM2** Update $\nu_{ki}^{(L+1)}$ and $u_{ki}^{(L+1)}$ as follows:

$$\nu_{ki}^{(L+1)} = \begin{cases} 1, & (J_k^\nu < J_k^u \wedge i = p_k) \\ 0, & (\text{otherwise}) \end{cases}$$

$$u_{ki}^{(L+1)} = \begin{cases} 1, & \left(J_k^\nu \geq J_k^u \wedge ( i = q_k^1 \vee i = q_k^2)\right) \\ 0. & (\text{otherwise}) \end{cases}$$

$p_k$, $q_k^1$, $q_k^2$, $J_k^\nu$, and $J_k^u$ are calculated by (3), (5), (6), (4), and (7), respectively, with $\nu_{ti}^{(L)}$ and $u_{ti}^{(L)}$.

**RHNM3** If the stop criterion satisfies, finish. Otherwise $L := L + 1$ and back to **RHNM2**.

---

## 4   RFNM

### 4.1   Objective Function and Optimal Solutions

In the above section, we proposed RHNM algorithm. In the algorithm, an object $x_k$ belongs to just two boundaries if $x_k$ does not belong to any lower approximation, since $u_{ki} \in \{0,1\}$ and the objective function (2) is linear for $u_{ki}$. Therefore, in this section we propose RFNM algorithm to make $x_k$ belong to two or more boundaries if $x_k$ does not belong to any lower approximation.

We consider the following objective function of RFNM:

$$J(N, U) = \underline{w} \sum_{i=1}^{c} \sum_{k=1}^{n} \sum_{t=1}^{n} \nu_{ki}\nu_{ti}r_{kt} + \overline{w} \sum_{i=1}^{c} \sum_{k=1}^{n} \sum_{t=1}^{n} u_{ki}u_{ti}r_{kt}$$
$$+ \lambda^{-1} \sum_{i=1}^{c} \sum_{k=1}^{n} u_{ki} \log u_{ki}, \tag{8}$$

here $\underline{w} + \overline{w} = 1$. $r_{kt}$ means a dissimilarity between $x_k$ and $x_t$. The last entropy term means fuzzification of $u_{ki}$ and makes the objective function nonlinear for $u_{ki}$. Hence, the value of the optimal solution on $u_{ki}$ which minimizes the objective function (8) is in $[0, 1)$.

We assume the following conditions for $\nu_{ki}$ and $u_{ki}$:

$$\nu_{ki} \in \{0, 1\}, \qquad u_{ki} \in [0, 1).$$

From (C1), (C2), and (C3) in Subsection 2.1, we can derive the following constraints:

$$\sum_{i=1}^{c} \nu_{ki} \in \{0,1\}, \quad \sum_{i=1}^{c} u_{ki} \in \{0,1\},$$

$$\sum_{i=1}^{c} \nu_{ki} = 1 \iff \sum_{i=1}^{c} u_{ki} = 0.$$

From the above constraints, we can derive the following relation for any $k$:

$$\sum_{i=1}^{c} \nu_{ki} = 0 \iff \sum_{i=1}^{c} u_{ki} = 1 \tag{9}$$

It is obvious that these relations are equivalent to (C1), (C2), and (C3) in Subsection 2.1.

Same as RHNM, optimal solutions to $\nu_{ki}$ and $u_{ki}$ can be obtained by comparing the following two cases for each $x_k$.

**Case 1:** Let assume that $x_k$ belongs to the lower approximation $\underline{A_{p_k}}$. $p_k$ is derived as follows:

The objective function $J$ can be rewritten as follows:

$$J(N,U) = \underline{w} \sum_{i=1}^{c} \underline{J_i} + \overline{w} \sum_{i=1}^{c} \sum_{l=1}^{n} \sum_{t=1}^{n} u_{li} u_{ti} r_{lt} + \lambda^{-1} \sum_{i=1}^{c} \sum_{k=1}^{n} u_{ki} \log u_{ki},$$

here

$$\underline{J_i} = \left( 2\nu_{ki} \sum_{t=1}^{n} \nu_{ti} r_{kt} + \sum_{l=1,l \neq k}^{n} \sum_{t=1,t \neq k}^{n} \nu_{li} \nu_{ti} r_{lt} \right).$$

Note that $r_{kk} = 0$ and $r_{kt} = r_{tk}$. Therefore,

$$p_k = \arg\min_i \sum_{t=1}^{n} \nu_{ti} r_{kt}. \tag{10}$$

It means the following relations:

$$\nu_{ki} = \begin{cases} 1, & (i = p_k) \\ 0, & (\text{otherwise}) \end{cases}$$

$$u_{ki} = 0. \quad (\forall i)$$

In this case, the value of the objective function can be calculated as follows:

$$J(N,U) = \underline{w}\left(2\sum_{t=1}^{n}\nu_{tp_k}r_{kt} + \sum_{i=1}^{c}\sum_{l=1,l\neq k}^{n}\sum_{t=1,t\neq k}^{n}\nu_{li}\nu_{ti}r_{lt}\right)$$

$$+ \overline{w}\sum_{i=1}^{c}\sum_{l=1}^{n}\sum_{t=1}^{n}u_{li}u_{ti}r_{lt} + \lambda^{-1}\sum_{i=1}^{c}\sum_{l=1}^{n}u_{li}\log u_{li}$$

$$= 2J_k^{\nu} + J_c,$$

here

$$J_k^{\nu} = \sum_{t=1}^{n}\left(\underline{w}\nu_{tp_k} + \sum_{i=1}^{c}\overline{w}u_{ki}u_{ti}\right)r_{kt} + \lambda^{-1}\sum_{i=1}^{c}u_{ki}\log u_{ki}$$

$$= \underline{w}\sum_{t=1}^{n}\nu_{tp_k}r_{kt}, \tag{11}$$

$$J_c = \underline{w}\sum_{i=1}^{c}\sum_{l=1,l\neq k}^{n}\sum_{t=1,t\neq k}^{n}\nu_{li}\nu_{ti}r_{lt} + \overline{w}\sum_{i=1}^{c}\sum_{l=1,l\neq k}^{n}\sum_{t=1,t\neq k}^{n}u_{li}u_{ti}r_{lt}$$

$$+ \lambda^{-1}\sum_{i=1}^{c}\sum_{l=1,l\neq k}^{n}u_{li}\log u_{li}.$$

**Case 2:** Let assume that $x_k$ belongs to the boundaries of two or more one clusters. The objective function $J$ is convex for $u_{ki}$, hence we can derive an optimal solution to $u_{ki}$ using a Lagrange multiplier.

In this case, the constraint 9, then we can introduce the following Lagrange function:

$$L(N,U) = J(N,U) + \sum_{k=1}^{n}\eta_k\sum_{i=1}^{c}(u_{ki} - 1).$$

We partially differentiate $L$ by $u_{ki}$ and get the following equation:

$$\frac{\partial L}{\partial u_{ki}} = 2\overline{w}\left(\sum_{t=1,t\neq k}^{n}u_{ti}r_{kt} + u_{ki}r_{kk}\right) + \lambda^{-1}(1 + \log u_{ki}) + \eta_k$$

$$= 2\overline{w}\sum_{t=1,t\neq k}^{n}u_{ti}r_{kt} + \lambda^{-1}(1 + \log u_{ki}) + \eta_k.$$

From $\frac{\partial L}{\partial u_{ki}} = 0$, we obtain the following relation:

$$u_{ki} = \exp\left(-\lambda(\eta_k + D_{ki}) - 1\right), \tag{12}$$

where

$$D_{ki} = 2\overline{w} \sum_{t=1,t\neq k}^{n} u_{ti}r_{kt} = 2\overline{w} \sum_{t=1}^{n} u_{ti}r_{kt}. \tag{13}$$

From the constraint (9) and the above equation (12), we get the following equation:

$$\sum_{i=1}^{c} u_{ki} = \sum_{i=1}^{c} \exp\left(-\lambda(\eta_k + D_{ki}) - 1\right) = 1,$$

$$\exp\left(-\lambda\eta_k - 1\right) = 1/\sum_{j=1}^{c} \exp\left(-\lambda D_{kj}\right).$$

We then obtain the following optimal solution:

$$u_{ki} = \frac{\exp\left(-\lambda D_{ki}\right)}{\sum_{j=1}^{c} \exp\left(-\lambda D_{kj}\right)}.$$

It means the following relations:

$$\nu_{ki} = 0, \quad (\forall i)$$

$$u_{ki} = \frac{\exp\left(-\lambda D_{ki}\right)}{\sum_{j=1}^{c} \exp\left(-\lambda D_{kj}\right)} \quad (\forall i).$$

In this case, the value of the objective function can be calculated as follows:

$$J(N,U) = \underline{w}\sum_{i=1}^{c}\sum_{l=1}^{n}\sum_{t=1}^{n} \nu_{li}\nu_{ti}r_{lt}$$

$$+ \overline{w}\sum_{i=1}^{c}\left(2u_{ki}\sum_{t=1}^{n} u_{ti}r_{kt} + \sum_{l=1,l\neq k}^{n}\sum_{t=1,t\neq k}^{n} u_{li}u_{ti}r_{lt}\right)$$

$$+ \lambda^{-1}\left(\sum_{i=1}^{c} u_{ki}\log u_{ki} + \sum_{i=1}^{c}\sum_{l=1,l\neq k}^{n} u_{li}\log u_{li}\right)$$

$$= 2J_k^u + J_c,$$

here

$$J_k^u = \sum_{t=1}^{n}\left(\sum_{i=1}^{c} \underline{w}\nu_{ki}\nu_{ti} + \sum_{i=1}^{c} \overline{w}u_{ki}u_{ti}\right)r_{kt} + \frac{1}{2}\lambda^{-1}\sum_{i=1}^{c} u_{ki}\log u_{ki}$$

$$= \sum_{i=1}^{c} u_{ki}\left(\sum_{t=1}^{n} \overline{w}u_{ti}r_{kt} + (2\lambda)^{-1}\log u_{ki}\right). \tag{14}$$

In comparison with $J_k^\nu$ and $J_k^u$, we determine $\nu_{ki}$ and $u_{ki}$ as follows:

$$\nu_{ki} = \begin{cases} 1, & (J_k^\nu < J_k^u \wedge i = p_k) \\ 0, & \text{(otherwise)} \end{cases}$$

$$u_{ki} = \begin{cases} \frac{\exp(-\lambda D_{ki})}{\sum_{j=1}^c \exp(-\lambda D_{kj})}, & (J_k^\nu \geq J_k^u) \\ 0. & \text{(otherwise)} \end{cases}$$

### 4.2   RFNM Algorithm

From the above discussion, we show the RFNM algorithm as Algorithm 2. The proposed algorithm is also constructed based on iterative optimization.

---

**Algorithm 2. RFNM**

---

**RFNM1** The iteration number $L = 0$. Give $r_{kt}$ and set initial values of $\nu_{ki}^{(0)}$ and $u_{ki}^{(0)}$.
**RFNM2** Update $\nu_{ki}^{(L+1)}$ and $u_{ki}^{(L+1)}$ as follows:

$$\nu_{ki}^{(L+1)} = \begin{cases} 1, & (J_k^\nu < J_k^u \wedge i = p_k) \\ 0, & \text{(otherwise)} \end{cases}$$

$$u_{ki}^{(L+1)} = \begin{cases} \frac{\exp(-\lambda D_{ki})}{\sum_{j=1}^c \exp(-\lambda D_{kj})}, & (J_k^\nu \geq J_k^u) \\ 0. & \text{(otherwise)} \end{cases}$$

$p_k$, $D_{ki}$, $J_k^\nu$, and $J_k^u$ are calculated by (10), (13), (11), and (14), respectively, with $\nu_{ti}^{(L)}$ and $u_{ti}^{(L)}$.
**RFNM3** If the stop criterion satisfies, finish. Otherwise $L := L + 1$ and back to **RFNM2**.

---

## 5   Kernelized RNM

In this section, we try to kernelize RNM. As mentioned above, kernel functions are a powerful tool to classify a dataset into some clusters with nonlinear boundaries.

### 5.1   Kernel Functions

To kernelize the ENM, we first define some symbols to introduce kernel functions. Mapping from the pattern space $\Re^p$ to high dimensional feature space $\Re^s$ is expressed as $\phi : \Re^p \to \Re^s$ $(p \ll s)$. Each datum in feature space is denoted by $\phi(x_k) = x_k^\phi = (x_{k1}^\phi, \ldots, x_{ks}^\phi)^T \in \Re^s$, and dataset $X^\phi = \left\{ x_1^\phi, \ldots, x_n^\phi \right\}$ is given.

The kernel function $K : \Re^p \times \Re^p \to \Re$ satisfies the following relation:

$$K(x, y) = \langle \phi(x), \phi(y) \rangle.$$

From Mercer's theorem [24], $K$ is a continuous symmetric nonnegative definite kernel if and only if mapping $\phi$ exists and satisfies the above relation. Note that $\phi$ is not explicit. With $K$, we can easily calculate the inner product of data in high dimensional feature space.

## 5.2  KRNM Algorithm

Given the above preparation, we consider a Kernelized Rough Non Metric model (KRNM) using the following relation:

$$
\begin{aligned}
r^{\phi}_{kt} &= \|x^{\phi}_k - x^{\phi}_t\|^2 \\
&= \langle x^{\phi}_k, x^{\phi}_k \rangle - 2\langle x^{\phi}_k, x^{\phi}_t \rangle + \langle x^{\phi}_t, x^{\phi}_t \rangle \\
&= K(x_k, x_k) - 2K(x_k, x_t) + K(x_t, x_t).
\end{aligned}
$$

Kernelized RHNM (KRHNM) and Kernelized RFNM (KRFNM) algorithms is the same as Algorithms 1 and 2 by substituting $r^{\phi}_{kt}$ for (3), (4), (5), (6), (7), (10), (11), (13), and (14) instead of $r_{kt}$. For example,

$$
u_{ki} = \frac{\exp\left(-\lambda D^{\phi}_{ki}\right)}{\sum_{j=1}^{c} \exp\left(-\lambda D^{\phi}_{kj}\right)},
$$

where

$$
D^{\phi}_{ki} = 2 \sum_{t=1, t\neq k}^{n} u_{ti} r^{\phi}_{kt} = 2 \sum_{t=1}^{n} u_{ti} r^{\phi}_{kt}.
$$

We show KRNM algorithm as Algorithm 3.

---

**Algorithm 3.** KRNM (KRHNM and KRFNM)

---

**KRNM1** The iteration number $L = 0$. Give $r_{kt}$ and set initial values of $\nu^{(0)}_{ki}$ and $u^{(0)}_{ki}$.

**KRNM2** Update $\nu^{(L+1)}_{ki}$ and $u^{(L+1)}_{ki}$ by (3), (4), (5), (6), and (7) in case of KRHNM, or by (10), (11), (13), and (14) in case of KRFNM, with $\nu^{(L)}_{ti}$ and $u^{(L)}_{ti}$ by substituting $r^{\phi}_{kt}$ instead of $r_{kt}$.

**KRNM3** If the stop criterion satisfies, finish. Otherwise $L := L + 1$ and back to RFNM2.

---

## 6  Conclusion

In this paper, we first constructed two type of RNM, that is, RHNM and RFNM. In the both algorithms, clusters are represented by rough sets and each cluster consists of lower and upper approximation. It is considered that RHNM is more simple than RFNM so that it is expected that the calculation cost of RHNM is less than RFNM, while RFNM is more flexible than RFNM in the meaning that each object allow to belong to two or more upper approximation. Second, we kernelized the proposed methods by introducing kernel functions.

We have mainly developed a theoretical discussion in this paper. So it cannot be said that the verification of the proposed algorithm through numerical examples is enough. We thus must do that through numerical examples in a future paper.

**Acknowledgment.** We would like to gratefully and sincerely thank Professor Sadaaki Miyamoto of the University of Tsukuba, and Associate Professor Yuchi Kanzawa of Shibaura Institute of Technology for their advice.

# References

1. Bezdek, J.C.: Pattern Recognition with Fuzzy Objective Function Algorithms. Plenum, New York (1981)
2. Bezdek, J.C., Keller, J., Krisnapuram, R., Pal, N.R.: Fuzzy Models and Algorithms for Pattern Recognition and Image Processing. The Handbooks of Fuzzy Sets Series (1999)
3. Roubens, M.: Pattern classification problems and fuzzy sets. Fuzzy Sets and Systems 1, 239–253 (1978)
4. Bezdek, J.C., Davenport, J.W., Hathaway, R.J.: Clustering with the Relational $c$-Means Algorithms using Different Measures of Pairwise Distance. In: Juday, R.D. (ed.) Proceedings of the 1988 SPIE Technical Symposium on Optics, Electro-Optics, and Sensors, vol. 938, pp. 330–337 (1988)
5. Hathaway, R.J., Davenport, J.W., Bezdek, J.C.: Relational Duals of the $c$-Means Clustering Algorithms. Pattern Recognition 22(2), 205–212 (1989)
6. Endo, Y.: On Entropy Based Fuzzy Non Metric Model –Proposal, Kernelization and Pairwise Constraints. Journal of Advanced Computational Intelligence and Intelligent Informatics 16(1), 169–173 (2012)
7. Lingras, P., Peters, G.: Rough clustering. In: Proceedings of the 17th International Conference on Machine Learning (ICML 2000), pp. 1207–1216 (2011)
8. Pawlak, Z.: Rough Sets. International Journal of Computer and Information Sciences 11(5), 341–356 (1982)
9. Inuiguchi, M.: Generalizations of Rough Sets: From Crisp to Fuzzy Cases. In: Tsumoto, S., Słowiński, R., Komorowski, J., Grzymała-Busse, J.W. (eds.) RSCTC 2004. LNCS (LNAI), vol. 3066, pp. 26–37. Springer, Heidelberg (2004)
10. Pawlak, Z.: Rough Classification. International Journal of Man-Machine Studies 20, 469–483 (1984)
11. Hirano, S., Tsumoto, S.: An Indiscernibility-Based Clustering Method with Iterative Refinement of Equivalence Relations. Journal of Advanced Computational Intelligence and Intelligent Informatics 7(2), 169–177 (2003)
12. Lingras, P., West, C.: Interval Set Clustering of Web Users with Rough $K$-Means. Journal of Intelligent Information Systems 23(1), 5–16 (2004)
13. Mitra, S., Banka, H., Pedrycz, W.: Rough-Fuzzy Collaborative Clustering. IEEE Transactions on Systems Man, and Cybernetics, Part B, Cybernetics 36(5), 795–805 (2006)

14. Maji, P., Pal, S.K.: Rough Set Based Generalized Fuzzy $C$-Means Algorithm and Quantitative Indices. IEEE Transactions on System, Man and Cybernetics, Part B, Cybernetics 37(6), 1529–1540 (2007)
15. Peters, G.: Rough Clustering and Regression Analysis. In: Yao, J., Lingras, P., Wu, W.-Z., Szczuka, M.S., Cercone, N.J., Ślęzak, D. (eds.) RSKT 2007. LNCS (LNAI), vol. 4481, pp. 292–299. Springer, Heidelberg (2007)
16. Mitra, S., Barman, B.: Rough-Fuzzy Clustering: An Application to Medical Imagery. In: Wang, G., Li, T., Grzymala-Busse, J.W., Miao, D., Skowron, A., Yao, Y. (eds.) RSKT 2008. LNCS (LNAI), vol. 5009, pp. 300–307. Springer, Heidelberg (2008)
17. Miyamoto, S., Mukaidono, M.: Fuzzy $c$-Means as a Regularization and Maximum Entropy Approach. In: Proc. of the 7th International Fuzzy Systems Association World Congress (IFSA 1997), vol. 2, pp. 86–92 (1997)
18. Miyamoto, S., Umayahara, K., Mukaidono, M.: Fuzzy Classification Functions in the Methods of Fuzzy c-Means and Regularization by Entropy. Journal of Japan Society for Fuzzy Theory and Systems 10(3), 548–557 (1998)
19. Vapnik, V.N.: Statistical Learning Theory. Wiley, New York (1998)
20. Vapnik, V.N.: The nature of Statistical Learning Theory, 2nd edn. Springer, New York (2000)
21. Endo, Y., Haruyama, H., Okubo, T.: On Some Hierarchical Clustering Algorithms Using Kernel Functions. In: IEEE International Conference on Fuzzy Systems, #1106 (2004)
22. Hathaway, R.J., Huband, J.M., Bezdek, J.C.: A Kernelized Non-Euclidean Relational Fuzzy $c$-Means Algorithm. Neural, Parallel and Scientific Computation 13, 305–326 (2005)
23. Miyamoto, S., Kawasaki, Y., Sawazaki, K.: An Explicit Mapping for Fuzzy c-Means Using Kernel Function and Application to Text Analysis. In: IFSA/EUSFLAT 2009 (2009)
24. Mercer, J.: Functions of Positive and Negative Type and Their Connection with the Theory of Integral Equations. Philosophical Transactions of the Royal Society A 209, 415–446 (1909)

# An Efficient Reasoning Method
# for Dependencies over Similarity
# and Ordinal Data

Radim Belohlavek[1], Pablo Cordero[2], Manuel Enciso[3],
Angel Mora[2], and Vilem Vychodil[1]

[1] Data Analysis and Modeling Laboratory (DAMOL), Dept. Computer Science,
Palacky University, Olomouc, Czech Republic
{radim.belohlavek,vychodil}@acm.org
[2] Dept. Applied Mathematics, University of Málaga, Spain
{pcordero,amora}@uma.es
[3] Dept. Languages and Computer Sciences, University of Málaga, Spain
enciso@lcc.uma.es

**Abstract.** We present a new axiomatization of logic for dependencies
in data with grades, including ordinal data and data in an extension of
Codd's model that takes into account similarity relations on domains.
The axiomatization makes possible an efficient method for automated
reasoning for such dependencies that is presented in the paper. The presented method of automatic reasoning is based on a new simplification
equivalence which allows to simplify sets of dependencies while retaining
their semantic closures. We include two algorithms for computing closures and checking semantic entailment from sets of dependencies and
present experimental comparison showing that the algorithms based on
the new axiomatization outperform the algorithms proposed in the past.

## 1   Introduction

We present a complete axiomatization of a logic for dependencies in data with
grades and an efficient automated reasoning method based on this axiomatization. The dependencies are expressed by formulas of the form

$$A \Rightarrow B, \tag{1}$$

where $A$ and $B$ are graded sets of attributes, such as

$$\{^{0.2}/y_1, y_2\} \Rightarrow \{^{0.8}/y_3\}. \tag{2}$$

Such formulas have two different kinds of semantics (two interpretations) whose
entailment relations coincide. First, the semantics given by object-attribute data
with grades [5] in which (2) means: every object that has attribute $y_1$ to degree
at least 0.2 and attribute $y_2$ to degree 1 (i.e. fully possesses $y_2$), has also attribute
$y_3$ to degree at least 0.8. Second, the semantics given by ranked tables over domains with similarities (an extension of Codd's model of relational data) in which
(2) means: every two tuples that are similar on attribute $y_1$ to degree at least 0.2

V. Torra et al. (Eds.): MDAI 2012, LNAI 7647, pp. 408–419, 2012.

and are equal on attribute $y_2$ are similar on attribute $y_3$ to degree at least 0.8. We assume that the set of degrees forms a partially ordered set and is equipped with particular aggregation operations, notably many valued conjunction and implication. If 0 and 1 are the only degrees, the first interpretation coincides with the well-known attribute dependencies in binary data, called attribute implications (meaning that presence of certain attributes implies presence of other attributes), and the second one with functional dependencies in the ordinary Codd's model.

The rules used in this paper serve several purposes. From the point of view of knowledge acquisition, the rules represent important if-then patterns that can be derived from data and are capable of representing various if-then dependencies that are present in the data. For instance, [5] shows that each object-attribute data table representing ordinal (graded) dependencies between objects and their attributes (features) can be characterized by a base of rules like (1). Similar situation applies to data tables with similarities over domains and ordinal ranks as in [8,9]. A base is an irreducible set of rules valid in the data such that all other dependencies that are valid in the data can be derived from the base. Thus, bases of rules like (1) are concise representations of knowledge inferred from data. In order to gain more knowledge from a base, one has to come up with an efficient inference system and this paper contributes to this area—it shows an algorithm for determining whether a formula follows (and to what degree of satisfaction) from a collection of formulas. In a broader context, the paper shows for the first time an efficient automated prover for reasoning about dependencies in data involving grades.

## 2  Preliminaries

We assume that the set of degrees, such as 0.2 or 0.8 in (2), is partially ordered and equipped with particular aggregation operations. Such structures are known from fuzzy logic [13,14,15,17], aggregation theory [16], and have been utilized in various models for combination of ordinal information [11]. In particular, we denote the set of degrees by $L$ and assume that it forms an algebraic structure $\mathbf{L} = \langle L, \wedge, \vee, \otimes, \rightarrow, \diagdown, ^*, 0, 1 \rangle$ such that $\langle L, \wedge, \vee, 0, 1 \rangle$ is a complete lattice, $\langle L, \otimes, 1 \rangle$ is a commutative monoid, and

- $\otimes$ and $\rightarrow$ satisfy the following adjointness property:
  for all $a, b, c \in L$, $a \otimes b \leq c$ if and only if $a \leq b \rightarrow c$;
- $\diagdown$ and $\vee$ satisfy the following adjointness property:
  for all $a, b, c \in L$, $a \diagdown b \leq c$ if and only if $a \leq b \vee c$;
- $^*$ is a unary operation (so-called hedge) satisfying: for all $a, b \in L$,
  $1^* = 1$, $a^* \leq a$, $(a \rightarrow b)^* \leq a^* \rightarrow b^*$, and $a^{**} = a^*$.

We recall that the above conditions mean that $\langle L, \wedge, \vee, \otimes, \rightarrow, 0, 1 \rangle$ forms a complete residuated lattice [14,17] and $\langle L, \wedge, \vee, \diagdown, 1 \rangle$ is a Brouwerian algebra (or equivalently, its dual $\langle L, \vee, \wedge, \diagdown, 0 \rangle$ is a Heyting algebra, which implies that the lattice is distributive). $\otimes$ and $\rightarrow$ are interpreted as a many-valued conjunction and implication; $\diagdown$ as a many-valued non-implication (used for set difference); and $^*$ as an intensifying hedge such as "very true", see [18].

The most commonly used set $L$ is the real unit interval $L = [0, 1]$ (or its finite subchains), in which case $\wedge$ and $\vee$ are the minimum and the maximum, $\otimes$ and $\rightarrow$ a left-continuous t-norm and its residuum, respectively, and $\setminus$ is given by

$$x \setminus y = \begin{cases} x & \text{if } x > y, \\ 0 & \text{otherwise.} \end{cases} \tag{3}$$

Two important, boundary cases of hedges are identity and so-called globalization (i.e. $1^* = 1$ and $x^* = 0$ for all $1 \neq x \in L$). We use the usual notions of **L**-sets, graded subsethood, and define if-then formulas like (2) and their interpretation in a general way (cf. early approaches like [23]), see [5] for details.

## 3    FASL Logic

In this section, we introduce the alternative axiomatization and prove its completeness. The axiomatization forms a theoretical base for the automated prover introduced in Section 4. The proposed axiomatization has the following benefit over the Armstrong-like [2] axiomatizations from [4,5]: the rules can always be applied to all formulas, meaning there is no restriction on the form of the formulas that appear in the input part of the inference rules. This property makes the rules suitable for sequential execution by an automated prover.

### 3.1    New Axiomatic System

In [4], the authors presented an axiomatic system for reasoning with formulas (1) that is syntactico-semantically complete w.r.t. the two kinds of semantics described in the previous section. The system consists of three deduction rules,

[Ax] $\vdash AB \Rightarrow A$ (Axiom)

[Cut] $A \Rightarrow B,\ BC \Rightarrow D \vdash AC \Rightarrow D$ (Cut)

[Mul] $A \Rightarrow B \vdash c^* \otimes A \Rightarrow c^* \otimes B$ (Multiplication)

where $A, B, C, D \in L^Y$ and $c \in L$. In [Ax] and [Cut], we use the convention of writing $BC$ instead of $B \cup C$, etc., and in [Mul], we use $a \otimes B$ to denote so-called $a$-multiple of $B \in L^Y$ which is an **L**-set such that $(a \otimes B)(y) = a \otimes B(y)$ for all $y \in Y$ (i.e., the degrees to which $y \in Y$ belongs to $B$ is multiplied by a constant degree $a \in L$).

As usual, if $\mathcal{R}$ is an axiomatic system (like that containing the rules [Ax], [Cut], and [Mul]), a formula $A \Rightarrow B$ is said to be *provable* from a theory $T$ by using $\mathcal{R}$, denoted by

$$T \vdash_{\mathcal{R}} A \Rightarrow B$$

if there is a sequence $\varphi_1, \ldots, \varphi_n$ such that $\varphi_n$ is $A \Rightarrow B$, and for each $\varphi_i$ we either have $\varphi_i \in T$ or $\varphi_i$ is inferred (in one step) from some of the preceding formulas using some inference rule in $\mathcal{R}$. The results in [4] have shown among other things that $\mathcal{R}$ consisting of [Ax], [Cut], and [Mul] is complete in the following sense:

**Theorem 1** *Let* **L** *and* $Y$ *be finite. Then for every set* $T$ *of formulas,* $T \vdash_{\mathcal{R}} A \Rightarrow B$ *if and only if* $T \models A \Rightarrow B$. $\qquad\qquad\square$

The rule [Cut] is powerful but it is not directly suitable for automated deduction. We now present a new syntactico-semantically complete axiomatic system which overcomes this drawback by replacing [Cut] by a new rule, called *rule of simplification* (denoted [Sim]). The new system consists of the following rules:

[Ax] $\vdash AB \Rightarrow A$

[Sim] $A \Rightarrow B,\ C \Rightarrow D \vdash A(C - B) \Rightarrow D$

[Mul] $A \Rightarrow B \vdash c^* \otimes A \Rightarrow c^* \otimes B$

where $A, B, C, D \in L^Y$ and $c \in L$. The new system is called FASL (Fuzzy Attribute Simplification Logic). The main motivation for introducing a new axiomatic system is to obtain a system that may be used for an efficient system of automated reasoning with formulas (1). Unlike [Cut], the new simplification rule [Sim] can be applied to any pair of formulas which makes it more suitable for automated provers.

## 3.2 Completeness

In this section, we prove completeness of the new axiomatic system. We start by recalling the following concepts.

When reasoning with degrees, theories are naturally conceived as L-sets of formulas [12,17,22], leaving theories as ordinary sets of formulas as particular cases. In our case, given a theory $T$, the degree $T(A \Rightarrow B)$ can been seen as the degree to which we assume the validity of $A \Rightarrow B$. For the following concepts and results, see e.g. [4,5,6]. Let $T$ be a theory. The set of all *models* of $T$ is defined by

$$\mathrm{Mod}(T) = \{\mathcal{D} \mid T(A \Rightarrow B) \le \|A \Rightarrow B\|_{\mathcal{D}} \text{ for all } A, B \in L^Y\},$$

i.e. $\mathrm{Mod}(T)$ is the set of all ranked tables in which every $A \Rightarrow B$ is true at least to the degree prescribed by $T$. The degree $\|A \Rightarrow B\|_T$ to which $A \Rightarrow B$ *semantically follows* from a theory $T$ is defined by

$$\|A \Rightarrow B\|_T = \bigwedge\nolimits_{\mathcal{D} \in \mathrm{Mod}(T)} \|A \Rightarrow B\|_{\mathcal{D}},$$

i.e. it may be seen as the degree to which $A \Rightarrow B$ is true in every model of $T$. If $\|A \Rightarrow B\|_T = 1$, we write

$$T \models A \Rightarrow B.$$

In the particular case when $T$ is crisp ($T(C \Rightarrow D)$ may only be 0 or 1), we get that $T \models A \Rightarrow B$ iff $\|A \Rightarrow B\|_{\mathcal{D}} = 1$ for all $\mathcal{D} \in \mathrm{Mod}(T)$. Also note that if $T$ is crisp, it may be regarded as a set of formulas, i.e. we write $C \Rightarrow D \in T$ instead of $T(C \Rightarrow D) = 1$. The following lemma [5] shows a technical trick due to which one may restrict to crisp theories only.

**Lemma 1.** *For any $A, B \in L^Y$, $c \in L$, and a ranked table $\mathcal{D}$,*

$$c \le \|A \Rightarrow B\|_{\mathcal{D}} \text{ if and only if } \|A \Rightarrow c \otimes B\|_{\mathcal{D}} = 1.$$

*As a consequence, for any theory $T$,*

$$\mathrm{Mod}(T) = \mathrm{Mod}(c(T)) \quad \text{and} \quad \|A \Rightarrow B\|_T = \|A \Rightarrow B\|_{c(T)},$$

*where $c(T)$ is the crisp theory defined by*

$$c(T) = \{A \Rightarrow T(A \Rightarrow B) \otimes B \mid A, B \in L^Y \text{ and } T(A \Rightarrow B) \otimes B \neq \emptyset\}. \quad (4)$$

Lemma 1 shows that the crisp theory $c(T)$ has the same models and consequences as $T$. Therefore, in what follows, we only consider crisp theories and, for simplicity, only entailment to degree 1. The following theorem shows that our new axiomatic system, consisting of [Ax], [Sim], and [Mul] is sound and complete.

**Theorem 2 (Completeness)** *Let $S$ be the axiomatic system given by* [Ax], [Sim], *and* [Mul]. *Let* **L** *and $Y$ be finite, let $T$ be a set of formulas. Then $T \vdash_S A \Rightarrow B$ if and only if $T \models A \Rightarrow B$.* $\qquad\square$

*Remark 1.* In addition to Theorem 2, which asserts an "ordinary-style completeness" (formula is provable if and only if it is entailed), a stronger theorem may be proved, asserting a "graded-style completeness": Using an appropriately defined concept of degree of provability in our system, one may show that a degree to which $A \Rightarrow B$ is provable from $T$ equals the degree to which $A \Rightarrow B$ is entailed from $T$. We omit the stronger theorem in this paper.

Recall that a deduction rule is called derivable in a given axiomatic system if the output formula of the rule is provable from the input formulas of the rule. The following assertion shows important derivable rules:

**Lemma 2.** *The following deduction rules are derivable in FASL: Let $A, B, C, D \in L^Y$ and $c \in L$. Then,*

> [Dec] $\{A \Rightarrow BC\} \vdash A \Rightarrow B$;                   *(Decomposition)* [1]
>
> [Com] $\{A \Rightarrow B, C \Rightarrow D\} \vdash AC \Rightarrow BD$.           *(Composition)*

Using [Dec] and [Com], we can obtain observations how certain formulas can be equivalently replaced by other formulas while retaining the semantic entailment. We call theories $T_1$ and $T_2$ equivalent, denoted by $T_1 \equiv T_2$, if the set of derivable formulas from both theories coincide. Using Lemma 2, we get the following observation.

**Theorem 3** *Let $A, B, C, D \in L^Y$. The following equivalences can be obtained from* [Ax] + [Sim].

> **(DeEq)** *Decomposition Equivalence:* $\{A \Rightarrow B\} \equiv \{A \Rightarrow B - A\}$;
>
> **(UnEq)** *Union Equivalence:* $\{A \Rightarrow B, A \Rightarrow C\} \equiv \{A \Rightarrow BC\}$;
>
> **(SiEq)** *Simplification Equivalence: If $A \subseteq C$ then*
>          $\{A \Rightarrow B, C \Rightarrow D\} \equiv \{A \Rightarrow B, A(C - B) \Rightarrow D - B\}$.

The previous equivalences, read from left to right, enable us to remove redundant information in the formulas. Namely, the sets on the right-hand sides can be seen as equivalent simplifications of the sets on the left-hand sides (simplified either in terms of the number of formulas as in case of (**UnEq**) or in terms of the number of elements in formulas as in case of the other equivalences).

---

[1] In the literature, [Dec] is also called the rule of projectivity.

*Remark 2.* If **L** is the two-element Boolean algebra, then (**SiEq**) becomes the ordinary simplification equivalence that has been utilized in an efficient prover for ordinary functional dependencies which is studied in [1,10,20,21].

## 4 Automated Reasoning Method

In this section, we utilize the inference rules and equivalences obtained in Section 3 in an automated prover. The role of the prover is twofold. First, given a theory $T$ and a formula $A \Rightarrow B$ of the above-mentioned forms, the prover is able to check whether $A \Rightarrow B$ is provable from $T$. Due to the completeness of our inference system, it means the prover tests whether $||A \Rightarrow B||_{\mathcal{D}} = 1$ for any model $\mathcal{D} \in \mathrm{Mod}(T)$. Needless to say, checking $||A \Rightarrow B||_{\mathcal{D}} = 1$ for any model $\mathcal{D} \in \mathrm{Mod}(T)$ directly (i.e., following the definition) is not possible since there are infinitely many pairwise different models of $T$. As we shall see later in the section, the prover is capable of checking more than the (semantic) entailment to degree 1. Indeed, it can be used to check (by means of finding a proof) that $||A \Rightarrow B||_{\mathcal{D}} \geq a$ for a given degree $a \in L$ and any model $\mathcal{D} \in \mathrm{Mod}(T)$, checking thus a lower bound for $||A \Rightarrow B||_T$. Moreover, the prover can be used to compute a syntactic closure of a given $A \in L^Y$ with respect to a theory $T$. Using this notion and taking advantage of the previous results, we show that the prover described in this section fully characterizes degrees of semantic entailment $||A \Rightarrow B||_T$ using inclusion degrees [5] and computed closures of $A$. Therefore, the automated prover presented in this section is an important (and simple) algorithm for determining degrees of semantic entailment which are (by definition) degrees of satisfaction in an infinite class of models.

### 4.1 Generalized Simplification Equivalence

We start by showing that for any crisp theory $T$, the fact $T \vdash A \Rightarrow B$ can equivalently be expressed by provability using formulas with empty antecedents derived from $A \Rightarrow B$, i.e., formulas of the form $\emptyset \Rightarrow C$ where $C \in L^Y$.

**Theorem 4** *If $T$ is a crisp theory, then for any $A \Rightarrow B$, we have $T \vdash A \Rightarrow B$ iff $T \cup \{\emptyset \Rightarrow A\} \vdash \emptyset \Rightarrow B$.* □

Notice that Theorem 4 can be seen as an analogy of the classic deduction theorem known from propositional logic. Indeed, the classic deduction theorem of propositional logic says that $T \vdash \varphi \Rightarrow \psi$ if and only if $T \cup \{\varphi\} \vdash \psi$. Using the fact that any propositional formula $\chi$ is equivalent to $\overline{1} \Rightarrow \chi$ where $\overline{1}$ denotes a tautology (e.g., $\overline{1}$ stands for $\vartheta \Rightarrow \vartheta$), the classic deduction theorem can be equivalently restated as $T \vdash \varphi \Rightarrow \psi$ if and only if $T \cup \{\overline{1} \Rightarrow \varphi\} \vdash \overline{1} \Rightarrow \psi$ which is close to the form in Theorem 4. Note that from the point of view of interpreting $\emptyset \Rightarrow A$ in ordinal data, $\emptyset \Rightarrow A$ can be seen as a formula saying "attributes from $A$ are (unconditionally) present". Using Theorem 1 and (4), we can extend Theorem 4 from crisp theories to arbitrary theories (**L**-sets of formulas) as follows:

**Corollary 1.** *For any $A \Rightarrow B$, the following are equivalent:*

(i) $a \leq \|A \Rightarrow B\|_T$,

(ii) $c(T) \vdash A \Rightarrow a \otimes B$,

(iii) $c(T) \cup \{\emptyset \Rightarrow A\} \vdash \emptyset \Rightarrow a \otimes B$,

*where $c(T)$ is the crisp theory given by (4).*     □

Using Theorem 3, we can prove the following assertion which is fundamental for the simplification procedure described below.

**Theorem 5** *The following equivalence can be obtained from FASL rules:*

**(gSiEq)** *Generalized Simplification Equivalence:*

$$\{\emptyset \Rightarrow A, U \Rightarrow V\} \equiv \{\emptyset \Rightarrow A', U - A' \Rightarrow V - A'\},$$

*where $A, U, V \in L^Y$ and $A' = A \cup (S(U, A)^* \otimes V)$.*

*Proof (sketch).* The assertion is proved using the fact that for all $A, U, V \in L^Y$, we get $\{\emptyset \Rightarrow A, U \Rightarrow V\} \vdash \emptyset \Rightarrow A(S(U, A)^* \otimes V)$, where $A(S(U, A)^* \otimes V)$ denotes $A \cup (S(U, A)^* \otimes V)$. The fact can be obtained as a corollary of the preceding observations. As a particular case of the observation, we get $\{U \Rightarrow V\} \vdash \emptyset \Rightarrow S(U, \emptyset)^* \otimes V$.     □

In particular, the algorithms employ the following corollary which follows immediately from the generalized simplification equivalence from Theorem 5.

**Corollary 2.** *The following equivalences hold.*

**(gSiUnEq)** *If $U - A' = \emptyset$ then $\{\emptyset \Rightarrow A, U \Rightarrow V\} \equiv \{\emptyset \Rightarrow A'V\}$;*

**(gSiAxEq)** *if $V - A' = \emptyset$ then $\{\emptyset \Rightarrow A, U \Rightarrow V\} \equiv \{\emptyset \Rightarrow A'\}$,*

*where $A, U, V \in L^Y$ and $A' = A \cup (S(U, A)^* \otimes V)$.*     □

### 4.2   The Algorithm

The automated prover is based on the algorithm shown in Figure 1. In the body of the while-loop, the algorithm maintains a formula of the form $\emptyset \Rightarrow A_i$. This formula shall be called a *guide*.

The algorithm applies the simplification rules from Theorem 5 and Corollary 2 using which the crisp theory $T_{i+1}$ is reduced. Either the number of formulas in $T_{i+1}$ is reduced using **(gSiUnEq)** and **(gSiAxEq)** or the number of formulas in $T_{i+1}$ remains the same but $T_{i+1}$ is modified by replacing $U \Rightarrow V$ by $U - A' \Rightarrow V - A'$ using **(gSiEq)** which means a reduction of antecedents and consequents in $U \Rightarrow V$. This can be seen as removing a particular type of redundancy from $T_{i+1}$. Also note that the sequence $A_1, A_2, \ldots$ of L-sets which appear in the guide during the computation is nondecreasing, i.e., $A_1 \subseteq A_2 \subseteq \cdots$ At the end of the computation, the algorithm returns the last $A_i$ considered. We will show that the returned result plays an important role in determining degrees to which particular formulas follow from $T$.

We now focus on the basic properties of the algorithm. In order to prove its soundness, we introduce the following notion.

---

**Input**: $T$ (theory), $A$ (**L**-set of attributes)
**Output**: $A^+$ (closure of $A$ with respect to $c(T)$)

  **begin**
1     Let $A_0 := \emptyset$, $T_0 := \emptyset$, and $i := 0$;
2     Let $A_1 := A$;
3     Apply (**DeEq**) to every formula in $c(T)$ obtaining $T_1$;
4     **while** ($A_i \neq A_{i+1}$ *or* $T_i \neq T_{i+1}$) **do**
        $i := i + 1$; $A_{i+1} := A_i$; $T_{i+1} := T_i$;
        Modify $T_{i+1}$ and $A_{i+1}$ applying
           (**gSiEq**), (**gSiUnEq**), and (**gSiAxEq**) to
           $\emptyset \Rightarrow A_i$ and each $U \Rightarrow V \in T_i$;
     **end**
5     Return $A_i$;
  **end**

**Fig. 1.** The core algorithm of the automated prover

---

**Definition 1.** *Let $A \in L^Y$ and $T$ be a crisp theory. The closure of $A$ (with respect to $T$), denoted by $A^+$, is the greatest **L**-set in $Y$ such that $T \vdash A \Rightarrow A^+$.*

Note that $A^+$ from Definition 1 is in fact a *syntactic* closure of $A$ with respect to $T$ since it is defined by means of syntactic entailment $\vdash$. The existence and uniqueness of $A^+$ has been shown in [5]. Moreover, from Theorem 1 and observations from [5], it follows that $A^+$ coincides with the semantic closure of $A$ with respect to $T$. The following assertion shows that $A^+$ is the value returned by the algorithm in Figure 1.

**Theorem 6** *For any theory $T$ and $A \in L^Y$, the algorithm from Figure 1 finishes after finitely many steps. Moreover, the returned value is equal to $A^+$ computed with respect to $c(T)$ given by (4).*

*Proof (sketch).* Tarski's fixed-point theorem ensures that the algorithm finishes because the sequence of the sets $A_i$ is growing in $L^Y$. So, since both $Y$ and $L$ are finite, the algorithm achieves a fixed point in a finite number of steps. Moreover, when the algorithm finishes, we can show (details are postponed to a full version of this paper) that for all $a \in L$ and $U \Rightarrow V \in c(T)$ if $a^* \otimes U \subseteq A_i$ then $a^* \otimes V \subseteq A_i$. The latter condition implies that $A_i$ is equal to $A^+$. $\square$

An important application of the algorithm from Figure 1 is a characterization of semantic entailment provided by the next corollary. This is due to the previous observations from [7] and the fact that syntactic closures coincide with semantic closures. Hence, we get the following characterization:

**Corollary 3.** *For any theory $T$ and $A, B \in L^Y$,*

$$\|A \Rightarrow B\|_T = S(B, A^+), \tag{5}$$

*where $A^+$ is returned by the algorithm from Figure 1.* $\square$

Thus, according to Corollary 3, in order to determine $||A \Rightarrow B||_T$, it suffices to run the algorithm from Figure 1 and compute the inclusion degree $S(B, A^+) \in L$ of $B$ in the closure $A^+$ of $A$. Therefore, as we have outlined in the beginning of this section, the automated prover is capable of deciding $||A \Rightarrow B||_T = a$ by means of finitely many syntactic manipulations with the crisp counterpart $c(T)$ of $T$.

The presented algorithm can also be used to check a lower bound for $||A \Rightarrow B||_T$ as it is shown by the following corollary:

**Corollary 4.** *For any theory $T$, degree $a \in L$, and $A, B \in L^Y$, the following are equivalent:*

    *(i)  The algorithm in Figure 1 returns $A^+$ such that $a \otimes B \subseteq A^+$,*

    *(ii)  $c(T) \vdash A \Rightarrow a \otimes B$,*

    *(iii)  $a \leq ||A \Rightarrow B||_T$.*            □

*Remark 3.* Note that Corollary 3 and 4 represent two applications of the core algorithm from Figure 1. They demonstrate a versatility of the pseudocode from Figure 1 which represents several algorithms resulting by a slight modification of the pseudocode (appending additional test conditions). In addition to computing the syntactic closure which, for the particular case $L = \{0, 1\}$, is a classic topic in database algorithms, it is capable of determining degrees of semantic entailment as in Corollary 3 and lower bounds of entailment degrees as in Corollary 4. In the latter case, the pseudocode can be made more efficient by modifying the halting condition (details are postponed to a full version of the paper).

## 5    Complexity And Performance

This section presents complexity analysis and empirical comparison of the presented algorithm with algorithm GRADEDCLOSURE which represents a graded generalization of the algorithm CLOSURE known from database systems [19].

### 5.1    Complexity Analysis

Regarding the complexity of the algorithm, the number of times in which Theorem 5 and Corollary 2 are applied is lower than $\alpha \cdot n$ where $\alpha$ is the cardinality of $L^* = \{c \in L \mid c^* = c\}$ (a constant) and $n$ equals $Sz(T)$, the size of the input $T$, which is defined as follows:

- For any $A \in L^Y$, put $Sz(A) = \{y \in Y \mid A(y) > 0\}$, i.e., $Sz(A)$ is the number of elements from $Y$ which belong to $A$ to nonzero degrees;
- for $A \Rightarrow B$, put $Sz(A \Rightarrow B) = Sz(A) + Sz(B)$;
- for a theory $T$, put

$$Sz(T) = \begin{cases} \sum_{A \Rightarrow B \in T} Sz(A \Rightarrow B), & \text{if } T \text{ is crisp,} \\ Sz(c(T)), & \text{otherwise.} \end{cases}$$

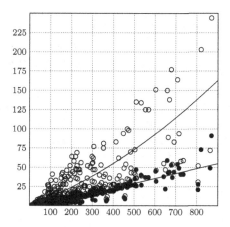

**Fig. 2.** Graded Closure vs. FASL Algorithm from Fig. 1

## 5.2  Experimental Evaluation

In this section, we show results of experimental evaluation of the algorithm. Note that in case of the classic algorithms for computing closures, the asymptotic worst-case complexity analysis is usually accompanied by experimental analysis to further illustrate their performance. For instance, while the famous LINCLO-SURE introduced in [3] has a linear-time complexity compared to CLOSURE whose complexity is quadratic, in most cases LINCLOSURE is slower due to the overhead caused by a construction of temporary data structures. Therefore, we present an experimental comparison showing how the present algorithm performs compared to a graded counterpart of CLOSURE from [5].

In the experiment, we use the following structures of truth degrees. We take for $L$ the following subset of the real unit interval

$$L = \{0, 0.1, 0.2, \ldots, 0.9, 1\}$$

with the natural ordering. Moreover, $\otimes$ and $\rightarrow$ are the restrictions of the Łukasiewicz operations to $L$. Thus,

$$a \otimes b = \max(a + b - 1, 0),$$
$$a \rightarrow b = \min(1 - a + b, 1),$$

for all $a, b \in L$. Furthermore, we consider * (hedge) given by

$$a^* = \begin{cases} 1, & \text{if } a = 1, \\ 0.5, & \text{if } 0.5 \leq a < 1, \\ 0, & \text{otherwise.} \end{cases}$$

for all $a \in L$. Altogether, $\mathbf{L} = \langle L, \wedge, \vee, \otimes, \rightarrow, \diagdown, {}^*, 0, 1 \rangle$ with $\diagdown$ given by (3) and $\wedge$ and $\vee$ being minimum and maximum, respectively, is a residuated structure of degrees with the desired properties, see Section 2.

The algorithm has been implemented in Java and tested together with GRADED CLOSURE from [5] with randomly generated sets of formulas. The sets have been generated with up to 1 200 pairwise different attributes and consist of up to 1 200 formulas. Sets of similar sizes can be obtained as bases of mid-scale ordinal data.

The results are shown in Figure 2. The $x$-axis represents sizes of sets $T$ of formulas measured by $Sz$ as in the previous subsection (the units are in thousands). The $y$-axis contains executions times in milliseconds. The graph shows that on average, FASL (the algorithm from this paper whose results are denoted by "•") outperforms GRADED CLOSURE (denoted by "○").

## 6   Conclusions

We have presented a sound and complete axiomatization of logic for dependencies in data with grades. The dependencies we use are formalized using residuated structures of grades and are in the form of implications between graded sets. The dependencies have two basic interpretations as (i) functional dependencies in ranked data tables over domains with similarities, and (ii) dependencies between attributes in ordinal data. Contrary to the mainstream approaches to axiomatizations of such rules which are based on the rule of cut (and related rules, e.g., the transitivity rule), a major role in the presented axiomatization is played by the simplification rule which is more suitable for designing an automated prover. Based on the axiomatization, we have presented an algorithm which can be used as an automated prover checking whether a dependency follows from a collection of dependencies (at least) to a prescribed degree. Moreover, the algorithm can be used for computing closures which can be used to determine degrees of entailment from a collection of dependencies. We have shown that the algorithm is sound and complete and provide empirical evidence that in most cases it outperforms algorithms proposed in the past.

**Acknowledgments.** Supported by Grants No. P103/11/1456 of the Czech Science Foundation, No. TIN2011-28084 of the Science and Innovation Ministry of Spain, and No. P09-FQM-5233 of the Junta de Andalucía. DAMOL is supported by project reg. no. CZ.1.07/2.3.00/20.0059 of the European Social Fund in the Czech Republic.

## References

1. Aguilera, G., Cordero, P., Enciso, M., Mora, A., de Guzmán, I.P.: A Non-explosive Treatment of Functional Dependencies Using Rewriting Logic. In: Bazzan, A.L.C., Labidi, S. (eds.) SBIA 2004. LNCS (LNAI), vol. 3171, pp. 31–40. Springer, Heidelberg (2004)
2. Armstrong, W.: Dependency structures of data base relationships. In: IFIP Congress, pp. 580–583 (1974)

3. Beeri, C., Bernstein, P.A.: Computational problems related to the design of normal form relational schemas. ACM Trans. Database Syst. 4, 30–59 (1979)
4. Belohlavek, R., Vychodil, V.: Axiomatizations of fuzzy attribute logic. In: IICAI 2005, pp. 2178–2193 (2005)
5. Bělohlávek, R., Vychodil, V.: Attribute Implications in a Fuzzy Setting. In: Missaoui, R., Schmidt, J. (eds.) ICFCA 2006. LNCS (LNAI), vol. 3874, pp. 45–60. Springer, Heidelberg (2006)
6. Bělohlávek, R., Vychodil, V.: Data Tables with Similarity Relations: Functional Dependencies, Complete Rules and Non-redundant Bases. In: Li Lee, M., Tan, K.-L., Wuwongse, V. (eds.) DASFAA 2006. LNCS, vol. 3882, pp. 644–658. Springer, Heidelberg (2006)
7. Belohlavek, R., Vychodil, V.: Properties of models of fuzzy attribute implications. In: SCIS & ISIS 2006, pp. 291–296 (2006)
8. Belohlavek, R., Vychodil, V.: Logical Foundations for Similarity-Based Databases. In: Chen, L., Liu, C., Liu, Q., Deng, K. (eds.) DASFAA 2009. LNCS, vol. 5667, pp. 137–151. Springer, Heidelberg (2009)
9. Belohlavek, R., Vychodil, V.: Query systems in similarity-based databases: logical foundations, expressive power, and completeness. In: ACM SAC, pp. 1648–1655 (2010)
10. Cordero, P., Enciso, M., Mora, A., de Guzmán, I.P.: SL$_{FD}$ logic: Elimination of data redundancy in knowledge representation. In: Garijo, F.J., Riquelme, J.-C., Toro, M. (eds.) IBERAMIA 2002. LNCS (LNAI), vol. 2527, pp. 141–150. Springer, Heidelberg (2002)
11. Fagin, R.: Combining fuzzy information: an overview. SIGMOD Record 31(2), 109–118 (2002)
12. Gerla, G.: Fuzzy Logic. Mathematical Tools for Approximate Reasoning. Kluwer Academic Publishers, Dordrecht (2001)
13. Goguen, J.A.: The logic of inexact concepts. Synthese 19(3–4), 325–373 (1969)
14. Gottwald, S.: A Treatise on Many-Valued Logics. Studies in Logic and Computation, vol. 98. Research Studies Press, Baldock (2000)
15. Gottwald, S.: Mathematical fuzzy logics. The Bulletin of Symbolic Logic 14(2), 210–244 (2008)
16. Grabisch, M., Marichal, J.L., Mesiar, R., Pap, E.: Aggregation Functions. Cambridge University Press, Cambridge (2009)
17. Hájek, P.: Metamathematics of Fuzzy Logic. Trends in Logic: Studia Logica Library. Kluwer Academic Publishers, Dordrecht (1998)
18. Hájek, P.: On very true. Fuzzy Sets and Systems 124(3), 329–333 (2001)
19. Maier, D.: The theory of relational databases. Computer software engineering series. Computer Science Press (1983)
20. Mora, A., Aguilera, G., Enciso, M., Cordero, P., Perez de Guzman, I.: A new closure algorithm based in logic: SLFD-Closure versus classical closures. Inteligencia Artificial, Revista Iberoamericana de IA 10(31), 31–40 (2006)
21. Mora, A., Cordero, P., Enciso, M., Fortes, I., Aguilera, G.: Closure via functional dependence simplification. International Journal of Computer Mathematics 89(4), 510–526 (2012)
22. Pavelka, J.: On fuzzy logic I, II, III. Mathematical Logic Quarterly 25, 45–52 (1979)
23. Raju, K.V.S.V.N., Majumdar, A.K.: Fuzzy functional dependencies and lossless join decomposition of fuzzy relational database systems. ACM Transactions on Database Systems (TODS) 13, 129–166 (1988) ACM ID: 42344

# Author Index

Abril, Daniel   210
Amgoud, Leila   282
Antunes, Cláudia   329
Armengol, Eva   258

Batet, Montserrat   173
Beliakov, Gleb   35
Belohlavek, Radim·   294, 408
Brax, Christoffer   185

Cabrerizo, Francisco J.   90
Cai, Zhiping   222
Čaklović, Lavoslav   102
Cardin, Marta   139
Casas-Roma, Jordi   197
Castellà-Roca, Jordi   161
Clivillé, Vincent   341
Confalonieri, Roberto   282
Cordero, Pablo   408

Dahlbom, Anders   185
Damez, Marc   234
De Baets, Bernard   306
de Jonge, Dave   282
Del Vasto-Terrientes, Luis   78
Díaz, Susana   306
d'Inverno, Mark   282
Domingo-Ferrer, Josep   367
Dubois, Didier   127

Elouedi, Zied   382
Enciso, Manuel   408
Endo, Yasunori   394
Ennaceur, Amel   382

Flaminio, Tommaso   23
Fujimoto, Katsushige   115

García-Cerdaña, Àngel   258
Godo, Lluís   23

Hamasuna, Yukihiro   394
Hazelden, Katina   282
Heki, Ayako   394

Herrera-Joancomartí, Jordi   197
Herrera-Viedma, Enrique   90

Inuiguchi, Masahiro   11, 270

James, Simon   35

Kawamura, Shizuya   270
Konecny, Jan   294
Kroupa, Tomáš   23
Kuhr, Tomas   246

Lefevre, Eric   382
Lesot, Marie-Jeanne   234
Long, Jun   222
Lopez, Beatriz   318
López-Gijón, Javier   90

Martínez Arqué, Néstor   149
Mesiar, Radko   13
Miyamoto, Sadaaki   1, 355
Montes, Susana   306
Mora, Angel   408
Murillo, Javier   210
Murillo, Javier   318

Narukawa, Yasuo   56
Nettleton, David F.   149

Ogryczak, Wlodzimierz   66
Osicka, Petr   294
Osman, Nardine   282

Pérez, Ignacio J.   90
Perny, Patrice   66
Pla, Albert   318
Prade, Henri   127, 282

Revault d'Allonnes, Adrien   234
Rico, Agnés   127

Sánchez, David   161, 173
Sierra, Carles   282
Silva, Andreia   329
Slowinski, Roman   78
Sugeno, Michio   115

Takumi, Satoshi    355
Torra, Vicenç    197, 210

Valet, Lionel    341
Valls, Aida    78
Viejo, Alexandre    161, 173
Vychodil, Vilem    246, 408

Weng, Paul    66

Xia, Geming    222

Yee-King, Matthew    282
Yin, Jianping    222
Yoshida, Yuji    45

Zhao, Wentao    222
Zielniewicz, Piotr    78